ASPIRE
SUCCEED
PROGRESS

# Complete Mathematics for Cambridge Secondary 1

# 3

Deborah Barton

Oxford excellence for Cambridge Secondary 1

OXFORD

Great Clarendon Street, Oxford OX2 6DP. United Kingdom

Oxford University Press is a department of the University of Oxford. It furthers the University's objective of excellence in research, scholarship, and education by publishing worldwide. Oxford is a registered trade mark of Oxford University Press in the UK and in certain other countries.



First published 2013



British Library Cataloguing in Publication Data

Data available

ISBN 978-0-19-913710-7

10 9 8 7 6

Paper used in the production of this book is a natural, recyclable product made from wood grown in sustainable forests. The manufacturing process conforms to environmental regulations of the country of origin.

Printed in Great Britain by Bell and Bain Ltd., Glasgow

**Acknowledgements**

® IGCSE is the registered trademark of Cambridge International Examinations.

The publishers would like to thank the following for permissions to use their photographs:

**p7:** Dmitry Kalinovsky/Shutterstock; **p8:** George Dolgikh/Shutterstock; **p12:** Aistov Alexey/Shutterstock; **p13:** Zts/Dreamstime; **p16:** Jakub Pavlinec/Shutterstock; **p19:** Nicky Rhodes/Shutterstock; **p21:** Photostogo; **p38:** REX/Patrick Frilet; **p43:** Set of 100 Snap Cubes® from Learning Resources® – www.LearningResources.co.uk; **p59:** Pressmaster/Shutterstock; **p70:** iPics/Shutterstock.com; **p71:** Aaron Amat/Shutterstock; **p71:** OUP; **p71:** Winai Tepsuttinun/Shutterstock; **p74:** rook76/Shutterstock.com; **p74:** © Travel Pictures Travel; **p76:** Dja65/Shutterstock; **p79:** Squaredpixels/ Istock; **p80:** Sylvie Bouchard/Shutterstock; **p90:** AFP/Getty Images; **p97:** imagebroker.net/SuperStock ; **p100:** Gregor Kervina/Shutterstock. com; **p102:** Jonathan Larsen/Shutterstock.com; **p106:** Sharon Day/ Shutterstock; **p110:** NERAMIT SISA/Shutterstock; **p119:** Eric Gevaert/ Shutterstock; **p126:** Philip Lange/Shutterstock; **p154:** © Oliver Knight/ Alamy; **p161:** Oleksiy Mark/Shutterstock; **p170:** fcknimages/Istock; **p176:** © Godfer/Dreamstime.com; **p185:** michaket/Shutterstock; **p194:** Oxford Scientific; **p299:** Jorg Hackemann/Shutterstock; **p206:** OUP; **p208:** Fancy Collection/SuperStock ; **p209:** © Purestock/Alamy; **p214:** BeckyBrockie/Istock; **p215:** tavi/Shutterstock; **p218:** auremar/ Shutterstock; **p264:** courtneyk/Istock; **p268:** Peter Baxter/Shutterstock. com; **p274:** vovan/Shutterstock; **p274:** Ene/Shutterstock; **p274:** Ene/ Shutterstock; **p274:** Trailexplorers/Shutterstock; **p275:** Tryphosa/Alamy; **p276:** Dorling Kindersley/Getty Images; **p280:** Pavel L Photo and Video/ Shutterstock; **p293:** UNIVERSITY OF DUNDEE/SCIENCE PHOTO LIBRARY; **p298:** Nikolai Pozdeev/Shutterstock; **p298:** Nikolai Pozdeev/Shutterstock

**Cover image:** ©2010 Elizabeth Papadopoulos/Getty Images

Illustrations by Ian West and Q2A Media

# Contents

## Contents

## About this book

This book follows the Cambridge Secondary 1 Mathematics curriculum framework for Cambridge International Examinations in preparation for the Cambridge Checkpoint assessments. It has been written by a highly experienced teacher, examiner and author.

This book is part of a series of nine books. There are three student textbooks, each covering stages 7, 8 and 9 and three homework books written to closely match the textbooks, as well as a teacher book for each stage.

The books are carefully balanced between all the content areas in the curriculum framework: number, algebra geometry, measure, data handling and problem solving. Some of the questions in the exercises and the investigations within the book are underpinned by the final framework area: problem solving – providing a structure for the application of mathematical skills.

Features of the book:

- **Objectives** – taken from the Cambridge Secondary 1 curriculum framework.
- **What's the point?** – providing rationale for inclusion of topics in a real-world setting.
- **Chapter Check in** – to assess whether the student has the required prior knowledge.
- **Notes and worked examples** – in a clear style using accessible English and culturally appropriate material.
- **Exercises** – carefully designed to gradually increase in difficulty, providing plenty of practice and techniques.
- **Considerable variation in question style** – encouraging deeper thinking and learning, including open questions.
- **Comprehensive practice** – plenty of initial practice questions followed by varied questions for stretch, challenge, cross-over between topics and links to the real world with questions set in context.
- **Extension questions** – providing stretch and challenge for students:
  - questions with a box e.g. [1] provide challenge for the average student
  - questions with a filled box e.g. **1** provide extra challenge for more able students.
- **Technology boxes** – direct students to websites for review material, fun games and challenges to enhance learning.
- **Investigation and Game boxes** – providing extra fun, challenge and interest.
- **Full colour with modern artwork** – to engage students and maintain their interest.
- **Consolidation examples and exercises** – providing review material on the chapter.
- **Summary and Check out** – providing a quick review of the chapter's key points aiding revision and enabling you to assess progress.
- **Review exercises** – provided every six chapters with mixed questions covering all chapters.
- **Bonus chapter** – the work from Chapter 19 is not in the Cambridge Secondary 1 curriculum. It is in the Cambridge IGCSE® curriculum and is included to stretch and challenge more able students.

**A note from the author:**

If you don't already love maths as much as I do, I hope that after working through this book you will enjoy it more. Maths is more than just learning concepts and applying them. It isn't just about right and wrong answers. It is a wonderful subject full of challenges, puzzles and beautiful proofs. Studying maths develops your analysis and problem-solving skills and improves your logical thinking – all important skills in the workplace.

Be a responsible learner – if you don't understand something, ask or look it up. Be determined and courageous. Keep trying without giving up when things go wrong. No-one needs to be 'bad at maths'. Anyone can improve with hard work and practice in just the same way sports men and women improve their skills through training. If you are finding work too easy, say. Look for challenges, then maths will never be boring.

Most of all, enjoy the book. Do the 'training', enjoy the challenges and have fun!

**Deborah Barton**

# 1 Fractions and indices

## Objectives

- Consolidate writing a fraction in its simplest form by cancelling common factors.
- Add, subtract, multiply and divide fractions, interpreting division as a multiplicative inverse, and cancelling common factors before multiplying or dividing.
- Use the order of operations, including brackets and powers.
- Use positive, negative and zero indices and the index laws for multiplication and division of positive integer powers.

## What's the point?

Fractions are useful in many different areas of life. We can use fractions when decorating. We could mix a tin of white paint with a quarter of a tin of red paint to make pink, or we may use two thirds of a tin of paint to decorate a room. We can use fractions in farming. A farmer could use three quarters of his land for crops and one quarter for animals.

## Before you start

### You should know ...

**1** Fractions can be simplified by dividing.
*For example*:

$$\frac{14}{35} \xrightarrow{\div 7} \frac{2}{5}$$

(numerator ÷ 7, denominator ÷ 7)

### Check in

**1** Simplify:

**a** $\frac{14}{20}$ **b** $\frac{16}{40}$ **c** $\frac{32}{48}$

**d** $\frac{15}{75}$ **e** $\frac{8}{96}$ **f** $\frac{18}{162}$

| | |
|---|---|
| **2** How to convert between mixed numbers and improper fractions.<br>*For example:*<br>$\frac{8}{5}$ means $8 \div 5 = 1$ remainder 3, so $\frac{8}{5} = 1\frac{3}{5}$<br>$3\frac{1}{4} = \frac{(3 \times 4) + 1}{4}$, so $3\frac{1}{4} = \frac{13}{4}$ | **2 a** Change these improper fractions to mixed numbers:<br>**i** $\frac{23}{20}$ **ii** $\frac{18}{5}$ **iii** $\frac{73}{10}$<br>**b** Change these mixed numbers to improper fractions:<br>**i** $1\frac{3}{10}$ **ii** $3\frac{7}{8}$ **iii** $5\frac{2}{3}$ |
| **3** The meaning of indices.<br>*For example:*<br>$5 \times 5 = 5^2$ or five squared $= 25$<br>$2 \times 2 \times 2 = 2^3$ or two cubed $= 8$<br>$p \times p \times p \times p \times p = p^5$ using index notation | **3 a** Calculate:<br>**i** $4^2$ **ii** $3^3$ **iii** $2^4$<br>**b** Write $y \times y \times y \times y \times y \times y \times y$ using index notation. |

## 1.1 Working with fractions

Before you start working with fractions you need to be comfortable cancelling fractions by dividing numerators and denominators by common factors. You also need to be able to convert between mixed numbers and improper fractions. These puzzles will help you practise these skills.

**Puzzles**

### Equivalent fraction puzzles

**1** I am equivalent to $\frac{21}{45}$ and my numerator is a prime number. What am I?

**2** I am equivalent to $\frac{64}{104}$ and my denominator is a prime number. What am I?

**3** I am equivalent to $\frac{45}{105}$ and my numerator and denominator are both prime numbers. What am I?

**4** I am equivalent to $\frac{3}{8}$ and the product of my numerator and denominator is 216. What am I?

**5** I am equivalent to $1\frac{2}{3}$ and as an improper fraction the sum of my numerator and denominator is 40. What am I?

**6** I am equivalent to $\frac{60}{204}$ and my denominator is less than 20. What am I?

**7** I am equivalent to $1\frac{3}{4}$ and as an improper fraction my numerator is 9 more than my denominator. What am I?

**8** I am equivalent to $\frac{5}{9}$ and my numerator is 72 less than my denominator. What am I?

**9** I am equivalent to $\frac{52}{65}$ and my numerator is smaller than 20 and a multiple of 3. What am I?

**10 a** When writing a fraction in its simplest form by dividing the numerator and denominator by a common factor, either the numerator or denominator or both will be prime. True or false?

**b** What are the first eight prime numbers?

### Adding and subtracting fractions

Fractions with common denominators are easily added or subtracted.
*For example:*

$$\frac{3}{8} + \frac{1}{8} = \frac{4}{8} = \frac{1}{2}$$

$$\frac{4}{5} - \frac{1}{5} = \frac{3}{5}$$

Fractions without common denominators need to be changed to equivalent fractions with the same denominators.

**EXAMPLE 1**

Work out: **a** $\frac{3}{5} + \frac{1}{4}$ **b** $\frac{3}{8} - \frac{2}{7}$

**a** $\frac{3}{5} + \frac{1}{4} = \frac{12}{20} + \frac{5}{20}$
$= \frac{17}{20}$

**b** $\frac{3}{8} - \frac{2}{7} = \frac{21}{56} - \frac{16}{56} = \frac{5}{56}$

Addition and subtraction of mixed numbers is done in a similar manner.

**EXAMPLE 2**

Work out: **a** $1\frac{3}{4} + 2\frac{1}{5}$ **b** $3\frac{2}{5} - 1\frac{3}{4}$

**a** $1\frac{3}{4} + 2\frac{1}{5} = 3 + \frac{3}{4} + \frac{1}{5} = 3 + \frac{15}{20} + \frac{4}{20}$

$= 3\frac{19}{20}$

**b** $3\frac{2}{5} - 1\frac{3}{4} = \frac{17}{5} - \frac{7}{4}$ Change to improper fractions.

$= \frac{68}{20} - \frac{35}{20}$ Common denominator is 20

$= \frac{33}{20}$

$= 1\frac{13}{20}$ Change to a mixed number.

## Exercise 1A

**1** Work out:

**a** $\frac{1}{4} + \frac{1}{4}$ **b** $\frac{2}{5} + \frac{1}{5}$

**c** $\frac{5}{8} - \frac{3}{8}$ **d** $\frac{3}{4} + \frac{1}{4}$

**e** $\frac{4}{9} - \frac{2}{9}$ **f** $\frac{5}{9} - \frac{5}{9}$

**2** Calculate:

**a** $\frac{1}{2} + \frac{1}{4}$ **b** $\frac{1}{3} + \frac{1}{2}$

**c** $\frac{2}{3} + \frac{1}{5}$ **d** $\frac{2}{7} + \frac{3}{8}$

**e** $\frac{2}{9} + \frac{1}{3}$ **f** $\frac{2}{9} + \frac{1}{6}$

**3** Calculate:

**a** $\frac{3}{5} - \frac{1}{4}$ **b** $\frac{5}{8} - \frac{1}{4}$

**c** $\frac{3}{7} - \frac{2}{9}$ **d** $\frac{4}{7} - \frac{3}{8}$

**e** $\frac{3}{4} - \frac{1}{2}$ **f** $\frac{11}{12} - \frac{2}{3}$

**4** Work out:

**a** $3\frac{1}{3} + 2\frac{1}{4}$ **b** $3\frac{1}{2} - 1\frac{1}{4}$

**c** $7\frac{2}{3} + 1\frac{3}{4}$ **d** $4\frac{3}{5} - 2\frac{2}{7}$

**e** $8\frac{1}{2} - 3\frac{2}{3}$ **f** $6\frac{3}{5} + 4\frac{1}{2}$

**g** $5\frac{2}{3} - 2\frac{3}{4}$ **h** $5\frac{1}{4} - 1\frac{4}{5}$

**i** $4\frac{1}{5} + 2\frac{3}{4} - 2\frac{1}{2}$ **j** $10\frac{3}{5} - 4\frac{3}{8} + 2\frac{7}{20}$

**5** Write down three fractions, with different denominators, that add up to make $4\frac{5}{24}$.

**6**

Aakesh bought $3\frac{1}{4}$ kg of oranges. He gave his sister $1\frac{2}{3}$ kg.

What was the mass of the oranges he had left?

**7** A plank of wood is 3 m in length. How long will it be if I cut $\frac{5}{8}$ m of wood from it?

**8** Karen spent $\frac{4}{7}$ of her money on Monday.

She spent $\frac{1}{3}$ of her money on Tuesday. What fraction of money did she

**a** spend on Monday and Tuesday?

**b** have left?

**9** A container holds $4\frac{1}{2}$ litres of water.

How much water is left in the container if Jerry drinks $2\frac{7}{8}$ litres?

**10** Fill in the missing numbers:

**a** $\square + 3\frac{1}{3} - 2\frac{1}{5} = 5\frac{4}{15}$

**b** $4\frac{1}{2} - \square + 3\frac{1}{10} = 2\frac{9}{10}$

**11** Find the value of the letters in the following:

**a** $1\frac{a}{5} + 2\frac{3}{10} = 4\frac{1}{10}$ **b** $1\frac{1}{3} + b\frac{5}{6} = 4\frac{1}{6}$

**c** $6\frac{c}{10} + 1\frac{3}{5} = 8\frac{1}{2}$ **d** $3\frac{7}{9} + \frac{7}{d} = 4\frac{1}{6}$

**12** Put these fractions into groups of three so that each group has a total of $4\frac{1}{2}$.

$2\frac{2}{3}$ $1\frac{3}{5}$ $2\frac{4}{5}$ $2\frac{3}{4}$ $2\frac{5}{12}$ $\frac{2}{10}$ $1$ $\frac{3}{20}$ $1\frac{1}{12}$ $1\frac{11}{24}$ $1\frac{1}{2}$ $\frac{3}{8}$

## INVESTIGATION

A **unit fraction** is a fraction with a numerator of 1, for example $\frac{1}{2}, \frac{1}{6}, \frac{1}{13}$.
The ancient Egyptians didn't write fractions with a numerator greater than 1. Instead they would write them as a sum of two or more *different* unit fractions. For example they wouldn't write $\frac{2}{3}$; instead they would write $\frac{2}{3}$ as $\frac{1}{2} + \frac{1}{6}$. (They wouldn't use $\frac{1}{3} + \frac{1}{3}$ as this involves repeating a unit fraction.)
Investigate different non-unit fractions. Can all non-unit fractions be written as a sum of different unit fractions?

## Multiplying and dividing fractions

To multiply fractions you just need to multiply the numerators and the denominators.
*For example:*

$$\frac{2}{3} \times \frac{2}{5} = \frac{2 \times 2}{3 \times 5} = \frac{4}{15}$$

$$1\tfrac{1}{2} \times \frac{3}{4} = \frac{3}{2} \times \frac{3}{4} = \frac{3 \times 3}{2 \times 4} = \frac{9}{8} = 1\tfrac{1}{8}$$

You know that $3^2$ means $3 \times 3$; so we square fractions in the same way.
Often you can simplify the fractions, by cancelling common factors, before doing the multiplication.
*For example:*

$$\frac{2}{\cancel{3}_1} \times \frac{\cancel{3}^1}{8} = \frac{2^1}{\cancel{3}_1} \times \frac{\cancel{3}^1}{\cancel{8}_4} = \frac{1 \times 1}{1 \times 4} = \frac{1}{4}$$

$$4\tfrac{2}{5} \times \frac{7}{33} = \frac{22}{5} \times \frac{7}{33} = \frac{\overset{2}{\cancel{22}}}{5} \times \frac{7}{\underset{3}{\cancel{33}}} = \frac{2 \times 7}{5 \times 3} = \frac{14}{15}$$

You can also use the idea of cancelling when you are multiplying more than two fractions together.

**EXAMPLE 3**

Calculate: **a** $1\tfrac{1}{15} \times \frac{3}{8} \times 10$ **b** $\left(1\tfrac{4}{5}\right)^2$

**a** $1\tfrac{1}{15} \times \frac{3}{8} \times 10$

$= \frac{16}{15} \times \frac{3}{8} \times \frac{10}{1}$

Change to improper fractions.

$= \frac{\overset{2}{\cancel{16}}}{15} \times \frac{3}{\underset{1}{\cancel{8}}} \times \frac{10}{1}$

$= \frac{2}{\underset{3}{\cancel{15}}} \times \frac{3}{1} \times \frac{\overset{2}{\cancel{10}}}{1}$

$= \frac{2}{\underset{1}{\cancel{3}}} \times \frac{\overset{1}{\cancel{3}}}{1} \times \frac{2}{1}$

$= \frac{2}{1} \times \frac{1}{1} \times \frac{2}{1}$

$= 4$

**b** $\left(1\tfrac{4}{5}\right)^2$

$= 1\tfrac{4}{5} \times 1\tfrac{4}{5}$

$= \frac{9}{5} \times \frac{9}{5}$

$= \frac{81}{25}$

$= 3\tfrac{6}{25}$



Division of fractions is a little more tricky.
Remember:

$$9 \div 3 = 9 \times \frac{1}{3} = 3$$

$$8 \div 2 = 8 \times \frac{1}{2} = 4$$

Notice that the operation

$\div 3$ is the same as $\times \frac{1}{3}$

$\div 2$ is the same as $\times \frac{1}{2}$

In general, division by a number is the same as multiplying by the number's **inverse** or **reciprocal**.

Under multiplication, the reciprocal of

$3$ is $\frac{1}{3}$ $\left(3 \times \frac{1}{3} = 1\right)$

$\frac{1}{2}$ is $2$ $\left(\frac{1}{2} \times 2 = 1\right)$

$\frac{3}{4}$ is $\frac{4}{3}$ $\left(\frac{3}{4} \times \frac{4}{3} = 1\right)$

Using this idea, all divisions of fractions can be turned into multiplications.

## EXAMPLE 4

Calculate: **a** $\frac{3}{5} \div \frac{2}{3}$ **b** $2\frac{1}{4} \div 1\frac{3}{8}$

**a** $\frac{3}{5} \div \frac{2}{3} = \frac{3}{5} \times \frac{3}{2}$

$= \frac{9}{10}$



**b** $2\frac{1}{4} \div 1\frac{3}{8} = \frac{9}{4} \div \frac{11}{8}$

$= \frac{9}{4} \times \frac{8}{11}$

$= \frac{9}{{}_1\not{4}} \times \frac{\not{8}^2}{11}$

$= \frac{18}{11} = 1\frac{7}{11}$

$= \dfrac{\frac{3}{4} \times \frac{3}{4}}{\frac{17}{5} - \frac{7}{4}} = \dfrac{\frac{9}{16}}{\frac{68}{20} - \frac{35}{20}}$

$= \dfrac{\frac{9}{16}}{\frac{33}{20}}$

$= \frac{9}{16} \div \frac{33}{20}$

$= \frac{9}{16} \times \frac{20}{33}$

$= \frac{\not{9}^{3}}{\not{16}_{4}} \times \frac{\not{20}^{5}}{\not{33}_{11}} = \frac{3}{4} \times \frac{5}{11}$

$= \frac{15}{44}$

The order of operations (BIDMAS) also applies to fractions. Remember in calculations we do:

Brackets first (note long dividing lines act like brackets)
Then Indices (or powers)
Then Division and Multiplication
Then Addition and Subtraction

## EXAMPLE 5

Calculate:

$$\frac{\left(\frac{3}{4}\right)^2}{3\frac{2}{5} - 1\frac{3}{4}}$$

## Exercise 1B

**1** Work out:

**a** $3 \times \frac{1}{2}$ **b** $\frac{1}{2} \times \frac{1}{2}$

**c** $\frac{3}{4} \times \frac{1}{2}$ **d** $\frac{3}{4} \times \frac{2}{3}$

**e** $\frac{7}{8} \times \frac{2}{5}$ **f** $\frac{3}{4} \times \frac{4}{7}$

**g** $\frac{6}{11} \times \frac{4}{9}$ **h** $\frac{5}{8} \times \frac{4}{5}$

**2** Calculate:

**a** $1\frac{1}{2} \times 3$ **b** $1\frac{1}{2} \times 1\frac{1}{2}$

**c** $2\frac{1}{4} \times 1\frac{1}{4}$ **d** $3\frac{2}{3} \times 2\frac{1}{2}$

**e** $2\frac{1}{2} \times 2\frac{1}{2}$ **f** $4\frac{3}{4} \times 1\frac{7}{8}$

**g** $3\frac{1}{3} \times 4\frac{3}{4}$ **h** $5\frac{2}{5} \times 3\frac{7}{11}$

**3** Calculate:

**a** $\frac{1}{2} \div \frac{1}{4}$ **b** $\frac{1}{2} \div \frac{1}{3}$

**c** $\frac{3}{4} \div \frac{1}{4}$ **d** $\frac{7}{8} \div \frac{3}{4}$

**e** $\frac{7}{8} \div \frac{2}{5}$ **f** $\frac{3}{4} \div \frac{4}{5}$

**g** $\frac{6}{11} \div \frac{4}{9}$ **h** $\frac{3}{14} \div \frac{4}{7}$

**4** Work out:

**a** $1\frac{1}{4} \div 2\frac{1}{2}$ **b** $2\frac{1}{2} \div 1\frac{1}{4}$

**c** $3\frac{1}{2} \div 1\frac{1}{3}$ **d** $2\frac{2}{3} \div 1\frac{3}{4}$

**e** $4\frac{2}{5} \div 1\frac{2}{3}$ **f** $4\frac{3}{7} \div 1\frac{6}{7}$

**g** $5\frac{4}{5} \div 2\frac{4}{7}$ **h** $3\frac{3}{4} \div 6\frac{2}{3}$

**5** Work out:

**a** $\frac{7}{10}$ of $8\frac{1}{3}$ m **b** $\frac{3}{5}$ of $40\frac{1}{2}$ kg

**c** $\frac{4}{5}$ of $30\frac{2}{3}$ ml **d** $\frac{2}{3}$ of $3\frac{1}{4}$ tonnes

**e** $\frac{14}{15} \times \frac{3}{8} \times \frac{5}{7}$ **f** $\frac{7}{8} \times \frac{4}{21} \times \frac{4}{5}$

**g** $\left(\frac{7}{8}\right)^2$ **h** $1\frac{1}{5} \times \left(\frac{2}{3}\right)^2$

**i** $1\frac{1}{5} \times \frac{2^2}{3}$

**6** Work out:

**a** $(1\frac{1}{2})^2$ **b** $2\frac{1}{4} \times 1\frac{1}{2} \times \frac{1}{3}$

**c** $6 \times \left(1\frac{2}{3} + 1\frac{1}{2}\right)$ **d** $\left(2\frac{5}{6} + 3\frac{1}{4}\right) \div 4$

**e** $3\frac{3}{4} - 1\frac{7}{8} + \frac{1}{2}$ **f** $\frac{5}{8} - \frac{7}{9} + \frac{1}{3}$

**g** $1\frac{4}{5} + 2\frac{3}{4} \times 1\frac{1}{3}$ **h** $1\frac{1}{2} - 4\frac{1}{8} \div 2\frac{3}{4}$

**i** $1\frac{1}{7} \times \left(2\frac{3}{4} + 1\frac{1}{3}\right)$

**j** $\frac{2}{7} \times \left(1\frac{3}{4} + 1\frac{2}{5}\right) + \frac{1}{10}$

**7** Calculate:

**a** $2\frac{7}{10} \div 1\frac{7}{15} - 1\frac{1}{2}$ **b** $\frac{3}{8}$ of $\left(1\frac{1}{2}\right)^2$

**c** $3\frac{1}{8} \times \frac{4}{5} \div 1\frac{1}{2}$ **d** $2\frac{2}{3} \div 2\frac{2}{5} \times 1\frac{1}{2}$

**e** $\dfrac{3\frac{1}{2}}{1\frac{1}{2} + 2\frac{1}{4}}$ **f** $\dfrac{2\frac{1}{2} \times \frac{7}{8}}{6 - \frac{3}{4}}$

**g** $\dfrac{3\frac{7}{8} - 1\frac{5}{6}}{2\frac{3}{4} - 1\frac{2}{3}}$ **h** $\dfrac{6\frac{7}{9} - 4\frac{8}{9}}{3\frac{1}{2} + 1\frac{3}{5}}$

**i** $1\frac{7}{8} \times \frac{4}{5} + 3\frac{1}{2} \times 1\frac{5}{7} - 7\frac{1}{2}$

**j** $1\frac{3}{4} + 2\frac{1}{5} \times \frac{5}{11} - 3 \div 1\frac{1}{4}$

**k** $\left(1\frac{3}{5} - 1\frac{1}{20}\right) \times \left(1\frac{2}{3} + 1\frac{4}{5} - 2\frac{8}{15}\right)$

**8**

A packet of biscuits has a mass of $1\frac{1}{3}$ kg.
Jason eats $\frac{2}{5}$ of the packet.
What is the mass of biscuits left?

**9** A rectangle measures $4\frac{1}{4}$ cm by $2\frac{4}{5}$ cm. What is the area of the rectangle?

$4\frac{1}{4}$ cm

$2\frac{4}{5}$ cm

**10** The area of a rectangular field is $130\frac{2}{3}$ m$^2$.
What is the field's length if its width is $10\frac{1}{2}$ m?

**11**

$\frac{3}{4}$-litre bottles of washing-up liquid are taken from a container holding $164\frac{1}{2}$ litres.

**a** How many bottles can be filled from the container?

**b** When 75 bottles have been removed from the container, what fraction of the original amount of liquid remains in the container?

**12** Work out:

**a** $^{-}1\frac{3}{8} \div \frac{2}{5}$ **b** $^{-}3\frac{3}{5} \times \frac{5}{18}$

**c** $^{-}1\frac{3}{4} \div 8$ **d** $\left(^{-}1\frac{4}{5}\right)^2$

**13** What do the letters stand for in the following?

**a** $a \times 1\frac{1}{4} = 3\frac{1}{4}$ **b** $b \div 3\frac{1}{2} = \frac{9}{28}$

**c** $2\frac{1}{c} \times 3\frac{1}{4} = 7\frac{7}{12}$ **d** $2\frac{3}{7} \div d = \frac{3}{7}$

**e** $(1\frac{2}{3})^2 \times e = 1$ **f** $f^2 = 2\frac{14}{25}$

**14** Fill in the missing numbers:

**a** $1\frac{1}{4} \times 40 = \square \div \frac{2}{5}$

**b** $20 \div 1\frac{7}{8} = 5\frac{1}{3} \times \square$

**c** $\square \div \frac{1}{2} = \frac{1}{5} \times \square$

**d** There are many possible answers to part **c**. Find some more answer pairs and write down the connection between them.

## INVESTIGATION

**a** Find as many pairs of fractions as you can with a product of $\frac{1}{2}$.

**b** Find as many pairs of fractions as you can that when divided give $\frac{1}{2}$.

## TECHNOLOGY

Need help with fractions?
Visit the website

www.coolmath.com

There is a complete course on fractions in the prealgebra section. Choose the lessons you need further help on.
Don't forget to do the questions!

## INVESTIGATION

Scientific calculators usually have a fractions key which looks like:

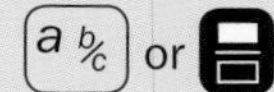

Examples of how to use it are as follows:

To simplify $\frac{35}{45}$, key in [3] [5] [a b/c] [4] [5] [=]

You should see an answer of $\frac{7}{9}$.

To do $4\frac{1}{2} \times \frac{3}{5}$, you need to work out how to type mixed numbers into your calculator, or change $4\frac{1}{2}$ into an improper fraction.

Try:

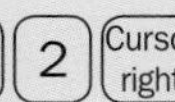
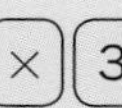

You should see an answer of $\frac{27}{10}$ or $2\frac{7}{10}$ or 2.7.
(Some calculators will let you set up whether the answer will be improper, a mixed number or a decimal – see what yours can do.)

Or you could do $4\frac{1}{2} \times \frac{3}{5}$ as:

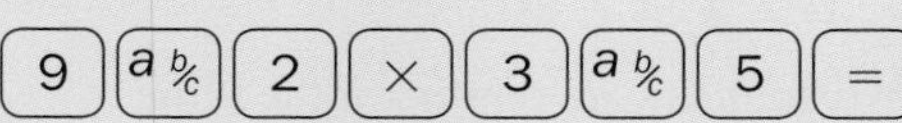


Your calculator will add and subtract fractions too, and you can use calculator brackets.

There are many different makes of calculators. Find out how to do calculations with fractions on yours. You may need to look at the instructions book, or you can ask your teacher. Repeat some of the questions from Exercises 1A and 1B using your calculator.

## 1.2 Indices

A short way of writing

$3 \times 3 \times 3 \times 3$ is $3^4$

$3^4$ is 3 raised to the power 4.

In the same way,

$4^3$ is 4 raised to the power 3.

$4^3 = 4 \times 4 \times 4 = 64$

The **power** is also called the **index**.

Notice that

$4^3 \times 4^2 = 4 \times 4 \times 4 \times 4 \times 4 = 4^5$

That is, $4^3 \times 4^2 = 4^{3+2} = 4^5$

- So long as the two numbers are powers of the same number you can multiply them by adding their indices. Using symbols,
  $a^m \times a^n = a^{m+n}$

**EXAMPLE 6**

Work out:

**a** $2^3 \times 2^4$ **b** $2^2 \times 2^5 \times 2^7$

**c** $7 \times 7^4$

**a** $2^3 \times 2^4 = 2^{3+4} = 2^7$

**b** $2^2 \times 2^5 \times 2^7 = 2^{2+5+7} = 2^{14}$

**c** $7 \times 7^4 = 7^5$

Notice also that

$$4^3 \div 4^2 = \frac{4 \times \cancel{4} \times \cancel{4}}{\cancel{4} \times \cancel{4}} = 4^1$$

That is $4^3 \div 4^2 = 4^{3-2} = 4^1$

- When two numbers are powers of the same number, you can divide them by subtracting their indices. Using symbols,
  $a^m \div a^n = a^{m-n}$

**EXAMPLE 7**

Work out:

**a** $5^7 \div 5^3$ **b** $8^{12} \div 8^{10}$

**a** $5^7 \div 5^3 = 5^{7-3} = 5^4$

**b** $8^{12} \div 8^{10} = 8^{12-10} = 8^2$

### Exercise 1C

**1** Copy and complete:

**a** $3^4 = 3 \times 3 \times \ldots = 81$

**b** $5^4 = 5 \times 5 \times \ldots = \square$

**c** $2^7 = 2 \times 2 \times \ldots = \square$

**d** $25 = 5^{\square} = 5 \times \ldots$

**e** $49 = 7^{\square} = 7 \times \ldots$

**f** $8 = 2^{\square} = 2 \times \ldots$

**g** $81 = 3^{\square} = 3 \times \ldots$

**h** $10\,000 = 10 \times 10 \times \ldots = 10^{\square}$

**i** $1\,000\,000 = 10 \times 10 \times \ldots = 10^{\square}$

**2** Copy and complete:

**a** $3^2 \times 3^4 = 3 \times 3 \times \ldots = 3^{\square}$

**b** $2^3 \times 2^5 = 2 \times 2 \times \ldots = 2^{\square}$

**c** $7^5 \times 7 = 7 \times 7 \times \ldots = 7^{\square}$

**d** $5^6 \times 5^4 = 5 \times 5 \times \ldots = 5^{\square}$

**3** Simplify:

**a** $6^2 \times 6^3 \times 6^5$ **b** $7 \times 7^{10} \times 7^{12}$

**c** $3^2 \times 3^{10} \times 3^5$ **d** $10 \times 10 \times 10^3$

**4** Simplify these, if possible, leaving the answer in index form. If it is not possible, explain why.

**a** $2^6 \times 2^7$ **b** $2^3 \times 3^2$

**c** $3^2 \times 4^2 \times 5^2$ **d** $4^8 \times 4^2 \times 4$

**5** Copy and complete:

**a** $\frac{3^4}{3^2} = \frac{3 \times 3 \times \ldots}{3 \times \ldots} = 3^{\square}$

**b** $\frac{5^7}{5^3} = \frac{5 \times 5 \times \ldots}{5 \times \ldots} = 5^{\square}$

**c** $\frac{7^{10}}{7^4} = \frac{7 \times 7 \times \ldots}{7 \times \ldots} = 7^{\square}$

**6** Simplify:

**a** $\frac{2^6}{2^3}$ **b** $\frac{3^7}{3^4}$ **c** $\frac{4^8}{4^3}$

**d** $\frac{7^{10}}{7^5}$ **e** $\frac{9^7}{9}$ **f** $\frac{5^{12}}{5^8}$

**7** Simplify, leaving your answer in index form.

**a** $6^3 \div 6^2$ **b** $5^7 \div 5^4$

**c** $12^6 \div 12^3$ **d** $7^5 \div 7$

**e** $20^9 \div 20^4$ **f** $q^2 \div q$

**g** $b^6 \div b^5$ **h** $y^4 \div y^m$

**i** $8p^7 \div 2p^3$ **j** $6x^6 \div 2x^2$

**k** $10m^{10} \div 2m^2$

## Zero and negative indices

What does $5^0$ mean?

$5^2 \div 5^2 = 25 \div 25 = 1$, but
$5^2 \div 5^2 = 5^{2-2} = 5^0$
So $5^0 = 1$

Using symbols,

$$a^0 = 1$$

Indices can also be negative.
For example,

$$6^3 \div 6^5 = \frac{6 \times 6 \times 6}{6 \times 6 \times 6 \times 6 \times 6} = \frac{1}{6 \times 6} = \frac{1}{6^2}$$

but $6^3 \div 6^5 = 6^{3-5} = 6^{-2}$

so $6^{-2} = \frac{1}{6^2}$

Using symbols,

$$a^{-n} = \frac{1}{a^n}$$

## Using the order of operations, including brackets and powers

The order of operations (BIDMAS) also applies to indices. Remember, in calculations we do:

Brackets first (note long dividing lines act like brackets)
Then Indices (or powers)
Then Division and Multiplication
Then Addition and Subtraction

**EXAMPLE 8**

Simplify: $16 \times 2^{-3}$

$$16 \times 2^{-3} = 16 \times \frac{1}{2^3} = 16 \times \frac{1}{2 \times 2 \times 2}$$
$$= 16 \times \frac{1}{8} = 2$$

**EXAMPLE 9**

Simplify: $\frac{(3^6 \times 3^4)^2}{3^2 \times 3^3}$

### Exercise 1D

**1** Copy and complete:

**a** $2^3 \div 2^3 = 2^{\square} = 1$
**b** $3^5 \div 3^5 = 3^{\square} = \square$
**c** $7^2 \div 7^2 = 7^{\square} = \square$
**d** $9^{10} \div 9^{10} = 9^{\square} = \square$

**2** Write down the value of:

**a** $5^0$ **b** $7^0$ **c** $(^-2)^0$
**d** $8^0$ **e** $\left(\frac{1}{2}\right)^0$ **f** $m^0$

**3** Copy and complete:

**a** $4^5 \div 4^6 = \frac{4 \times \ldots}{4 \times \ldots} = \frac{1}{\square}$

**b** $4^5 \div 4^6 = 4^{\cdots} = 4^{-1}$

**c** $\frac{1}{\square} = 4^{-1}$

**4** Write as a fraction:

**a** $7^{-1}$ **b** $3^{-1}$ **c** $4^{-1}$

**d** $8^{-1}$ **e** $5^{-2}$ **f** $3^{-3}$

**5** Write in index form:

**a** $\frac{1}{6}$ **b** $\frac{1}{2}$ **c** $\frac{1}{20}$

**d** $\frac{1}{100}$ **e** $\frac{1}{4^2}$ **f** $\frac{1}{5^3}$

**6** Find: **a** $5^6 \div 5^8$ **b** $8^4 \div 8^7$

**7** By considering $4^7 \div 4^9$, show that $4^{-2} = \frac{1}{4^2}$.

**8** By considering $9^5 \div 9^8$, show that $9^{-3} = \frac{1}{9^3}$.

**9** Look carefully at the first two examples below. Then copy and complete the list.

$$0.1 = \frac{1}{10} = 10^{-1}$$

$$0.01 = \frac{1}{100} = 10^{-2}$$

$$0.001 = \frac{1}{1000} = \square$$

$$0.0001 = \square = \square$$

$$0.00001 = \square = \square$$

## INVESTIGATION

A 3 × 3 × 3 cube made up of 27 smaller cubes is painted green on its outside.

How many of the smaller cubes are unpainted?

How many of the smaller cubes have 1 face painted? 2 faces painted? 3 faces painted?

What about a 4 × 4 × 4 cube made up of 64 smaller cubes?

Copy and complete the table.

| Cube | Number of faces painted on smaller cubes | | | |
|---|---|---|---|---|
| | None | 1 | 2 | 3 |
| 2 × 2 × 2 | | | | |
| 3 × 3 × 3 | | | | |
| 4 × 4 × 4 | | | | |
| 5 × 5 × 5 | | | | |

Can you spot any rules?

What about a 13 × 13 × 13 cube?

## Exercise 1E

**1** A piece of ribbon is 50 cm long. How many $10\frac{1}{2}$ cm lengths of ribbon can be cut from it? How much ribbon is left over?

**2** I walk $\frac{2}{3}$ km to school, then $2\frac{1}{2}$ km to the store, then $1\frac{1}{4}$ km back home. How far do I walk altogether?

**3** On each bounce a ball rises to $\frac{2}{3}$ of its height at the start of the bounce. To what height will it rise after the fourth bounce, if it was originally dropped from a height of 81 cm?

**4** A bottle of juice holds $2\frac{1}{4}$ litres. How many glasses of juice can be poured from the bottle if each glass holds $\frac{3}{8}$ litre?

**5** A $\frac{1}{2}$-litre jug is filled with milk. It is used to fill two cups, one holding $\frac{1}{6}$ litre the other $\frac{2}{7}$ litre. How much milk remains in the jug?

**6** Calculate the exact value of $\dfrac{3\frac{2}{3} - \frac{1}{7}}{2\frac{3}{7}}$.

**7** **a** How many quarters in 16?
**b** How many thirds in 3?
**c** What is a tenth of 10?
**d** What is three tenths of 3?

**8** **a** Find a tenth of three quarters of $800.
**b** Find three quarters of a tenth of $800.

**9** Leon has made a mistake in his homework.

The question is: $2\frac{7}{45} \times 35$

Leon has written:

$$2\frac{7}{45} \times 35 = 2\frac{7}{9} \times 7$$

$$= \frac{25}{9} \times \frac{7}{1}$$

$$= \frac{175}{9}$$

$$= 19\frac{4}{9}$$

What mistake has he made?

# Consolidation

### Example 1

Work out:

**a** $3\frac{1}{4} + 2\frac{2}{5}$ **b** $4\frac{1}{2} \times 2\frac{2}{3}$ **c** $2\frac{1}{2} \div 1\frac{3}{4}$

**a** $3\frac{1}{4} + 2\frac{2}{5}$

$= 5 + \frac{1}{4} + \frac{2}{5}$

$= 5 + \frac{5}{20} + \frac{8}{20}$

$= 5 + \frac{13}{20} = 5\frac{13}{20}$

**b** $4\frac{1}{2} \times 2\frac{2}{3}$

$= \frac{9}{2} \times \frac{8}{3}$

$= \frac{\overset{3}{\cancel{9}}}{\underset{1}{\cancel{2}}} \times \frac{\overset{4}{\cancel{8}}}{\underset{1}{\cancel{3}}}$

$= \frac{3 \times 4}{1 \times 1} = 12$

**c** $2\frac{1}{2} \div 1\frac{3}{4}$

$= \frac{5}{2} \div \frac{7}{4}$

$= \frac{5}{2} \times \frac{4}{7}$

$= \frac{20}{14} = 1\frac{6}{14} = 1\frac{3}{7}$

### Example 2

Work out $\dfrac{(1\frac{1}{3})^2}{2\frac{2}{5} + 1\frac{1}{3}}$

$$\frac{(1\frac{1}{3})^2}{2\frac{2}{5} + 1\frac{1}{3}} = \frac{(\frac{4}{3})^2}{\frac{12}{5} + \frac{4}{3}}$$

$$= \frac{\frac{16}{9}}{\frac{36}{15} + \frac{20}{15}} = \frac{\frac{16}{9}}{\frac{56}{15}}$$

$$= \frac{16}{9} \div \frac{56}{15} = \frac{16}{9} \times \frac{15}{56}$$

$$= \frac{\overset{2}{\cancel{16}}}{\underset{3}{\cancel{9}}} \times \frac{\overset{5}{\cancel{15}}}{\underset{7}{\cancel{56}}} = \frac{2}{3} \times \frac{5}{7} = \frac{10}{21}$$

### Example 3

Simplify:

**a** $2^7 \times 2^3$ **b** $\dfrac{4^3 \times 4^4}{4^7}$

**c** $\dfrac{(2^3 \times 2^4)^2}{2^{16}}$

**a** $2^7 \times 2^3 = 2^{7+3} = 2^{10}$

**b** $\dfrac{4^3 \times 4^4}{4^7} = \dfrac{4^7}{4^7} = 4^{7-7} = 4^0 = 1$

**c** $\dfrac{(2^3 \times 2^4)^2}{2^{16}} = \dfrac{(2^7)^2}{2^{16}} = \dfrac{2^7 \times 2^7}{2^{16}} = \dfrac{2^{14}}{2^{16}}$

$= 2^{14-16} = 2^{-2} = \dfrac{1}{2^2} = \dfrac{1}{4}$

## Exercise 1

**1** Work out:

**a** $\frac{1}{2} + \frac{2}{3}$ **b** $\frac{3}{8} + \frac{4}{7}$ **c** $2\frac{3}{5} + 1\frac{2}{7}$

**d** $\frac{6}{7} - \frac{2}{3}$ **e** $6\frac{1}{2} - 2\frac{5}{8}$

**2** Work out:

**a** $\frac{3}{5} \times \frac{2}{3}$ **b** $2\frac{1}{2} \times 3\frac{1}{3}$ **c** $\frac{6}{7} \div \frac{2}{3}$

**d** $\frac{4}{5} \div 1\frac{2}{3}$ **e** $4\frac{3}{4} \div 2\frac{2}{5}$

**3** Work out:

**a** $1\frac{3}{4} + 1\frac{3}{7} \times 2\frac{1}{3}$ **b** $\dfrac{2\frac{3}{4}}{2\frac{1}{4} + 1\frac{1}{3}}$

**c** $1\frac{1}{3} \times \left(4\frac{1}{4} - 3\frac{3}{8}\right) - \frac{5}{6}$ **d** $\dfrac{1\frac{1}{3} + \frac{2}{7}}{1\frac{1}{7} \times 2\frac{5}{6}}$

**e** $\dfrac{(1\frac{1}{2})^2}{2\frac{1}{5} - 1\frac{1}{4}}$

**4** Simplify:

**a** $2^4 \times 2^2$ **b** $5^8 \div 5^2$

**c** $4^8 \div 4^{12}$ **d** $4^7 \times 4^3 \times 4^2$

**e** $4^4 \times 4^3 \div 4^{10}$ **f** $\dfrac{9^{20}}{(9 \times 9^9)^2}$

**g** $\dfrac{(6^2 \times 6^3)^2}{6^9}$ **h** $\dfrac{(8^4 \times 8^5)^2}{(8^{20} \div 8^2)}$

**5** A piece of cloth is $4\frac{1}{2}$ m long. Amandeep used $\frac{2}{5}$ of it. What length did he use?

**6** It takes $3\frac{3}{4}$ hours to plaster a wall. One third of this time is spent mixing the plaster. How long does the actual plastering take?

**7** Janice bought $6\frac{1}{2}$ metres of cloth. She used $1\frac{1}{4}$ metres. What fraction did she use?

**8** $7\frac{1}{2}$ kg of coffee are put into $1\frac{1}{4}$ kg bags.
How many bags will be needed?

**9** A tin holds $5\frac{1}{2}$ litres of oil.
How many $\frac{2}{3}$-litre cans can be filled from it?

**10** A piece of wood is $4\frac{1}{3}$ metres long.
How many $1\frac{1}{4}$-metre strips can be cut from it?

**11** A man's work day is $8\frac{1}{2}$ hours. He spends $3\frac{1}{4}$ hours dealing with customers and the remaining time working in his office.

**a** What fraction of his work day does he spend with customers?

**b** What fraction of the whole day is he in his office?

(**Hint:** think about how many hours there are in a *whole* day.)

**12** A man spends $\frac{2}{7}$ of his weekly wage on food and $\frac{2}{5}$ on rent. If he spends \$200 on food, how much does he spend on rent?

**13** What are the numbers $p$, $q$, $r$ and $s$ if

**a** $2^p = 4^q = 16^2 = 256$

**b** $3^r = 9^s = 81^2 = 6561$

**14** Yafeu says that if he starts with a number and squares it he ends up with the same number he started with. What could his number be? Is there more than one answer?

## Summary

### You should know ...

**1** How to add and subtract fractions.
*For example:*

$$\frac{3}{4} + \frac{1}{3} = \frac{9}{12} + \frac{4}{12} = \frac{13}{12} = 1\frac{1}{12}$$

**2** How to multiply and divide fractions.
*For example:*

$$\frac{3}{5} \times \frac{10}{12} = \frac{\overset{1}{\cancel{3}}}{\underset{1}{\cancel{5}}} \times \frac{\overset{2}{\cancel{10}}}{\underset{4}{\cancel{12}}} = \frac{2}{4} = \frac{1}{2}$$

$$3\frac{3}{4} \div \frac{5}{8} = \frac{15}{4} \div \frac{5}{8} = \frac{15}{4} \times \frac{8}{5} = \frac{\overset{3}{\cancel{15}}}{\underset{1}{\cancel{4}}} \times \frac{\overset{2}{\cancel{8}}}{\underset{1}{\cancel{5}}} = \frac{3}{1} \times \frac{2}{1} = 6$$

**3** How to multiply and divide with indices, and about the zero index.
*For example:*

$3^2 \times 3^4 = 3^{(2+4)} = 3^6$

$6^5 \div 6^2 = 6^{(5-2)} = 6^3$

$4^0 = 1$

### Check out

**1** Work out:

**a** $\frac{1}{2} + \frac{1}{4}$ **b** $\frac{3}{5} + \frac{1}{3}$

**c** $\frac{2}{3} - \frac{1}{4}$ **d** $\frac{7}{8} - \frac{3}{4}$

**e** $1\frac{4}{5} + \frac{3}{4}$ **f** $2\frac{2}{3} - 1\frac{3}{5}$

**2** Work out

**a** $\frac{1}{2} \times \frac{3}{4}$ **b** $\frac{3}{4} \div \frac{1}{2}$

**c** $\frac{3}{5} \times \frac{2}{3}$ **d** $1\frac{1}{2} \div \frac{1}{4}$

**e** $2\frac{1}{2} \times 2\frac{1}{2}$ **f** $2\frac{3}{4} \div 1\frac{1}{4}$

**g** $\left(1\frac{2}{9}\right)^2$

**h** $1\frac{3}{4} \times \left(\frac{1}{8} + \frac{1}{2}\right)$

**3** Work these out. Write your answers in index form.

**a** $2^2 \times 2^3$ **b** $5^4 \times 5^7$

**c** $7^8 \times 7^3$ **d** $9^5 \times 9^2$

**e** $4^5 \div 4^2$ **f** $3^3 \div 3^3$

**g** $2^8 \div 2^4$ **h** $7^0$

# 2 Expressions and formulae

## Objectives

- Know the origins of the word *algebra* and its links to the work of the Arab mathematician Al'Khwarizmi.
- Simplify or transform algebraic expressions by taking out single-term common factors.
- Use index notation for positive integer powers; apply the index laws for multiplication and division to simple algebraic expressions.
- Construct algebraic expressions.
- Add and subtract simple algebraic fractions.
- Expand the product of two linear expressions of the form $x \pm n$ and simplify the corresponding quadratic expression.
- Substitute positive and negative numbers into expressions and formulae.
- Derive formulae and, in simple cases, change the subject; use formulae from mathematics and other subjects.

## What's the point?

Relationships between two variables are commonly expressed in algebraic form, as equations. For example, the equation $D = v \times t$ relates distance travelled, $D$, by a car to its average speed, $v$, and the time taken, $t$.

## Before you start

### You should know ...

**1** How to add and subtract negative numbers.
*For example:*
$8 + {}^-3 = 8 - 3 = 5$
$8 - {}^-3 = 8 + 3 = 11$

### Check in

**1** Work out:

**a** $6 + {}^-2$
**b** ${}^-2 - 3$
**c** ${}^-3 - {}^-6$
**d** ${}^-6 + 5$
**e** ${}^-3 + {}^-6 - {}^-2$

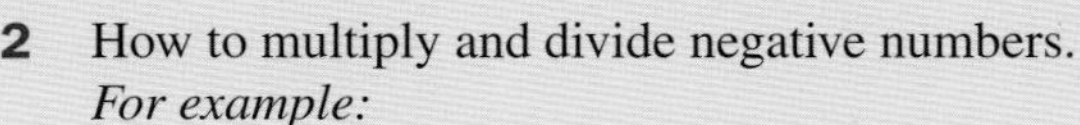

| | |
|---|---|
| **2** How to multiply and divide negative numbers.<br>*For example:*<br>$6 \times {}^{-}3 = {}^{-}18$<br>${}^{-}6 \times {}^{-}3 = 18$<br>$6 \div {}^{-}3 = {}^{-}2$<br>${}^{-}6 \div {}^{-}3 = 2$ | **2** Calculate:<br>**a** ${}^{-}3 \times 5$<br>**b** ${}^{-}2 \times {}^{-}6$<br>**c** $\frac{{}^{-}8}{4}$<br>**d** ${}^{-}15 \div {}^{-}3$<br>**e** $\frac{{}^{-}3 + 5}{{}^{-}2}$ |
| **3** How to substitute numbers for letters.<br>*For example:*<br>if $x = 10$ and $y = 6$<br>then $2x + 3y = 2 \times 10 + 3 \times 6$<br>$= 20 + 18$<br>$= 38$ | **3** If $a = 5$ and $b = 3$, find:<br>**a** $ab$<br>**b** $2a + b$<br>**c** $a + 2b$<br>**d** $ab + 2b$ |
| **4** How to work with indices.<br>*For example:*<br>$7^2 \times 7^3 = 7^{2+3} = 7^5$<br>$8^5 \div 8^2 = 8^{5-2} = 8^3$ | **4** Simplify:<br>**a** $9^5 \times 9^4$<br>**b** $8^{10} \div 8^2$<br>**c** $5^6 \div 5^5$<br>**d** $3^6 \times 3^2$ |
| **5** The lowest common multiple (LCM) is the smallest multiple common to two or more numbers.<br>*For example:*<br>The LCM of 3 and 4 is 12. | **5** Find the LCM of:<br>**a** 2 and 5<br>**b** 2 and 14<br>**c** 4 and 6 |

## 2.1 The origins of algebra and algebraic expressions

Muhammad ibn Musa Al'Khwarizmi was a mathematician who lived more than a thousand years ago. He worked in Baghdad at the 'House of Wisdom', where philosophical articles were written and translated. He worked with Hindu-Arabic numbers and was among the first to use zero as a placeholder. He wrote the first book about algebra, called *Al-Kitab al-mukhtasar fi hisab al-jabr wa-l-muqabala* or 'The Compendious Book on Calculation by Completion and Balancing'. The word 'algebra' comes from the *al-jabr* of the book's title, and Al'Khwarizmi is known as the 'father of algebra'.

Algebra involves using letters to stand for numbers, allowing us to manipulate letters in a similar way to numbers.

$3 + 3 + 3 + 3 = 4 \times 3 = 12$
$a + a + a + a = 4 \times a = 4a$

$3 \times 3 \times 3 \times 3 = 3^4 \quad = 81$
$a \times a \times a \times a = a^4$

$4a$ and $a^4$ are **algebraic expressions**. This just means they contain letters to stand for numbers.

**EXAMPLE 1**

Construct an expression for the perimeter of this rectangle.

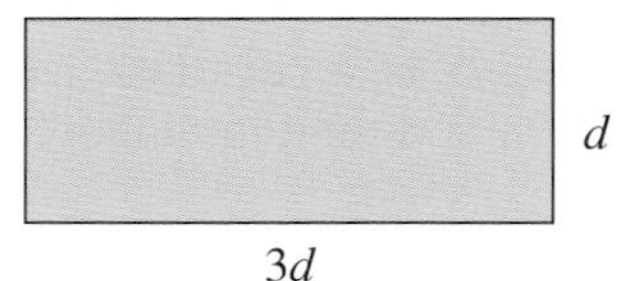


Perimeter $= 3d + d + 3d + d$

$= 8d$

If $d = 2\,\text{cm}$, then

perimeter $= 8 \times 2\,\text{cm} = 16\,\text{cm}$

### Exercise 2A

**1** Construct an expression for the perimeter of each shape.

**a**

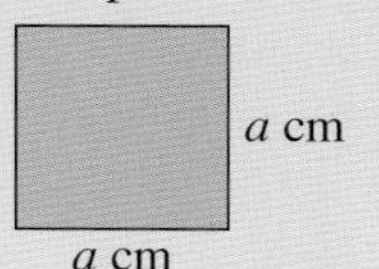


**b**

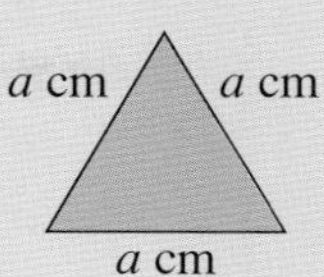


**c**

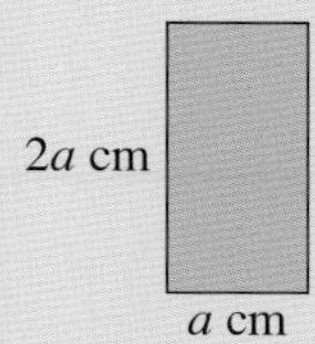


**2** Find the perimeter of each shape in Question **1**, when:

**a** $a = 5$ **b** $a = 7$ **c** $a = 12$

**3** Derive a formula for the perimeter, $P$, of each rectangle.

**a**

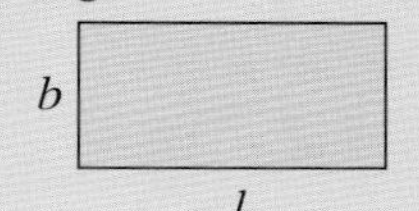


**b**

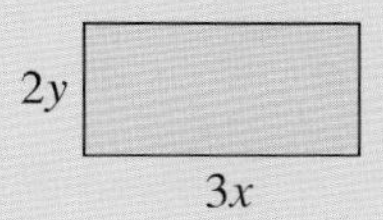


**4** Which of these formulae can be used for the perimeter of the rectangle in Question **3 a**?

**a** $P = l + b + l + b$

**b** $P = 2l + 2b$

**c** $P = 2(l + b)$

**5** Find the perimeter, $P$, of the rectangle in Question **3 a**, when:

**a** $l = 6, b = 3$ **b** $l = 12, b = 5$

**6** Find the perimeter, $P$, of the rectangle in Question **3 b**, when:

**a** $x = 5, y = 2$ **b** $x = 10, y = 7$

**7** Write in a shorter way:

**a** $3x + 2x$ **b** $7y + 9y$

**c** $5z + 7z + 8z$ **d** $9x - 3x$

**e** $12y - 7y$ **f** $9z - 6z - 2z$

**8** Find the value of each expression in Question **7** when $x = 3, y = 8$ and $z = 12$.

**9** The perimeter of a rectangle is given by the formula

$$P = 2(l + w)$$

where $l$ is the length and $w$ is the width of the rectangle.

What is the perimeter of a rectangle with:

**a** $l = 6\,\text{cm}, w = 4\,\text{cm}$

**b** $l = 9\,\text{cm}, w = 5\,\text{cm}$?

**10** Derive a formula for $A$, the area of these rectangles.

**a**

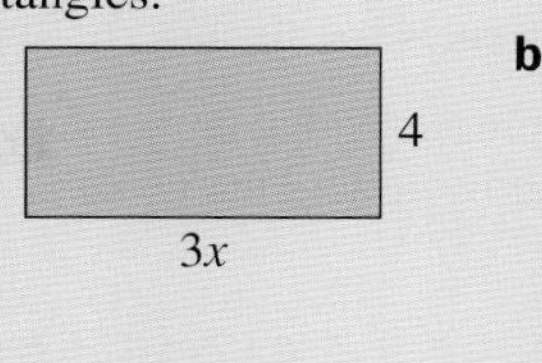


**b**

6

$2y$

**11** Find the area, $A$, of the rectangles in Question **10** if

**i** $x = 7$ and $y = 1$

**ii** $x = 3$ and $y = 1.5$

**12** To convert a Celsius temperature, $C$, to a Fahrenheit temperature, $F$, the formula

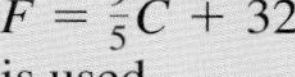

$F = \frac{9}{5}C + 32$

is used.

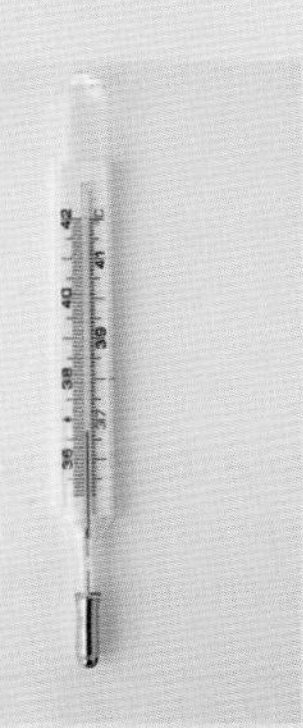

**a** Water boils at 100 °C. What Fahrenheit temperature is this?

**b** What Celsius temperature is 32 °F?

## 2.2 Simplifying and indices

In algebra, multiplication signs are often missed out.

For example:

$12 \times x = 12x$

$3 \times p \times q = 3pq$

$2 \times 3 \times p \times r = 6pr$

$2 \times 5m \times 10n = 100mn$

$\frac{1}{3} \times 5x \times (^{-}9z) = {}^{-}15xz$

It may help you to use the longer form when simplifying fractions.

**EXAMPLE 2**

$$\frac{3x^3y^2z^2}{x^2y^3z} = \frac{3 \times x \times x \times x \times y \times y \times z \times z}{x \times x \times y \times y \times y \times z}$$

$$= \frac{3 \times x \times \cancel{x} \times \cancel{x} \times \cancel{y} \times \cancel{y} \times z \times \cancel{z}}{\cancel{x} \times \cancel{x} \times y \times \cancel{y} \times \cancel{y} \times \cancel{z}}$$

$$= \frac{3 \times x \times z}{y}$$

$$= \frac{3xz}{y}$$

## Using indices with algebra

In Chapter 1 you learned about the rules for indices. You can use the same rules for algebra.

You can multiply expressions that are powers of the same number or letter by adding the powers.
*For example:*

$$a^5 \times a^3 = (a \times a \times a \times a \times a) \times (a \times a \times a)$$
$$= a^{5+3} = a^8$$

Similarly, you can divide expressions that are powers of the same number or letter by subtracting the powers.

$$\frac{a^5}{a^3} = \frac{a \times a \times \cancel{a} \times \cancel{a} \times \cancel{a}}{\cancel{a} \times \cancel{a} \times \cancel{a}} = a^{5-3} = a^2$$

The rules for indices are

$$a^m \times a^n = a^{m+n}$$
$$a^m \div a^n = a^{m-n}$$
$$a^0 = 1$$
$$a^{-n} = \frac{1}{a^n}$$

**Note that negative powers go beyond what you are expected to do at stage 9 and are included here as extension work.**

**EXAMPLE 3**

Simplify:

**a** $p \times p^4$
**b** $m^7 \div m^3$
**c** $3x^3 \times 4x^2$
**d** $y^2 \div y^{10}$

**a** $p \times p^4 = p^{1+4} = p^5$
**b** $m^7 \div m^3 = m^{7-3} = m^4$
**c** $3x^3 \times 4x^2 = 3 \times 4 \times x^3 \times x^2 = 12 \times x^{3+2}$
$= 12x^5$
**d** $y^2 \div y^{10} = y^{2-10} = y^{-8} = \frac{1}{y^8}$

**EXAMPLE 4**

Simplify $\frac{a^2bc^3}{a^5b^2} \times \frac{ab^3c^2}{bc^4}$

Multiply the two numerators together and the two denominators together:

$$\frac{a^2bc^3ab^3c^2}{a^5b^2bc^4}$$
$$= \frac{\cancel{a} \times \cancel{a} \times \cancel{b} \times \cancel{c} \times \cancel{c} \times \cancel{c} \times \cancel{a} \times \cancel{b} \times \cancel{b} \times b \times \cancel{c} \times c}{\cancel{a} \times \cancel{a} \times \cancel{a} \times a \times a \times \cancel{b} \times \cancel{b} \times \cancel{b} \times \cancel{c} \times \cancel{c} \times \cancel{c} \times \cancel{c}}$$
$$= \frac{bc}{a^2}$$

Alternatively, using the rules for indices:

$$\frac{a^2bc^3}{a^5b^2} \times \frac{ab^3c^2}{bc^4} = \frac{a^{2+1} \times b^{1+3} \times c^{3+2}}{a^5 \times b^{2+1} \times c^4}$$
$$= \frac{a^3b^4c^5}{a^5b^3c^4}$$
$$= a^{3-5} \times b^{4-3} \times c^{5-4}$$
$$= a^{-2} \times b \times c = \frac{bc}{a^2}$$

### Exercise 2B

**(Note: Questions 6 and 7 are extension work.)**

**1** Simplify:

**a** $a^2 \times a^3$ **b** $q^5 \times q^3$
**c** $r^4 \times r^6$ **d** $s^{10} \times s^5$
**e** $p \times p^2 \times p^5$ **f** $j^3 \times j^7 \times j^9$
**g** $a^m \times a^n$ **h** $p^a \times p^b$
**i** $m^a \times m^b \times m^c$ **j** $x^a \times x^m \times x^c$

**2** Simplify, leaving your answer in index form.

**a** $q^2 \div q$ **b** $b^8 \div b^5$
**c** $y^8 \div y^4$ **d** $8p^7 \div 2p^3$
**e** $6x^6 \div 2x^2$ **f** $14y^{10} \div 7y^5$
**g** $\frac{10n^6}{2n^3}$ **h** $\frac{50x^3}{5x^5}$

**3** Simplify:

**a** $2a^3 \times 4a^4$ **b** $3c^2 \times c^3 \times 2c^4$
**c** $12p^3 \div 2p^2$ **d** $10q^5 \div 5q$

**4** $(a^3)^4 = a^3 \times a^3 \times a^3 \times a^3$
$= a^{3+3+3+3} = a^{12}$

Use the method above to simplify:

**a** $(x^2)^3$ **b** $(x^3)^2$
**c** $(x^3)^3$ **d** $(2x^2)^4$
**e** $(x^{-2})^4$ **f** $(3x^2)^4$

**5** Write $(a^m)^n$ in another way.

**6** Simplify:

**a** $x^2 \times x^{-3} \times x^4$ **b** $p^7 \times p^5 \div p^3$
**c** $y^4 \times y^{-3} \times y^{-5}$ **d** $q^5 \times q^{-4} \div q^{-2}$

**7** Write with positive indices only:

**a** $p^{-5}$ **b** $x^{-2} \times x^{-5} \times x^4$
**c** $q^{-3} \div q^4$ **d** $y^2 \times y^{-5} \div y^{-3}$
**e** $(k^2 \times k^{-3}) \div (k^4 \div k^{-1})$

**8** Simplify:

a $lm^3n^2 \div mn^2$ b $x^5y^2z^4 \div y^3z^3$

**9** Simplify:

a $\frac{p^3q}{pq^4}$ b $\frac{x^4yz^3}{xy^2z^4}$

c $\frac{k^4l^2m^5}{kl^2m^3}$

**10** Simplify:

a $2(p^2q)^3$ b $\frac{(pq)^3}{p^2q^3}$

c $\frac{(3p^2q)^2}{3pq^2}$ d $\frac{4p^2(q^2r)^3}{3pr \times 2qr^2}$

**11** Copy out the expressions below. Draw lines to show which are equivalent. One has been done for you.

$x^2 \times x^3$ $\frac{x^{10}}{x^3} \times \frac{x^2}{x \times x}$

$\frac{x^4}{x^2 \times x} \times x^2$ $x^{11} \times x^2 \div x^{10}$

$x^4 \times x^2 \times x$ $x^7 \div x^2$

$x^5 \div x^8$ $x^4 \div \frac{(x^4)^2}{x}$

**12** Put these expressions in order, starting with the lowest index.

$\frac{(p^5 \times p^3)^2}{(p \times p^3)}$ $\frac{p^{30}}{(p^2 \times p^8)^2}$ $\frac{(p^4 \times p^2)^3}{p^5}$

**13** Simplify:

a $(P^3)^2$ b $x^2 \times x^3 \div x^4$

c $A^5 \div A^3 \times A^4$ d $(m \times m^4)^2$

e $\frac{x^2 \times x^7}{x^4}$ f $\frac{(y^2 \times y^3)^2}{(y \times y^2)^3}$

g $(3T^2 + T^2) \div T$ h $\left(\frac{h^8}{h^2}\right)^2 \times h^3$

i $g^3 + (g^4 + g^4) \div 2g$

**TECHNOLOGY**

For some further work on this, go to www.onlinemathlearning.com/multiply-divide-expressions.html

An expression is a collection of **terms** separated by plus or minus signs.

**Like terms** are terms with the same combination of letters, raised to the same power.

For example, in

$3ab + 5ab + 6$

$3ab$ and $5ab$ are like terms.

Similarly,

$2a$ and $5a$
$ab^2$ and $10ab^2$
$xy^3$ and $6xy^3$

are like terms.

You can simplify expressions with several terms by collecting like terms together.

**EXAMPLE 5**

Simplify:

a $3a + 4b - 2a + 5b$

b $x + 3y + 7x - 12x$

a $3a + 4b - 2a + 5b$
$= (3a - 2a) + (4b + 5b)$
$= a + 9b$

b $x + 3y + 7x - 12x$
$= 3y + (x + 7x - 12x)$
$= 3y - 4x$

You can simplify more complicated expressions in the same way.

**EXAMPLE 6**

Simplify:

a $3pq - 8pq + 4pq$

b $xy + 2ab - 5xy + 10ab$

c $2n^2 + m^2 - n^2 + 2m^2$

a $3pq - 8pq + 4pq = {}^-pq$

b $xy + 2ab - 5xy + 10ab$
$= (xy - 5xy) + (2ab + 10ab)$
$= {}^-4xy + 12ab$

c $2n^2 + m^2 - n^2 + 2m^2$
$= (2n^2 - n^2) + (m^2 + 2m^2)$
$= n^2 + 3m^2$

## Exercise 2C

**1** Simplify:

**a** $3x + 5x$  **b** $8a + 2a$
**c** $14b - 2b$  **d** $^{-}7y + 4y$
**e** $8a + 3a - 2a$  **f** $7b - 3b - b$
**g** $4p - 7p - 5p$  **h** $3ab + 5ab + 9ab$
**i** $2a^2 + 7a^2$  **j** $9b^2 - 2b^2$
**k** $x^2 + 3x^2 + 7x^2$  **l** $8y^3 - 2y^3 - 4y^3$

**2** Collect like terms together and simplify:

**a** $4a + 3a + 7b + 5b$
**b** $7a + 3b + 4a + 2b$
**c** $4a + 9b - 2a - 3b$
**d** $6x - 3y - 2x + 7y$
**e** $x - y - 8x - 9y$
**f** $5xy + 2z - 3xy - 5z$
**g** $3ab + pq - ab + 5pq$
**h** $a^2 + b^2 + 6a^2 - 3b^2$
**i** $4a^3 - a - a^3 + 5a$

**3** Simplify the expressions:

**a** $4a + 3a + 7b + 2a$
**b** $4p + 9q - 2p - 3q$
**c** $6z^3 - z - 7z^3 + 4z$
**d** $a^2 + b^2 + 6a^2 - 3b^2$
**e** $p^2 + 2pq + 6pq - 4p^2$

**4** Simplify:

**a** $2 \times 3a$  **b** $5b \times 4$
**c** $4a \times 3b$  **d** $7a \times 2a$
**e** $3 \times 2b \times 4c$  **f** $2a \times 5b \times 3c$
**g** $^{-}3 \times 7y$  **h** $(^{-}6p) \times (^{-}5)$
**i** $(^{-}3q) \times 4p$  **j** $(^{-}5y) \times (^{-}3y)$
**k** $3a \times 2a \times a$
**l** $(^{-}2x) \times 5x \times (^{-}4x)$

**5** Simplify:

**a** $6a \div 2$  **b** $8b \div 4b$
**c** $5a \div 2a$  **d** $4pq \div q$
**e** $6ab \div 2ab$  **f** $8pq \div 2q$
**g** $10 \div 2x$  **h** $3 \div 6y$
**i** $a^2b \div ab$  **j** $a^2b^2 \div ab^2$
**k** $7ab^4 \div 3a^3b$  **l** $4x^2y^2z^2 \div xy^3z^2$

**[6]** Simplify the expressions:

**a** $mn + 4xy + 17xy - mn + xy$
**b** $3p^2q + lm + 14p^2q + 6lm$
**c** $pqr + abx + mny + 5abx$
**d** $15ab - 29ab + 4pq + ab$
**e** $x^2y^2 - 4xy + 13xy + 3x^2y^2$

**[7]** Find the value of each expression when $m = 2, n = 3$ and $p = 4$.

**a** $5m + 7n + 2p$  **b** $2m \times 3n \times p$
**c** $5mnp$  **d** $m^2n$
**e** $pm^3$  **f** $p^2 + n^2$
**g** $mn + 2p$  **h** $p^3n \div 6$

**[8]** Simplify:

**a** $4x^2 + 7x^2$
**b** $2mn + 5mn - 3mn$
**c** $4l^2m^2 - 7l^2m^2$
**d** $8pqr - 9pqr - 15pqr$
**e** $^{-}3 \times 5m$
**f** $6a \times 11b$
**g** $7a^2 \times 3a^3$
**h** $(^{-}4pq) \times (^{-}2p^2q^3)$
**i** $10m^2 \div 2$
**j** $7a^2b^3 \div 2ab$
**k** $4x^5 \div 7x^8$
**l** $2lm^2n^3 \div 3l^3m^2n$

## 2.3 Expanding brackets

$3 \times (x + 5y + 4z)$ can be written as $3x + 15y + 12z$.

Each term inside the brackets is multiplied by the term outside. The brackets have been removed. This is called **expanding** the brackets.

**EXAMPLE 7**

Expand the brackets.

**a** $3(a + 2b + 5c)$
$= 3a + 6b + 15c$

**b** $^{-}3y(z - 2w) = (^{-}3y) \times z - (^{-}3y) \times 2w$
$= {^{-}3yz} + 6yw$

**c** $2x(y + 4z) - 3y(z - 2w)$
$= 2xy + 8xz - 3yz + 6yw$

## Exercise 2D

**1** Expand the brackets.

**a** $2(a + b)$  **b** $3(x + 2y)$
**c** $4(3p - 5q)$  **d** $5a(b + c)$
**e** $3p(q - 2r)$  **f** $5(a - 2b + 3c)$
**g** $2x(y - 2 + 3w)$  **h** $^{-}3(2l + m)$
**i** $^{-}2l(m - 5n)$  **j** $p(p + 3q - 5r)$
**k** $^{-}4y(3x - 2)$  **l** $4x(x^2 - x + 1)$

**2** Expand the brackets and simplify:

- **a** $3(x + y) + 2(x - y)$
- **b** $5(l + 2m) - (m - 2l)$
- **c** $6(p - q) + 4(r - q - p)$
- **d** $4a(b + c - a) + 2c(3a + b)$
- **e** $^{-}(m - 5) + 4m(m + 2)$
- **f** $^{-}3(mn + m) + 4n(2 - m)$

**3** Expand the brackets and simplify:

- **a** $3(x + y) + 2(x - y)$
- **b** $4(x + y) + 3(2x + 5y)$
- **c** $8(a - b) - 3(2a - b)$
- **d** $6(x - 3) + 5(x + 2)$
- **e** $x(x^2 + 1) + 2x(x^2 + 5)$
- **f** $7a - 4a(b + 3)$

**4** Expand the brackets and simplify:

- **a** $2(x - 3y) + 3(y - 2z) + 4(z - 5x)$
- **b** $3m(2 + n) + 5n(1 - 3m) + 12mn$
- **c** $x(x^2 - 1) + x^2(x + 2) - x(5 - x^2)$

**5** Expand the brackets and simplify:

- **a** $m(2 + n) + n(m - 3) - (mn - m)$
- **b** $^{-}(x^2 - 1) - x^2(4 - x) + 5(6 - x^3)$
- **c** $pq(p^2 + 1) + pq(1 + p^{-2})$
- **d** $xy(x^{-3} + x) + 5xy(3 + y^{-3})$
- **e** $p^{-2}(p^3 - p^4) + p^{-1}(pq + p^2a)$
- **f** $\frac{1}{3}p^2(p - p^{-1}) + \frac{5}{3}p(p^2 + p)$

$p^{-2} = \frac{1}{p^2}$

## 2.4 Factorising expressions

The opposite of expanding brackets is to put brackets in. This is called **factorising**.

For example:

3 is the highest common factor of the terms $3l$, $6m$ and $9n$.
So $3l + 6m + 9n = 3(l + 2m + 3n)$

$x$ is a common factor of the terms $ax$ and $bx$.
So $ax + bx = x(a + b)$

$x$ is a common factor of the terms $x^2$ and $3x$.
So $x^2 + 3x = x(x + 3)$
$x$ and $(x + 3)$ are the factors of $x^2 + 3x$.

$2m$ is one factor of $2lm + 6mn + 10mp$.
$(l + 3n + 5p)$ is the other factor.

### Exercise 2E

**1** Copy and complete:

- **a** $3x + 3y = 3(\quad)$
- **b** $5a - 5b = 5(\quad)$
- **c** $4x + 4y + 4z = 4(\quad)$
- **d** $6a - 6b + 6c = 6(\quad)$
- **e** $2x + 6y = 2(\quad)$
- **f** $8a - 4b = 4(\quad)$
- **g** $3x + 6y + 9z = 3(\quad)$
- **h** $25a - 10b - 5c = 5(\quad)$

**2** Copy and complete:

- **a** $ax + ay = a(\quad)$
- **b** $pa - pb = p(\quad)$
- **c** $px + py + pz = p(\quad)$
- **d** $ra - rb + rc = r(\quad)$
- **e** $qx + 3qy = q(\quad)$
- **f** $5sa - sb = s(\quad)$
- **g** $2tx + 5ty + tz = t(\quad)$
- **h** $7la - 4lb - lc = l(\quad)$

**3** Copy and complete:

- **a** $px + qx = (\quad)x$
- **b** $as - bs = (\quad)s$
- **c** $px + qx + rx = (\quad)x$
- **d** $ra - sa + ta = (\quad)a$
- **e** $3ly + 2my = (\quad)y$
- **f** $6fh - 5gh = (\quad)h$
- **g** $4xt + 9yt + zt = (\quad)t$
- **h** $21g - 7mg - 3ng = (\quad)g$

**4** Copy and complete:

- **a** $3m + 5mn + m^2 = m(\quad)$
- **b** $2p + 3pr + p^2 = p(\quad)$
- **c** $6l + 2lm + 2l^2 = 2l(\quad)$
- **d** $5rs + 50rs^2 + 15r^2s = 5rs(\quad)$

**5** Factorise:

- **a** $2a + 2b$ **b** $3a - 3b$
- **c** $4x + 12y$ **d** $9p - 6q$
- **e** $px + py$ **f** $ra - rb$
- **g** $7sx + 4sy$ **h** $2ta - 7tb$
- **i** $xa + xb + xc$ **j** $la - lb - lc$
- **k** $4rx + 5ry + rz$ **l** $pa - 6pb + 8pc$

**6** Factorise:

- **a** $lx + mx$ **b** $an - bn$
- **c** $7py + 2qy$ **d** $rt - 5st$
- **e** $pt + qt + rt$ **f** $an + bn - cn$
- **g** $5lx + mx + 2nx$ **h** $4kg - 2lg - mg$

**7** Factorise:

- **a** $4p + 2pr + 6pz$
- **b** $5m + 15mp + 25mg$
- **c** $9sr + 3s + 6s^2$
- **d** $4lm + 2mn + 8pmn$
- **e** $ab^2 + 6ab^3 + 2a^2b^2$
- **f** $2x^2y^2 + 3xy^2 + x^2y^2$

(**Hint**: simplify first)

**8** Factorise:

| | | | |
|---|---|---|---|
| **a** | $x^2 + 3x$ | **b** | $y^2 - 5y$ |
| **c** | $2z^2 + 3z$ | **d** | $4m^2 - m$ |
| **e** | $x^3 + 2xy$ | **f** | $4y^2z - y^3$ |
| **g** | $ab^2 + a^2b$ | **h** | $x^2yz^2 - xyz^3$ |
| **i** | $\pi r^2 + 2\pi rh$ | **j** | $2lm^2 + 8l^2m$ |
| **k** | $x^4 + x^3 + x^2$ | **l** | $32y + 16y^3 + 8y^5$ |

**9** Factorise:

**a** $abc^2 + ab^2 + a^2b$
**b** $p^3q^2r + p^2q^2r^2 + pq^2r^3$
**c** $7axy + 14bxy + 21cxy$
**d** $8x^6 + 16x^4 + 48x^3$
**e** $2lmp - lm + 5lm^2$
**f** $f^4g^2 - 6f^2g^3 + 2fg^4$
**g** $5abcd + 35bcde$
**h** $24k^2lm^2n - 32kl^2m^2n^3$
**i** $16abcx - 28bcdx - 20cdex$

**10** Expand the brackets, simplify, and factorise:

**a** $3(x - 5) + 4(x - 2) - 5$
**b** $x(y + 3) + 2x(4 - y) - 10xy$
**c** $a(b + c) + b(c - a) + 3c$
**d** $lm(5l + m) + 2l^2(m - 3) + 6l(m^2 + l)$

**11** Factorise each expression.

**a** $x^2y + 3xy + 4xy^2$
**b** $2rs + 18rst + 8r^2s^2 + 10rst^2$
**c** $14mn^2 + 2mn + 8m^2n + 8m^2n^2$
**d** $\frac{1}{3}g^2h + \frac{5}{3}g^3h + 2g^3h^3$
**e** $x^2y + 2xy^2 + x^2y^3 + x^4y^4$

You can often factorise an expression as the product of two expressions. This is known as **factorising by grouping**.

**EXAMPLE 8**

Factorise

**a** $ax + bx + ay + by$
**b** $4ny - 5py + 5px - 4nx$

**a** $ax + bx + ay + by = x(a + b) + y(a + b)$
$= (x + y)(a + b)$

**b** $4ny - 5py + 5px - 4nx$
$= y(4n - 5p) - x(4n - 5p)$
$= (y - x)(4n - 5p)$

4n−5p is a common factor.

## Exercise 2F

**1** Copy and complete:

**a** $px + qx + py + qy = x(\quad) + y(\quad)$
$= (x + y)(\quad)$

**b** $ax - bx + ay - by = x(\quad) + y(\quad)$
$= (x + y)(\quad)$

**c** $ut^2 + us + vt^2 + vs = u(\quad) + v(\quad)$
$= (u + v)(\quad)$

**d** $2lx + mx - my - 2ly = x(\quad) - y(\quad)$
$= (x - y)(\quad)$

**2** Factorise pairs of terms and then write the expression as a product of two expressions:

**a** $rx + sy + sx + ry$
**b** $sy + 2ry + 2rx + sx$
**c** $by + 3ay + bx + 3ax$
**d** $2ap + 2bp + 2qb + 2qa$
**e** $8ap + 2bq + 2pb + 8aq$

**3** Factorise:

**a** $6t^2y + xz + 3t^2x + 2yz$
**b** $2ax - 3by + 3bx - 2ay$
**c** $mp^2 - 2mq - 2nq + np^2$
**d** $gx - 4hy + 2gy - 2hx$
**e** $a^2l - b^2m + b^2l - a^2m$

## 2.5 Algebraic fractions

When adding fractions the numerators can be added when the denominators are the same:

$$\frac{2}{7} + \frac{3}{7} = \frac{4}{7}$$

The same applies when the numerators or denominators include letters:

$$\frac{w}{6} + \frac{p}{6} = \frac{w + p}{6}$$

The same method is used for subtracting:

$$\frac{3}{t} - \frac{m}{t} = \frac{3 - m}{t}$$

If the denominators are not the same then you must first make them the same, using equivalent fractions to find a common denominator:

$$\frac{1}{3} + \frac{2}{7} = \frac{7}{21} + \frac{6}{21} = \frac{13}{21}$$

21 is the common denominator here. The same method is used in algebra.

**EXAMPLE 9**

Simplify: **a** $\frac{2}{b} + \frac{b}{7}$ **b** $\frac{x}{5} + \frac{4}{y}$

**a** First, give the fractions a common denominator, $7b$:

$$\frac{2}{b} + \frac{b}{7} = \frac{14}{7b} + \frac{b^2}{7b} = \frac{14 + b^2}{7b}$$

**b** $\frac{x}{5} + \frac{4}{y} = \frac{xy}{5y} + \frac{20}{5y} = \frac{xy + 20}{5y}$

You can work out more complex problems in the same way.

**EXAMPLE 10**

Simplify $\frac{2}{(x-1)} + \frac{3}{(x-2)}$

$$\frac{2}{(x-1)} + \frac{3}{(x-2)}$$

$$= \frac{2(x-2)}{(x-1)(x-2)} + \frac{3(x-1)}{(x-1)(x-2)}$$

$$= \frac{2(x-2) + 3(x-1)}{(x-1)(x-2)}$$

$$= \frac{2x - 4 + 3x - 3}{(x-1)(x-2)} = \frac{5x - 7}{(x-1)(x-2)}$$

**Note that Example 10 goes beyond what you are expected to do at stage 9 and is included here as extension work.**

## Exercise 2G

**(Note: Question 9 is extension work.)**

**1** Work out:

**a** $\frac{3}{4} + \frac{1}{8}$ **b** $\frac{2}{3} + \frac{1}{4}$

**c** $\frac{5}{6} - \frac{1}{3}$ **d** $\frac{5}{7} - \frac{2}{3} + \frac{1}{4}$

**2** Simplify:

**a** $\frac{a}{5} + \frac{a}{3}$ **b** $\frac{2a}{9} + \frac{a}{5}$

**c** $3a + \frac{2a}{3}$ **d** $\frac{3a}{4} - \frac{a}{6} + \frac{a}{3}$

**e** $\frac{a}{7} - \frac{2a}{3}$ **f** $\frac{5a}{21} + \frac{a}{7} + \frac{a}{3}$

**3** Simplify:

**a** $\frac{2x}{3} - \frac{y}{6}$ **b** $\frac{3y}{11} + \frac{x}{3}$

**c** $\frac{y}{2} + \frac{2x}{5} + \frac{z}{2}$ **d** $\frac{x}{4} + \frac{y^2}{2}$

**e** $\frac{3x}{2} + \frac{5y^2}{2} - \frac{y}{4}$ **f** $\frac{x^2}{9} + \frac{2y}{7}$

**4** Simplify:

**a** $\frac{2}{7b} + \frac{3}{14b}$

**b** $\frac{2}{y} - \frac{1}{8y}$

**c** $\frac{6}{pq} + \frac{r}{2pq}$

**d** $x + \frac{2x}{5}$

**e** $p - \frac{3p}{8}$

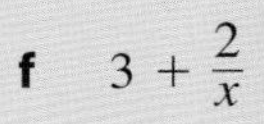

**f** $3 + \frac{2}{x}$

**g** $5 - \frac{4}{y}$

**5** Simplify:

**a** $\frac{p}{3q} + \frac{2q}{p}$ **b** $\frac{5}{m} + \frac{4l}{3m}$

**c** $\frac{5}{4} + \frac{3z}{x}$ **d** $\frac{1}{pq} + \frac{3}{r}$

**e** $\frac{xy}{z} + \frac{az}{y}$ **f** $\frac{ab}{5c} + \frac{bc}{4a}$

**6** The mean of $p$ numbers is 48. The mean of another $q$ numbers is 51. Express, as a single fraction, the mean of $p + q$ numbers.

**7** Avril cycles 10 km at $x$ km/h and then 14 km at $y$ km/h. Express the time for the journey as a single fraction.

**8** $b$ bananas cost \$1.45 and $a$ apples cost \$2.35.

**a** How much does one banana cost?

**b** How much does one apple cost?

**c** Write as a single fraction the cost of one banana and one apple.

**9** Simplify:

**a** $\frac{(x+3)}{4} + \frac{(x-1)}{5}$

**b** $\frac{(x-2)}{5} - \frac{(2x-1)}{7}$

**c** $\frac{4}{(x+1)} + \frac{3}{(x+2)}$

**d** $\dfrac{5}{(x+3)} + \dfrac{3}{(x-1)}$

**e** $\dfrac{1}{(x-1)} + \dfrac{1}{(x+1)}$

**f** $\dfrac{6}{(x+1)} - \dfrac{4}{(x-3)}$

## 2.6 The product of two linear expressions

Look at this rectangle.

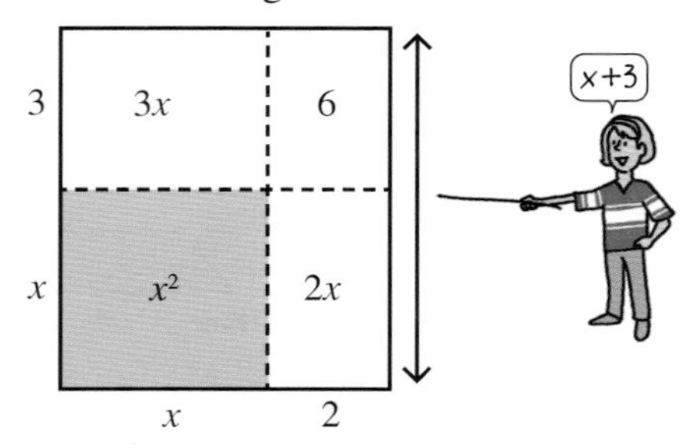


The area of the big rectangle

$= $ sum of the four smaller rectangles

$= x^2 + 2x + 3x + 6$

$= x^2 + 5x + 6$

Also, the area of the big rectangle $= (x + 2) \times (x + 3)$

So, $(x + 2) \times (x + 3) = x^2 + 5x + 6$

This shows how the product of two expressions in brackets can be found.

Another way is to use the **distributive law**:

$(x + 3) \times (x + 2)$

$= x \times (x + 2) + 3 \times (x + 2)$

$= x^2 + 2x + 3x + 6$

$= x^2 + 5x + 6$

Notice that each term in one bracket is multiplied by each term in the other bracket. The box below shows this:

$(x + 3) \times (x + 2)$

$= x^2 + 2x + 3x + 6$

$= x^2 + 5x + 6$

When multiplying two linear expressions like this, you need to remember to collect the x terms together to simplify the expression formed.

The expression formed, here $x^2 + 5x + 6$, has $x^2$ in it. An expression like this, where the highest power of $x$ is 2, is called a **quadratic** expression. You can find out more about these in Chapter 19.

### Exercise 2H

In this exercise, simplify all answers.

**1** By considering the areas of the rectangles, simplify the products:

**a**

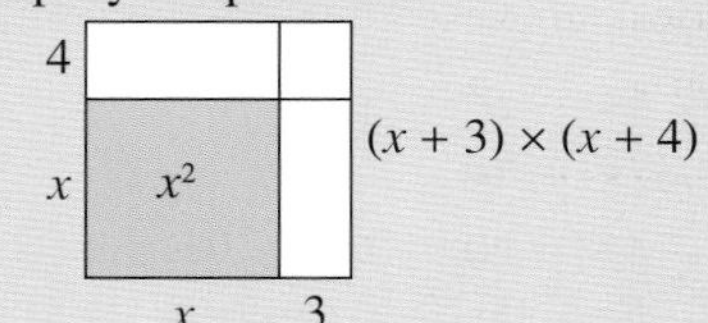


**b**

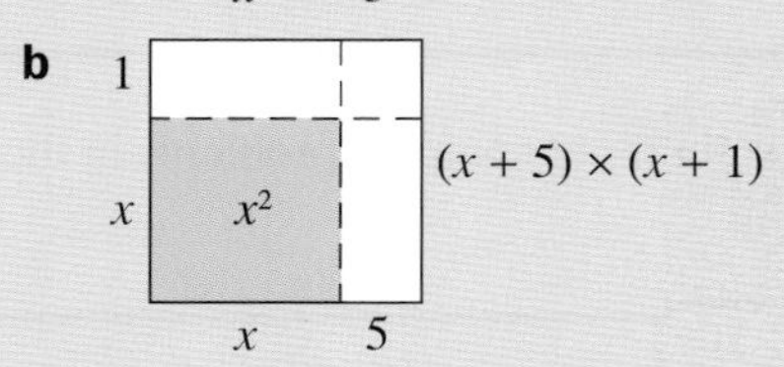


**2** Use the distributive law to work out:

**a** $(x + 3) \times (x + 8)$

**b** $(x + 6) \times (x + 2)$

**c** $(x + 3) \times (x + 5)$

**d** $(x + 7) \times (x + 2)$

**e** $(x + 3) \times (x + 1)$

**f** $(x + 1) \times (x + 2)$

**3** Simplify these expressions:

**a** $(x + 7) \times (x - 2)$

**b** $(x + 5) \times (x - 3)$

**c** $(x + 2) \times (x - 5)$

**d** $(x + 8) \times (x - 7)$

**e** $(x + 3) \times (x - 7)$

**f** $(x - 3) \times (x + 2)$

**Note:**

$(x - 3) \times (x - 4)$ can be written as $(x - 3)(x - 4)$, with the $\times$ sign left out.

**4** Simplify each expression, as in Question **3**.

**a** $(x - 3)(x - 4)$ **b** $(x - 2)(x - 5)$

**c** $(x - 7)(x - 3)$ **d** $(x - 9)(x - 7)$

**e** $(x - 2)(x - 2)$ **f** $(x - 1)(x - 1)$

**5** Simplify the expressions:

**a** $(x + 3)(x + 9)$ **b** $(x + 4)(x - 5)$

**c** $(x - 6)(x + 7)$ **d** $(x - 4)(x - 7)$

**e** $(x - 4)(x - 11)$ **f** $(x + 3)(x - 10)$

**6** Fatima and Aisha were both working out the answer to $(x + 3)^2$

Fatima wrote: $(x + 3)^2 = x^2 + 9$

Aisha wrote: $(x + 3)^2 = (x + 3)(x + 3)$
$= x^2 + 3x + 3x + 9 = x^2 + 6x + 9$

Whose working is correct?

7 Simplify:

**a** $(x + 5)^2$ **b** $(x + 4)^2$

**c** $(x - 1)^2$ **d** $(7 - x)^2$

**e** $(p + 9)^2$ **f** $(4 + t)^2$

8 Write down an expression for the area of each shape.

**a**

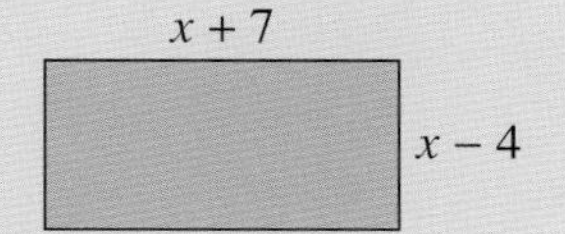


**b**

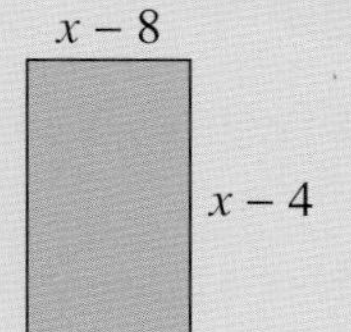


**c**

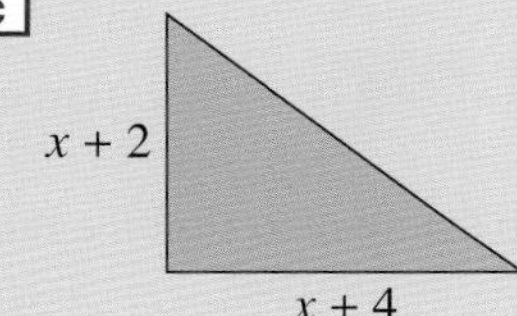


9 Check that your answers for Question **5** are correct, by letting $x = 2$.

**10** Use the distributive law to explain why $(x + a)(x + b) = x^2 + (a + b)x + ab$

**11** Write the product $(a + b)(c + d)$ without brackets.

## Challenge

1 Show that $(x + 2)(x + 5)(x + 3) = x^3 + 10x^2 + 31x + 30$

2 Expand $(x - 1)(x + 1)$ and simplify your answer. Show how you can use this result to help you work out $49 \times 51$.

## TECHNOLOGY

If you want some more help on this or you want to try some harder questions, go to

www.maths.com/algebra/expand-brackets/one-set-of-double-brackets

There are lots of examples and videos to watch, and a mixture of easy and harder questions to try.

## 2.7 Substitution into expressions and formulae

When substituting into expressions and formulae, letters are replaced by numbers, multiplication signs need to be put back in, and the order of operations (BIDMAS) needs to be followed.

**EXAMPLE 11**

Using the formula $A = 100t - 5p^2$, find $A$ when $p = {}^{-}3$ and $t = 0.5$

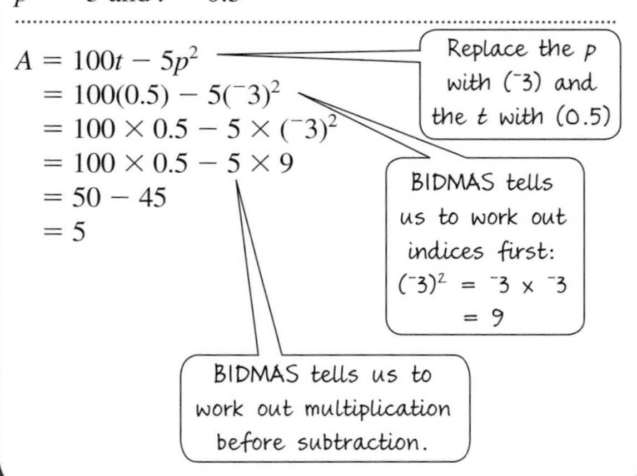


### Exercise 2I

1 If $m = {}^{-}4$, $x = 2$, $y = {}^{-}5$ and $r = 3$, find the value of:

**a** $10(mx + y)$ **b** ${}^{-}3x - y + r$

**c** $5my - 6r$ **d** $\frac{5xmr}{y}$

**e** $\frac{5x - 2y}{my}$ **f** $x^2(m + r)$

2 Repeat Question **1**, this time using $m = 0.5$, $x = 0.3$, $y = 2$ and $r = 0.8$.

3 Find the value of:

**a** $x^2 + 5$ when $x = 0.2$

**b** $1000T^3$ when $T = 0.1$

**c** $9 + m^2$ when $m = {}^{-}3$

**d** $10r^2 + 2r - 4$ when $r = 0.2$

**e** $4W^2 + 1 + W$ when $W = \frac{1}{2}$

**f** $10 - 200v^2 - 10v$ when $v = 0.4$

**g** $5p + 2p^2 + 20$ when $p = {}^{-}10$

4 Copy these expressions into your exercise book:

$6xy$ $4x + y^2$ $(xy)^2$ $48x^2 - y^2 - y$

Tick (✓) which expressions have the same value when $x = 0.5$ and $y = {}^{-}6$.

**5** Using $v = u + at$, find v when

**a** $u = 10$, $a = 2.5$ and $t = 30$

**b** $u = 70$, $a = {}^{-}3.4$ and $t = 20$

**6** Using $s = ut + \frac{1}{2}at^2$, find $s$ when

**a** $u = 25$, $a = 0.7$ and $t = 10$

**b** $u = 80$, $a = {}^{-}3$ and $t = 5$

**7** Using $A = \pi x^2 - \pi y^2$, find $A$ when $x = 0.3$ and $y = 0.2$. (Use $\pi = 3.142$)

**8** $V = \pi r^2 h$ gives the volume of a cylinder with radius $r$ and height $h$.
$V = \frac{4\pi r^3}{3}$ gives the volume of a sphere with radius $r$.
Which of these two solids has the smallest volume?

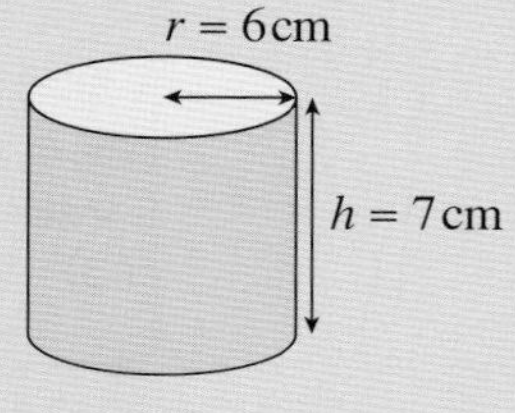


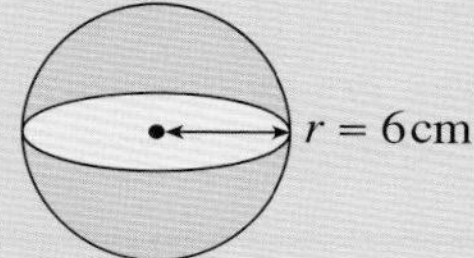


**9** The distance, $d$, between two sets of coordinates, $(x_1,y_1)$ and $(x_2,y_2)$, can be found using the formula
$d = \sqrt{(x_2 - x_1)^2 + (y_2 - y_1)^2}$. What sort of triangle joins the points (0, 3), (3, 9) and (6, 0)?

## INVESTIGATION

Heron of Alexandria (also known as Hero) was an ancient Greek mathematician and engineer who lived around 2000 years ago. He worked out a way of finding the area of a triangle from its side lengths.

**Heron's** (or **Hero's**) **formula** states that the area, $A$, of a triangle with side lengths $a$, $b$ and $c$ is
$A = \sqrt{s(s - a)(s - b)(s - c)}$, where s is half the perimeter (the 'semiperimeter') of the triangle:

$$s = \frac{a + b + c}{2}$$

Using the right-angled triangle with side lengths 3, 4 and 5, work out the area of the triangle using $\frac{1}{2}bh$, then work it out using Heron's formula. Check that the two answers agree. Investigate this further.

## 2.8 Changing the subject of a formula

The formula for the circumference of a circle is $C = 2\pi r$.

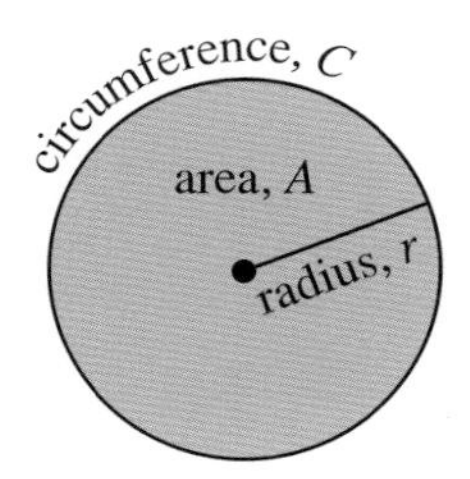


How would you find the radius, $r$, given the circumference, $C$?

You can find the circumference of a circle using a function machine:

$r \rightarrow$ [$\times \pi$] $\xrightarrow{\pi r}$ [$\times 2$] $\xrightarrow{2\pi r}$ $C$ (circumference)

You can use the reverse machine to find the radius:

$r \xleftarrow{\frac{C}{2\pi}}$ [$\div \pi$] $\xleftarrow{\frac{C}{2}}$ [$\div 2$] $\leftarrow C$

That is, $r = \dfrac{C}{2\pi}$

$r$ is now the **subject** of the formula.

A function machine can also be used to find the radius, $r$, of a circle given its area, $A$.

Remember, the area of a circle, $A = \pi r^2$

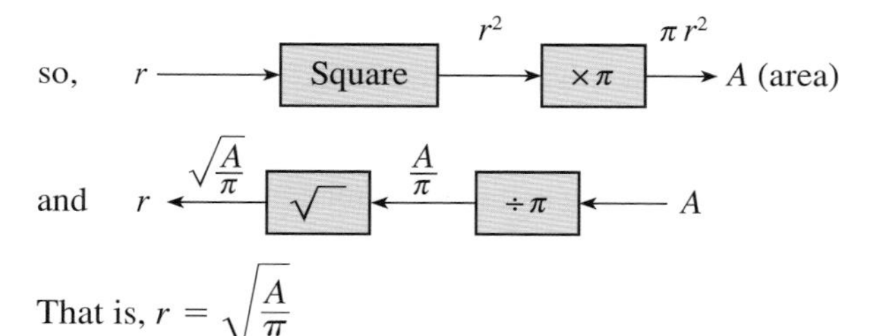


That is, $r = \sqrt{\dfrac{A}{\pi}}$

## EXAMPLE 12

Make $P$ the subject of the formula
$I = P \times R \times T$

The formula can be shown as a function machine:

$P \rightarrow \boxed{\times R} \rightarrow \boxed{\times T} \rightarrow I$

$P$ can be found by reversing the machine:

$P \leftarrow \boxed{\div R} \leftarrow \boxed{\div T} \leftarrow I$

Which gives the formula

$$P = \frac{I}{T \times R}$$

Notice that in Example 12 we do not write $I \div T \div R$ or $\frac{I \div T}{R}$. It is common for students to write 'two-tiered' fractions such as $\frac{\frac{I}{T}}{R}$, which are messy and should be avoided. It is worth remembering that dividing by $x$ is the same as multiplying by $\frac{1}{x}$. This is why, in Example 12, dividing $\frac{I}{T}$ by $R$ is shown as multiplying the $T$ by $R$: $\frac{I}{T \times R}$.

## Exercise 2J

**1** Using function machines to help you, rearrange each formula to make the letter in brackets the subject:

**a** $t = u + v$ $(v)$ **b** $d = h + y$ $(y)$
**c** $x = f - 3$ $(f)$ **d** $m = 2c + g$ $(c)$
**e** $\frac{b}{n} = s$ $(b)$ **f** $\frac{de}{f} = a$ $(d)$
**g** $q = 3(x - f)$ $(x)$

**2** The simple interest formula is $I = \frac{P \times R \times T}{100}$

**a** Draw a function machine for finding $I$ when starting with $T$.
**b** Draw the reverse machine, and complete the equation $T = \frac{\square \times \square}{\square \times \square}$

**3** The formula for converting temperature from Fahrenheit to Celsius is $C = \frac{5}{9}(F - 32)$.

**a** Draw a function machine to show how you can find C, starting with a value for $F$.
**b** Use your machine to find $C$ when $F$ is
**i** 41° **ii** 59°
**iii** 86° **iv** 212°

**4 a** Draw the reverse machine for Question **3 a**.
**b** Use this to write down the rearranged formula with $F$ as its subject.
**c** Find $F$ when $C$ is **i** 50° **ii** 80°

**5** Draw a function machine to show how to find:

**a** $V$ using $V = l \times b \times h$, starting with $l$
**b** $S$ using $S = 2\pi rh$, starting with $r$
**c** $v$ using $v = u + at$, starting with $t$
**d** $V$ using $V = \pi r^2 h$, starting with $r$
**e** $S$ using $S = \pi r(2h + r)$, starting with $h$
**f** $s$ using $s = t(u + \frac{1}{2}at)$, starting with $a$

**6** Use the reverse machine for each part of Question **5** to rewrite the formula. Make the letter you originally started with the subject.

**7** The time, $T$, for each complete swing of a pendulum of length $l$ is given by the formula

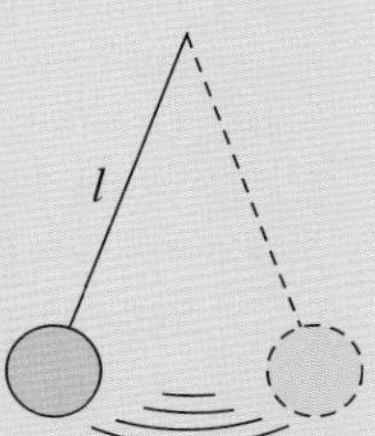


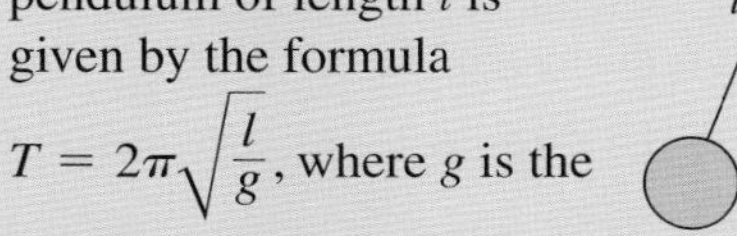

$T = 2\pi\sqrt{\frac{l}{g}}$, where $g$ is the acceleration due to gravity.
$T$ is the subject of the formula.

**a** This function machine is for finding $T$ when you start with $l$.
Copy and complete the machine.

**b** Taking $\pi$ as 3.142, use your machine to find $T$ when $l = 20$ and $g = 980$.

**8 a** Draw the reverse machine for Question **7 a** and show that the formula can be rearranged as

$$l = \left(\frac{T}{2\pi}\right)^2 \times g$$

**b** If $T = 1.571$ and $g = 32$, use the formula in part **a** to find $l$. (Take $\pi$ as 3.142)

**9** $x \rightarrow \boxed{\times m} \rightarrow \boxed{\div c} \rightarrow y$

**a** Use this machine to write down a formula with $y$ as its subject.
**b** Use the reverse machine to rewrite the formula with $x$ as its subject.

**10** $V = \frac{1}{3}\pi r^2 h$.

**a** Draw a function machine to show how to find:

**i** $V$, starting with $r$

**ii** $r$, starting with $V$

**b** Find $r$ if $h = 2.1$ cm and $V = 100$ cm$^3$. (Take $\pi$ as 3.142)

**11 a** Derive the formula for the area, $A$, of this rectangle:

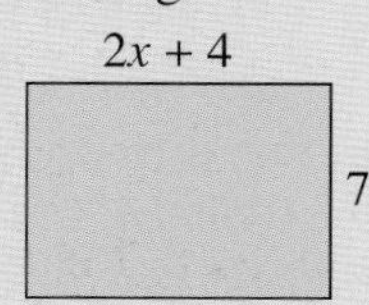


**b** Rearrange the formula to make $x$ the subject.

**c** Use this rearranged formula to find $x$ when the area, $A$, is 63

**d** What is the length of the rectangle in part **c**?

**12 a** Derive the formula for the perimeter, $P$, of this triangle:

$2x - 3$

$x - 3$

$x + 4$

**b** Rearrange this formula to make $x$ the subject.

**c** Use this rearranged formula to find $x$ when the perimeter, $P$, is 30.

**d** What are each of the side lengths of the triangle in part **a**?

**13** $V = \pi h(R^2 - r^2)$.

**a** Find $V$, when $h = 2$ cm, $R = 7.5$ cm and $r = 2.5$ cm, taking $\pi = 3.142$

**b** Draw a function machine to show how to find $V$ when starting with $R$.

**c** Use the reverse machine to rewrite the formula to give $R$ in terms of $V$, $h$, $r$ and $\pi$.

**14 a** Draw a function machine to show how to find $y$; starting with $x$:

**i** $y = px + q$

**ii** $y = k(x - l)$

**iii** $y = r(x - s) + t$

**iv** $y = m(nx + l)$

**b** Use the reverse machine to rearrange the formula, making $x$ the subject.

**15** Rearrange each formula, making $r$ the subject:

**a** $A = rh$ **b** $A = 2\pi rh$

**c** $V = \pi r^2 h$ **d** $S = \pi h(R + r)$

**e** $V = \frac{4}{3}\pi r^3$ **f** $A = P\left(1 + \frac{r}{100}\right)$

## Using the balance method

It is not always possible to use a function machine to change the subject of a formula, particularly when the chosen letter appears more than once.

Remember that a formula is an equation. By keeping the equation balanced you can rearrange the formula to change the subject.

**EXAMPLE 13**

Make $R_1$ the subject of the formula $V = I(R_1 + R_2)$.

*Step 1:* Divide both sides of the equation by $I$:

$$\frac{V}{I} = (R_1 + R_2)$$

*Step 2:* Subtract $R_2$ from both sides:

$$\frac{V}{I} - R_2 = R_1$$

*Step 3:* Turn the equation around:

$$R_1 = \frac{V}{I} - R_2$$

Take care when the letter you want to make the subject is negative or in the denominator of a fraction. Example 14 shows you how to avoid mistakes. It is easier if the letter you are trying to make the subject is positive. It is also easier if the letter is in the numerator of a fraction.

**EXAMPLE 14**

Make $x$ the subject of these formulae:

**a** $M = 4(t - x)$

**b** $g = \frac{f}{x} + y$

**c** $P = \frac{2}{5}x - e$

**a** $M = 4(t - x)$

*Step 1:* Divide both sides by 4:

$$\frac{M}{4} = t - x$$

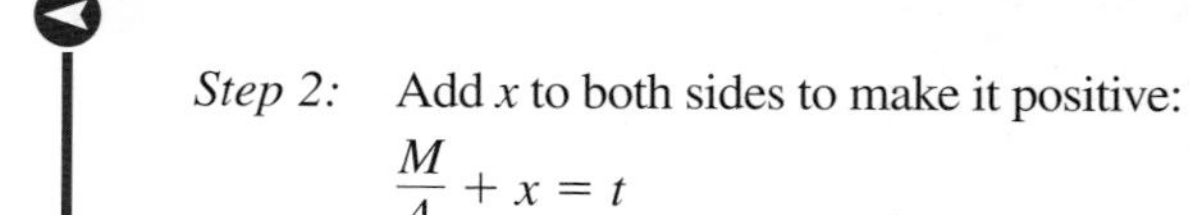

*Step 2:* Add $x$ to both sides to make it positive:

$$\frac{M}{4} + x = t$$

*Step 3:* Subtract $\frac{M}{4}$ from both sides:

$$x = t - \frac{M}{4}$$

**b** $g = \frac{f}{x} + y$

*Step 1:* Subtract y from both sides:

$$g - y = \frac{f}{x}$$

*Step 2:* Multiply both sides by $x$ so it is no longer in the denominator:

$$x(g - y) = f$$

*Step 3:* Divide both sides by $(g - y)$,:

$$x = \frac{f}{g - y}$$

Notice that when $(g - y)$ is in the denominator brackets are no longer necessary because of the long fraction line.

**c** $P = \frac{2}{5}x - e$

*Step 1:* Add $e$ to both sides:

$$P + e = \frac{2}{5}x$$

*Step 2:* Divide by $\frac{2}{5}$. This is the same as multiplying by $\frac{5}{2}$:

$$\frac{5}{2}(P + e) = x$$

Don't forget the brackets.

*Step 3:* You can leave your answer like this or you can write the $x$ first:

$$x = \frac{5(P + e)}{2}$$

This means exactly the same thing as $\frac{5(P = e)}{2} = x$

## Exercise 2K

**1** Using the balance method, make $y$ the subject of each formula.

**a** $C = x + y$ **b** $T = 3x + 2y$

**c** $P = \frac{3y}{4}$ **d** $S = \frac{2}{3}y + 4$

**2** Rearrange the formula $V = I(R_1 + R_2)$ so that the subject is

**a** $R_2$ **b** $I$

**3** Rearrange the formula $I = \frac{PRT}{100}$ so that the subject is

**a** $R$ **b** $P$

**4** If $P = P_0(1 + \alpha t)$, rearrange the formula so that the subject is

**a** $P_0$ **b** $\alpha$ **c** $t$

**5** Rearrange the formula and make the given letter the subject:

**a** $l$ in $A = l \times b$

**b** $r$ in $C = 2\pi r$

**c** $I$ in $V = IR$

**d** $L$ in $S = \pi rL$

**e** $f$ in $v = u + ft$

**f** $x$ in $y = mx + c$

**g** $T$ in $P = \frac{RT}{V}$

**h** $R$ in $I = \frac{E}{R}$

**i** $m$ in $P = \frac{m(v - u)}{t}$

**j** $r$ in $V = \pi r^2 h$

**k** $u$ in $v^2 = u^2 + 2as$

**l** $x$ in $r^2 = (x - a)^2 + y^2$

**m** $g$ in $T = 2\pi\sqrt{\frac{l}{g}}$

**6** Make $r$ the subject of each formula:

**a** $S = 4\pi r^2$ **b** $V = 3\pi rh$

**c** $A = \frac{1}{2}h(R + r)$ **d** $V = \pi h(R^2 - r^2)$

**7** Make $t$ the subject of each formula:

**a** $s = \frac{(u + v)t}{2}$ **b** $u = v - ft$

**c** $R = \frac{PV}{mt}$ **d** $T = k(\alpha t + \beta)$

**8** $P = \frac{k}{d} + mv$

Rearrange the formula so that the subject is:

**a** $k$ **b** $m$ **c** $v$ **d** $d$

# Consolidation

**Example 1**

Simplify these expressions:

**a** $6a - 4b + 2y - 5b$
$= 6a - 4b - 5b + 2y$
$= 6a - 9b + 2y$

**b** $\dfrac{2x^3y^2}{6x^4yz^4}$

$$= \frac{2 \times x \times x \times x \times y \times y}{6 \times x \times x \times x \times x \times y \times z \times z \times z \times z}$$

$$= \frac{\overset{1}{\cancel{2}} \times \cancel{x} \times \cancel{x} \times \cancel{x} \times \cancel{y} \times y}{\underset{3}{\cancel{6}} \times \cancel{x} \times \cancel{x} \times \cancel{x} \times x \times \cancel{y} \times z \times z \times z \times z}$$

$$= \frac{y}{3 \times x \times z \times z \times z \times z} = \frac{y}{3xz^4}$$

**Example 2**

Simplify $\dfrac{3}{a} - \dfrac{4}{b}$

Common denominator is $ab$:

$$\frac{3}{a} - \frac{4}{b} = \frac{3b}{ab} - \frac{4a}{ab}$$

$$= \frac{3b - 4a}{ab}$$

**Example 3**

Simplify:

**a** $x^4 \times x^5$
$= x^{4+5} = x^9$

**b** $a^6 \div a^4$
$= a^{6-4} = a^2$

**Example 4**

Expand and simplify:

**a** $(x + 3)(x - 5)$
$= x^2 + 3x - 5x - 15$
$= x^2 - 2x - 15$

**b** $(x - 3)^2$
$= (x - 3)(x - 3)$
$= x^2 - 3x - 3x + 9$
$= x^2 - 6x + 9$

**Example 5**

Factorise:

**a** $4x - 12y$
$= 4 \times x - 4 \times 3y$
$= 4(x - 3y)$

**b** $3y^2 - 4y + 5y^3$
$= y \times 3y - y \times 4 + y \times 5y^2$
$= y(3y - 4 + 5y^2)$

**Example 6**

Using the formula $M = 10y + 3r^2$, find $M$ when $y = 2.2$ and $r = {}^-2$.

$M = 10y + 3r^2$
$= 10(2.2) + 3({}^-2)^2$
$= 10 \times 2.2 + 3 \times ({}^-2)^2$
$= 10 \times 2.2 + 3 \times 4$
$= 22 + 12$
$= 34$

**Example 7**

Make $x$ the subject of these formulae:

**a** $y = 3(t + x)$

$\dfrac{y}{3} = t + x$ [divide both sides by 3

$\dfrac{y}{3} - t = x$ [subtract $t$ from both sides

**b** $M = \dfrac{P}{x} - r$

$M + r = \dfrac{P}{x}$ [add $r$ to both sides

$x(M + r) = P$ [multiply by $x$ to clear the denominator

$x = \dfrac{P}{M + r}$ [divide both sides by $M + r$

## Exercise 2

**1** Simplify:

**a** $x^2 \times x^3$ **b** $y^2 \times y^6$
**c** $2x^3 \times x^4$ **d** $3x^4 \times x^5$
**e** $x^3 \times x^4 \times x^2$ **f** $y^2 \times y^3 \times y^6$

**2** Simplify:

**a** $x^5 \div x^4$ **b** $y^5 \div y^5$
**c** $x^7 \div x^3$ **d** $y^8 \div y^3$
**e** $x^2 \times x^7 \div x^4$ **f** $y^3 \times y^4 \div y^5$

**3** Simplify:

**a** $3x - 2y + 4x + 5y$
**b** $6x - 3y - 5x - 5y$
**c** $6 - 2x - 3y - 5x$
**d** $(4xy^3)^2$
**e** $\dfrac{3x^3y^2}{6x^2y^4}$
**f** $\dfrac{z^5x^3y^5}{(x^2y)^2z^3}$

**4** Simplify:

**a** $y^4 \times y^6$
**b** $a^8 \div a^4$
**c** $x^8 \div x^{10}$
**d** $m^7 \times m^3 \times m^{-2}$
**e** $p^4 \times p^3 \div p^{10}$
**f** $7x^4 \times 3x^5$
**g** $\dfrac{8m^5}{2m^7}$
**h** $\dfrac{q^2 \times q^{10}}{(q \times q^9)^2}$
**i** $\dfrac{(p^2 \times p^3)^2}{p^8}$
**j** $\dfrac{(R^2 \times R^4)^3}{R^{10} \div R^2}$

**5** Expand and simplify:

**a** $(x + 6)(x + 2)$
**b** $(x + 3)(x + 5)$
**c** $(x + 4)(x - 7)$
**d** $(x - 2)(x + 8)$
**e** $(x - 4)(x - 3)$
**f** $(x - 10)(x - 1)$
**g** $(x + 5)^2$
**h** $(x - 9)^2$

**6** Factorise:

**a** $3x + 9y$ **b** $2a - 4b$
**c** $6x - 12y$ **d** $14x - 7y$
**e** $15x + 18y$ **f** $6x - 24y$
**g** $6x + 72y$ **h** $ax + ay$

**7** Factorise:

**a** $ax + 3ay$ **b** $an - 3am$
**c** $6rx + 2ry$ **d** $3ax - 18ay$
**e** $5mn - 5mp$ **f** $4pq - 3pr$
**g** $6r + 4pr - 2qr$ **h** $3ab - 3a + ac$

**8** Write down a formula for

**a** the area, $A$
**b** the perimeter, $P$, of these shapes:

**i**

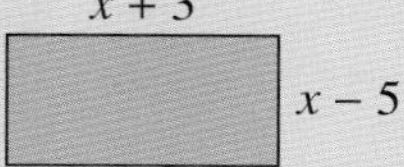


**ii**

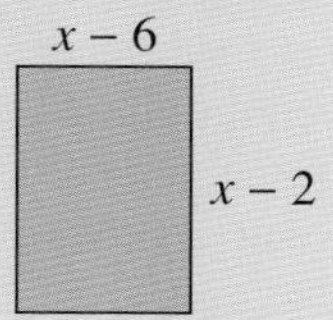


**c** Use your formulae to find the areas and perimeters of the shapes when $x = 10$.

**9** Write as a single fraction:

**a** $\dfrac{d}{4} + \dfrac{d}{6}$
**b** $\dfrac{3x}{5} + \dfrac{x}{4}$
**c** $\dfrac{2m}{3} - \dfrac{m}{10}$
**d** $\dfrac{y}{3} + \dfrac{2y}{5} + \dfrac{y}{4}$
**e** $\dfrac{3}{m} + \dfrac{1}{2m}$
**f** $1 - \dfrac{1}{x}$
**g** $\dfrac{3}{4b} + \dfrac{2}{3c}$
**h** $\dfrac{2}{(x + 1)} + \dfrac{1}{(x + 3)}$

**10** Rearrange each formula to make the letter in brackets the subject:

**a** $b = 2v + p$ $(v)$
**b** $T = y - h$ $(h)$
**c** $x = \dfrac{m}{3} + y$ $(m)$
**d** $Y = 2(r + P)$ $(r)$
**e** $\dfrac{b}{n} = j$ $(n)$
**f** $h = \dfrac{d + s}{y}$ $(s)$

**11** Make $h$ the subject of these formulae:

**a** $R = F - \dfrac{h}{20}$ **b** $\sqrt{\dfrac{x}{2} + \dfrac{h}{3}} = y$
**c** $A = 2\pi r^2 h$ **d** $\dfrac{1}{x} + \dfrac{1}{y} = \dfrac{1}{h}$

**12** The volume of a sphere, $V$, is given by the formula $V = \frac{4}{3}\pi r^3$, where $r$ is the radius of the sphere.

**a** Make $r$ the subject of this formula.

**b** What is the radius of a sphere with volume $200\,\text{cm}^3$?

**13** I started with a number, $n$. I subtracted 10 from this number and then multiplied the result by 7, and got the same number as I do when I add 18 to $n$ and multiply the result by 3. What is $n$?

**14** Find the value of

**a** $2x^2$ when $x = 0.5$

**b** $10m^2 + 5y$ when $m = {}^-0.1$ and $y = 4$

**c** $4 + y^3$ when $y = {}^-2$

**d** $50h^2 + 10h - f$ when $h = 0.4$ and $f = 3$

**e** $200g^2 - 20g$ when $g = 0.3$

**f** $3kw + 20w^2 + k$ when $k = {}^-10$ and $w = 0.1$

# Summary

| You should know ... | Check out |
|---|---|
| **1** How to simplify algebraic expressions by combining like terms.<br>*For example:*<br>$3xy + 4z - 2xy + z$<br>$= xy + 5z$ | **1** Simplify:<br>**a** $7x + 2x$<br>**b** $x^3 + 3x^3$<br>**c** $a - b - 2b$<br>**d** $3b^3 - 2b + b^3 - 4b$ |
| **2** The rules for indices.<br>$a^m \times a^n = a^{m+n}$<br>$a^m \div a^n = a^{m-n}$<br>$a^0 = 1$ | **2** Work out:<br>**a** $b^3 \times b^5$<br>**b** $3c^5 \div c^3$<br>**c** $\frac{3a^2b}{2ab}$<br>**d** $\frac{5a^3b^4c^2}{b^2c^5}$ |
| **3** How to use the distributive law to simplify an expression.<br>*For example:*<br>$6(3x - 2y) + 2(x - y)$<br>$= 6 \times 3x - 6 \times 2y + 2 \times x - 2 \times y$<br>$= 18x - 12y + 2x - 2y$<br>$= 20x - 14y$ | **3** Expand the brackets and simplify:<br>**a** $4(2x - 3)$<br>**b** $6(2x + 7y)$<br>**c** $3x(2 - 4y)$<br>**d** $6x^2(3 - 4x)$<br>**e** $3x(1 + 2y) - 2y(1 - 2x)$<br>**f** $4y(1 - y) + 3y(2 - y)$ |
| **4** How to factorise an expression.<br>*For example:*<br>$3ax + 4ay = a(3x - 4y)$ | **4** Factorise:<br>**a** $20x + 15y$<br>**b** $3x - x^2$<br>**c** $4xy - x^2$<br>**d** $6x + 72$ |

**5** How to form equations to solve problems.
*For example:*
In four years' time James will be five times as old as he is now. What is his present age?

Let his present age be $x$ years.
In four years' time his age will be $x + 4$
Five times as old as his present age is $5x$

so $\quad x + 4 = 5x$

$-x]\quad 4 = 4x$

$\div 4]\quad 1 = x$

James is 1 year old now.

**5** The width of a rectangle is $w$ cm. Its length is 4 cm more than its width. If the perimeter of the rectangle is 28 cm, find its length and width.

---

**6** How to change the subject of a formula.
*For example:*
Make $x$ the subject of $y = m\sqrt{x} + c$

$$y = m\sqrt{x} + c$$

$-c]\quad y - c = m\sqrt{x}$

$\div m]\quad \frac{y - c}{m} = \sqrt{x}$

square] $\quad x = \left(\frac{y - c}{m}\right)^2$

**6** Make $x$ the subject of these equations:

**a** $y = mx + c$
**b** $y^2 = x^2 + c^2$
**c** $y = \frac{x + c}{p - c}$

---

**7** How to expand the product of two linear expressions and simplify the resulting quadratic expression.
*For example:*

Expand and simplify $(x + 7)(x - 2)$

$(x + 7)(x - 2)$
$= x^2 + 7x - 2x - 14$
$= x^2 + 5x - 14$

**7** Expand and simplify:

**a** $(x + 5)(x + 1)$
**b** $(x - 3)(x + 4)$
**c** $(x + 2)(x - 6)$
**d** $(x - 3)(x - 8)$
**e** $(x + 5)^2$

---

**8** How to add and subtract simple algebraic fractions by finding a common denominator.
*For example:*

$$\frac{d}{4} + \frac{3}{2y} = \frac{dy + 6}{4y}$$

**8** Write as a single fraction:

**a** $\frac{e}{3} + \frac{2e}{5}$ **b** $\frac{2y}{3} - \frac{y}{4}$

**c** $\frac{5}{x} + \frac{1}{2x}$ **d** $\frac{1}{4} + \frac{d}{8p}$

---

**9** How to substitute numbers into an expression or formula and find the result following the rules of BIDMAS.
*For example:*

Using $A = \pi x^2 - \pi y^2$, find $A$ when $x = 5$ and $y = 2.4$.

Using $\pi = 3.142$,
$A = 3.142 \times 5^2 - 3.142 \times 2.4^2$
$= 3.142 \times 25 - 3.142 \times 5.76$
$= 78.55 - 18.09792$
$= 60.45208$

**9** Using the formula $s = ut + \frac{1}{2}at^2$, find $s$ when $u = 20$, $a = 2.5$ and $t = 15$.

# Shapes and mathematical drawings

## Objectives

- Draw 3D shapes on isometric paper.
- Analyse 3D shapes through plans and elevations.
- Identify reflection symmetry in 3D shapes.
- Use a straight edge and compasses to:
  - construct the perpendicular from a point to a line and the perpendicular from a point on a line
  - inscribe squares, equilateral triangles, and regular hexagons and octagons by constructing equal divisions of a circle.
- Use bearings (angles measured clockwise from the North) to solve problems involving distance and direction.
- Make and use scale drawings and interpret maps.

## What's the point?

Some of the elements of design are form, shape and symmetry. Knowledge of shapes and mathematical drawings informs us about these and enables architects to design attractive buildings.

## Before you start

### You should know ...

**1** How to measure angles using a protractor.
*For example:*

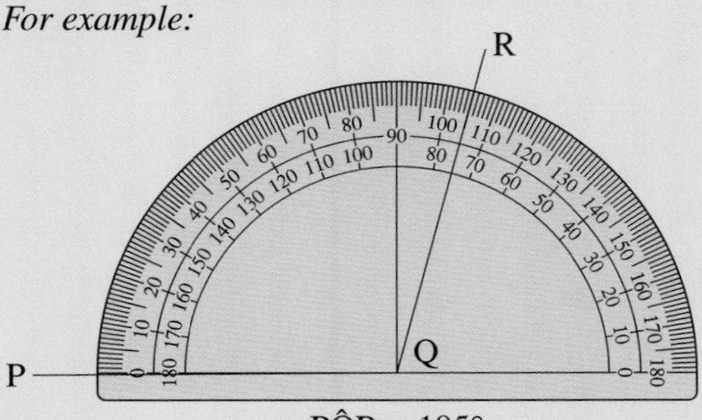


$P\hat{Q}R = 105°$

### Check in

**1** Measure these angles with your protractor:

**a**

**b**

**2** How to find lines of symmetry.
*For example:*
A regular hexagon has six lines of symmetry.

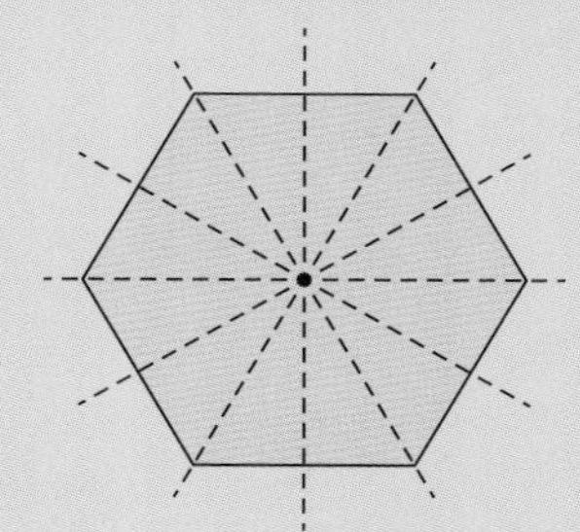

**3** How to construct the perpendicular bisector of a line segment using a pair of compasses and a ruler.
*For example:*
Construct the perpendicular bisector of line PQ.

Draw an arc with centre P and radius more than $\frac{1}{2}$ PQ.
Draw another arc, centre Q, with the same radius.

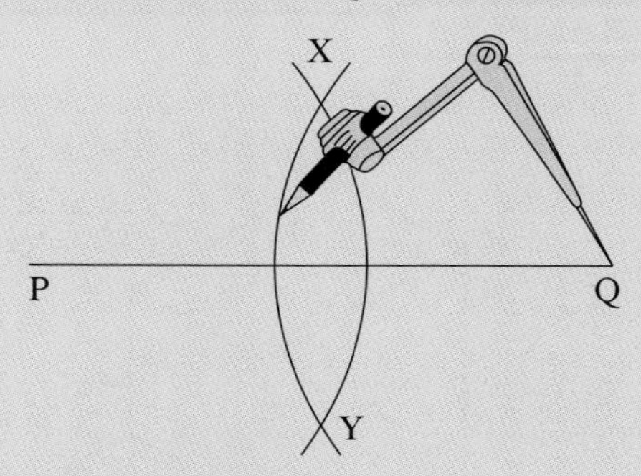


The arcs meet at X and Y.
Join XY.

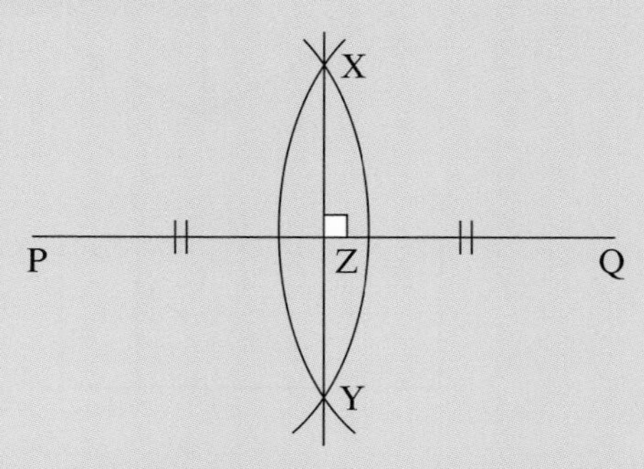


XY is **perpendicular** to PQ and bisects PQ at the midpoint Z.
PZ = QZ

**2** How many lines of symmetry do these shapes have?

**a** a rectangle
**b** a square
**c** a regular pentagon
**d**

**3** Make a rough, enlarged copy of this diagram:

A

B

Construct the perpendicular bisector of the line segment AB.

## 3.1 Three-dimensional (3D) shapes

### Isometric drawings

Here is a drawing of a building:

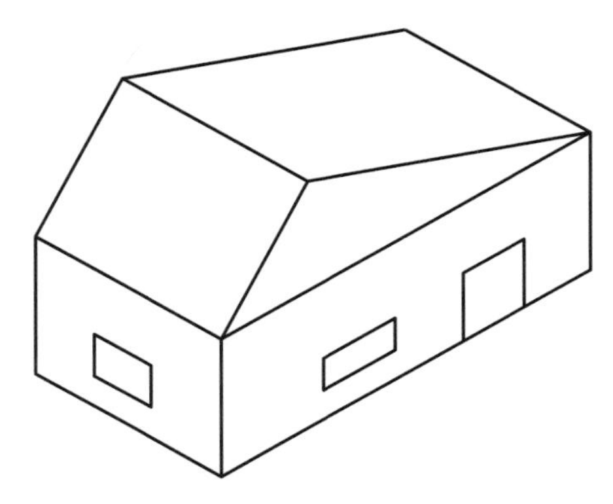

The basic shape of this building can be drawn on **isometric** paper:

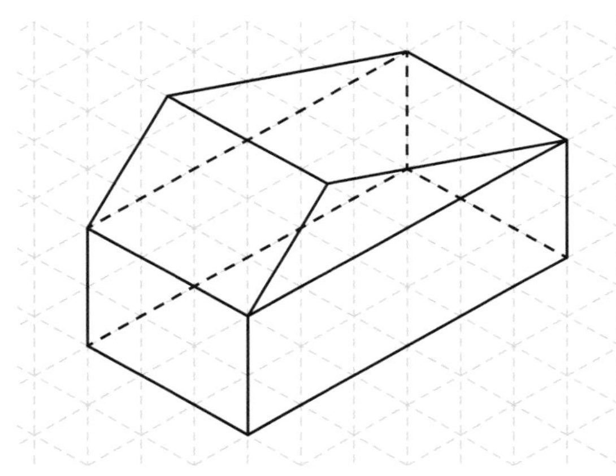

An **isometric drawing** is made on a grid consisting of equilateral triangles instead of squares. In isometric drawings, all horizontal lines should be parallel to each other. All vertical lines should be parallel to each other and drawn vertically on the paper. Often in isometric drawings hidden edges are shown as dashed lines, as here, or they are not shown at all, as in the next drawing, of a cube.

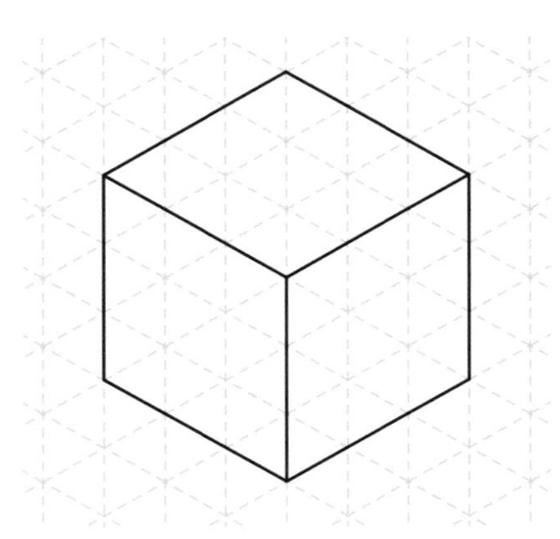

### Plans and elevations

You can also represent a three-dimensional (3D) shape using two-dimensional (2D) drawings of it. These are called '**plans**' and '**elevations**', and they are projections of a 3D object onto a 2D surface. The plan is the view from directly above the object, while elevations are views of the object from the front, back or sides.

**EXAMPLE 1**

Draw the plan, front elevation and side elevation of this object:

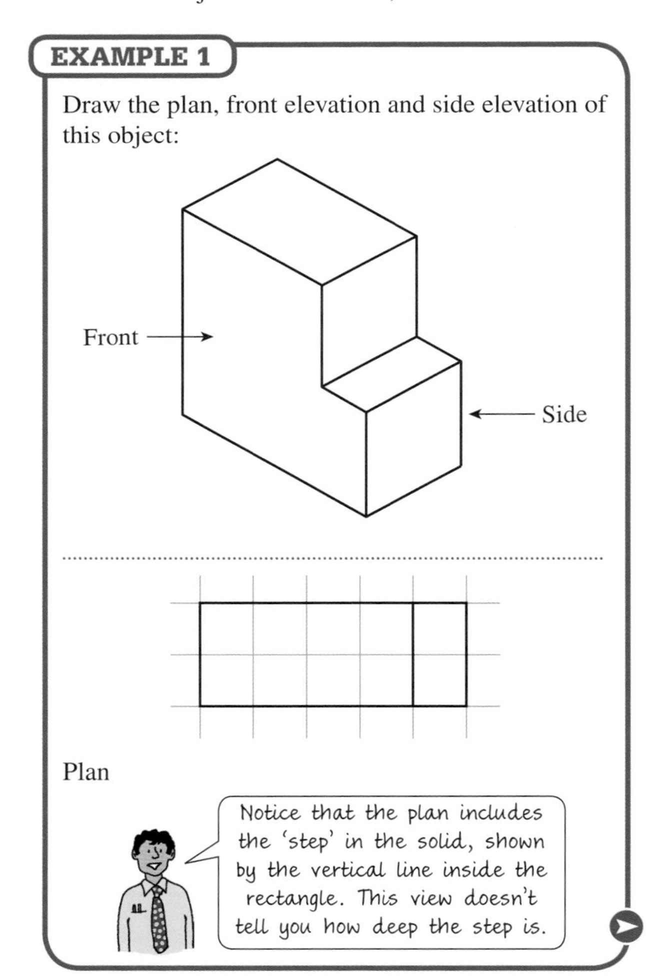

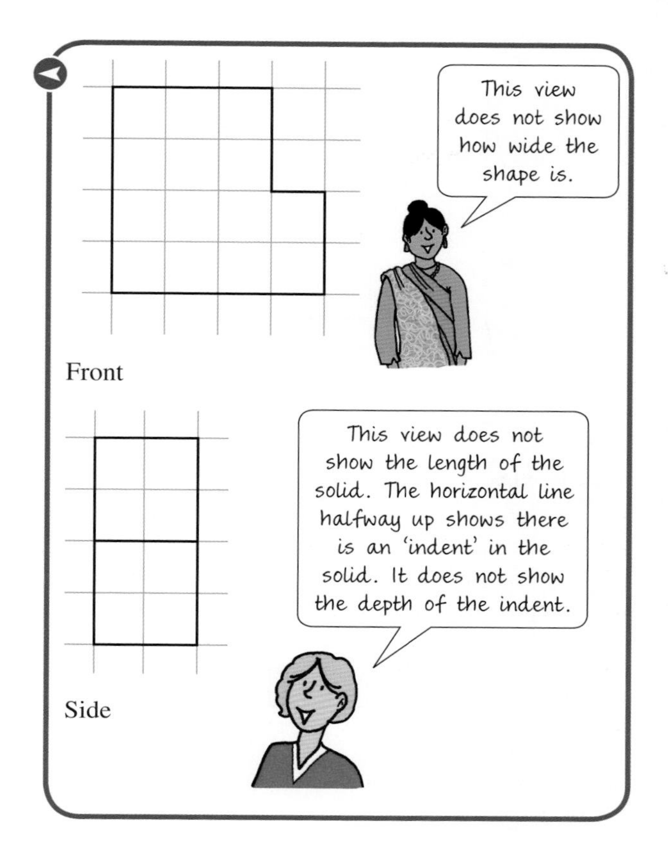


Notice that, in each of the 2D diagrams in Example 1 (the plan and the two elevations) some information is missing. All three diagrams are therefore required to represent the 3D shape.

Also note that, if the side elevation had been drawn from the left-hand side rather than the right-hand side, the line across the middle would be a dashed line like this, indicating that the edge is hidden from view:

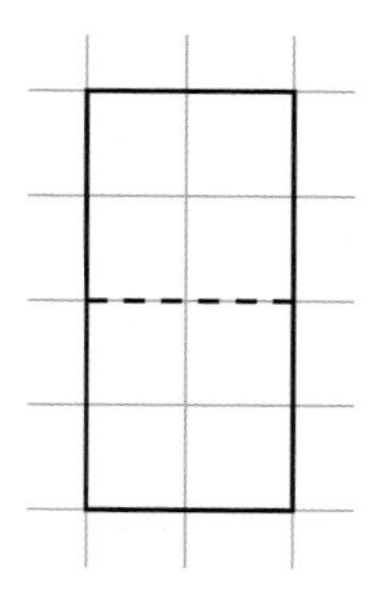

## Exercise 3A

For Questions **1–5**, use isometric paper to draw the 3D objects represented by the plans and elevations.

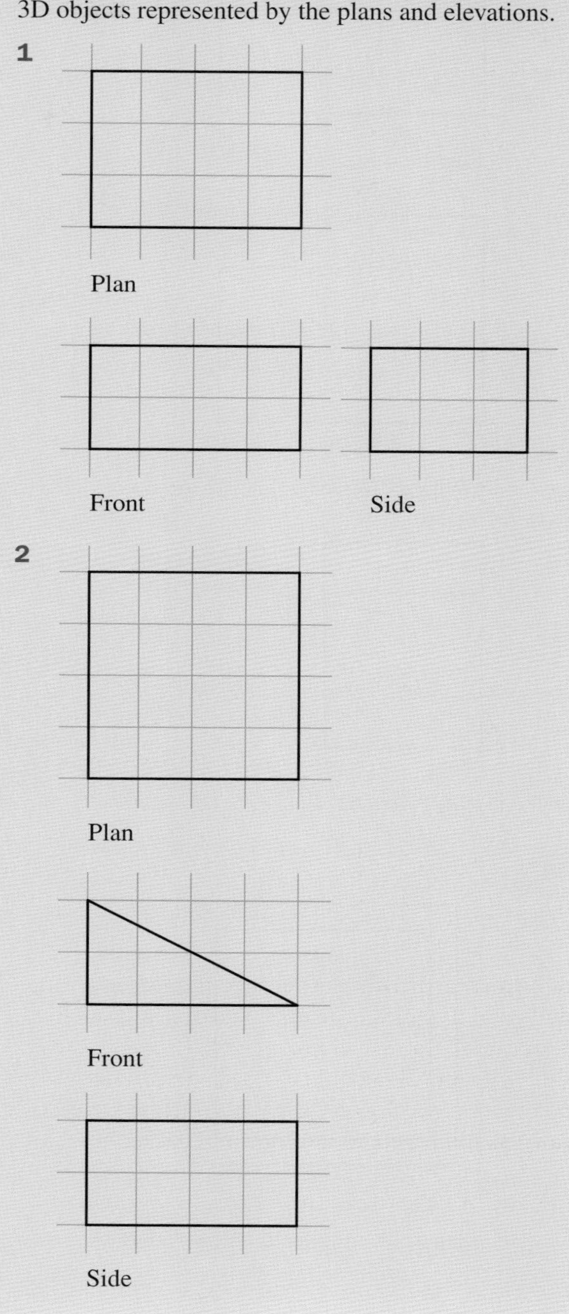

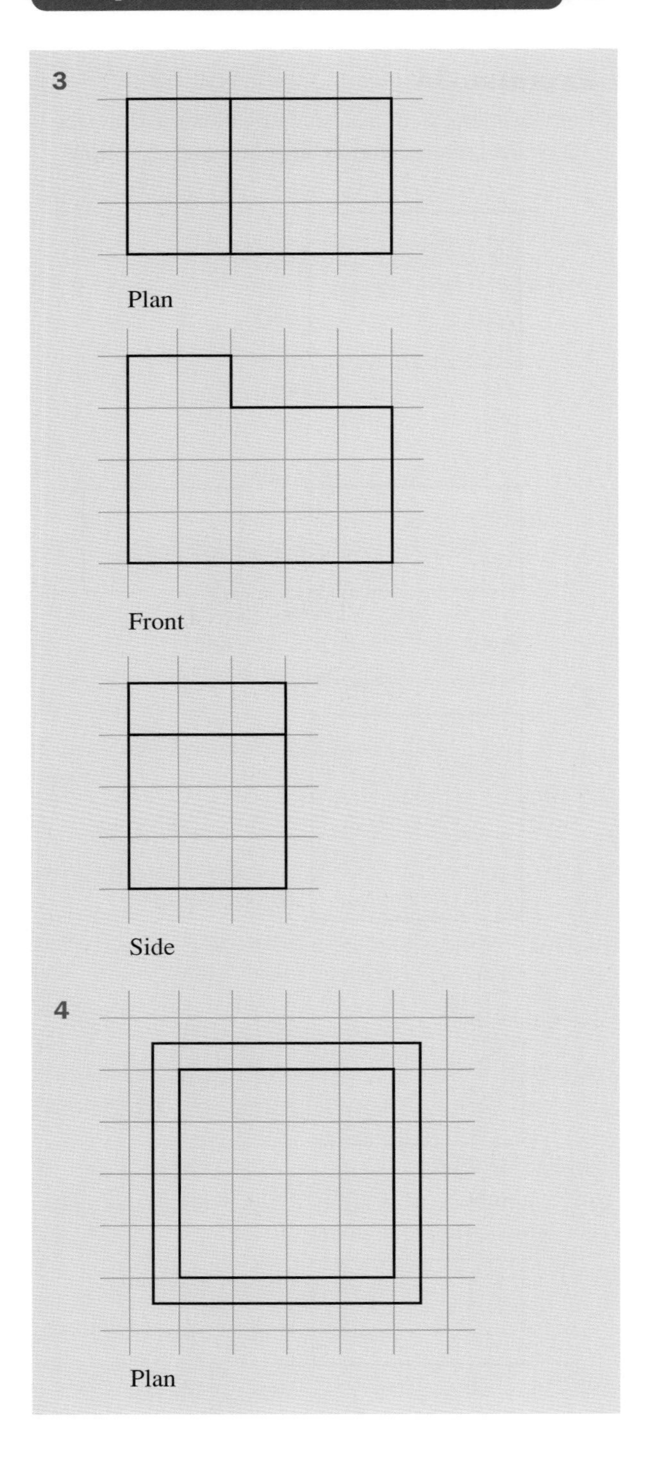
3
Plan
Front
Side
4
Plan

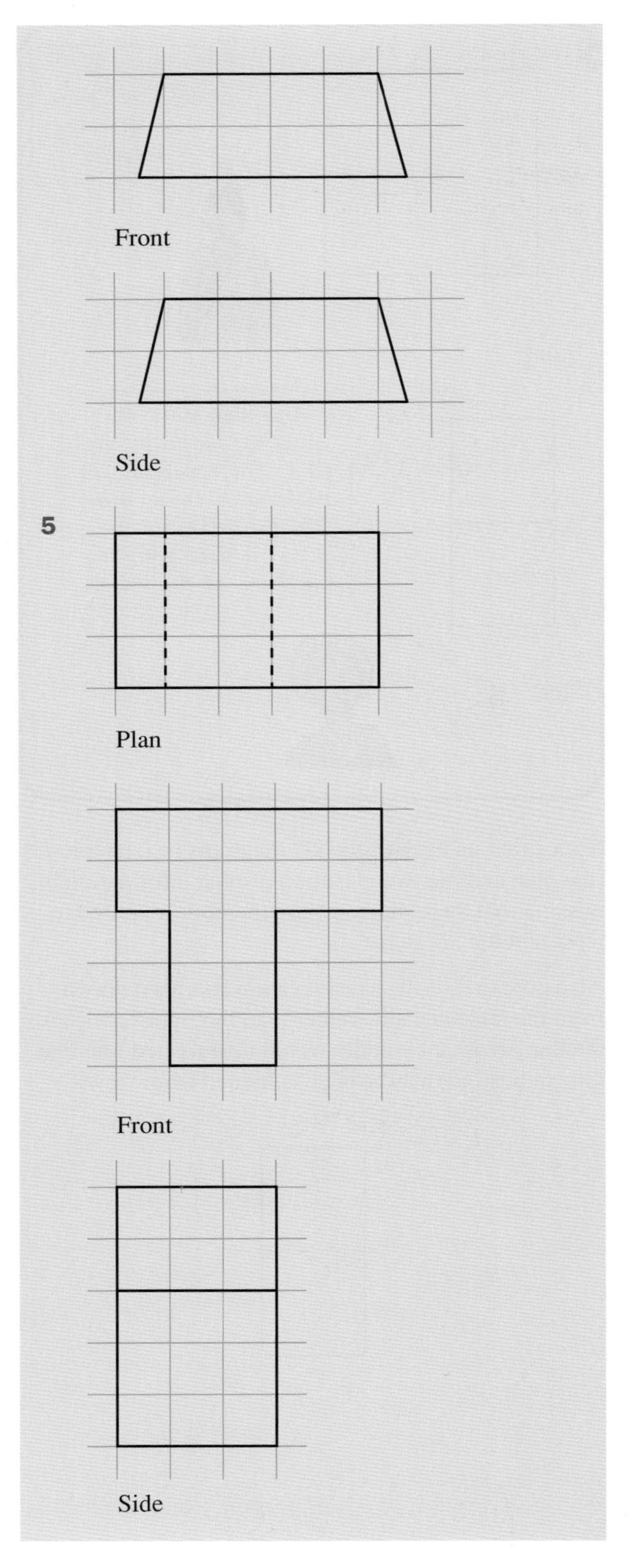
Front
Side
5
Plan
Front
Side

For Questions **6–10**, draw the plan, front elevation and side elevation for each 3D shape. Note that it is conventional for the plan to be at the top, the front elevation directly below the plan and the side elevation to be to the right of the front elevation, as in Question **1**, if you have space.

**6**

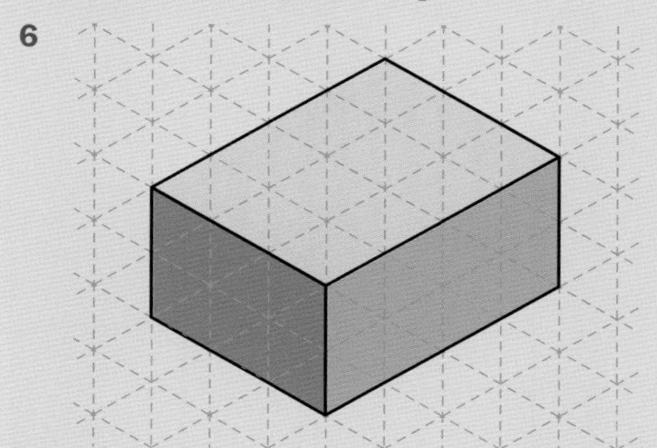

**7**

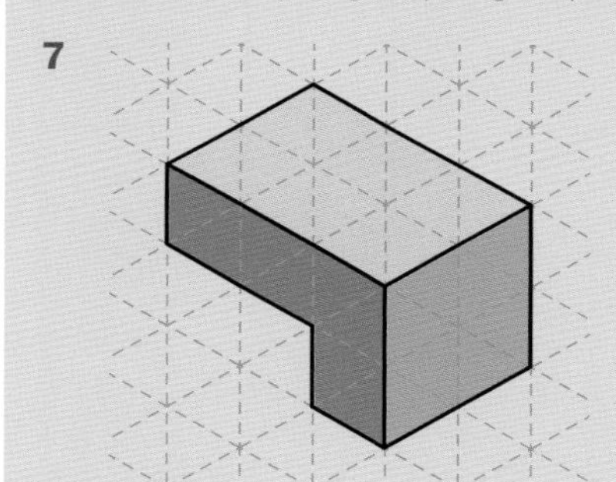

**8**

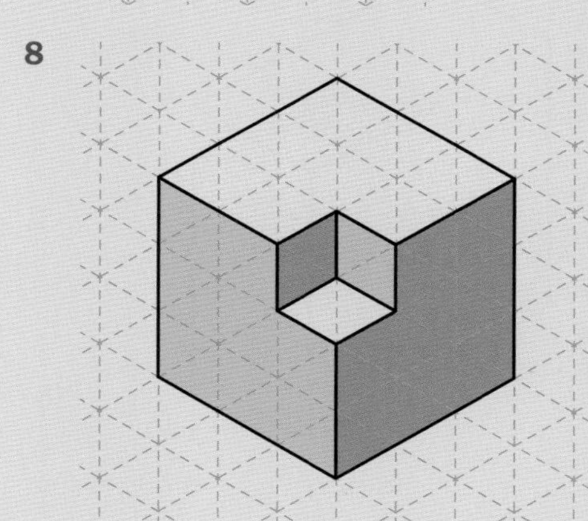

**9**

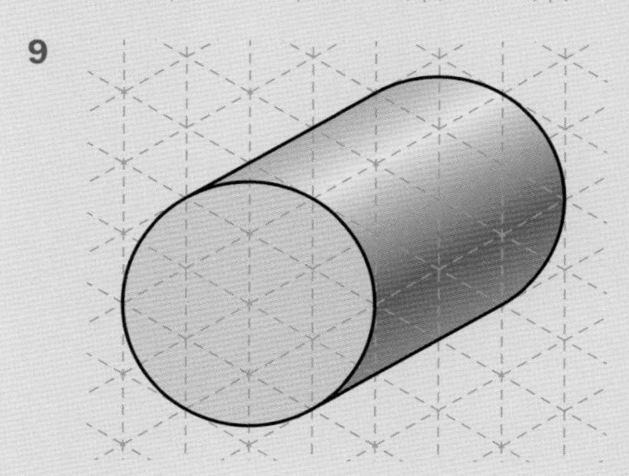

**10** Use the descriptions in the table to identify the solids.

| | Plan | Front elevation | Side elevation |
|---|---|---|---|
| **a** | Square | Square | Square |
| **b** | Circle | Rectangle | Rectangle |
| **c** | Rectangle | Triangle | Rectangle |

## ACTIVITY

You will need multilink cubes for this Activity.

Make a 3D model with multilink cubes.
To keep it easy, limit the number of cubes to 5.
To make it harder, increase the number of cubes allowed.

Draw the plan and elevations for your model.

Ask your neighbour to try to make the model using the plan and elevations you drew.

Were they right?

Repeat this for other models.

## 3D symmetry

Isometric paper is also useful for showing the symmetrical properties of 3D shapes.

In 2D shapes you learnt about lines of symmetry.

A rectangle has two lines of reflection symmetry that split the shape into two congruent (identical) halves:

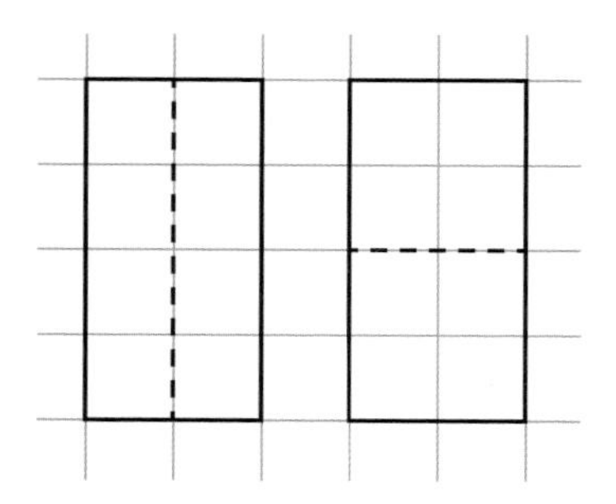

A cuboid has three **planes** of reflection symmetry that split the shape into two congruent halves. A 'plane' is a flat surface.

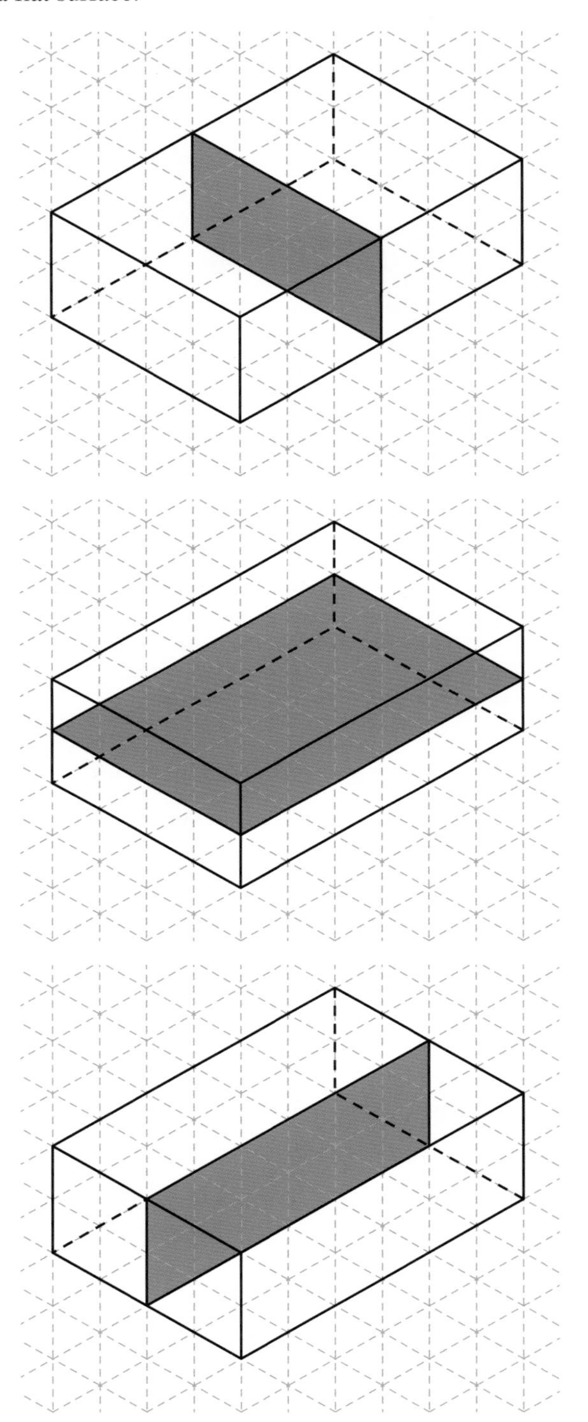

A 3D shape has reflection symmetry if it has one or more planes of symmetry.

## Exercise 3B

For Questions **1**–**4**, copy the 3D shapes onto isometric paper and draw the planes of symmetry. (For some questions you will need to make more than one copy of the shape.)

**1**

**2**

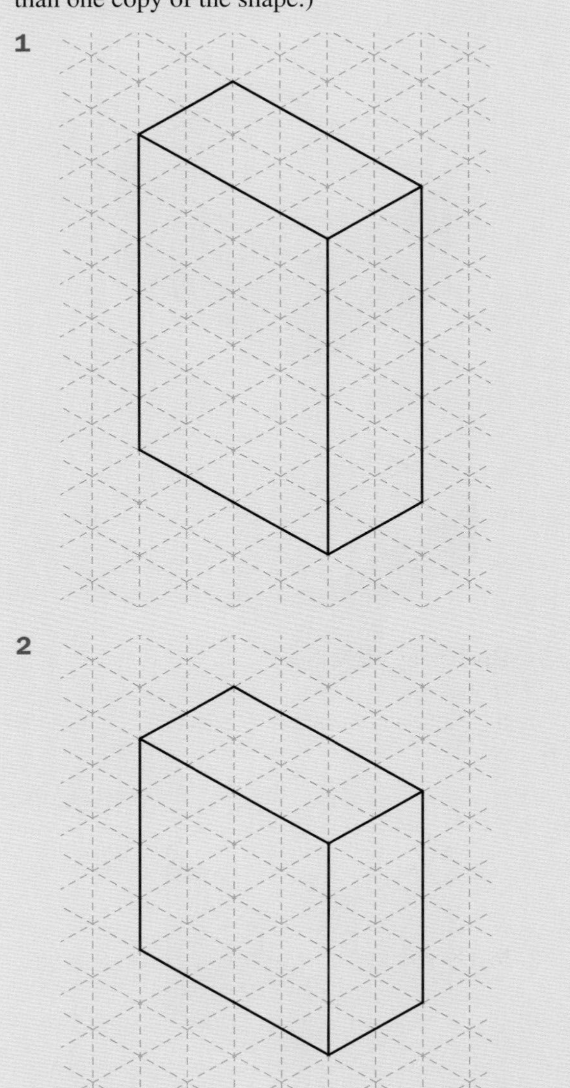

**3**

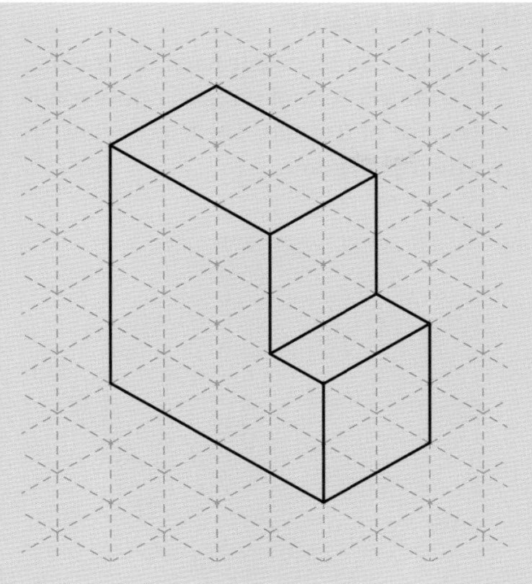

**4**

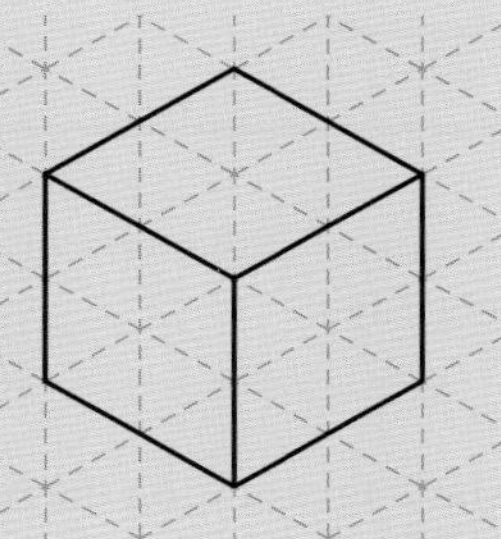

**5** How many planes of symmetry do these shapes have?

**a** Equilateral triangular prism

**b** Regular pentagonal prism

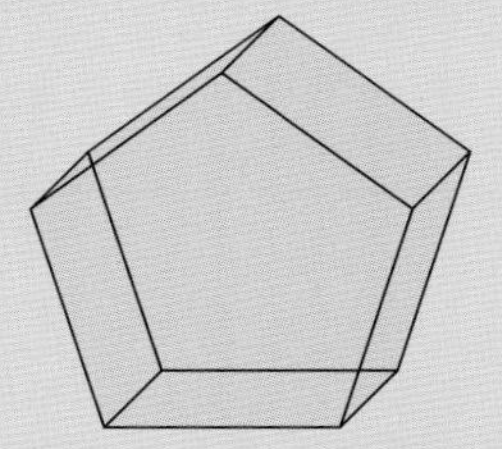

**c** Regular hexagonal prism

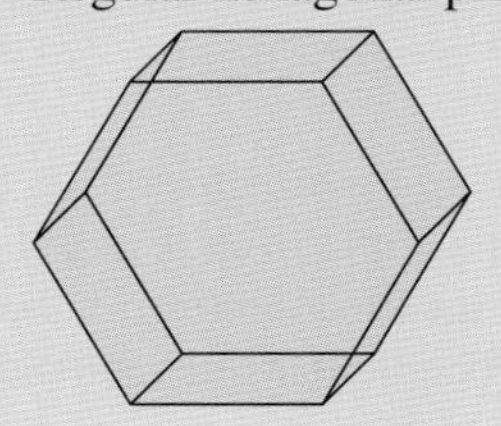

**6** Using your answers to Questions **2** and **5**, how do you find the number of planes of symmetry of a prism with a regular cross section?

**7** How many planes of symmetry are there in a cylinder?

# 3.2 Constructions

## Constructing the perpendicular from a point to a line

Sometimes you want to draw a perpendicular line from a point, X, to a line.

To do this, with your compasses draw arcs centred at X that cut the line.

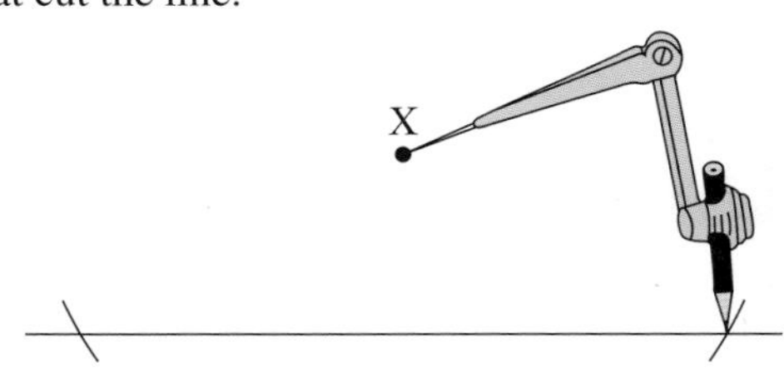


Then draw arcs centred where the previous arcs cut the line.

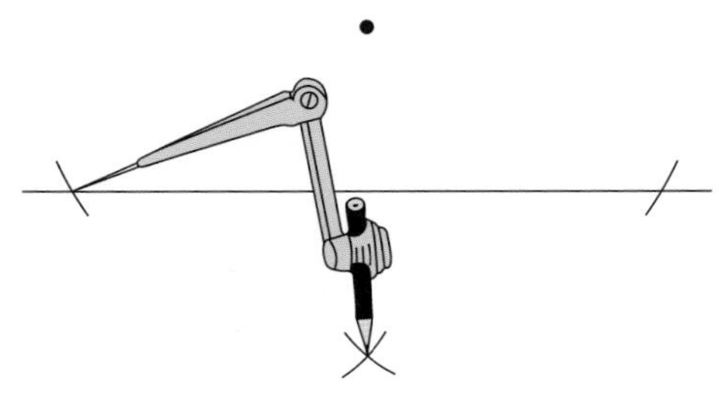

Finally, with your ruler join the point X to where the arcs intersect.

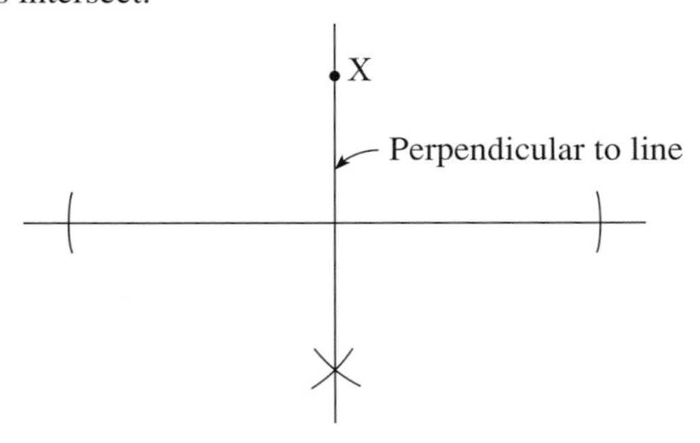

## Constructing a perpendicular line from a point on a line

You can construct an angle of 90° at a point X on a line using a pair of compasses and a ruler. First, draw two arcs centred at X to cut the line at A and B.

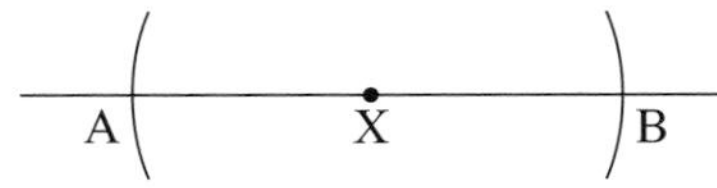


Increase the radius of your compasses. Draw arcs centred at A and B to intersect each other at P and Q. Join PQ.

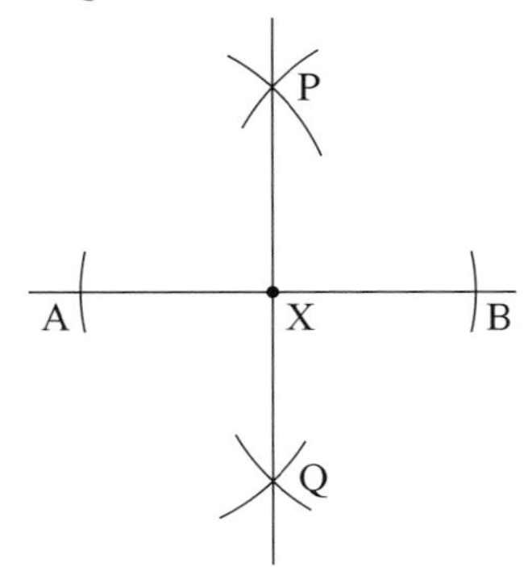


Angle AXP = 90°

### Exercise 3C

**1** Draw a line AB and a point C which is about 6 cm above the line. Construct a perpendicular from C to the line AB.

**2** **a** Draw a triangle.
**b** Construct a perpendicular from each corner of the triangle to its opposite side.
**c** What do you notice?

**3** Repeat Question **2** for four other triangles.

**4** **a** Draw a line AB = 8 cm.
**b** Construct a 90° angle at A.
**c** Draw the line AC = 6 cm, where $B\hat{A}C = 90°$.
**d** Join the points B and C.
**e** Measure the line BC.

**5** **a** Draw a line PQ = 7 cm.
**b** Construct angles of 90° at both P and Q.
**c** On the perpendicular through P mark the point S so that it is 5 cm from P.
**d** Draw a perpendicular through S to meet the perpendicular through Q at R.
**e** What is the shape PQRS?
**f** Measure the length RP.

## Inscribed polygons

Regular polygons can be constructed by drawing circles. All of the vertices of a regular polygon lie on the circumference of a circle. These polygons are called **inscribed polygons**.

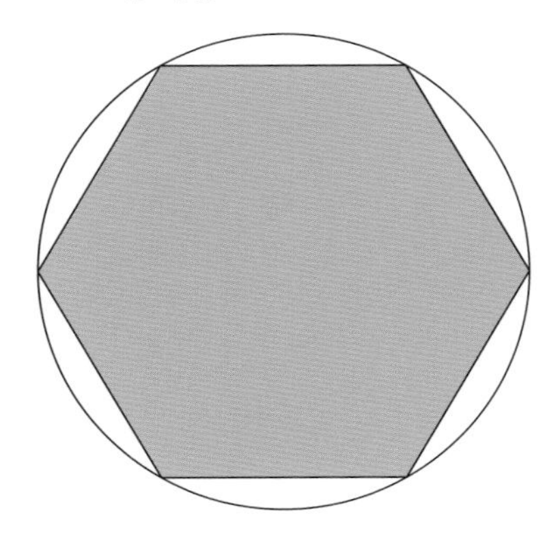

**EXAMPLE 2**

Construct an inscribed square, with a pair of compasses and a straight edge.

**Step 1:**
Draw a straight line. Mark a point C on the line, somewhere near the middle. Remember, if you are asked to construct with a straight edge, this means you are not allowed to measure!

**Step 2:**
Draw a circle with centre C using your compasses. Label the points where the circle intersects the line as A and B.

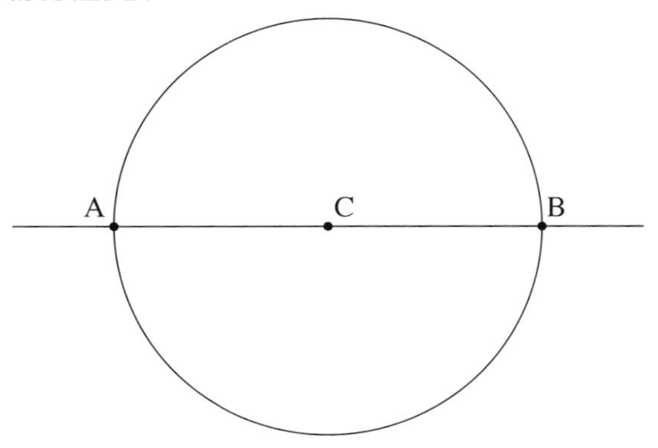


**Step 3:**
Construct the perpendicular bisector of AB by drawing arcs above the circle from A and B and joining their point of intersection to C. Label the points where this perpendicular line intersects with the circle as X and Y.

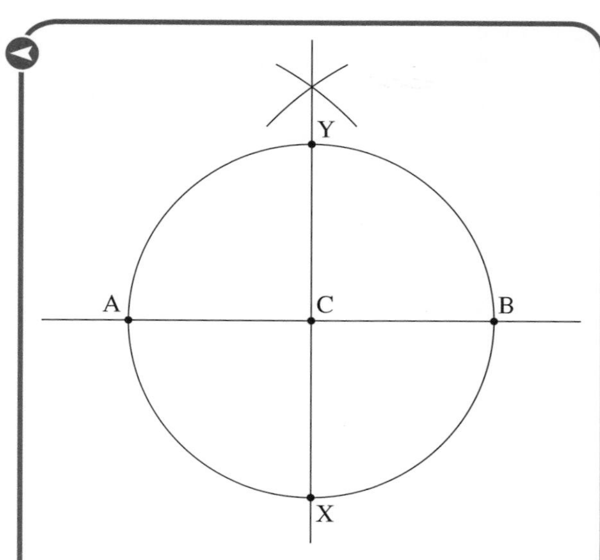


**Step 4:**
Join A to X, X to B, B to Y and Y to A to finish the construction of the square.

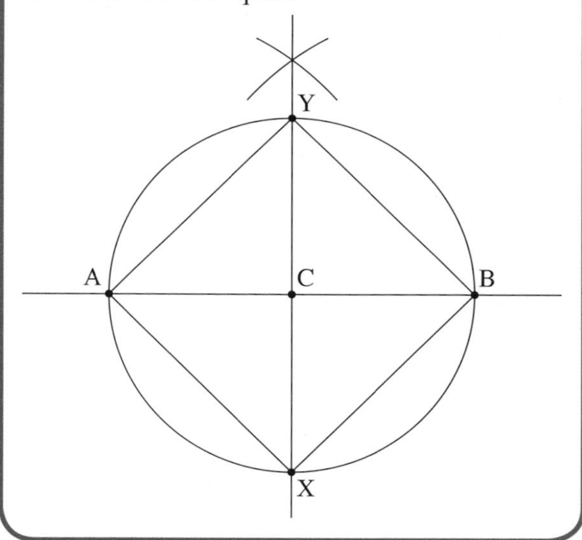


To construct a regular hexagon, remember that it is made up from 6 equilateral triangles:

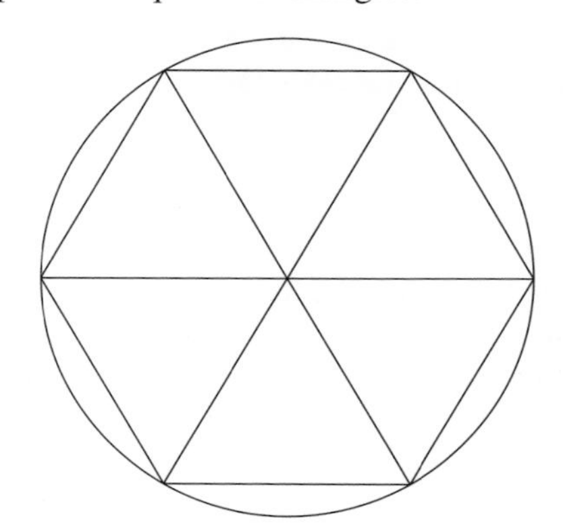

### EXAMPLE 3

Construct an inscribed regular hexagon

**Step 1:**
Draw a circle. Keep the compasses set to the radius of the circle and with the point of the compasses anywhere on the circumference draw an arc intersecting the circumference. Label the point of intersection A.

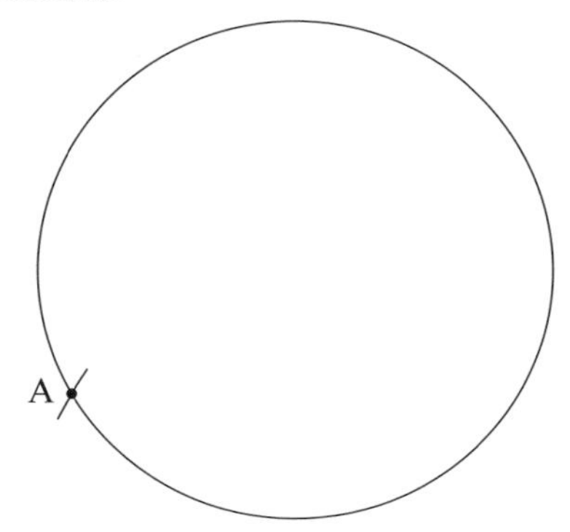


**Step 2:**
With the point of the compasses on A and still set to the radius of the circle, draw the next arc along. Label it B.

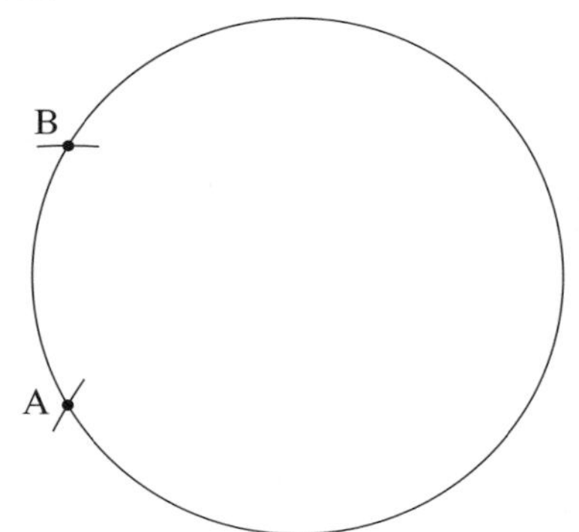


**Step 3:**
Repeat Step 2 to construct points C, D, E and F.

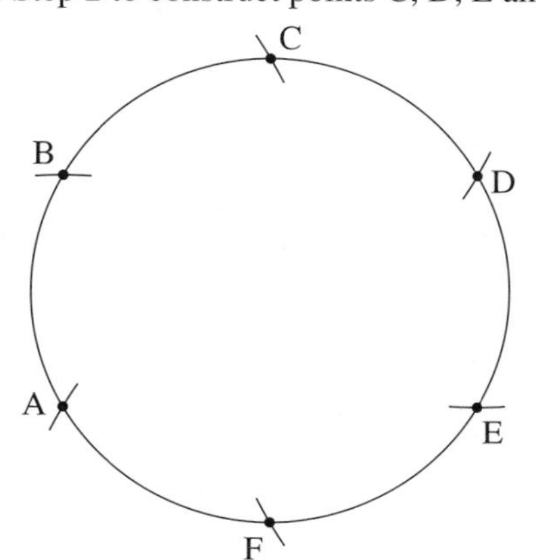

**Step 4:**
Join A to B, B to C, C to D, D to E, E to F and F to A to finish the construction of the hexagon.

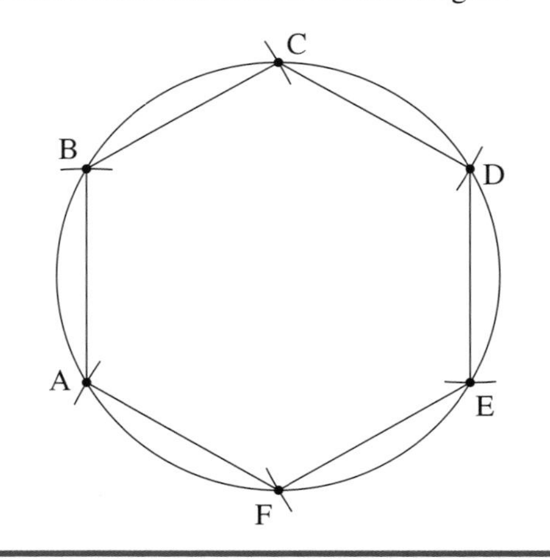


## Exercise 3D

**1** Construct an inscribed equilateral triangle by following Example 3, Steps 1–3, then, for Step 4, joining A to C, C to E and E to A.

**2** Construct an inscribed octagon by following Example 2, Steps 1–3, then constructing the perpendicular bisectors of AY and AX.
You should have a diagram like this:

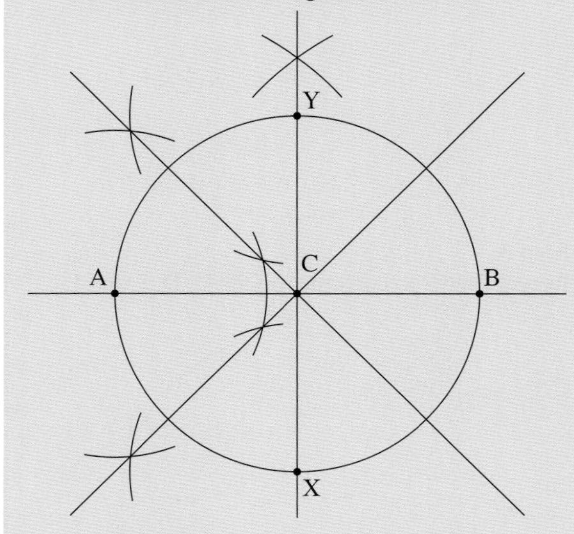


Where these new perpendicular bisectors intersect the circle draw four points. Join the eight points on the circumference to complete your inscribed octagon.

**3** Construct the geometric patterns shown. Remember, you can only use your ruler for drawing straight edges, and not for measuring.

**a**

**b**

**c**

**4** Design your own geometric pattern using inscribed shapes.

### INVESTIGATION

A polygon is inscribed inside a circle with radius 10 cm.

If it is a square it will have area 200 $cm^2$.

What will the area of the polygon be if it is

**a** a hexagon

**b** a triangle?

**Hint:** You may want to look ahead to Chapter 9 to help you with this.

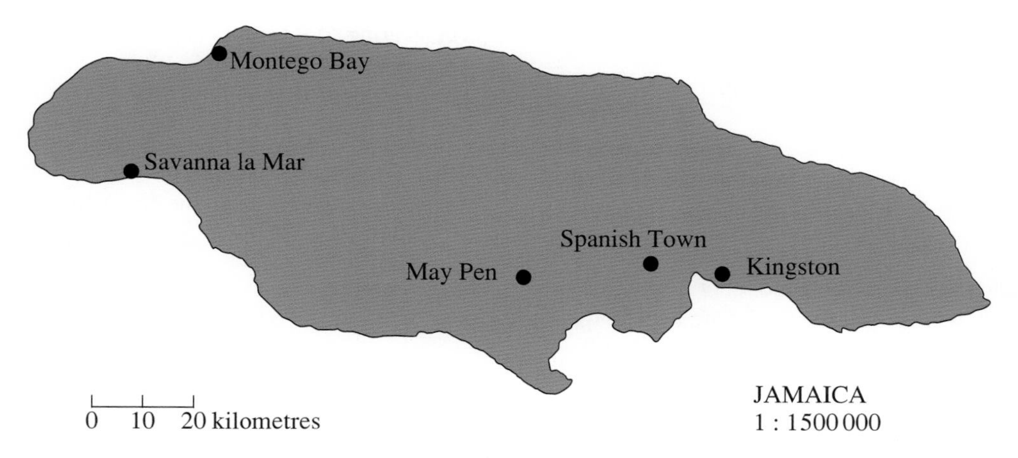


## 3.3 Maps and scale drawings

Maps are always drawn to a **scale**. The scale of the sketch map of Jamaica shown here is 1 : 1 500000. This is the ratio of a length on the map to the actual distance.

If you measure the distance on the map between May Pen and Spanish Town it is 1.8 cm.

$$
\begin{aligned}
\text{Actual distance} &= 1.8\,\text{cm} \times 1\,500\,000 \\
&= 270\,000\,\text{cm} \\
&= 27\,000\,\text{m} \\
&= 27\,\text{km}
\end{aligned}
$$

Of course, this is the direct distance. The road distance may well be longer.

**EXAMPLE 4**

A road map is drawn to a scale 1 : 50000
Two towns are 5 km apart.
How far apart are they on the map?

5 km = 5 × 1000 m = 5000 m
5000 m = 5000 × 100 cm = 500000 cm
The scale is 1 : 50000
500000 cm ÷ 50000 = 10 cm
The towns are 10 cm apart on the map.

### Exercise 3E

**1** The distance on the map between Kingston and Spanish Town is 1 cm.
What is the distance in kilometres?

**2** Use a ruler to measure, in centimetres, the distance from Kingston to:
**a** Montego Bay **b** May Pen
What are the actual distances, in kilometres?

**3** Use the sketch map to find, to the nearest kilometre, the greatest length of the island.

**4** The distance from Kingston to Savanna la Mar is approximately 120 km. What is the distance on a map with a scale of 1 : 500000?
Check your answer by measuring.

**5** A more detailed map has a scale of 1 : 50000.
Find the actual distance between two places if the distance on the map is:
**a** 1 cm **b** 5 cm **c** $\frac{1}{2}$ cm
Find the map distance if the actual distance is:
**d** 5 km **e** 10 km **f** 1 km

**6** On a map the scale is written 5 cm : 1 km.
**a** What is the distance, in metres, represented by 1 cm on the map?
**b** What is the distance, in centimetres, represented by 1 cm on the map?
**c** Express the scale as a ratio 1 : $n$.

**7** On a road map the scale can be shown as 1 : 100000 or 1 cm : 1 km.
Match these scales:

| | |
|---|---|
| 1 : 50000 | 1 cm : 1 km |
| 1 : 100000 | 2 cm : 25 km |
| 1 : 1 250000 | 10 cm : 1 km |
| 1 : 10000 | 1 cm : 4 km |
| 1 : 4000000 | 2 cm : 1 km |
| 1 : 400000 | 1 cm : 40 km |

**8** What distance does 1 cm represent on a scale of:
**a** 1 : 50000 **b** 1 : 200000
**c** 1 : 5000 **d** 1 : 1250000
**e** 1 : 5000000?

**9** The scale on a road map is $1:25000$.

**a** What is the distance, in metres, represented by 3 cm?

**b** What is the area of a field represented on the map by a rectangle 3 cm long and 4 cm wide?

**c** What is the area of the field in hectares? $(10000\,\text{m}^2 = 1\,\text{ha})$

**10** Sheldon makes a scale model of his bedroom. The dimensions of the model are 8 cm long, 6 cm wide and 4 cm high. He uses a scale of $1:50$.

**a** What is the area of the floor in the model?

**b** What is the actual area of Sheldon's bedroom floor, in $\text{cm}^2$?

**c** Copy and complete:
Area in the model : actual area
= 48 : □
= 1 : □

**d** What is the volume of the room in the scale model?

**e** What is the actual volume of Sheldon's bedroom?

**f** Copy and complete:
Volume in the model : actual volume
= 192 : □
= 1 : □

**g** How do the ratios of the areas in part **c** and the volumes in part **f** compare to the ratio of the lengths $(1:50)$?

- In general, if the ratio of the lengths is $1:a$ then the ratio of the areas is $1:a^2$ and the ratio of the volumes is $1:a^3$

## 3.4 Bearings

The position of an object relative to another object is called its **bearing**.

The bearing is given as a three-figure angle, such as 045°. It is measured in a *clockwise* direction from North.

For example:

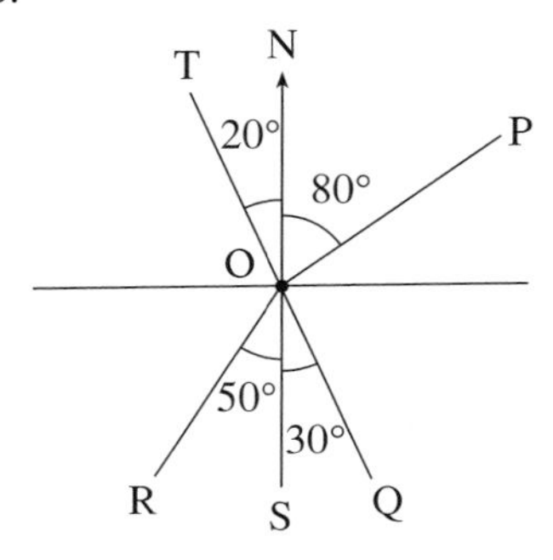


The bearing of P from O is 080°
The bearing of Q from O is 150°
The bearing of R from O is 230°
The bearing of T from O is 340°

When we write the bearing of A from B, this is short for the bearing of A from the North line at B.

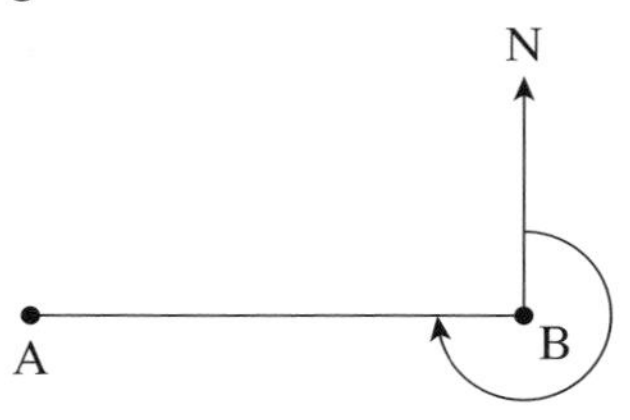


The bearing of B from A is different! Make sure you understand the difference.

**EXAMPLE 5**

A ship travels from Chennai, India on a bearing of 147°, for a distance of 100 km. Show this in a diagram using 1 cm to represent 20 km.

The distance the ship has travelled East and South of its starting point can be found from the scale drawing.

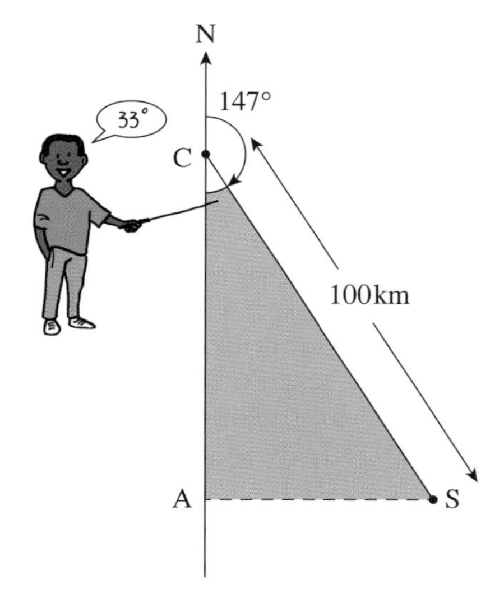


You can see that:

AS ≈ 2.7 cm
CA ≈ 4.2 cm

So the ship has travelled $2.7 \times 20 = 54$ km East and $4.2 \times 20 = 84$ km South.

## Exercise 3F

1. Make a diagram for each of these bearings:
   - **a** 145°
   - **b** 210°
   - **c** 330°
   - **d** 085°
   - **e** 305°
   - **f** 010°
   - **g** 258°
   - **h** 340°
   - **i** 220°
   - **j** 250°

2. By first making a scale drawing, find the distance travelled East and South by a ship sailing on a bearing of:
   - **a** 132° for 100 km
   - **b** 151° for 50 km
   - **c** 165° for 20 km.

3. By first making a scale drawing, find the distance travelled West and South by a ship sailing on a bearing of:
   - **a** 210° for 100 km
   - **b** 236° for 50 km
   - **c** 261° for 20 km.

4. Measure the bearing of A from B in each of these diagrams. ('From B' means from the North line at B.)

**a**

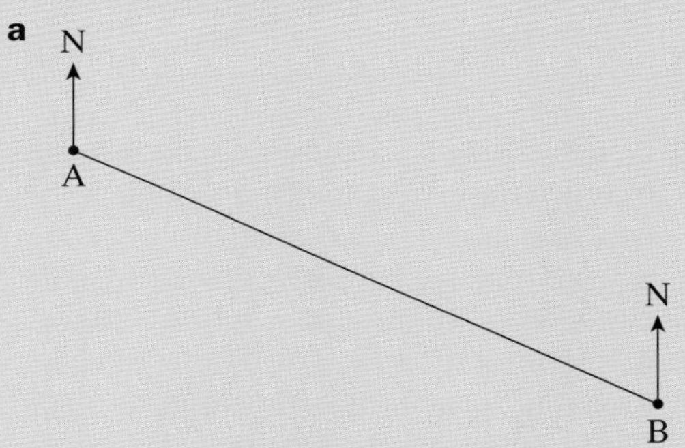


**b**

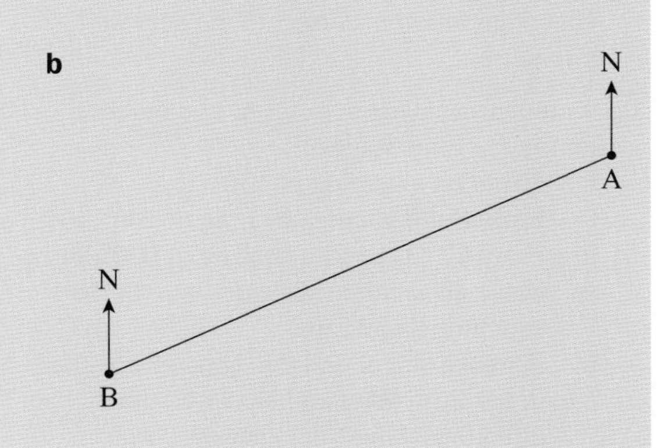


**c**

**d**

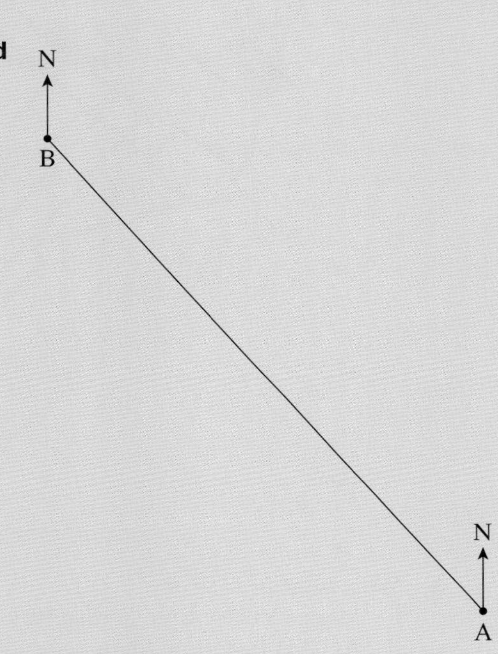


5. In each of the diagrams in Question **4**, find the bearing of B from A.

6. Can you see a relationship between your answers to Question **4** and your answers to Question **5**?

**7** The map shows part of Abu Dhabi.

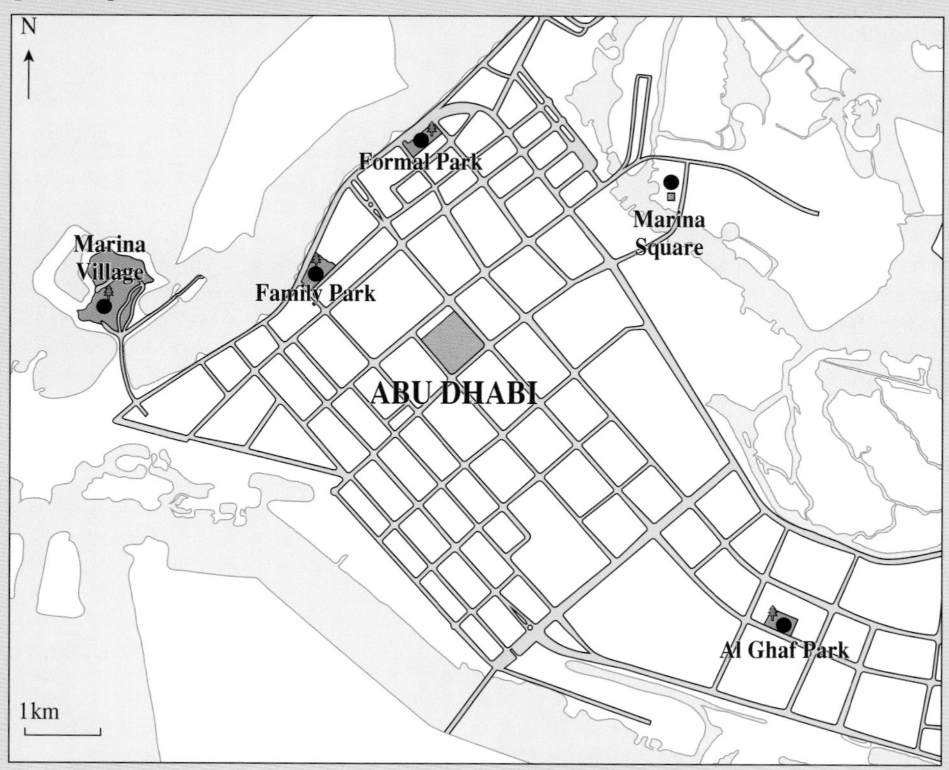


**a** What is the bearing of

- **i** Marina Village from Formal Park
- **ii** Al Ghaf Park from Marina Square?

**b** The scale on this map is 1 : 100 000.
By measuring on the map, find the distance, in metres, from

- **i** Formal Park to Marina Village
- **ii** Marina Square to Al Ghaf Park.

**8** After sailing from port, a ship finishes 10 km East and 20 km South of its starting point.

**a** Find the bearing on which the ship sailed by making a scale drawing.

**b** Use your drawing to find how far the ship sailed from port.

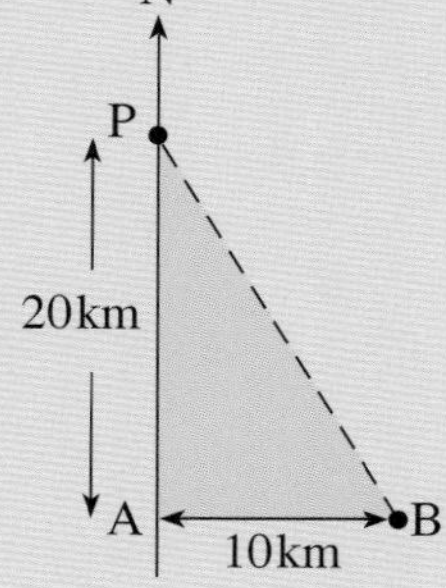


**9** Find the bearing on which a ship sails from port if it finishes:

- **a** 20 km East and 10 km South
- **b** 20 km West and 30 km South
- **c** 10 km West and 40 km North
- **d** 30 km West and 20 km North
- **e** 40 km East and 20 km South.

Hint: Use scale drawings.

**10** Find, by using a scale drawing, the distances travelled East–West and North–South by a ship sailing on a bearing of:

- **a** 066° for 10 km
- **b** 126° for 20 km
- **c** 256° for 40 km
- **d** 296° for 50 km
- **e** 320° for 25 km
- **f** 335° for 17 km.

Each time, give the direction of the distance (East or West, North or South).
Give your answers correct to the nearest kilometre.

**11** A submarine starts at point A, and makes a journey in three stages:

- **i** 40 km on a bearing of 130°, to B
- **ii** From B, 40 km on a bearing of 240°, to C
- **iii** From C, 43 km on a bearing of 030°, to D.

**a** Make an accurate scale drawing of the submarine's journey.

**b** For each stage, find the distance travelled in the East–West and North–South directions. Give your answers correct to the nearest kilometre.

**c** Find how far D is North or South and East or West of A.

# Consolidation

### Example 1

Draw the 3D shape represented by the plan and elevations.

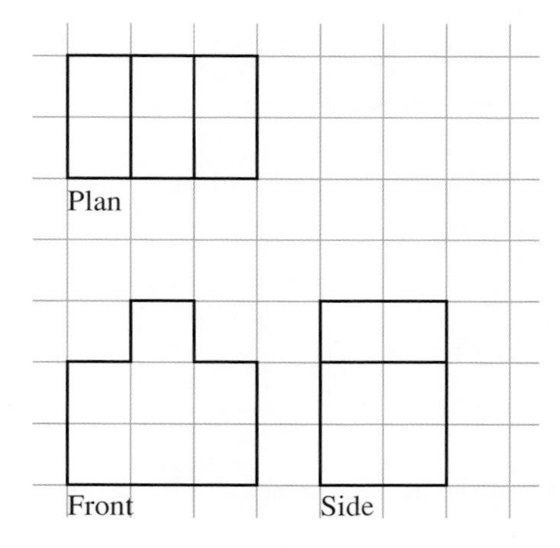


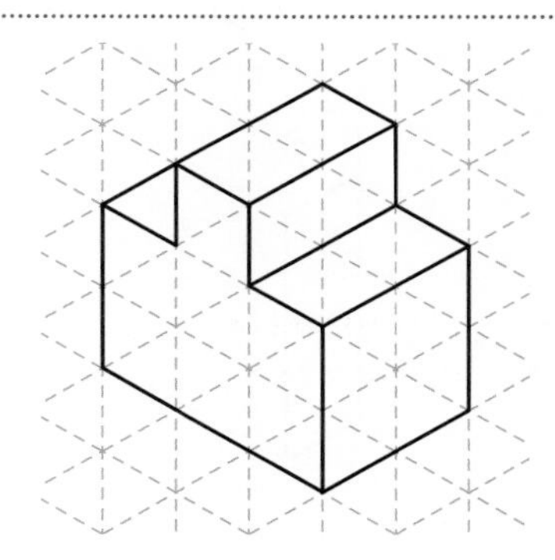

### Example 2

Draw the planes of symmetry of the shape in Example 1.

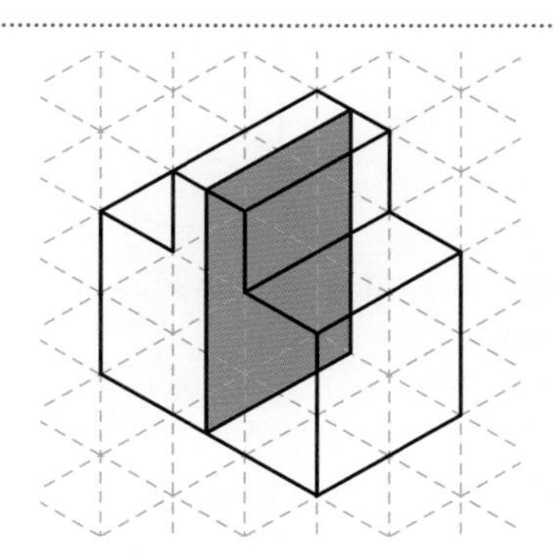

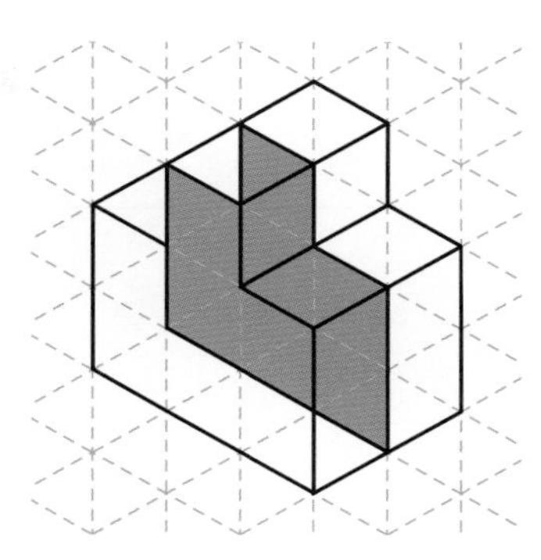

### Example 3

Construct the perpendicular line from P to MN.

.P

Draw arcs centred at P to cut MN:

.P

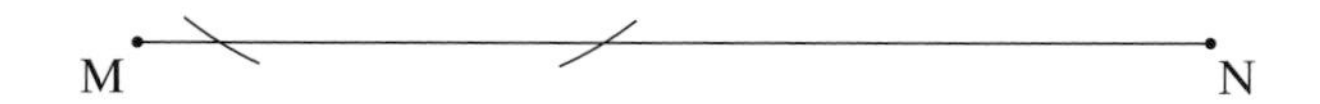


Using the intersections of these arcs with the line MN as centres, draw two intersecting arcs below MN:

.P

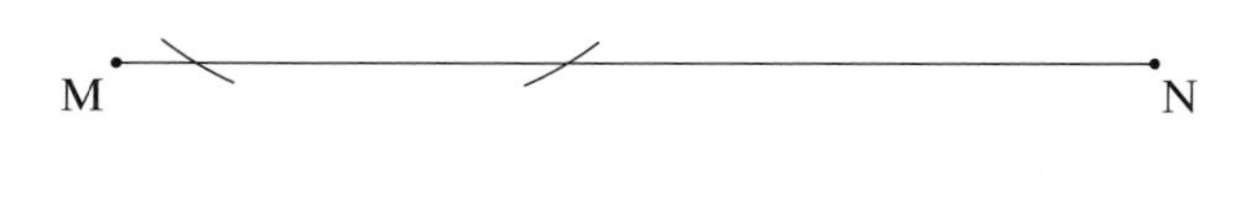

Join the point of intersection of the arcs to P:

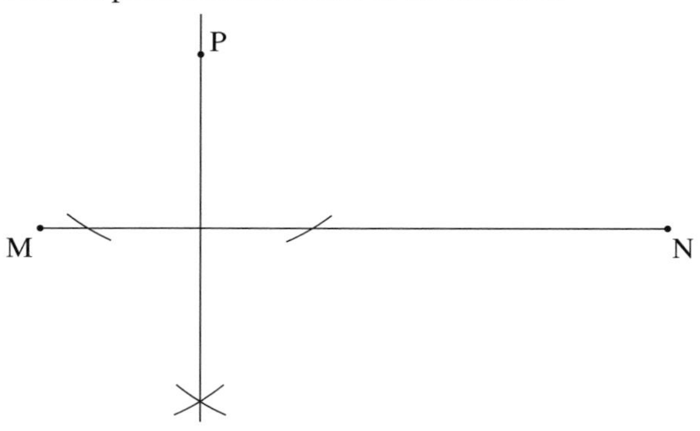


**Example 4**

What is the bearing of B from A?

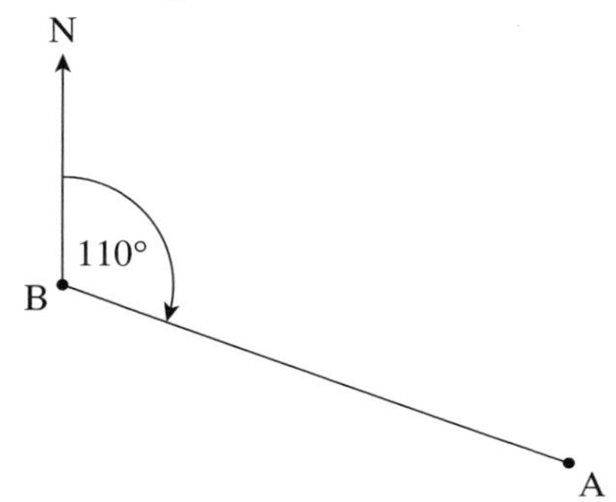


The bearing of B from A is 290°.

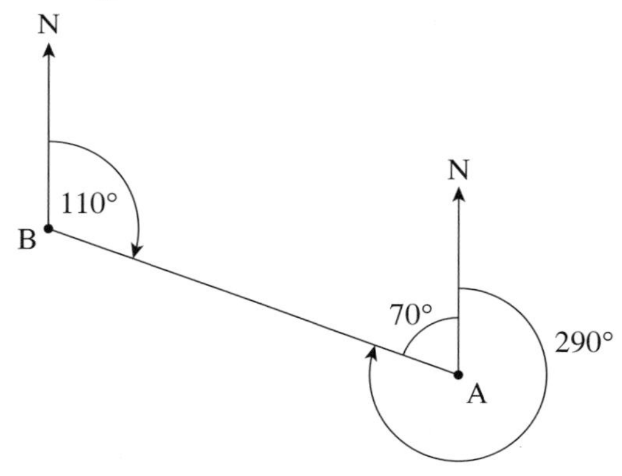


## Exercise 3

**1** Using isometric paper, draw the 3D shape represented in the plan and elevations.

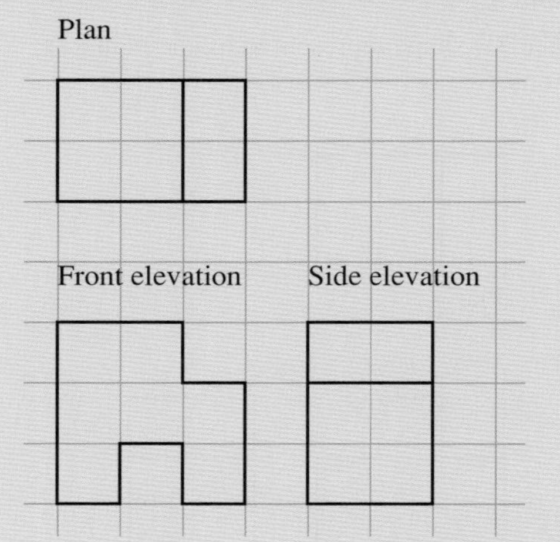


**2** Draw the plan and elevations for these shapes.

**a**

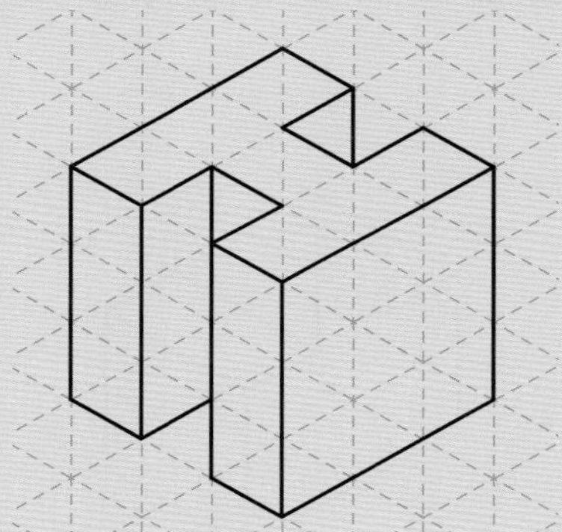

**b**

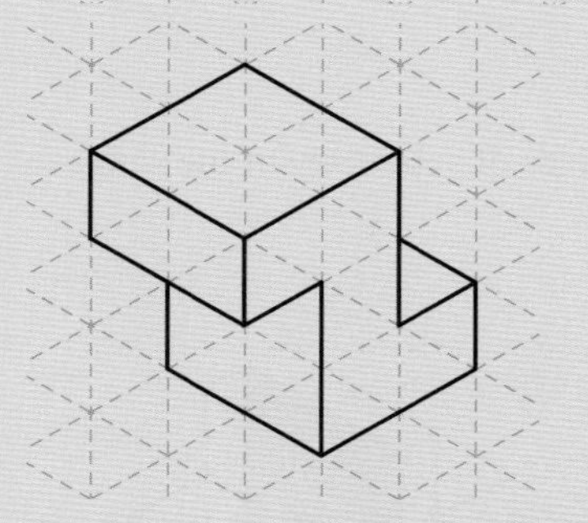

**3** What distance does 4.3 cm represent on a scale of:

**a** 1 : 10 000 **b** 1 : 25 000
**c** 1 : 500 000 **d** 1 : 200?

**4** The distance of 7 km would be represented by what length on a map with scale:

**a** 1 : 50 000 **b** 1 : 200 000
**c** 1 : 1 000 000 **d** 1 : 250 000?

**5** A map is drawn to a scale of 1 : 3000. Find the actual distance in metres between two towns which are 3.8 cm apart on the map.

**6** **a** What is the bearing of A from B?

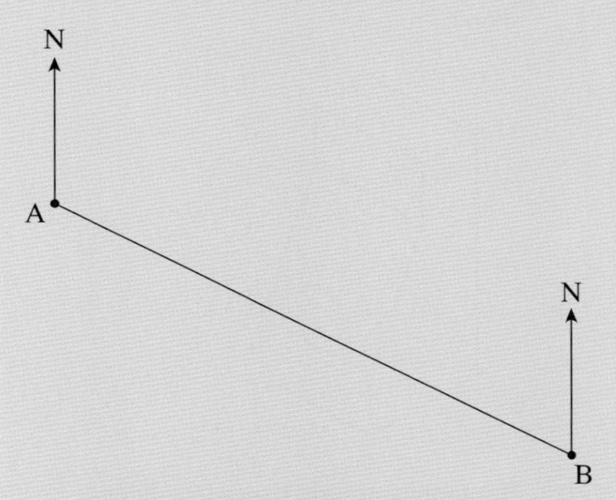


**b** What is the bearing of C from D?

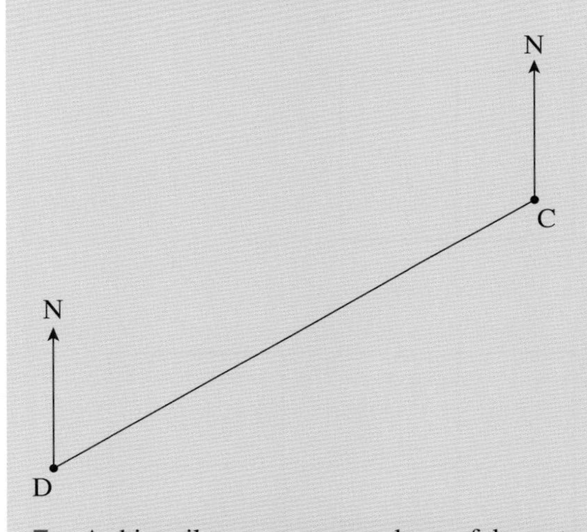


**7** A ship sails on a course made up of three stages:

**i** 10 km on a bearing of 130°
**ii** 20 km on a bearing of 215°
**iii** 40 km on a bearing of 330°

**a** Make a scale drawing of the ship's journey.
**b** Find for each stage the distance travelled to the nearest kilometre
  **i** in the East–West direction
  **ii** in the North–South direction.
**c** Find also the total distance travelled to the North and West of the ship's starting point.

**8** How many planes of symmetry do these shapes have?

**a**

**b**

**c**

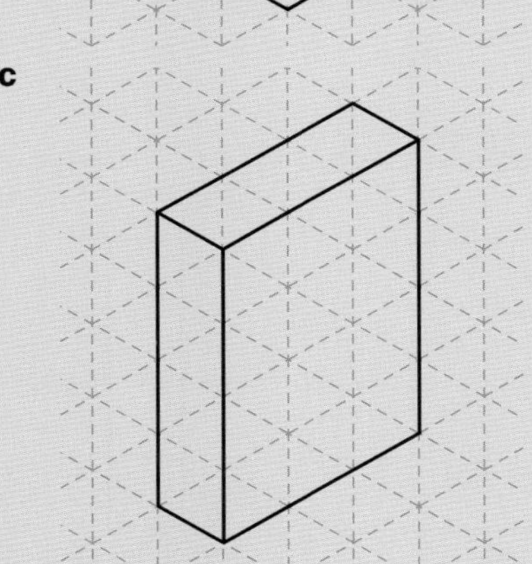

**9** Copy each diagram.
Construct a perpendicular line from point X to line AB.

**a**

**b**

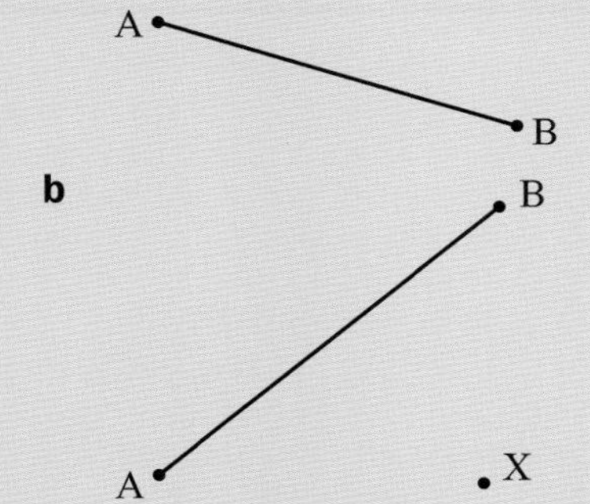

# Summary

## You should know ...

**1** How to draw 3D shapes using isometric paper and how to draw plans and elevations. You should also know about planes of symmetry.
*For example:*

This shape has one plane of symmetry. Its plan and elevations are:

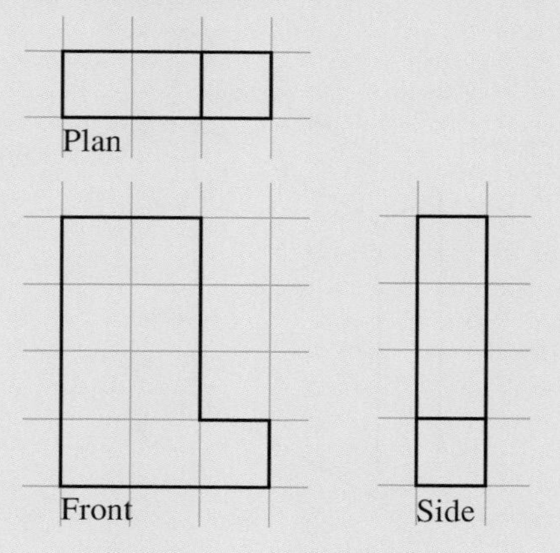


## Check out

**1 a** Draw the plan and elevations for this shape:

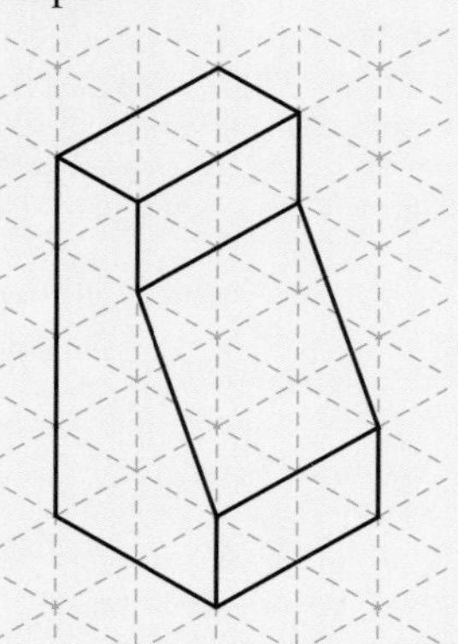

**b** Draw on isometric paper the shape with the plan and elevations shown here.

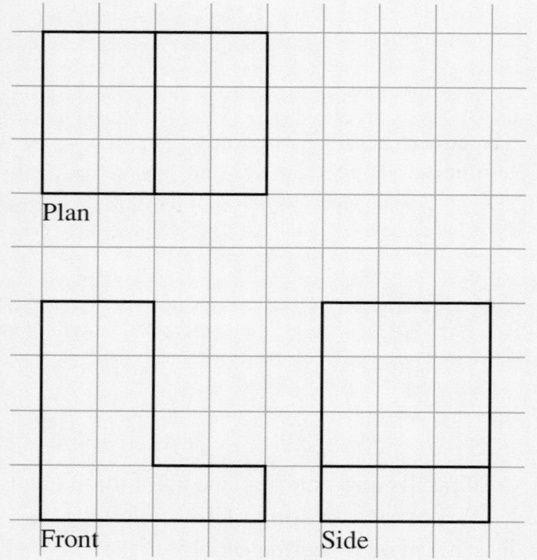


**c** How many planes of symmetry do the shapes in **a** and **b** have?

**2** How to use a straight edge and compasses to construct the perpendicular from a point to a line, and from a point on a line.

To construct the perpendicular from a point to a line: keeping your compasses at the same radius throughout, draw arcs on the line with the point X as centre; then use the intersections of these arcs with the line as centres to draw intersecting arcs below the line; join the intersection with point X.

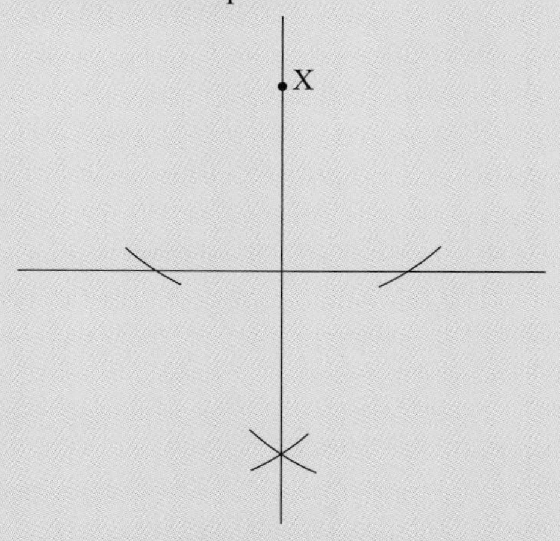


To construct the perpendicular from a point on a line: draw arcs on the line with the point X as centre; then increase the radius of the compasses and use the intersections of these arcs with the line as centres to draw intersecting arcs both above and below the line; join the intersections.

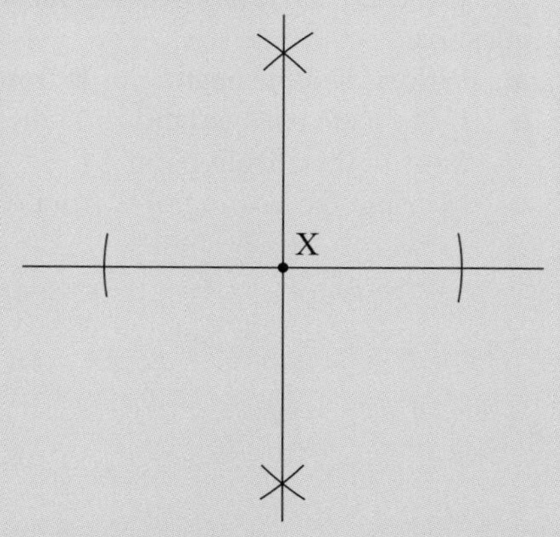


Remember not to rub out any construction arcs.

**2** **a** Make a rough copy of the diagram. Construct the perpendicular from the point X to the line.

• X

**b** Make a rough copy of the diagram. Construct the perpendicular to the line from the point X.

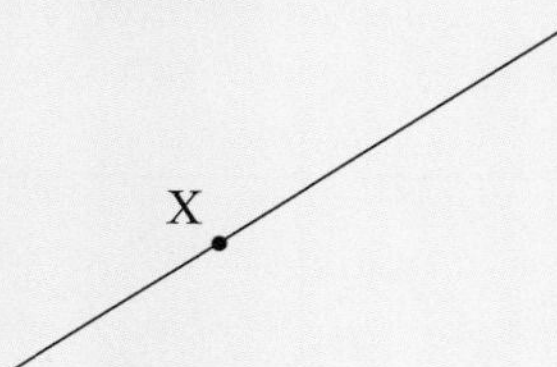

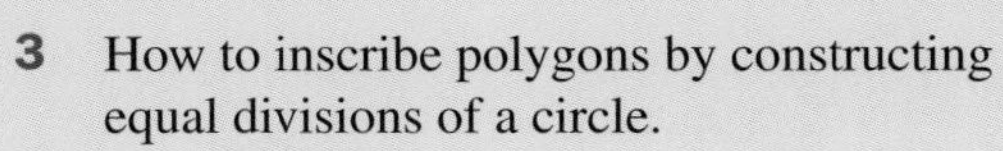

**3** How to inscribe polygons by constructing equal divisions of a circle.
*For example:*
An inscribed square, when constructed, should look like this (do not rub out construction arcs).

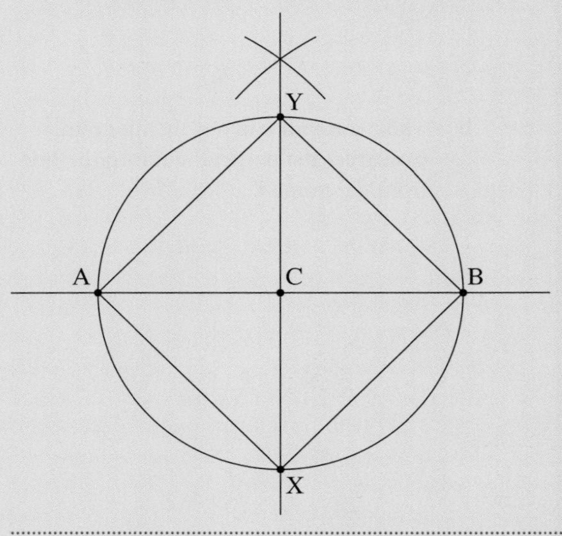


**4** How to use bearings to solve problems involving distance and direction and use scale drawings and maps.
Bearings are always three-digit angles measured clockwise from the North direction.

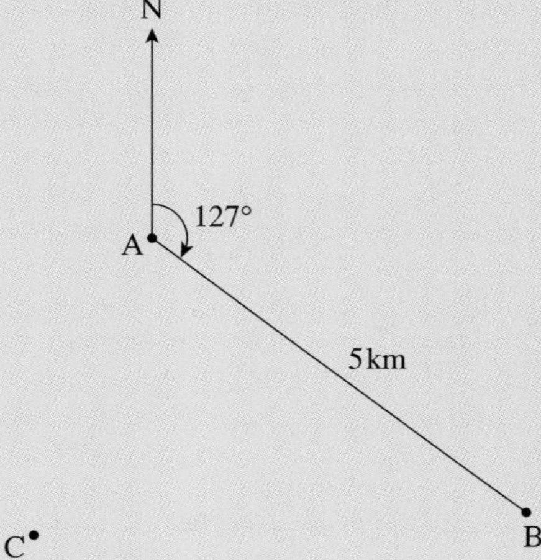


B is on a bearing of 127° from point A and 5 km from A.
Using the scale 1 cm = 1 km, since C is 3.5 cm from A then C is 3.5 km from A.

**3** Draw a circle using a pair of compasses. Inside the circle inscribe an equilateral triangle.

**4**

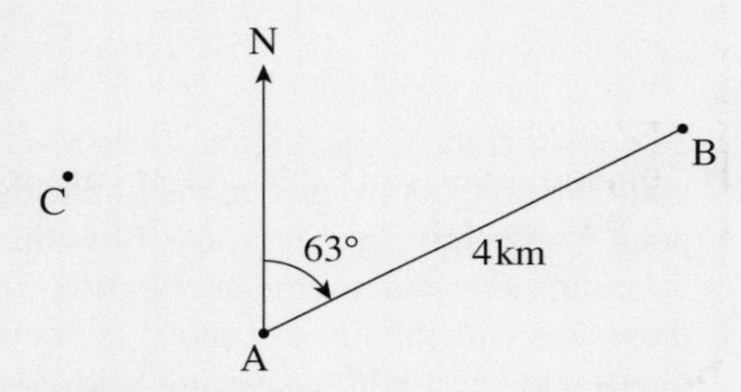


Use the scale diagram to answer these questions.

**a** Write down the bearing of B from A.

**b** If the scale used is 1 cm = 1 km, how far is point C from point A?

**c** Measure the bearing of C from A.

# Number

## Objectives

- Add, subtract, multiply and divide directed numbers.
- Estimate square roots and cube roots.
- Recognise the equivalence of 0.1, $\frac{1}{10}$ and $10^{-1}$; multiply and divide whole numbers and decimals by 10 to the power of any positive or negative integer.

## What's the point?

Being confident with working with numbers is important in many professions. For example, negative numbers are used in banking: money coming into an account is positive, while money leaving an account is negative, so that the balance can be clearly recorded. Square roots are used in engineering, to work out lengths of material for constructions. In this instance, it is useful if the engineer can estimate the square root so that he knows if his calculation is correct.

## Before you start

### You should know ...

**1** How to divide by a decimal.
*For example:*

$4 \div 0.5$ can be worked out using equivalent fractions:

$4 \div 0.5 = \frac{4}{0.5} = \frac{40}{5} = 8$

**2** The squares and square roots of integers up to 20 and the cubes and cube roots of integers up to 5.
*For example:*

$3^3 = 27$
$\sqrt{144} = 12$
$8^2 = 64$
$\sqrt[3]{125} = 5$

### Check in

**1** Work out:
**a** $6 \div 0.5$
**b** $4 \div 0.2$
**c** $8 \div 0.01$
**d** $1.2 \div 0.3$

**2** Work out:
**a** $4^3$
**b** $\sqrt{121}$
**c** $7^2$
**d** $\sqrt[3]{8}$

| | |
|---|---|
| **3** How to work out the square root of some decimals without a calculator.<br>*For example:*<br>$\sqrt{0.25} = 0.5$<br>$\sqrt{0.04} = 0.2$ | **3** Work out:<br>**a** $\sqrt{0.36}$<br>**b** $\sqrt{0.09}$<br>**c** $\sqrt{0.49}$<br>**d** $\sqrt{0.01}$ |

## 4.1 Directed numbers

Numbers can be both positive and negative. Negative numbers are commonly used to represent temperatures

**Directed numbers** are numbers which have direction as well as size. Direction can be positive, as in the case of temperatures above zero, or negative, as for temperatures below zero. Directed numbers that are also whole numbers are called **integers**, a few of which are shown here on a number line:

$^-5 \quad ^-4 \quad ^-3 \quad ^-2 \quad ^-1 \quad 0 \quad 1 \quad 2 \quad 3 \quad 4 \quad 5$

Addition and subtraction can be shown on the number line by movements to the right (positive direction) or to the left (negative direction).

**EXAMPLE 1**

Work out

**a** $^-3 + 5.4$ **b** $^-2 - 5.1$

**a**

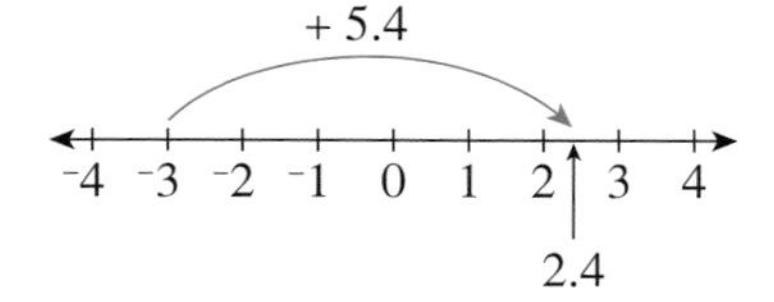


$^-3 + 5.4 = 2.4$

**b**

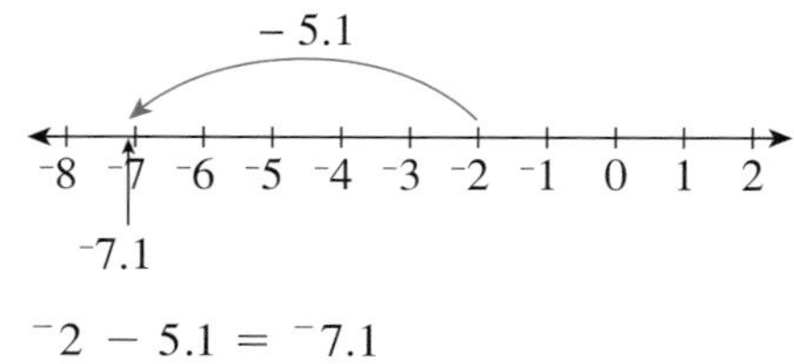


$^-2 - 5.1 = {}^-7.1$

You can also add negative numbers, for example:

$$4.3 + (^-5)$$

Notice that $4.3 + (^-5) = {}^-5 + 4.3$, as you can add numbers in any order.

On a number line:

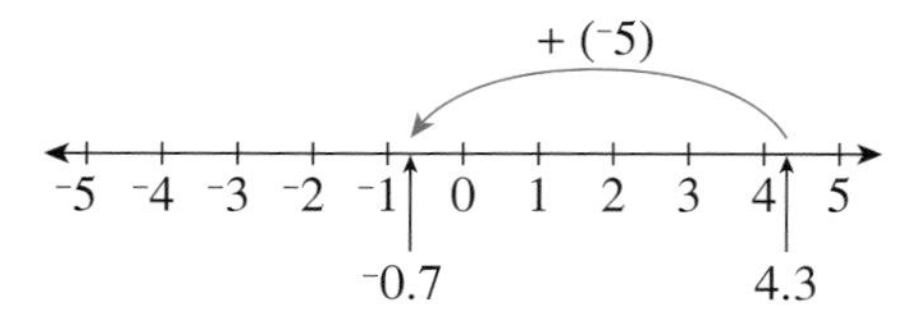


$$4.3 + (^-5) = {}^-0.7$$

So **adding a negative number is the same as subtracting a positive number**.

### Exercise 4A

**1** Draw diagrams to show that:

**a** $4 + (^-2) = 2$

**b** $^-3 + (^-2) = {}^-5$

**c** $4 + (^-6) = {}^-2$

**2** Work out:

**a** $^-3 + 4$ **b** $^-5 - 2$

**c** $^-4 - 3.4$ **d** $^-4 + 5.6$

**e** $3 - 7.1$ **f** $^-6 + 2.8$

**g** $2.7 - 4.6$ **h** $^-3.8 + 5.2$

**3** Use a number line to calculate:

**a** $4 + (^-3)$ **b** $^-3 + (^-5)$

**c** $4.6 + (^-2)$ **d** $^-6 + (^-2.4)$

**e** $1.7 + (^-3)$ **f** $8 + (^-9.2)$

**g** $2.6 + (^-4.8)$ **h** $^-5.1 + (^-4.8)$

**4** Copy and complete this addition table.

| + | $^-2.6$ | $^-1.8$ | $^-0.8$ | 2.9 |
|---|---|---|---|---|
| $^-2.3$ | | | | |
| $^-1.4$ | | $^-3.2$ | | |
| $^-0.6$ | | | | |
| 2.3 | | | | 5.2 |

**5** Work out:

**a** $3 - 5 - 2$
**b** $6 - 2 + 4$
**c** ${}^-3 - 2 - 1$
**d** ${}^-4 + ({}^-3) - 5$
**e** $4.1 + ({}^-3.1) - 3.9$
**f** $6.3 - 7.4 + ({}^-3.2)$
**g** ${}^-5.4 + 8.3 - 9.7$
**h** ${}^-3.2 + ({}^-7.7) - 6.5$
**i** $7.21 + ({}^-3.46) + 6.8$
**j** ${}^-5.36 + ({}^-4.2) + 3.84$

## Subtracting negative numbers

Look at the pattern in these subtractions:

$3 - 3 = 0$

$3 - 2 = 1$

$3 - 1 = 2$

$3 - 0 = 3$

$3 - {}^-1 = ?$

What do you think $3 - {}^-1$ should be? Since the answers increase by 1 each time, $3 - {}^-1 = 4$.

That is, **subtracting a negative number is the same as adding a positive number**.

### EXAMPLE 2

Work out

**a** $7 - {}^-3$ **b** ${}^-7 - {}^-3.2$

**a** $7 - {}^-3 = 7 + 3 = 10$

**b** ${}^-7 - {}^-3.2 = {}^-7 + 3.2 = {}^-3.8$

A number line is useful if you are not sure.

### EXAMPLE 3

Calculate

**a** ${}^-1.5 + {}^-2.8$ **b** ${}^-2.1 - {}^-1.3$

**a**

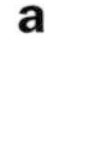

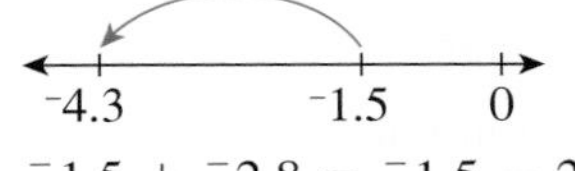


${}^-1.5 + {}^-2.8 = {}^-1.5 - 2.8 = {}^-4.3$

**b**

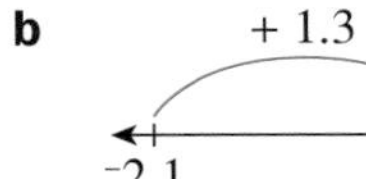


${}^-2.1 - {}^-1.3 = {}^-2.1 + 1.3 = {}^-0.8$

## Exercise 4B

**1** Work out:

**a** $6 - {}^-3$ **b** $5 - {}^-3$
**c** ${}^-4 - {}^-3.6$ **d** ${}^-5 - {}^-2.4$
**e** $1.7 - {}^-4$ **f** $1.3 - {}^-5$

**2** Use a number line to find:

**a** ${}^-1.3 + {}^-1.4$ **b** ${}^-1.3 - {}^-1.3$
**c** ${}^-1.5 - {}^-0.6$ **d** ${}^-2.3 + {}^-1.7$
**e** ${}^-4.1 + {}^-3.7$ **f** ${}^-5.3 - {}^-3.4$

**3** Calculate:

**a** $6 + {}^-4 - 2$ **b** $6 - {}^-4 - 2$
**c** $1 - 6 - {}^-2$ **d** $9 - {}^-3 + 3$
**e** $7 + {}^-6 - {}^-7.1$ **f** ${}^-6.8 + {}^-4 - {}^-6$
**g** ${}^-0.9 + 0.6 + {}^-0.3$
**h** ${}^-1.3 - {}^-1.1 - 1.2$
**i** ${}^-4.3 + {}^-1.7 + 1.4$
**j** ${}^-3.3 - {}^-4.1 + {}^-1.8$

**4** Copy and complete this subtraction table.

Second number (columns); First number (rows)

| – | -3.4 | -2.8 | -1 | 1.7 | 2.4 |
|---|---|---|---|---|---|
| -3 | 0.4 | | | | |
| -2.5 | | | -1.5 | | |
| -1 | | | | | |
| 1.7 | | | | 0 | |
| 2 | | | | | |

**5** Copy these diagrams and fill in numbers that make the subtraction tables work.

**a**

Second number (columns); First number (rows)

| – | | |
|---|---|---|
| | 4 | 7 |
| | 3 | 6 |

**b**

Second number (columns); First number (rows)

| – | | |
|---|---|---|
| | -1 | 2 |
| | -3 | 0 |

## Multiplying and dividing negative numbers

Multiplying by a negative number is straightforward. For example:

$${}^-5 \times 3 = {}^-5 + {}^-5 + {}^-5$$
$$= {}^-15$$

Similarly: $3 \times {}^-5 = {}^-5 \times 3 = {}^-15$

Notice that, if ${}^-5 \times 3 = {}^-15$

then $3 = {}^-15 \div {}^-5$

So **a negative number divided by a negative number gives a positive number**.

Also $^{-}5 = {}^{-}15 \div 3$

So **a negative number divided by a positive number gives a negative number**.

The rules for multiplication and division of negative numbers are:

| Multiply (×) Divide (÷) | Positive number | Negative number |
|---|---|---|
| Positive number | + | − |
| Negative number | − | + |

**EXAMPLE 4**

Work out

**a** $^{-}9.6 \div 8$

**b** $^{-}2.1 \times {}^{-}7$

**a** $8\overline{)9.6}$ = 1.2 — Do the calculation ignoring the signs.

Using the table of rules above, a negative number divided by a positive number gives a negative answer.

So $^{-}9.6 \div 8 = {}^{-}1.2$

**b** $2.1 \times 7 = 14.7$ — Do the calculation ignoring the signs.

Using the table of rules, a negative number multiplied by a negative number gives a positive answer.

So $^{-}2.1 \times {}^{-}7 = 14.7$

## Exercise 4C

**1** Work out:

**a** $3 \times {}^{-}4$  **b** $^{-}5 \times 3$

**c** $7 \times {}^{-}11$  **d** $^{-}3 \times {}^{-}2$

**e** $^{-}6 \times {}^{-}8.2$  **f** $^{-}3 \times 6.4$

**g** $({}^{-}4)^2$  **h** $({}^{-}2)^3$

**i** $({}^{-}1.2)^2$  **j** $({}^{-}1.1)^2$

**2** Work out:

**a** $18 \div {}^{-}3$  **b** $21 \div {}^{-}7$

**c** $^{-}14 \div 2$  **d** $^{-}9.6 \div {}^{-}12$

**e** $7.2 \div {}^{-}8$  **f** $^{-}4.8 \div {}^{-}6$

**3** Find the missing numbers.

**a** $\square \times 3 = {}^{-}9$  **b** $18 \div \square = {}^{-}3$

**c** $^{-}4 \times \square = 12$  **d** $^{-}5 \times \square = 35$

**e** $\square \div {}^{-}3 = {}^{-}0.9$  **f** $\square \div {}^{-}5 = 1.4$

**4** Copy and complete this multiplication grid. You may use a calculator for the blue shaded regions. If you use a calculator, don't forget to use the [±] or [−], rather than the subtraction key, to get a negative number.

| × | $^{-}1.4$ | $^{-}2.3$ | 3.5 | $^{-}1.79$ |
|---|---|---|---|---|
| 4.1 | | | | |
| $^{-}1.6$ | | | | |
| $^{-}2.6$ | | | | |
| $^{-}5.86$ | | | | |

**5** Copy and complete this division grid.

Second number (across); First number (down)

| ÷ | $^{-}5$ | 2 | $^{-}0.5$ |
|---|---|---|---|
| $^{-}2.5$ | | | |
| 3.5 | | | |
| $^{-}10$ | | | |

**6** Copy and complete these multiplication grids.

**a**

| × | | |
|---|---|---|
| | 6 | $^{-}15$ |
| | 8 | $^{-}20$ |

**b**

| × | | $^{-}4$ |
|---|---|---|
| | 10 | 8 |
| 6 | | |

**7** Copy and complete these division grids.

**a** Second number (across); First number (down)

| ÷ | | | $^{-}1$ |
|---|---|---|---|
| | $^{-}9$ | 6 | $^{-}18$ |
| | $^{-}6$ | 4 | |
| $^{-}12$ | 6 | | |

**b** Second number (across); First number (down)

| ÷ | | |
|---|---|---|
| | 14 | 7 |
| | $^{-}6$ | $^{-}3$ |

**8** Design your own multiplication and division grids with missing numbers. Ask your partner to solve them.

Visit the website
www.nrich.maths.org
and search for 'negative numbers'.
Have a go at the investigation about consecutive negative numbers.
It's quite a challenge!

## 4.2 Estimating square roots and cube roots

### Square roots

Previously you learned that, when you square an integer, you get a square number, for example $7^2 = 49$. 49 is a square number and 7 is the square root of 49.

You also learned that square roots can be negative. For example, $(^-7)^2 = 49$ so the square root of 49, or $\pm\sqrt{49}$, is 7 or $^-7$.

When you square a decimal the result is not a square number. For example, $1.4^2 = 1.96$, but 1.96 is not a square number as 1.4 is not an integer.

It is possible to estimate the square root of a number by looking at square numbers either side of it.

**EXAMPLE 5**

Estimate the square root of 28.

The closest square numbers either side of 28 are 25 and 36.

Look just at the positive answers to start with:
Since $\sqrt{25} = 5$ and $\sqrt{36} = 6$, $\sqrt{28}$ lies between 5 and 6.

28 is closer to 25 than 36, so the square root is going to be closer to 5 than to 6.

In fact, $\pm\sqrt{28} = 5.3$ or $^-5.3$, to 1 d.p.

Discuss with others how to make a good guess at the first decimal place for $\sqrt{28}$. How can you tell that it will be closer to 5.3 than to 5.1 or 5.5?

### Cube roots

Previously you learned that, when you cube an integer, you get a cube number, e.g. $2^3 = 8$.

8 is a cube number and 2 is the cube root of 8, or $\sqrt[3]{8}$ for short.

When you cube a decimal the result is not a cube number. For example, $1.2^3 = 1.728$, but 1.728 is not a cube number as 1.2 is not an integer.

It is possible to estimate the cube root of a number by looking at cube numbers either side of it, in a similar way to the method used for estimating square roots.

This list may help with the next example and exercise.

$1^3 = 1$
$2^3 = 8$
$3^3 = 27$
$4^3 = 64$
$5^3 = 125$
$6^3 = 216$
$7^3 = 343$
$8^3 = 512$
$9^3 = 729$
$10^3 = 1000$

**EXAMPLE 6**

Estimate $\sqrt[3]{200}$.

$\sqrt[3]{125} = 5$ and $\sqrt[3]{216} = 6$ so the answer lies between 5 and 6.
200 is 75 away from 125 and only 16 away from 216 so the answer is a lot closer to 6.

In fact, $\sqrt[3]{200} = 5.8$, to 1 d.p.

### Exercise 4D

Do **not** use a calculator for this exercise.

**1** Which of these estimates are obviously wrong? Why?
**a** $\sqrt{20} \approx 5.1$
**b** $\sqrt{3} \approx 1.7$
**c** $\sqrt{95} \approx 9.1$
**d** $\sqrt{50} \approx 6.9$

**2** Find **a** $\sqrt{49}$ **b** $\sqrt{64}$
**c** Use your answers to **a** and **b** to estimate $\sqrt{59}$ to 1 d.p.

**3** Find **a** $\sqrt{16}$ **b** $\sqrt{25}$
**c** Use your answers to **a** and **b** to estimate $^-\sqrt{21}$ to 1 d.p.

**4** Find **a** $\sqrt{144}$ **b** $\sqrt{169}$

**c** Use your answers to **a** and **b** to estimate $\sqrt{150}$ to 1 d.p.

**5** By considering the two closest square roots, estimate the following square roots, to 1 d.p.:

**a** $\pm\sqrt{32}$ **b** $\pm\sqrt{75}$

**c** $\pm\sqrt{6}$ **d** $\pm\sqrt{90}$

**e** $\pm\sqrt{110}$ **f** $\pm\sqrt{135}$

**6** Find **a** $\sqrt[3]{8}$ **b** $\sqrt[3]{27}$

**c** Use your answers to **a** and **b** to estimate $\sqrt[3]{10}$ to 1 d.p.

**7** By considering the two closest cube roots, estimate the following cube roots, to 1 d.p.:

**a** $\sqrt[3]{35}$ **b** $\sqrt[3]{4}$

**c** $\sqrt[3]{80}$ **d** $\sqrt[3]{150}$

**e** $\sqrt[3]{^-15}$ **f** $\sqrt[3]{^-400}$

**8** Find **a** $\sqrt{0.01}$ **b** $\sqrt{0.04}$

**c** Use your answers to **a** and **b** to estimate $\sqrt{0.03}$ to 2 d.p.

**9** Find **a** $\sqrt{0.25}$ **b** $\sqrt{0.36}$

**c** Use your answers to **a** and **b** to estimate $^-\sqrt{0.3}$ to 2 d.p.

**10** Estimate the side lengths of these squares, to the nearest millimetre.

**a**

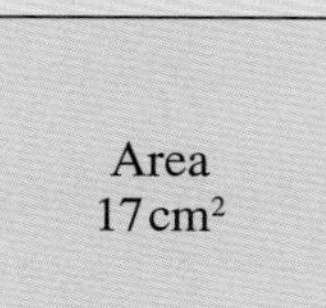


**b**

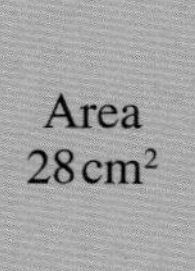


**11** Estimate the side lengths of these cubes, to the nearest millimetre.

**a**

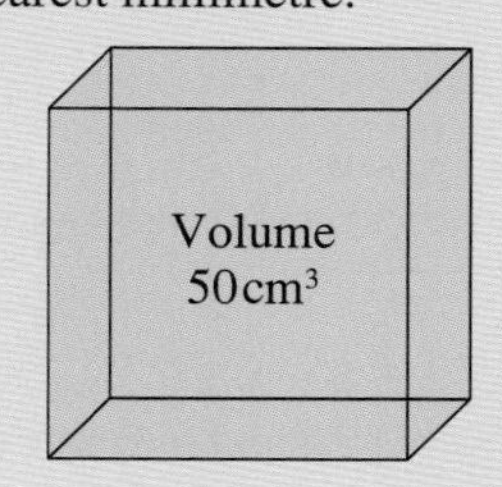


**b**

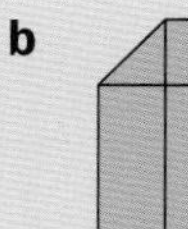

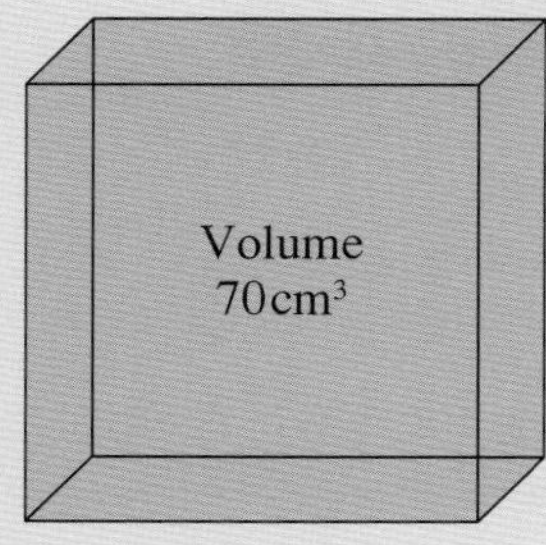


**12** Estimate the following square roots, to 2 d.p.:

**a** $\pm\sqrt{0.45}$

**b** $\pm\sqrt{0.92}$

**c** $\pm\sqrt{0.71}$

**d** $\pm\sqrt{0.2}$

**13** Estimate the positive solutions to these equations, to 1 d.p.:

**a** $x^2 = 70$

**b** $x^3 = 100$

**14** Farhad wanted to estimate the square root of 27. He wrote this working:

$\sqrt{36} = 6$ and $\sqrt{25} = 5$

36 is 11 more than 25

27 is 2 more than 25

So the square root must be $\frac{2}{11}$ more than 5 which is 5.2 to 1 d.p.

Is Farhad's method correct? Discuss it with your neighbour or teacher.

## Challenge

The examples here show the Babylonian method for finding the square root of a number. Note this requires you to be confident about dividing with decimals or working with fractions – the ancient Babylonians did not have calculators!

To estimate the square root of a number $N$, say $N = 120$, use this method:

**Example 1:** $\sqrt{120}$

| Guess | $\frac{N}{\text{guess}}$ | Average these numbers to give next guess |
|---|---|---|
| 12 | $\frac{120}{12} = 10$ | $(12 + 10) \div 2 = 11$ |
| 11 | $\frac{120}{11}$ (= 10.9090...) | $(11 + \frac{120}{11}) \div 2 = \frac{241}{22}$ = 10.954545... |

In fact, $\sqrt{120} = 10.9545$ to 4 d.p., so this estimate is *very* close after only two guesses!

**Example 2: $\sqrt{20}$**

| Guess | $\frac{N}{\text{guess}}$ | Average these numbers to give next guess |
|---|---|---|
| 5 | $\frac{20}{5} = 4$ | $\frac{5+4}{2} = 4.5$ |
| 4.5 or $\frac{9}{2}$ | $20 \div \frac{9}{2} = \frac{40}{9}$ (= 4.444...) | $\left(\frac{9}{2} + \frac{40}{9}\right) \div 2 = \frac{161}{36}$ = 4.4722... |

In fact, $\sqrt{20} = 4.4721$ to 4 d.p., so again the estimate is very close after just two guesses.

Try the Babylonian method to answer Question **4 c** of Exercise 4D.

## INVESTIGATION

Which of these statements are true?

**a** $\sqrt{16} + \sqrt{9} = \sqrt{(16 + 9)}$

**b** $\sqrt{16} \div \sqrt{9} = \sqrt{(16 \div 9)}$

**c** $\sqrt{16} - \sqrt{9} = \sqrt{(16 - 9)}$

**d** $\sqrt{16} \times \sqrt{9} = \sqrt{(16 \times 9)}$

Are any of these statements always true, for any pair of numbers?

## TECHNOLOGY

Go to the website:

http://nrich.maths.org/5955

Read the text (and listen to the audio) to find out about a method similar to long division used to work out the square root of a number. Try some numbers of your own.

## 4.3 Multiplying and dividing by powers of 10

In your work on indices in Chapter 1, you learned about positive indices. For example, you learned that $10^3 = 10 \times 10 \times 10$.

You learned about the zero index, for example $10^0 = 1$.

You also learned about negative indices, in particular that $a^{-n} = \frac{1}{a^n}$.

For example, $10^{-1} = \frac{1}{10^1} = \frac{1}{10} = 0.1$

Similarly, $10^{-2} = \frac{1}{10^2} = \frac{1}{100} = 0.01$

and $10^{-3} = \frac{1}{10^3} = \frac{1}{1000} = 0.001$

etc.

We can use this to help us multiply and divide by positive and negative powers of 10.

**EXAMPLE 7**

Work out:

**a** $3 \times 10^4$

**b** $420 \div 10^2$

**c** $530 \times 10^{-3}$

**d** $0.819 \div 10^{-5}$

---

**a** $3 \times 10^4 = 3 \times 10\,000 = 30\,000$

**b** $420 \div 10^2 = 420 \div 100 = 4.2$

**c** $530 \times 10^{-3} = 530 \times \frac{1}{10^3} = 530 \times \frac{1}{1000}$
$= 530 \div 1000 = 0.53$

**d** $0.819 \div 10^{-5} = 0.819 \div \frac{1}{10^5}$
$= 0.819 \div \frac{1}{100\,000} = 0.819 \times 100\,000$
$= 81\,900$

### Exercise 4E

**1** Copy the diagram below and match the numbers in the first column with their equivalents in the second column. One has been done for you.

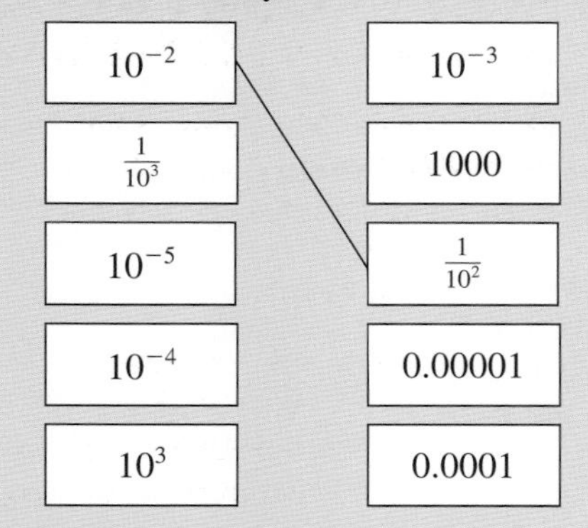


**2** Work out:

**a** $7 \times 10^3$ **b** $2.8 \times 10^4$

**c** $0.43 \times 10^5$ **d** $71 \times 10^2$

**3** Work out:

**a** $80\,300 \div 10^4$

**b** $42 \div 10^2$

**c** $510\,000 \div 10^5$

**d** $958\,000 \div 10^3$

**4** Copy and complete:

**a** $800 \times 10^{-3} = 800 \times \frac{1}{10^3} = 800 \times \frac{1}{\square}$
$= 800 \div \square = \square$

**b** $3500 \times 10^{-2} = 3500 \times \frac{1}{10^{\square}} = 3500 \times \frac{1}{\square}$
$= 3500 \div \square = \square$

**c** $7\,200\,000 \times 10^{-5} = 7\,200\,000 \times \frac{1}{\square}$
$= 7\,200\,000 \div \square = \square$

**d** $2 \times 10^{-1} = 2 \times \frac{1}{\square} = 2 \div \square = \square$

**5** Copy and complete:

**a** $0.074 \div 10^{-2} = 0.074 \div \frac{1}{10^2}$
$= 0.074 \div \frac{1}{\square} = 0.074 \times \square = \square$

**b** $0.008 \div 10^{-5} = 0.008 \div \frac{1}{10^{\square}}$
$= 0.008 \div \frac{1}{\square} = 0.008 \times \square = \square$

**c** $2.145 \div 10^{-3} = 2.145 \div \frac{1}{10^{\square}}$
$= 2.145 \div \frac{1}{\square} = 2.145 \times \square = \square$

**d** $0.3 \div 10^{-1} = 0.3 \div \frac{1}{10^{\square}} = 0.3 \div \frac{1}{\square}$
$= 0.3 \times \square = \square$

**6** Copy and complete these sentences:

**a** Multiplying by $10^{-2}$ is the same as multiplying by 0.01 or dividing by ......... .

**b** Dividing by $10^{-3}$ is the same as dividing by 0.001 or multiplying by ......... .

**c** Multiplying by $10^{-4}$ is the same as ......... by ......... .

**d** Dividing by $10^{-5}$ is the same as ......... by ......... .

**7** Work out:

**a** $90\,000 \times 10^{-5}$

**b** $2\,870\,000 \times 10^{-7}$

**c** $86\,300 \times 10^{-4}$

**d** $4.6 \times 10^{-2}$

**8** Work out:

**a** $0.00048 \div 10^{-5}$

**b** $0.021 \div 10^{-4}$

**c** $0.6 \div 10^{-2}$

**d** $0.0143 \div 10^{-3}$

**9** Work out:

**a** $3.4 \times 10^3$

**b** $71\,300 \div 10^4$

**c** $810\,000 \times 10^{-5}$

**d** $0.07 \div 10^{-3}$

**10** Here are some number cards:

| $10^2$ | $10^0$ | 100 | $10^{-3}$ | 0.0001 | 0.01 | 1000 |
|---|---|---|---|---|---|---|

Use four of these cards to make the statements below correct.
You cannot use a card more than once.

$7 \times \square = 7 \div \square$

$5.2 \div \square = 5.2 \times \square$

**11** Copy and complete:

**a** $7.1 \times 10^{\square} = 710\,000$

**b** $\square \div 10^3 = 249$

**c** $60\,000 \times 10^{\square} = 6$

**d** $0.02 \div 10^{\square} = 2$

**e** $\square \times 10^3 = 2840$

**f** $44\,000 \div 10^{\square} = 4.4$

**g** $\square \times 10^{-2} = 2.3$

**h** $\square \div 10^{-6} = 7000$

Don't forget the order of operations, BIDMAS.

**12** Work out:

**a** $7 \times 10^5 + 6 \times 10^3$

**b** $8.1 \times 10^4 - 25.4 \times 10^2$

**c** $12 \div 10^{-4} - 11 \times 10^3$

**d** $1700 \times 10^{-2} + 0.03 \div 10^{-3}$

## TECHNOLOGY

To extend this work **beyond stage 9**, find out about writing numbers in standard index form. Look at the website

www.bbc.co.uk/bitesize

and search for 'standard index form'.

# Consolidation

**Example 1**

Work out:

**a** $^-3.5 + 8$

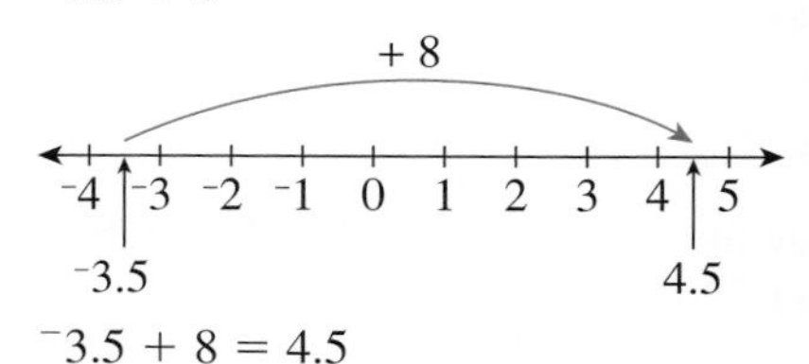


$^-3.5 + 8 = 4.5$

**b** $^-4.1 - 3.7$

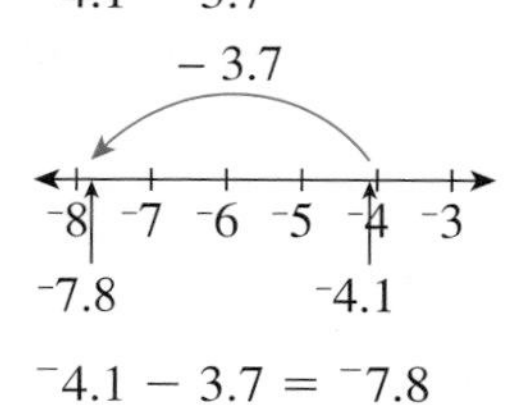


$^-4.1 - 3.7 = {}^-7.8$

**c** $2 + {}^-4.3$

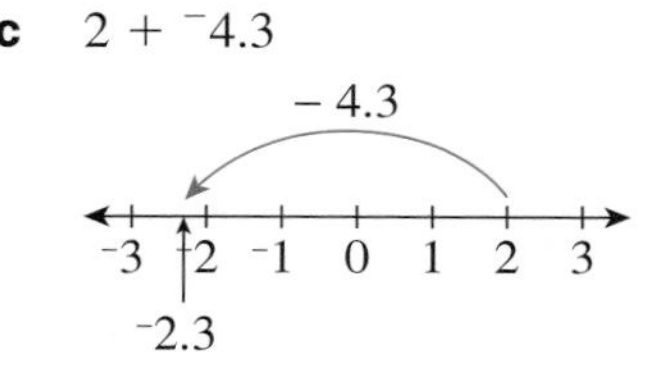


$2 + {}^-4.3 = 2 - 4.3 = {}^-2.3$

**d** $^-8 - {}^-10.6$

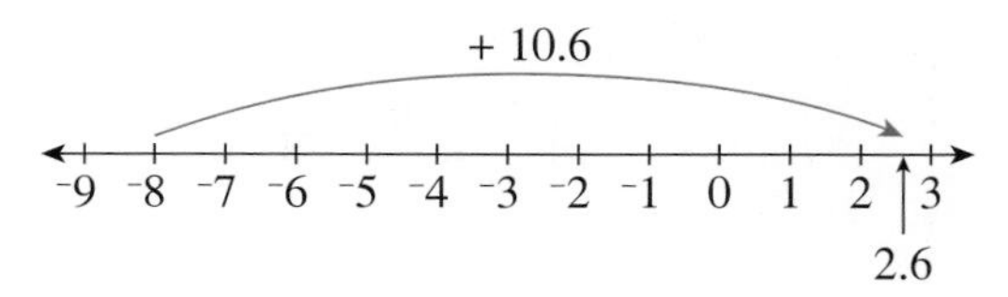


$^-8 - {}^-10.6 = {}^-8 + 10.6 = 2.6$

**Example 2**

Work out **a** $^-8.4 \div {}^-6$
**b** $^-3.7 \times 5$

**a** $^-8.4 \div {}^-6$

$$6\overline{)8.4} = 1.4$$

Do the calculation ignoring the signs.

A negative number multiplied or divided by a negative number gives a positive answer.
So $^-8.4 \div {}^-6 = 1.4$

**b** $^-3.7 \times 5$

$$\begin{array}{r} 3.7 \\ \times \quad 5 \\ \hline 18.5 \\ \hline {}_3 \end{array}$$

Do the calculation ignoring the signs.

A negative number multiplied or divided by a positive number gives a negative answer.
So $^-3.7 \times 5 = {}^-18.5$

**Example 3**

Estimate **a** $\pm\sqrt{78}$ **b** $\sqrt[3]{32}$ to 1 d.p.

**a** $\sqrt{64} = 8$ and $\sqrt{81} = 9$ so $\sqrt{78}$ lies between 8 and 9.
78 is closer to 81 than to 64 so the square root is going to be closer to 9 than to 8.
$\pm\sqrt{78} \approx 8.8$ or $^-8.8$ to 1d.p.

**b** $\sqrt[3]{27} = 3$ and $\sqrt[3]{64} = 4$ so $\sqrt[3]{32}$ lies between 3 and 4.
32 is closer to 27 than to 64 so the cube root is going to be closer to 3 than to 4.
$\sqrt[3]{32} \approx 3.2$ to 1 d.p.

**Example 4**

Work out:

**a** $2 \times 10^3$
**b** $63\,000 \div 10^4$
**c** $9800 \times 10^{-3}$
**d** $0.25 \div 10^{-2}$

**a** $2 \times 10^3 = 2 \times 1000 = 2000$
**b** $63\,000 \div 10^4 = 63\,000 \div 10\,000 = 6.3$
**c** $9800 \times 10^{-3} = 9800 \times \frac{1}{10^3} = 9800 \times \frac{1}{1000} = 9800 \div 1000 = 9.8$
**d** $0.25 \div 10^{-2} = 0.25 \div \frac{1}{10^2} = 0.25 \div \frac{1}{100} = 0.25 \times 100 = 25$

## Exercise 4

**1** Work out:

**a** $7 + {}^-8$ **b** $3 + {}^-9$
**c** $1 + {}^-3$ **d** $5 + {}^-11$
**e** $2 - 7$ **f** $6 - 13$
**g** $^-4 - 12$ **h** $^-9 - 13$
**i** $^-2 - ({}^-5)$ **j** $^-14 - ({}^-8)$

**2** Work out:

**a** $2 - 4.6$ **b** $^{-}7.1 + 5$
**c** $^{-}3 - 4.6$ **d** $^{-}4.7 + 5.3$
**e** $1.9 + {}^{-}3.7$ **f** $^{-}4 - {}^{-}0.87$
**g** $3.4 - {}^{-}2.6$ **h** $^{-}9.8 + {}^{-}2.4$

**3** Find the answer to:

**a** $7 \times {}^{-}9$ **b** $^{-}5 \times 3$
**c** $^{-}6 \times {}^{-}7$ **d** $^{-}9 \times {}^{-}5$
**e** $50 \div {}^{-}5$ **f** $^{-}72 \div {}^{-}9$
**g** $^{-}8 \div 4$ **h** $^{-}35 \div {}^{-}7$

**4** Work out:

**a** $2.3 \times {}^{-}5$
**b** $^{-}3.5 \times 3$
**c** $^{-}6 \times {}^{-}2.8$
**d** $^{-}8 \times {}^{-}7.2$
**e** $10 \times {}^{-}1.4$
**f** $^{-}100 \times {}^{-}0.23$
**g** $^{-}1.6 \times 0.8$
**h** $^{-}2.4 \times {}^{-}3.2$

**5** Find the answer to:

**a** $9 \div {}^{-}0.3$ **b** $1.6 \div {}^{-}4$
**c** $^{-}12 \div 0.4$ **d** $^{-}1.5 \div {}^{-}3$
**e** $\frac{1}{^{-}10}$ **f** $\frac{2.5}{^{-}5}$
**g** $\frac{^{-}35}{0.7}$ **h** $\frac{^{-}4.2}{^{-}6}$

**6** Estimate, to 1 d.p.:

**a** $\sqrt{46}$
**b** $\pm\sqrt{11}$
**c** $\sqrt{84}$
**d** $\sqrt{29}$
**e** $\pm\sqrt{156}$
**f** $^{-}\sqrt{18}$

**7** Estimate, to 1 d.p.:

**a** $\sqrt[3]{50}$
**b** $\sqrt[3]{10}$
**c** $\sqrt[3]{70}$
**d** $\sqrt[3]{140}$
**e** $\sqrt[3]{^{-}7}$
**f** $\sqrt[3]{^{-}101}$

**8** Estimate the side length of this square, to the nearest millimetre.

Area
$30\text{cm}^2$

**9** Estimate the side length of this cube, to the nearest millimetre.

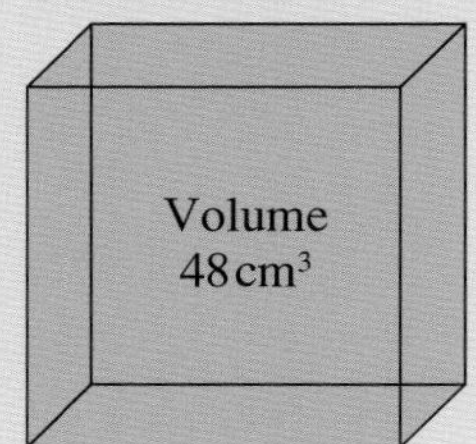


**10** Are these statements true or false?

**a** $10^4 = 40$
**b** $10^{-3} = 0.0001$
**c** $\frac{1}{10^5} = 100\,000$
**d** $0.02 = 10^{-2}$

**11** Work out:

**a** $4.2 \times 10^4$
**b** $0.3 \times 10^2$
**c** $5300 \div 10^3$
**d** $490\,000 \div 10^4$
**e** $0.0061 \times 10^5$
**f** $2\,100\,000 \div 10^6$
**g** $0.0001 \times 10^4$
**h** $7\,000\,000 \div 10^5$

**12** Work out:

**a** $20\,000 \times 10^{-4}$
**b** $6\,030\,000 \times 10^{-6}$
**c** $0.000021 \div 10^{-6}$
**d** $0.00435 \div 10^{-3}$
**e** $50 \times 10^{-3}$
**f** $2 \div 10^{-2}$
**g** $36\,000 \times 10^{-2}$
**h** $0.056 \div 10^{-4}$

# Summary

## You should know ...

**1** How to add and subtract negative numbers.

*For example:*

**a**

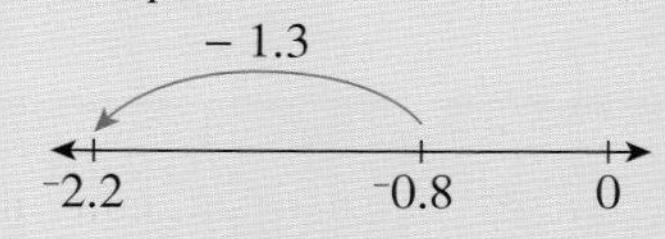


$^-0.8 + {}^-1.3 = {}^-0.8 - 1.3 = {}^-2.2$

**b**

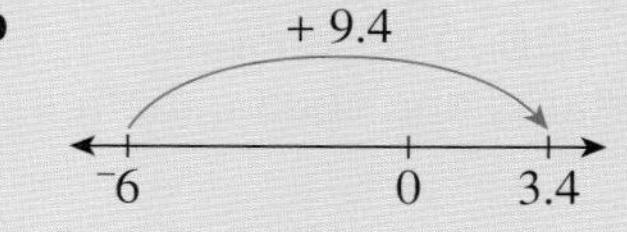


$^-6 - {}^-9.4 = {}^-6 + 9.4 = 3.4$

**2** How to multiply and divide negative numbers.

*For example:*

$^-16.8 \div {}^-4$

$4\overline{)16.8}$ = 4.2

Do the calculation ignoring the signs. A negative number divided by a negative number gives a positive answer.

So $^-16.8 \div {}^-4 = 4.2$

$4.3 \times {}^-5$

$$\begin{array}{r} 4.3 \\ \times \quad 5 \\ \hline 21.5 \\ \hline {}_1 \end{array}$$

Do the calculation ignoring the signs. A positive number multiplied by a negative number gives a negative answer.

So $4.3 \times {}^-5 = {}^-21.5$

**3** How to estimate square roots and cube roots.

*For example:*

Estimate $\pm\sqrt{39}$.

$\sqrt{36} = 6$ and $\sqrt{49} = 7$ so $\sqrt{39}$ lies between 6 and 7, 39 is closer to 36.

$\pm\sqrt{39} \approx 6.2$ or $^-6.2$ to 1d.p.

Estimate $\sqrt[3]{60}$.

$\sqrt[3]{27} = 3$ and $\sqrt[3]{64} = 4$ so $\sqrt[3]{60}$ lies between 3 and 4, 60 is closer to 64.

$\sqrt[3]{60} \approx 3.9$ to 1 d.p.

**4** How to multiply and divide whole numbers and decimals by 10 to the power of any positive or negative integer.

*For example:*

$2.1 \times 10^3 = 2.1 \times 1000 = 2100$

$350 \div 10^2 = 350 \div 100 = 3.5$

$7100 \times 10^{-4} = 7100 \times \frac{1}{10^4} = 7100 \div 10\,000 = 0.71$

$0.16 \div 10^{-5} = 0.16 \div \frac{1}{10^5} = 0.16 \times 100\,000 = 16\,000$

## Check out

**1** Work out:

**a** $^-3 + {}^-7.5$

**b** $^-2.4 + {}^-6 + 5.4$

**c** $^-5.3 - {}^-4.7$

**d** $1.4 - {}^-6 - 7.2$

**2** Work out:

**a** $^-23.8 \div 7$

**b** $5.1 \times {}^-3$

**c** $^-25.8 \div {}^-6$

**d** $^-2.7 \times {}^-4$

**3** Estimate, to 1 d.p.:

**a** $\sqrt{20}$

**b** $\sqrt[3]{15}$

**c** $\pm\sqrt{80}$

**d** $\sqrt[3]{45}$

**4** Work out:

**a** $2.3 \times 10^3$

**b** $2100 \div 10^2$

**c** $0.00047 \times 10^5$

**d** $19\,200\,000 \div 10^6$

**e** $340\,000 \times 10^{-4}$

**f** $0.00036 \div 10^{-5}$

**g** $290\,000 \times 10^{-3}$

**h** $0.041 \div 10^{-2}$

# 5 Measures

## Objectives

- Solve problems involving measurements in a variety of contexts.
- Convert between metric units of area, e.g. $mm^2$ and $cm^2$, $cm^2$ and $m^2$, and of volume, e.g. $mm^3$ and $cm^3$, $cm^3$ and $m^3$; know and use the relationship $1\,cm^3 = 1\,ml$.
- Know that land area is measured in hectares (ha), and that 1 hectare $= 10\,000\,m^2$; convert between hectares and square metres.

## What's the point?

Finding length, mass and capacity is of practical importance in our lives. Without such measures we could not ensure our football fields or cricket pitches were the correct size, or even know the amount of juice in a bottle.

## Before you start

### You should know ...

1 How to multiply and divide by 10, 100, 1000, etc.
*For example:*
$0.32 \times 100 = 32$
$40.7 \div 1000 = 0.0407$

2 The abbreviations for metric units of measurement.
*For example:*
$mm^2$ = millimetres squared (area)
$cm^3$ = centimetres cubed (volume)
m = metres (length)
kg = kilograms (mass)
t = tonnes (mass)
ml = millilitres (capacity)
ℓ = litres (capacity)
Other metric units include: $cm^2$ and $m^2$ (area), $mm^3$ and $cm^3$ (volume), mm, cm and km (length), and g (mass).

### Check in

1 Work out:
- **a** $5.6 \times 100$
- **b** $32.8 \div 10$
- **c** $714 \div 1000$
- **d** $0.00345 \times 1\,000\,000$

2
- **a** Which metric units of measurement from the list on the left are used for larger objects?
- **b** Which metric units of measurement from the list on the left are used for smaller objects?

## 5.1 Length, mass and capacity

### Units of length

You can measure length using millimetres (mm), centimetres (cm), metres (m) or kilometres (km). Remember, a centimetre is about the width of your little finger:

A metre is about one long stride.

Millimetres are used to measure very short lengths, while kilometres are used to measure very long distances. You should be able to walk 1 km in about 15 minutes.

Metric units of length are all related:

$$10\,\text{mm} = 1\,\text{cm}$$
$$100\,\text{cm} = 1\,\text{m}$$
$$1000\,\text{m} = 1\,\text{km}$$

You can convert one metric unit of length to another.

### Units of mass

You can measure mass using grams (g), kilograms (kg) or tonnes (t).

A large pineapple has a mass of about 1 kilogram:

Grams are used to measure small masses, while tonnes are used to measure very large masses. A large car has a mass of about 1 t.
Metric units of mass are all related:

$$1000\,\text{g} = 1\,\text{kg}$$
$$1000\,\text{kg} = 1\,\text{t}$$

You can convert one metric unit of mass to another.

### Units of capacity

Capacity is a measure of the quantity of liquid that will fit in a container. You can measure capacity in millilitres (ml), centilitres (cl) or litres (ℓ).

A typical measuring jug has a capacity of about 1 litre:

Millilitres are used for small capacities, while litres are used for large capacities. A teaspoon has a capacity of about 5 ml. Centilitres, which are between millilitres and litres, are sometimes used.

Metric units of capacity are all related, similar to the way millimetres, centimetres and metres are related:

$$10\,\text{ml} = 1\,\text{cl}$$
$$100\,\text{cl} = 1\,\ell$$
$$1000\,\text{ml} = 1\,\ell$$

You can convert one metric unit of capacity to another.

**EXAMPLE 1**

Convert:

**a** 1.35 m to cm
**b** 430 mm to m
**c** 2 350 000 g to t
**d** 3.4 ℓ to ml
**e** 12 cl to ml

**a** $1.35\,\text{m} = 1.35 \times 100\,\text{cm} = 135\,\text{cm}$
**b** $430\,\text{mm} = 430 \div 10\,\text{cm} = 43\,\text{cm}$
$43 \div 100\,\text{m} = 0.43\,\text{m}$
**c** $2\,350\,000\,\text{g} = 2\,350\,000 \div 1000\,\text{kg} = 2350\,\text{kg}$
$2350\,\text{kg} = 2350 \div 1000\,\text{t} = 2.35\,\text{t}$
**d** $3.4\,\ell = 3.4 \times 1000\,\text{ml} = 3400\,\text{ml}$
**e** $12\,\text{cl} = 12 \times 10\,\text{ml} = 120\,\text{ml}$

## Exercise 5A

**1** What is the best unit to measure the
**a** mass of a mouse
**b** height of a door
**c** distance from Wellington to New York
**d** length of your fingernail
**e** capacity of a water bottle
**f** distance around your school yard
**g** mass of an elephant
**h** capacity of a barrel?

**2** Write down two things that are
**a** about 1 m in length
**b** about 10 cm long
**c** 50 g in mass
**d** about 2 ℓ in capacity.

**3** Estimate, in centimetres,
**a** the length of your pen
**b** the width of your desk
**c** the width of your exercise book
Check your answers with your ruler.

**4** Convert to centimetres:
**a** 30 mm **b** 140 mm
**c** 2 m **d** 17 m
**e** 2.7 m **f** 4.35 m

**5** Convert to millilitres:
**a** 0.43 ℓ **b** 220 cl
**c** 5.4 ℓ **d** 0.3 cl

**6** Convert to kg:
**a** 400 g **b** 0.4 t
**c** 50 g **d** 3.41 t

**7** Convert to metres:
**a** 400 cm **b** 1890 cm
**c** 3 km **d** 29 km
**e** 6.4 km **f** 9.213 km

**8** Copy and complete:
**a** 0.03 kg = □ g
**b** 52.1 kg = □ g
**c** 5.3 ml = □ cl
**d** 0.42 ℓ = □ ml
**e** 2.14 t = □ kg
**f** 0.00031 t = □ g

**9** Write in metres:
**a** 1 m 50 cm **b** 3 m 25 cm
**c** 6 m 9 cm **d** 4 m 2 cm

**10** Write in millilitres:
**a** 3 ℓ 25 cl **b** 2 ℓ 14 cl
**c** 17 ℓ 8 cl **d** 20 ℓ 4 cl 3 ml

**11** Write in grams:
**a** 3 t 825 kg **b** 0.2 t 50 kg
**c** 0.004 t 7 g **d** 0.8 t 20 kg 35 g

**12** Convert:
**a** 215 m to km
**b** 5 km 20 m to km
**c** 5 m 16 cm 7 mm to mm
**d** 651 m 25 cm to km

**13** A piece of material is 0.7 m long. A 25-cm length and a 17-cm length are cut from it. How many metres of material are left?

**14** I walk to my friend's house, which is a journey of 2.8 km. Then I walk another 830 m to the shop. Finally, I walk 1.73 km to work. How many kilometres have I walked altogether?

**15** A small bottle of lemonade contains 350 ml. A large bottle of lemonade contains 2 ℓ. How many litres of lemonade are there in 4 small bottles and 2 large bottles?

**16** A bucket contains 5 ℓ of water. If 32 cl are poured out, what is the capacity of the water left in the bucket?

**17** A pile of waste material has a mass of 23 kg. If 14 230 g of this is removed, what mass of material is left?

## Prefixes

The metric measurements that we use today are largely based on the SI, or International System of Units ('SI' is an abbreviation of the French title, *Système international d'unités*).

Basic metric units are metres, grams and litres. To make basic units smaller or larger prefixes can be used, some of which – such as the 'centi-' in 'centimetre', 'milli-' in 'millilitre' or 'kilo-' in 'kilogram' – you have already met.

The tables below show more of these prefixes, most of which are not used for 'everyday' measurements. You may be familiar with the units 'megabytes', 'gigabytes' and 'terabytes', used in computing.

**Prefixes making the unit smaller than the basic unit**

| **Prefix** | deci- | centi- | milli- | micro- | nano- | pico- |
|---|---|---|---|---|---|---|
| **Symbol** | d | c | m | μ | n | p |
| **Times smaller** | ÷ 10 | ÷ 100 | ÷ 1000 | ÷ 1 000 000 | ÷ 1 000 000 000 | ÷ 1 000 000 000 000 |

**Prefixes making the unit larger than the basic unit**

| **Prefix** | deca- | hecto- | kilo- | mega- | giga- | tera- |
|---|---|---|---|---|---|---|
| **Symbol** | da | h | k | M | G | T |
| **Times larger** | × 10 | × 100 | × 1000 | × 1 000 000 | × 1 000 000 000 | × 1 000 000 000 000 |

The prefixes continue beyond those shown in these tables. Try to find some more.

When are the prefixes you have found used? Are they commonly used? If not, why not?

## 5.2 Area

### Units of area

Area is measured in millimetres squared ($mm^2$), centimetres squared ($cm^2$), metres squared ($m^2$) and kilometres squared ($km^2$).

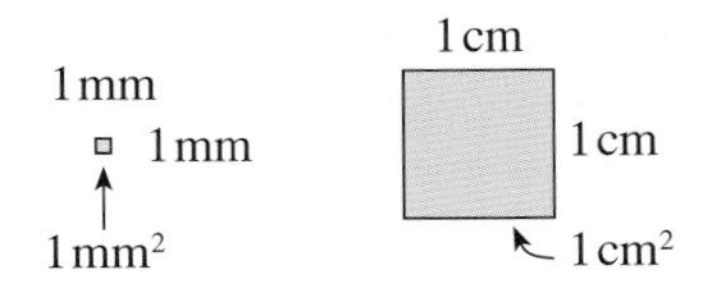


Millimetres squared and centimetres squared are used to find the area of small shapes.

The metric units of area are related:

1 cm = 10 mm (a 1 cm square equals a 10 mm square)

$1\ cm = 10\ mm$

$1\ cm^2 = 100\ mm^2$

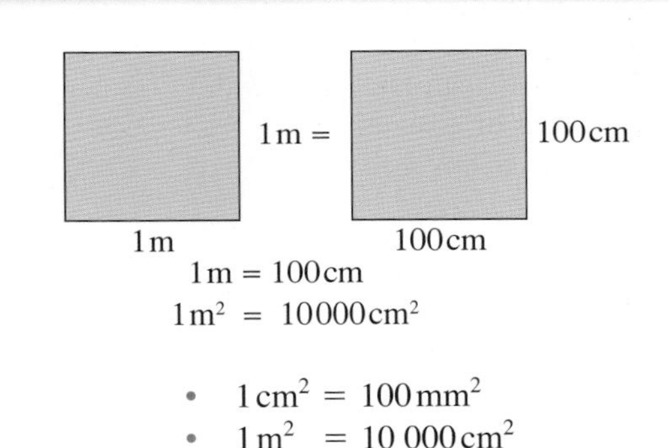


- $1\ cm^2 = 100\ mm^2$
- $1\ m^2 = 10\ 000\ cm^2$

### Exercise 5B

**1** Which unit of area would you use to measure:

**a** a page of your text book
**b** your thumbnail
**c** your classroom floor
**d** the country you live in
**e** the wing of a fly
**f** a football pitch?

2 Write down three objects that have an area of about
a $2\ m^2$ b $5\ cm^2$

3

Estimate, in centimetres squared, the area of:
a a page in your exercise book
b a postcard
c the postage stamp pictured
d your palm.

4 Write in millimetres squared:
a $2\ cm^2$ b $41\ cm^2$
c $3.6\ cm^2$ d $17.92\ cm^2$

5 Convert to centimetres squared:
a $300\ mm^2$ b $30\ mm^2$
c $193\ mm^2$ d $4824\ mm^2$

6 Convert:
a $5\ m^2$ to $cm^2$
b $60\ mm^2$ to $m^2$
c $8.2\ m^2$ to $mm^2$

7 Copy and complete:
a $0.004\ km^2 = \square\ m^2$
b $52\,000\,000\ m^2 = \square\ km^2$
c $0.021\ km^2 = \square\ m^2$
d $370\,000\ m^2 = \square\ km^2$

8 A table in a restaurant has a surface area of $25\,000\ cm^2$. If the restaurant has 30 of these tables, what is their total surface area, in $m^2$?

For further practice on converting metric measures go to:

www.sheppardsoftware.com/math.htm

Look at the Measuring section and try some of the games on metric measures.

For extension work, you can look at imperial measures too.

## Hectares

The unit used to measure the area of large plots of land is the hectare (ha). A **hectare** is a square with sides 100 m.

$1\ ha = 100\ m \times 100\ m = 10\,000\ m^2$

The area of an athletics track is about 1 ha.

It is also possible to convert between kilometres squared and hectares:
$100\ \text{hectares} = 100 \times 10\,000\ m^2 = 1\,000\,000\ m^2 = 1\ km^2$

## Exercise 5C

1 Convert to hectares:
a $20\,000\ m^2$
b $310\,000\ m^2$
c $600\ m^2$
d 5 million square metres

2 Convert to metres squared:
a 3 ha
b 4.71 ha
c 0.072 ha
d half a hectare

3 Given that 1 km = 1000 m, express
a $1\ km^2$ in $m^2$
b $1\ km^2$ in ha.

4 Convert to hectares:
a $3\ km^2$
b $0.1\ km^2$
c $2.4\ km^2$
d $32\ km^2$

5 Convert to kilometres squared:
a 400 ha b 23 ha
c 5000 ha d 54 100 ha

6 The area covered by a city measures 3.1 km by 4.7 km. Assuming it is roughly rectangular in shape, estimate the area of the city, in hectares.

7 The list shows the largest areas of water in five countries. Write them in order of size, smallest first.

Lake Taupo, New Zealand: 613 km$^2$
Wadi Shih Reservoir, United Arab Emirates: 5 000 000 m$^2$
Danau Toba, Indonesia: 11 3000 ha
Kainji Lake, Nigeria: 1243 km$^2$
Lake Nasser, Egypt: 525 000 ha

8 The diagram shows a piece of land shaped like an L.

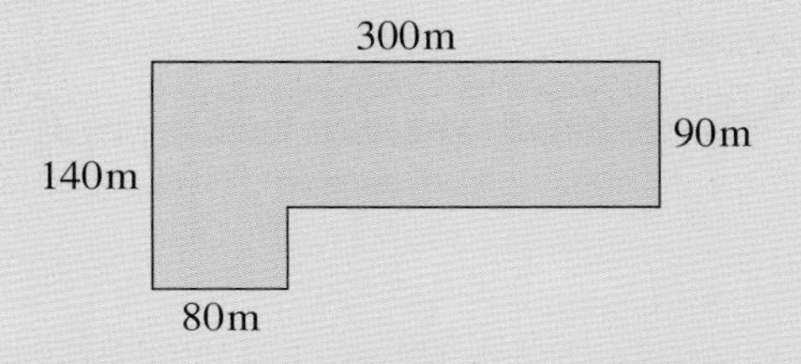


Find its area, giving your answer in hectares.

**INVESTIGATION**

What is the largest area of land, in hectares, that you can fence with 800 m of fence?

## 5.3 Volume

### Units of volume

You can find the area of a surface by covering it with squares and counting them:

In the same way you can find the volume of a space by filling it with cubes and counting them.

The two most commonly used units of volume are centimetres cubed (cm$^3$) and metres cubed (m$^3$):

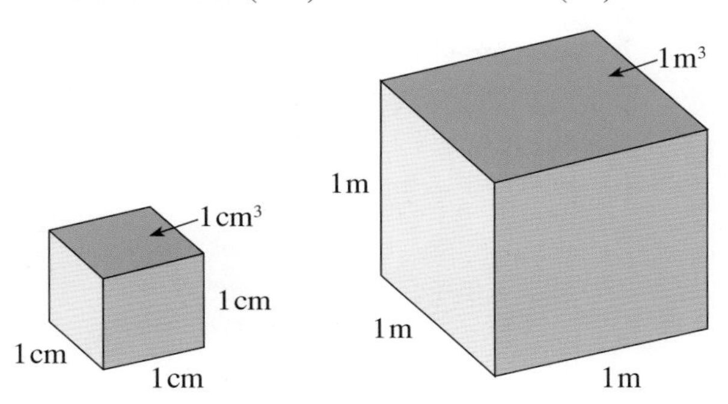


For very small units of volume you can use 1 mm$^3$:

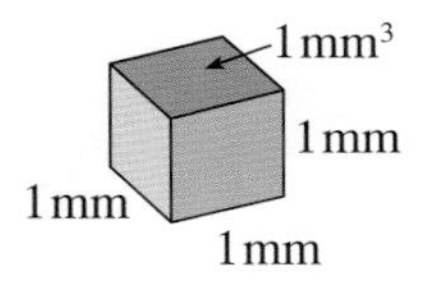


The metric units of volume are related:
1 cm = 10 mm,
so 1 cm$^3$ = 10 × 10 × 10 mm$^3$ = 1000 mm$^3$
1 m = 100 cm,
so 1 m$^3$ = 100 × 100 × 100 cm$^3$ = 1 000 000 cm$^3$

- 1 cm$^3$ = 1000 mm$^3$
- 1 m$^3$ = 1 000 000 cm$^3$

The amount of liquid or gas a container can hold is called the **capacity** of the container.

Capacity is measured in millilitres (ml) or litres (ℓ).

- 1000 cm$^3$ = 1 ℓ
- or 1 cm$^3$ = 1 ml

That is, a container with volume 1 cm$^3$ can hold 1 ml of liquid.

Similarly, a 50 cm$^3$ container can hold 50 ml.

## Exercise 5D

**1** Estimate the volume, in $cm^3$, of:

**a** an egg
**b** your little finger
**c** a reel of thread
**d** a pencil.

**2** Convert to $mm^3$:

**a** $3\,cm^3$ **b** $42\,cm^3$
**c** $0.6\,cm^3$ **d** $0.017\,cm^3$

**3** Convert to $cm^3$:

**a** $2000\,mm^3$ **b** $1700\,mm^3$
**c** $420\,mm^3$ **d** $38\,mm^3$

**4** Find the volume of these shapes, made of centimetre cubes

**i** in $cm^3$ **ii** in $mm^3$

**a**

**b**

**c**

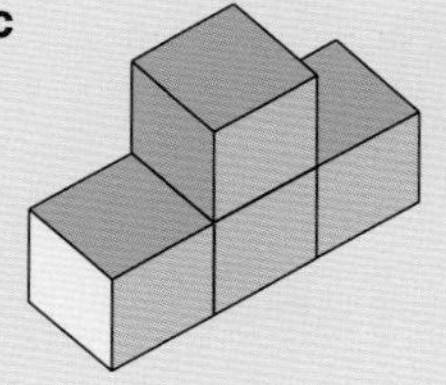

**d**

**5** What is the capacity, in millilitres, of these containers?

**a** a $300\,cm^3$ box
**b** a $450\,cm^3$ bottle
**c** a $175\,cm^3$ jar

**6** Convert to $m^3$:

**a** $71\,000\,000\,cm^3$ **b** $500\,000\,cm^3$
**c** $1\,400\,000\,cm^3$ **d** $72\,000\,cm^3$

**7** Convert to $cm^3$:

**a** $6\,m^3$ **b** $0.00287\,m^3$
**c** $32.4\,m^3$ **d** $0.00019\,m^3$

**8** Copy and complete:

**a** $32\,cm^3 = \square\,ml$
**b** $172\,ml = \square\,cm^3$
**c** $32\,000\,cm^3 = \square\,\ell$
**d** $0.4\,\ell = \square\,cm^3$
**e** $0.07\,m^3 = \square\,ml$
**f** $320\,000\,000\,ml = \square\,m^3$

**9** Jim stores food for his rabbit in a box with a volume of $3000\,cm^3$.
He stores food for his hamster in a box with a volume of $1\,050\,000\,mm^3$.
Which box is larger?

**10** A fish tank is 400 mm wide, 70 cm long and 0.3 m high. If it is filled right to the top, what is its capacity, in litres?

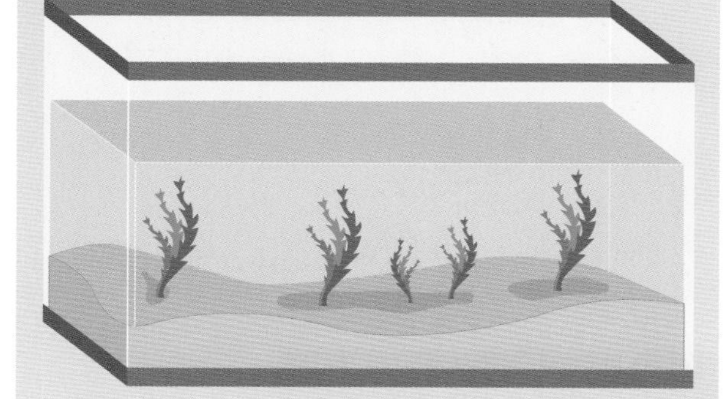

**11** A toilet has a flush capacity of 5 litres. To conserve water, a water-saving device with volume $500\,cm^3$ is placed in the water tank. What is the new, smaller flush capacity of the toilet, in millilitres?

### TECHNOLOGY

For further practice on problems involving metric measures go to

www.nrich.maths.org

Search for the investigation called 'Does this sound about right?' It's about a scientist who makes a set of estimates of various quantities – you need to consider whether or not the answers sound 'about right'.

## 5.4 Error in measurements

**Note that error is not mentioned specifically in the Cambridge Secondary 1 curriculum framework, and is included here as extension work.**

Look at these watches.

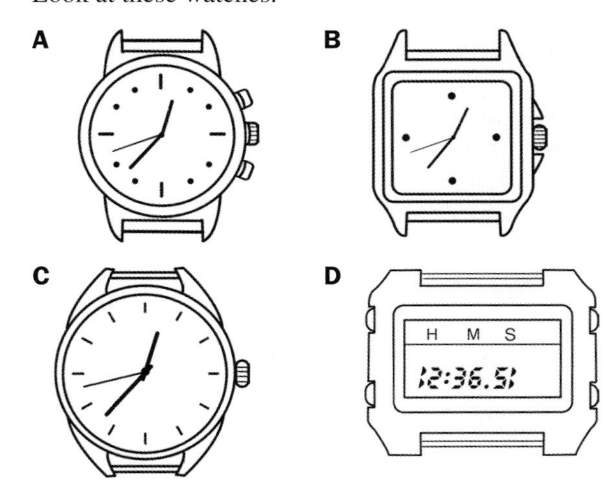


Which watch is easiest to read?

If you read the time as 12:36 when it is actually 12:37, there is an error of one minute in your reading.

### Error in measuring

In measuring, the word **error** has a special meaning. The error in a measurement is the difference between the result you record and the correct one.

This line is 3.7 cm long: ______________________

Its length is 4 cm, to the nearest cm.

When you read its length to the nearest cm, the error is 0.3 cm or 3 mm. ($4 - 3.7 = 0.3$)

### Exercise 5E (extension)

**1** Which of the watches at the start of the section would be accurate enough for:

- **a** measuring the fastest time for the 100 m final on school sports day
- **b** telling when maths class is nearly over
- **c** measuring the time a swimmer takes to swim a length of a pool
- **d** measuring how long your friend takes to run two miles
- **e** measuring how long you can hold your breath?

Give reasons for your answers.

**2** Which of these would you *not* measure with a ruler?

- **a** the thickness of a hair
- **b** the length of this book
- **c** the distance from the school to the post office
- **d** the width of your desk
- **e** the length of your foot

Explain your answers.

**3** Use a ruler to measure each line below. Write down the measurements correct to:

- **a** the nearest centimetre
- **b** the nearest millimetre.

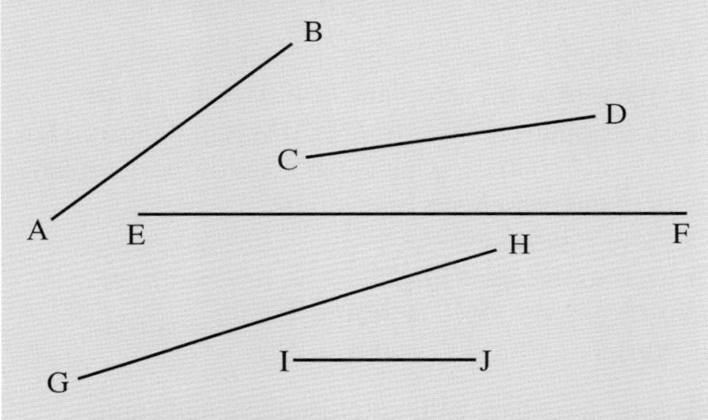


**4** Part of a centimetre ruler is shown magnified. Each small division represents 1 mm.

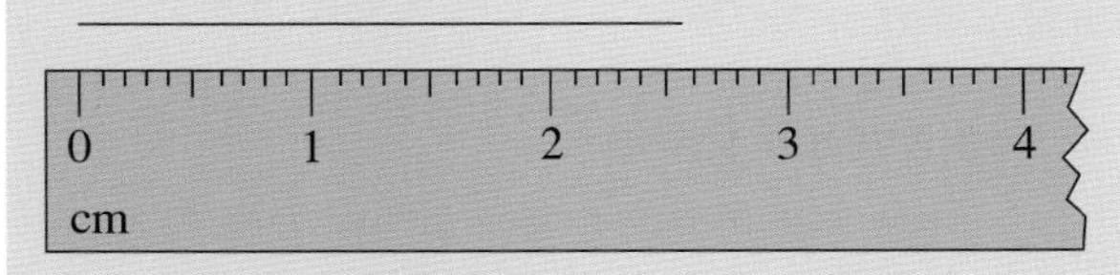


- **a** Look at the three lines above the ruler. Do they all look the same length?
- **b** How long is each line, to the nearest centimetre?

(**Note:** you need to use the ruler shown to measure the lines, not your own ruler.)

- **c** Were your answers for part **b** all the same?

Do you think your measurements include an error?

**5** If you are measuring to one decimal place, how long is the shortest line that could be recorded as:

**a** 5.6 cm **b** 4.1 km **c** 1.0 m?

# Consolidation

**Example 1**

**a** Convert 43 150 000 g to t
**b** Convert 28 000 $mm^2$ to $cm^2$
**c** Convert 0.053 $m^3$ to $cm^3$
**d** Convert 23 $cm^3$ to ml

**a** 43 150 000 g = 43 150 000 ÷ 1000 kg = 43 150 kg
43 150 kg = 43 150 ÷ 1000 t = 43.15 t
**b** 28 000 $mm^2$ = 28 000 ÷ 100 $cm^2$ = 280 $cm^2$
**c** 0.053 $m^3$ = 0.053 × 1 000 000 $cm^3$ = 53 000 $cm^3$
**d** 23 $cm^3$ = 23 ml

**Example 2**

The mass of a full aeroplane is 430 tonnes. If the passengers have a mass of 45 000 kg and the cargo has a mass of 17 000 kg, what is the mass of the aeroplane without the passengers and cargo?

Convert to the same units:
45 000 kg = 45 000 ÷ 1000 t = 45 t
17 000 kg = 17 000 ÷ 1000 t = 17 t

Mass of aeroplane without passengers and cargo
= 430 − 45 − 17 = 368 t

## Exercise 5

**1** Copy and complete:
**a** 0.2 kg = □ g
**b** 2800 mm = □ cm
**c** 3100 g = □ kg
**d** 240 cl = □ ml
**e** 320 000 cm = □ km
**f** 0.3 ℓ = □ ml
**g** 0.81 t = □ kg
**h** 0.00005 t = □ g

**2** A piece of wood is 1.3 m long. A 32 cm length and a 140 mm length are cut from it. How many metres of wood are left?

**3** Copy and complete:
**a** 87 $cm^3$ = □ ml
**b** 67 $cm^2$ = □ $mm^2$
**c** 950 000 $cm^3$ = □ $m^3$
**d** 0.6 $m^2$ = □ $cm^2$
**e** 6.2 ℓ = □ $cm^3$
**f** 1.8 $m^3$ = □ $cm^3$
**g** 2300 $mm^2$ = □ $cm^2$
**h** 39 ml = □ $cm^3$
**i** 0.49 ha = □ $m^2$

# Summary

### You should know ...

**1** How to convert between metric units of length, mass and capacity:
10 mm = 1 cm
100 cm = 1 m
1000 m = 1 km
1000 g = 1 kg
1000 kg = 1 t
10 ml = 1 cl
100 cl = 1 ℓ
1000 ml = 1 ℓ

**2** How to convert between metric units of area, volume and capacity:
100 $mm^2$ = 1 $cm^2$
10 000 $cm^2$ = 1 $m^2$
10 000 $m^2$ = 1 ha
1000 $mm^3$ = 1 $cm^3$
1 000 000 $cm^3$ = 1 $m^3$
1 $cm^3$ = 1 ml

### Check out

**1** Copy and complete:
**a** 28 000 g = □ kg
**b** 4 010 000 cm = □ km
**c** 0.032 kg = □ g
**d** 30 cl = □ ml
**e** 510 mm = □ cm
**f** 0.004 t = □ g
**g** 0.97 ℓ = □ ml
**h** 1.32 t = □ kg
**i** 72 300 ml = □ ℓ

**2** Copy and complete:
**a** 0.7 ha = □ $m^2$
**b** 3 ℓ = □ $cm^3$
**c** 450 $mm^2$ = □ $cm^2$
**d** 41 $cm^3$ = □ ml
**e** 850 $cm^2$ = □ $mm^2$
**f** 0.96 $m^3$ = □ $cm^3$
**g** 3 200 000 $cm^3$ = □ $m^3$
**h** 689 ml = □ $cm^3$
**i** 1.3 $m^2$ = □ $cm^2$

# 6 Planning, collecting and processing data

## Objectives

- Suggest a question to explore using statistical methods; identify the sets of data needed, how to collect them, sample sizes and degree of accuracy.
- Identify primary or secondary sources of suitable data.
- Design, trial and refine data collection sheets.
- Collect and tabulate discrete and continuous data, choosing suitable equal class intervals where appropriate.
- Calculate statistics and select those most appropriate to the problem.

## What's the point?

What is an average? How can you tell? Statistics help you to interpret data and, for example, help a company to determine what quantities of a product they should stock.

## Before you start

### You should know ...

**1** Whether data is discrete or continuous.

Discrete data can only take on certain values.
*For example:*
The number of children in a family is 0, 1, 2, 3, … .
You can't have 1.42 children.

Continuous data can take on any value.
*For example:*
The height of a person can be 1.4 m, or 1.47 m, or 1.471 m, and so on, depending on how accurately you can measure it.

### Check in

**1** Are the following types of data discrete or continuous?

**a** the speed of an aeroplane
**b** the number of people in a class
**c** a score out of 100 in a test
**d** the mass of a bicycle
**e** the time it takes to walk 100 m
**f** the length of a school hall

## 6.1 Planning and collecting data

In the modern world huge amounts of data are collected every week by many different people:

| Person needing data | Purpose |
| --- | --- |
| Store manager | Stock control |
| Pollster | Find out public opinion on an issue |
| Teacher | Determine student progress |

To collect data you need a suitable method of data collection. Two simple methods are:

- questionnaires
- data collection sheets.

To find answers to many everyday questions a **survey** is often used. For example, you may wish to find out about television viewing in your school. A survey question might be:

- For how long do you watch television each night?

There are four steps in carrying out a survey:

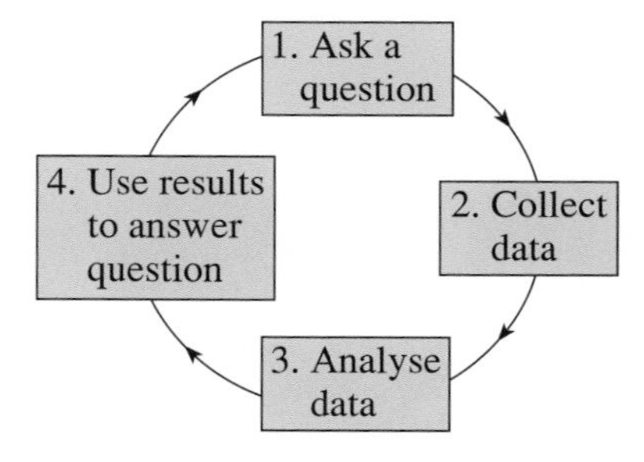


Often your results will prompt you to make a follow-up survey. In the case of television viewing, a follow-up question might be:

- 'Do you watch less television on a school night or a weekend night?'

You need to be able to suggest a question that can be explored using statistical methods. You can do this by imagining you have a particular job and thinking about what it would be useful for you to find out. For example, if you worked in a clothes shop your question might be

- 'Which clothes sell the best?'

Once you have decided on your question, you need to think about how you might collect information to answer it. You could collect the data yourself. Data collected this way is **primary data**. The advantage of collecting it yourself is that you can be fairly confident it is reliable and accurate. The disadvantage of collecting it yourself is that it could take a long time.

Another way is to use data that someone else has collected. This is called **secondary data**. Secondary data could be collected by someone you know, such as your teacher, or from the internet. The advantage of using secondary data is that someone else has done a lot of the work for you. The disadvantage is that you don't know how reliable or honest it is. You should always be careful with data collected from the internet, as there is a chance that it is biased or untrue.

With both primary and secondary data, you need to identify the sets of data you need. For example, to answer the question about which clothes sell best, you may want to collect all the sales figures for men's clothes and all the figures for women's clothes. Or you might collect the sales figures for t-shirts and the sales figures for trousers and skirts. You may also want to find out the answer to a follow-up question such as 'Which days of the week have the best sales figures?'

The shop owner should decide what they want to do with the data once it has been collected. For example, they may want to use the survey to decide what stock to buy more of, or which days of the week they will run sales promotions.

A difficult thing to decide is how much data to collect. Looking at all the sales since the shop first opened would involve handling a lot of data, so a **sample** of data is collected instead. You must decide what sample size to use.

The advantage of a smaller sample size is that the data can be collected quickly; a disadvantage is that the data can be less reliable. For instance, you might look at sales figures from just one week, or instead for a month or even a year to give a more accurate picture. The disadvantage of a larger sample is that it can take a long time and be expensive to collect.

You also need to think about what degree of accuracy you should use. For example, the shop owner could decide to round all figures to the nearest dollar.

## Exercise 6A

**1** For what purpose do you think the following people would need to collect data? Suggest some questions they could use.
- **a** Hotel managers
- **b** Tourist promotion officials
- **c** School principals
- **d** Hospital administrators
- **e** Politicians

**2** *The students in my class watch less than 2 hours of television each night.*
- **a** Decide what data to collect to test this statement.
- **b** Explain what sample size you are planning to use and why.
- **c** Explain what degree of accuracy you are planning to use and why.
- **d** Design a suitable data collection sheet.
- **e** Use your data collection sheet to collect the data.
- **f** Look at your data and decide whether the statement is true.

**3** *Do boys perform better in maths than girls?*
- **a** Decide what data to collect to test this question.
- **b** Explain what sample size you are planning to use and why.
- **c** Explain what degree of accuracy you are planning to use and why.
- **d** Design a suitable questionnaire.
- **e** Use your questionnaire to collect the data.
- **f** Study your data to decide the answer to the question.

**4** Frankie is collecting data to find out whether teenagers eat a healthy diet. He collects data in different ways. Are these sets of data primary or secondary?
- **a** He asks ten people in the dinner queue at school what they are buying for their lunch.
- **b** He asks the canteen staff for a list of what teenagers bought for their school dinners last month.
- **c** He looks on the internet for sales figures for different types of food supplied to school canteens.
- **d** He uses a data collection sheet to collect data from friends about whether they eat meat, fruit and vegetables every lunchtime.

**5** Suggest possible follow-up questions for the data collected for Questions **2** and **3**.

**6** Jalad thinks that the older you are, the more money you spend. Discuss how he could find out whether this is true.

Write about: possible questions, the sets of data needed, whether to use primary or secondary data, how to collect the data, sample sizes and degree of accuracy. As part of the discussion, give the advantages and disadvantages of each option you consider.

You can see that there are many different decisions that can be made when collecting data. There is no one method that works in all circumstances. A mistake commonly made when designing questionnaires or data collection sheets is to ask for data that is not required. Data is also often collected in a form that is not helpful for answering the particular question asked.

It is a good idea to trial your data collection sheet or questionnaire by conducting a **pilot survey**. This means you ask just a few people to complete the survey, to see if it is working the way you need it to work. Then you can refine (change and improve) your questionnaires and data collection sheets.

## Exercise 6B

**1** Safiya wants to find out whether the amount of exercise you do improves your general health. She wrote this data collection sheet, which was completed by five people in a pilot survey:

| Age | Health (1, 2, 3, 4 or 5) | How much exercise? |
|---|---|---|
| 14 | 2.5 | 2 |
| 25 | 4 | 100 |
| – | 3 | none |
| 51 | ? | ? |
| 34 | 0 | half |

- **a** Why do you think someone put a dash instead of writing their age?
- **b** Why is there a question mark in two of the columns?
- **c** Rewrite these three columns of the data collection sheet, to refine and improve them.
- **d** What other columns could you add to this data collection sheet to improve the survey? Or do you think that no further questions are needed? Would you remove any?

2 James wrote this data collection sheet to try to find out whether parents think that children today are spoilt, compared to when they were young themselves.

| 5 = agree strongly 1 = disagree strongly | | | | | |
|---|---|---|---|---|---|
| Dr Raphael says that children today have too much pocket money. What do you think? | 5 | 4 | 3 | 2 | 1 |
| Do you think children have more freedom? | 5 | 4 | 3 | 2 | 1 |
| Are children properly disciplined (told off)? | 5 | 4 | 3 | 2 | 1 |

**a** Write down all the things that you think are wrong with this data collection sheet.
**b** Rewrite this this data collection sheet to refine and improve it.

3 Look at your questionnaire and data collection sheets from Exercise 6A, Questions **2** and **3**. Can they be improved? If so, how and why?

4 Mr Thorne is deputy head at East Side Primary School. The school are thinking of changing the school timetable because the children don't seem to concentrate as well in the afternoon as they do in the morning.
Mr Thorne has designed a data collection sheet and used it in a pilot survey. Four students have completed the sheet:

| Age | Do you like lunchtime when it is? | Would you like lunch at a different time? | Lunch is currently at 12. We are thinking of moving it to 1 – what do you think? |
|---|---|---|---|
| 9 | yes | No | I would be too hungry at 1 |
| 7 | yes | I don't mind | I don't mind |
| 11 | yes | No | Will it still be an hour long? |
| 10 | yes | Depends when it would be | It would be better at 12.30 as that is half way. |

What improvements would you suggest for Mr Thorne's data collection sheet? Why?

# 6.2 Organising data

## Data types

There are two basic types of data:

- **discrete** and
- **continuous**.

Discrete data can only take definite values.
For example: clothes sizes – small, medium, large etc.

Continuous data can take any value.
For example: height, mass, time.

## Frequency tables

When you have collected your data, you need to organise it. A good way of organising discrete data is in a frequency table using a tally.

**EXAMPLE 1**

Make a frequency table for the test scores of a class of students:

3, 7, 6, 4, 2, 8, 8, 1, 10, 9
2, 5, 5, 6, 4, 7, 8, 6, 5, 8

| Score | Tally | Frequency |
|---|---|---|
| 1 | I | 1 |
| 2 | II | 2 |
| 3 | I | 1 |
| 4 | II | 2 |
| 5 | III | 3 |
| 6 | III | 3 |
| 7 | II | 2 |
| 8 | IIII | 4 |
| 9 | I | 1 |
| 10 | I | 1 |

The table shows that only four students scored less than 4 marks and six scored 8 or more.

## Exercise 6C

1. Here are the shoe sizes of 20 adults:

   6, 7, 6, 5, 6, 9, 8, 7, 10, 6,
   5, 8, 9, 10, 7, 6, 5, 9, 8, 7

   Show the data in a frequency table.

2. The favourite colours of 16 pupils were noted during a survey:

   red, blue, green, red, yellow, red, green, blue,
   blue, red, yellow, red, blue, blue, red, green

   Make a frequency table to display the data.

3. A 6-sided dice was rolled 30 times. Here are the scores:

   1, 3, 2, 6, 5, 4, 5, 2, 5, 3, 6, 1, 3, 1, 4,
   1, 4, 2, 1, 3, 6, 5, 2, 4, 3, 2, 1, 5, 1, 3

   Display the scores in a frequency table.

A **grouped** frequency table is often used for both continuous and discrete data.

### EXAMPLE 2

The heights of 25 boys, in centimetres, are:

103, 145, 138, 162, 149, 150, 175, 168, 138,
142, 161, 136, 125, 111, 143, 147, 159, 172,
165, 166, 133, 147, 152, 168, 171

Construct a grouped frequency table to show the data. Use groups of 100–109, 110–119, … 170–179.

| Group | Tally | Frequency |
|---|---|---|
| 100-109 | \| | 1 |
| 110-119 | \| | 1 |
| 120-129 | \| | 1 |
| 130-139 | \|\|\|\| | 4 |
| 140-149 | ~~\|\|\|\|~~ \| | 6 |
| 150-159 | \|\|\| | 3 |
| 160-169 | ~~\|\|\|\|~~ \| | 6 |
| 170-179 | \|\|\| | 3 |

You learned in Books 1 and 2 that we group data to help us look for trends and patterns.

Deciding what group size, or **class interval**, to use is not always easy. If the class interval is too large, the data loses a lot of accuracy. For instance, the data from Example 2 could have been grouped like this:

| Group | Frequency |
|---|---|
| 100-149 | 13 |
| 150-199 | 12 |

But grouping it like this shows very little.

If the class interval is too small, there are too many groups and often they will have no data or only one piece of data in them.

To begin with, aim for more than four and fewer than ten groups. This is just a rough guide – you will need to see if this makes sense for the data you are working with. Keep the class intervals equal for now (when you study Cambridge IGCSE® Maths you will find out when it may be appropriate to use different class intervals).

## Exercise 6D

1. The heights of 25 girls, in centimetres, are:

   101, 111, 159, 172, 132, 125, 113, 126, 138,
   142, 158, 107, 109, 117, 125, 104, 129, 121,
   143, 133, 168, 141, 121, 118, 141

   Display the data in a grouped frequency table, using the same groups as in Example 2.

2. Rainfall (mm) in London, UK is shown for the 30 days of November:

   21, 20, 0, 12, 1, 11, 0, 3, 1, 4,
   9, 1, 9, 0, 13, 3, 7, 18, 8, 4,
   7, 18, 0, 4, 1, 2, 0, 12, 2, 16

   Construct a grouped frequency table to show the data. Choose suitable groups.

3. The times taken by 25 students to complete a mathematical puzzle are shown in minutes:

   55, 22, 8, 13, 7, 9, 6, 8, 12, 10, 4, 9, 33,
   69, 18, 9, 14, 2, 5, 45, 21, 41, 5, 17, 31

   Make a grouped frequency table for the data using suitable groups.

## 6.3 Calculating statistics

Sometimes, instead of looking at a large set of numbers, it is more convenient to use a single number that is a good representation of all the data.
This number is an **average** or a **measure of central tendency**.

There are three commonly used averages:

- mean
- mode
- median.

The choice of which measure to use will depend on the circumstances.

### The mean

The most frequently used average is the mean. It is found by adding up all the data and dividing by the number of values.

- **Mean** $= \dfrac{\textbf{sum of data}}{\textbf{number of values}}$

**EXAMPLE 3**

A cricket batsman scored 35, 2, 71, 16 and 8 runs in five innings.

What is his mean score?

$$\text{Mean} = \frac{35 + 2 + 71 + 16 + 8}{5}$$
$$= \frac{132}{5}$$
$$= 26.4 \text{ runs}$$

You calculate the mean of a frequency distribution in the same way.

**EXAMPLE 4**

The number of goals scored by a football team in each game over a 40-game period is:

| No. of goals | 0 | 1 | 2 | 3 | 4 | 5 | 6 |
|---|---|---|---|---|---|---|---|
| Frequency (No. of games) | 3 | 7 | 6 | 5 | 12 | 7 | 0 |

What was the mean number of goals scored per game?

To find the total number of goals scored, you need to multiply the number of goals by the frequency. This is best done in a table:

| No. of goals ($x$) | Frequency ($f$) | No. of goals × Frequency ($fx$) |
|---|---|---|
| 0 | 3 | 0 × 3 = 0 |
| 1 | 7 | 1 × 7 = 7 |
| 2 | 6 | 2 × 6 = 12 |
| 3 | 5 | 3 × 5 = 15 |
| 4 | 12 | 4 × 12 = 48 |
| 5 | 7 | 5 × 7 = 35 |
| 6 | 0 | 6 × 0 = 0 |
| | 40 | 117 |

$$\text{Mean} = \frac{\text{total number of goals}}{\text{total number of matches}}$$
$$= \frac{117}{40}$$
$$= 2.93 \text{ (to 3 s.f.)}$$

The mean of a frequency distribution is sometimes written as

$$\textbf{Mean} = \frac{\Sigma fx}{\Sigma f}$$

where $x$ = value of each observation

$f$ = frequency

and Σ is the Greek letter 'sigma' meaning 'the sum of'.

So in Example 4:

$\Sigma f$ = the sum of the frequencies
= 40

$\Sigma fx$ = the sum of number of goals × frequency
= 117

### Exercise 6E

**1** Here are the total scores of two dice thrown together twenty times: 9, 2, 8, 6, 10, 7, 7, 4, 5, 8, 9, 12, 3, 10, 8, 11, 7, 4, 6, 9.
Calculate the mean score.

**2** A biologist takes a sample of 10 grasses and measures the stem length. His results, in centimetres, are: 30, 28, 32, 29, 25, 27, 31, 39, 33, 26.
Calculate the mean stem length.

**3** Two dice are thrown together 100 times. The following table is used to record the results and to calculate the mean:

| Score $x$ | Frequency $f$ | $fx$ |
|---|---|---|
| 2 | 1 | 2 |
| 3 | 4 | 12 |
| 4 | 7 | |
| 5 | 8 | |
| 6 | 12 | |
| 7 | 15 | |
| 8 | 16 | |
| 9 | 16 | |
| 10 | 12 | |
| 11 | 7 | |
| 12 | 2 | |

**a** Copy and complete the table.
**b** What is the mean score?

**4** A sample of 50 electric light bulbs was tested for length of life, and the results were:

| Hours | 80 | 81 | 82 | 83 | 84 | 85 | 86 |
|---|---|---|---|---|---|---|---|
| No. of bulbs | 1 | 5 | 11 | 18 | 8 | 4 | 3 |

Calculate the mean length of bulb life.

**5** In a game, a machine shows the numbers 0, 1, 2 or 3. An analysis of 100 games produces these results:

| Number | 0 | 1 | 2 | 3 |
|---|---|---|---|---|
| Frequency | 25 | 55 | 15 | 5 |

Calculate the mean of the numbers displayed.

**6** The table shows the lengths of 100 rods:

| Length (mm) | 196 | 197 | 198 | 199 | 200 |
|---|---|---|---|---|---|
| Frequency | 9 | 18 | 31 | 22 | 20 |

**a** Calculate the mean length.
**b** Calculate the mean length of the 80 rods that measure less than 200 mm.

**7** These are the scores for 20 throws of a dice:

| Score $x$ | 1 | 2 | 3 | 4 | 5 | 6 |
|---|---|---|---|---|---|---|
| Frequency $f$ | 3 | 5 | 6 | 3 | 1 | 2 |

**a** What is the value of $\Sigma f$?
**b** Calculate the value of $\frac{\Sigma fx}{\Sigma f}$.
What name is given to this measure?

**8** The mean number of children per family is 3. If there are 7 families, how many children are there in total?

**9** The mean of five numbers is 6. If four of the numbers are 8, 5, 7 and 9, what is the missing number?

**10** The mean of ten numbers is 5. When another number is added the mean is still 5. What is the extra number?

**11** In Class 9V there are 10 boys and 15 girls. The mean height of the ten boys is 158 cm. The mean height of the fifteen girls is 155 cm. What is the mean height of all the students in Class 9V?

## Means of grouped distributions

**Note that means of grouped distributions are not mentioned specifically in the Cambridge Secondary 1 curriculum framework, and are included here as extension work.**

### Using the mid-interval value

In the case of grouped frequency tables the **mid-interval** value is used to help estimate the mean.

For example, here is a frequency table recording the heights of 25 children:

| Height (cm) | Frequency |
|---|---|
| 140–144 | 1 |
| 145–149 | 3 |
| 150–154 | 11 |
| 155–159 | 7 |
| 160–164 | 2 |
| 165–169 | 0 |
| 170–174 | 1 |

The table shows that 11 children had heights in the class interval 150–154 cm. This interval includes all heights between 149.5 cm and 154.5 cm.

149.5 and 154.5 are the interval boundaries.

The mid-interval value is $\frac{150 + 154}{2} = 152$ cm.

The mean height of the children can be calculated using the mid-interval value. This will give an approximation to the mean, as it assumes that all 11 children have a height of 152 cm.

Use a table:

| Mid-interval value (cm) $x$ | Frequency $f$ | $fx$ |
|---|---|---|
| 142 | 1 | 142 |
| 147 | 3 | 441 |
| 152 | 11 | 1672 |
| 157 | 7 | 1099 |
| 162 | 2 | 324 |
| 167 | 0 | 0 |
| 172 | 1 | 172 |
| | 25 | 3850 |

$\Sigma f = 25$

$\Sigma fx = 3850$

$$\text{Mean} = \frac{\Sigma fx}{\Sigma f} = \frac{3850}{25} = 154$$

So the approximate mean height is 154 cm.

## Exercise 6F (extension)

**1** A census gives the following data for the ages of the population of a village:

| Age (years) | 0–9 | 10–19 | 20–29 | 30–39 |
|---|---|---|---|---|
| Number | 92 | 88 | 85 | 68 |

| Age (years) | 40–49 | 50–59 | 60–69 | 70–79 |
|---|---|---|---|---|
| Number | 55 | 52 | 42 | 18 |

**a** What is the mid-interval value of the class interval 20–29 years?

**b** Use mid-interval values to calculate an estimate for the mean age of the population.

**2** A group of students record the distances of their homes from school:

| Distance (km) | Number of students |
|---|---|
| Less than 1 | 10 |
| 1–2 | 15 |
| 2–3 | 7 |
| 3–4 | 2 |
| 4–5 | 1 |

**a** What is the mid-interval value of the class interval 2–3 km?

**b** Use mid-interval values to calculate an estimate for the mean distance of the students' homes from school.

Make up your own grouped frequency table that will give an estimated mean of 6.5.

## The mode

- The **mode** is the **most common** item in a distribution. It is the easiest average to find.

**EXAMPLE 5**

The shoe sizes of ten girls are:

6, 4, 5, 4, 2, 1, 7, 6, 3, 6

What is the mode?

The most frequent shoe size is 6, so the mode is 6.

For a frequency distribution the mode or modal class is the class interval with the highest frequency.

**EXAMPLE 6**

The masses of 100 school children are:

| Mass (kg) | Frequency |
|---|---|
| 30–34 | 4 |
| 35–39 | 7 |
| 40–44 | 23 |
| 45–49 | 30 |
| 50–54 | 16 |
| 55–59 | 11 |
| 60–64 | 5 |
| 65–69 | 4 |

What is the modal mass?

The highest frequency is 30.
The group with the highest frequency is 45–49 kg.
The modal mass is 45–49 kg.

Sometimes there is more than one mode or modal class. The data 2, 2, 3, 4, 5, 5, 8, 11 is bimodal as it has two modes, 2 and 5. Sometimes there is no mode. For example the data 6, 8, 9, 12, 20 has no mode.

## Exercise 6G

**1** The number of books in 30 students' bags are:

1, 2, 2, 1, 3, 1, 2, 4, 0, 1, 2, 2, 1, 0, 0,
1, 2, 3, 0, 2, 4, 2, 4, 2, 6, 5, 2, 5, 8, 2

What is the modal number of books?

**2** Find the modal class for Questions **1** and **2** of Exercise 6F.

**3** Find the modal class for this distribution:

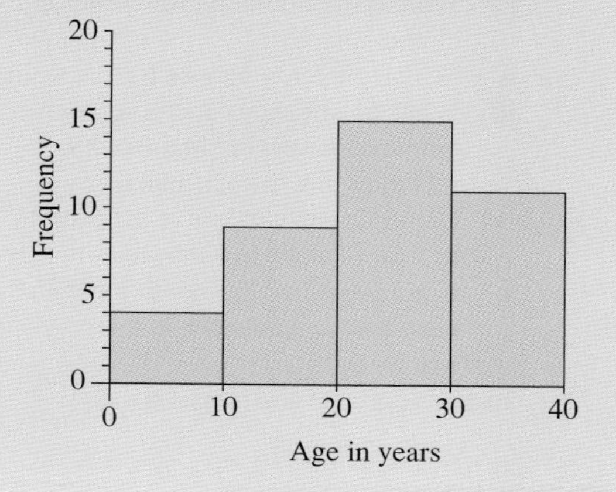


## The median

- When the data is arranged in ascending or descending order, the **median** is the **middle value**.

For example, the heights of eleven boys, in centimetres, are:

150, 146, 158, 165, 168, 170, 158, 154, 162, 180, 181

Written in ascending order they are:

146, 150, 154, 158, 158, 162, 165, 168, 170, 180, 181

The **median** height is the height of the middle boy – the sixth, that is, 162 cm:

146, 150, 154, 158, 158, (162), 165, 168, 170, 180, 181

median

Sometimes there are two middle values, so we take the median to be halfway between them.

For example, if there were only eight boys, with these heights:

146, 150, 154, $\underbrace{158, 160}$, 162, 165, 168

$$\text{median} = \frac{158 + 160}{2} = 159$$

So the median height is 159 cm.

## The range

Look at the scores of two groups of six students in a test:

| Group 1 | 1 | 0 | 0 | 10 | 9 | 10 |
|---|---|---|---|---|---|---|
| **Group 2** | 4 | 6 | 7 | 5 | 3 | 5 |

A student claims that both groups did equally well since:

$$\text{Mean for Group 1} = \frac{1 + 0 + 0 + 10 + 9 + 10}{6} = 5$$

$$\text{Mean for Group 2} = \frac{4 + 6 + 7 + 5 + 3 + 5}{6} = 5$$

However, the frequency distributions show that the performances of the two groups are very different. The mean does not completely describe the data.

The scores in Group 1 are much more dispersed, or spread out, than the scores in Group 2.

A single number can be found that gives a measure of this spread. One number you can use in the range.

Generally the larger the value of the range the more spread out the data.

**Range = highest value − lowest value**

In the example:

Range for Group 1 = 10 − 0 = 10

Range for Group 2 = 7 − 3 = 4

This indicates that the scores in Group 1 are widely spread and those in Group 2 are not so widely spread.

## Exercise 6H

**1** Find **i** the median
**ii** the range of the numbers:
- **a** 2, 3, 5, 7, 8
- **b** 6, 1, 4, 3, 9
- **c** 4, 4, 1, 4, 6, 2

**2** Find **i** the median
**ii** the range of each set of numbers:
- **a** 2, 5, 7, 9, 10, 11, 13
- **b** 4, 3, 6, 2, 1, 8, 4
- **c** 7, 2, 1, 7, 6, 9, 15, 13, 4, 9, 1
- **d** 5, 8, 12, 15, 10, 12, 17, 13
- **e** 3, 4, 9, 9, 6, 10, 12, 10, 8, 6, 10, 9

**3** The masses of five people are 70 kg, 64 kg, 58 kg, 80 kg, 78 kg.
- **a** What is the median mass?
- **b** What is the range?

Sometimes you have to be careful which average you use.

**EXAMPLE 7**

Here are the masses of nine athletes:

85 kg, 91 kg, 84 kg, 94 kg, 84 kg,
88 kg, 93 kg, 84 kg, 93 kg

**a** Find their median mass.
**b** Find the mode.
**c** Which of these is not a good average to use?

**a** First write the values in order:
84, 84, 84, 85, (88), 91, 93, 93, 94
middle value = median = 88 kg
**b** mode = most common = 84 kg
**c** The mode is not a good average to use for this data as 84 kg is also the lowest mass.

## Exercise 6I

**1** Here are the amounts of money that eight friends have managed to save over the course of a year:

$23, $31, $1602, $58, $39, $31, $33, $23

**a** What was the mean amount saved?
**b** Find the median.
**c** Which of your answers to parts **a** and **b** is not a good indicator of the average savings? Why?

**2** Here are the prices charged in eight different shops for a new watch strap:
$9, $4, $3, $5, $6, $9, $3, $9
**a** Work out the median price.
**b** Find the mode.
**c** What is the mean price?
**d** Which of your answers to parts **a**, **b** and **c** is not a good indicator of the average price of a new watch strap? Why?

**3** The heights of six friends are listed:

174 cm, 101 cm, 162 cm, 183 cm, 191 cm, 178 cm

**a** Find the mean height.
**b** Suggest a better average to use for this data.
**c** What is the value of the average you suggested in part **b**?

**4** **a** Decide which statistics – mean, median, mode or range – are most useful in these cases:
**i** What size of clothing to reorder in a clothes shop.
**ii** Which player to choose for the team.
**iii** What size of salary increase to give employees working in a factory, including the factory manager.
**iv** Comparing the masses of the packets of flour filled by two machines in a day.
**v** Comparing two classes to see which class performed better in their maths test.
**b** Justify your answers to part **a**.

### INVESTIGATION

Find data sets with the following properties:

1. Whatever number you remove the mode remains the same.
2. When you remove one number the mean stays the same.
3. When you remove one number the mean is double the original value.
4. When you remove a number the median and range remain the same.
5. When you remove a number the mean and range remain the same.

### TECHNOLOGY

If you want to learn about cumulative frequency curves and how these are used to estimate the median from grouped data, go to

www.mrbartonmaths.com

Scroll down to the search box and search for 'cumulative frequency'.

Note that cumulative frequency is not mentioned specifically in the Cambridge Secondary 1 curriculum framework, and is included here as extension work.

Also note that Mr Barton is no relation to the author!

# Consolidation

### Example 1

The masses, in grams, of 20 bars of soap made at a factory are:

134 137 132 134 135 135 134 133 135 136
136 134 134 137 136 132 133 134 134 135

Construct a frequency table for this data.

| Mass (g) | Tally | Frequency |
|---|---|---|
| 132 | II | 2 |
| 133 | II | 2 |
| 134 | ~~IIII~~ II | 7 |
| 135 | IIII | 4 |
| 136 | III | 3 |
| 137 | II | 2 |

### Example 2

The lifetimes of 100 electric light bulbs is shown in the table.

What is the modal lifetime?

| Lifetime (hours) | Frequency |
|---|---|
| 801-900 | 8 |
| 901-1000 | 12 |
| 1001-1100 | 51 |
| 1101-1200 | 23 |
| 1201-1300 | 6 |

The modal lifetime is 1001–1100 hours.

### Example 3

The table shows the heights, in centimetres, of 50 boys.

| Height | 148 | 149 | 150 | 151 | 152 | 153 | 154 | 155 | 156 |
|---|---|---|---|---|---|---|---|---|---|
| Frequency | 3 | 3 | 4 | 7 | 12 | 6 | 5 | 4 | 6 |

What is the mean height of the boys?

Mean height $= \frac{\Sigma fx}{\Sigma f}$

$$= \frac{(3 \times 148) + (3 \times 149) + (4 \times 150) + (7 \times 151) + (12 \times 152) + (6 \times 153) + (5 \times 154) + (4 \times 155) + (6 \times 156)}{3 + 3 + 4 + 7 + 12 + 6 + 5 + 4 + 6}$$

$$= \frac{444 + 447 + 600 + 1057 + 1824 + 918 + 770 + 620 + 936}{50}$$

$$= \frac{7616}{50} = 152.3\text{cm to 1 d.p.}$$

### Example 4

A local authority want to change the opening times of a swimming pool. What data could they collect to help them decide how the times should change, and how could they do this?

They could collect secondary data collected by the swimming pool staff on how many people paid to come in at different times during the day. They could look at the data for one week or one month.

They might also want to find out the ages of the people coming in at different times of the day. They might want to collect primary data, for example using this data collection sheet:

| Time | | |
|---|---|---|
| Day | | |
| Date | | |
| Number of people | | |
| Ages of people | | |

## Exercise 6

**1** Roll a die 30 times.
Construct a frequency table for the data you obtain.

**2** The masses of 24 children, in kilograms, are:

53 42 44 51 60 58 56 47
48 52 53 65 44 55 54 53
49 49 53 54 53 57 52 63

**a** Find the range.

**b** Construct a suitable grouped frequency table to show the data.

**3** Three weeks after planting, the heights, in centimetres, of 50 seedlings were:

| Height (cm) | Number of seedlings |
|---|---|
| 0-2.9 | 3 |
| 3.0-5.9 | 12 |
| 6.0-8.9 | 15 |
| 9.0-11.9 | 16 |
| 12.0-14.9 | 4 |

**a** What is the modal class?

**b** Estimate the mean height of a seedling.

**4** Conduct a survey to find out how many hours the students in your class spend watching television each week.

**a** Draw a grouped frequency table to show the data.

**b** Find the mean time spent watching television each week.

**c** Find the modal class.

**5** Do girls in your class spend more time each week on their homework than boys?

**a** What data do you need to collect?

**b** What different methods could you use to collect the data?

**c** What sample size would you use?

**d** How accurate does the data need to be?

**e** Conduct a survey to find out the answer to this question.

**f** Estimate the mean time spent per week on homework by both boys and girls.

**6**

At the 2012 Olympic Games in London, the distances thrown by the women's discus finalists are shown below.

| Distance (metres) | 58–59 | 60–61 | 62–63 | 64–65 | 66–67 | 68–69 |
|---|---|---|---|---|---|---|
| Frequency | 3 | 7 | 11 | 14 | 6 | 2 |

Estimate the mean distance a discus was thrown by a finalist.

**7** Janet is collecting data to find out whether her town is better at sport than the neighbouring town.

**a** She collects different sets of data. Are these sets of data primary or secondary?

**i** She looks at the football results in the newspapers.

**ii** She uses a data collection sheet to ask the opinions of a sample of 10 people from her town and 10 from the neighbouring town, to see which town they think is better at sport.

**iii** She looks at the records of trophies won by teams and sportspeople in each town.

**b** Janet uses this data collection sheet for her survey:

| Which town are you from? | How old are you? | Are you male or female? | Which town do you think is better at sport? |
|---|---|---|---|
| | | | |
| | | | |

**i** What problems are there with this survey?

**ii** Are there any irrelevant questions?

**8** For each of these situations, calculate the most appropriate average. Give your reasons for choosing that average.

**a** A carrot farmer picks 10 carrots at random and measures their lengths, in centimetres. The results are:

18 15 17 16 19 6 9 14 14 20

The supermarket only wants to sell the carrots if they are, on average, longer than 15 cm.

**b** A high jumper has recorded the height, in metres, of his best jumps at his last 9 competitions.

1.92 1.95 1.89 1.74 1.83 1.91 1.94 1.96 2.00

**c** 5 members of a family are weighed and their masses, in kilograms, are:

4 53 61 55 58

**d** A shop in Oxford, UK intends to reorder the clothes sizes that they have sold the most of, on average. The clothes they sell come in sizes 8, 10, 12, 14, 16, 18 and 20. These are the sales for one day:

| Clothes size | Number sold |
|---|---|
| 8 | 8 |
| 10 | 14 |
| 12 | 18 |
| 14 | 28 |
| 16 | 6 |
| 18 | 5 |
| 20 | 1 |

# Summary

## You should know ...

**1** How to show discrete data in a frequency table.
*For example:*
This list of scores:
0, 1, 1, 5, 2, 4, 3, 1, 2, 4, 0, 5, 1, 5, 4
can be shown in a frequency table:

| Score | 0 | 1 | 2 | 3 | 4 | 5 |
|---|---|---|---|---|---|---|
| Frequency | 2 | 4 | 2 | 1 | 3 | 3 |

## Check out

**1** Here are the numbers of catches taken by 20 cricketers during a season:

6, 7, 2, 0, 4, 5, 3, 1, 6, 5,
1, 9, 7, 1, 2, 3, 8, 9, 7, 6

Show the data in a frequency table.

---

**2** How to show continuous data in a grouped frequency table, choosing suitable class intervals.
*For example:*

It is sensible to use between about 4 and 10 class intervals, depending on the data.

| Mass (kg) | *f* |
|---|---|
| 40-49 | 1 |
| 50-59 | 5 |
| 60-69 | 8 |
| 70-79 | 6 |
| 80-89 | 4 |
| 90-99 | 2 |

**2** The lengths of a sample of 20 cucumbers are measured in centimetres. The data is shown below. Show this data in a grouped frequency distribution.

| | | | | |
|---|---|---|---|---|
| 13.8 | 29.8 | 20.0 | 31.4 | 25.0 |
| 15.3 | 22.0 | 34.3 | 22.4 | 15.1 |
| 27.2 | 26.4 | 19.0 | 28.5 | 13.0 |
| 29.0 | 23.1 | 34.7 | 26.0 | 27.3 |

---

**3** How to find the midpoint of a class interval.
*For example:*

In a table in which the class intervals are 100–104, 105–109, 110–114, … the midpoint of the second class is:

$$\frac{105 + 109}{2} = 107$$

**3** The table shows the masses, in kilograms, of some boys:

| Mass | *f* |
|---|---|
| 45-49 | 7 |
| 50-54 | 8 |
| 55-59 | 11 |
| 60-64 | 3 |

State the midpoint of each interval.

---

**4** Mean $= \dfrac{\text{sum of data}}{\text{number of values}}$

Mode is the most common value in a distribution.
Median is the middle value when the data is arranged in order.
*For example:* 3, 3, 4, 7, 8

Mean $= \dfrac{3 + 3 + 4 + 7 + 8}{5} = 5$

Mode = most common value = 3
Median = middle value = 4
If there are two middle values, the median is half way between these values.
If there are two or more most common values these are all modes.

**4** Here is a list of heights of plants:

4 cm, 6 cm, 9 cm, 10 cm,
5 cm, 3 cm, 2 cm, 1 cm,
8 cm, 11 cm, 9 cm, 4 cm

Find the:
**a** mean
**b** mode
**c** median.

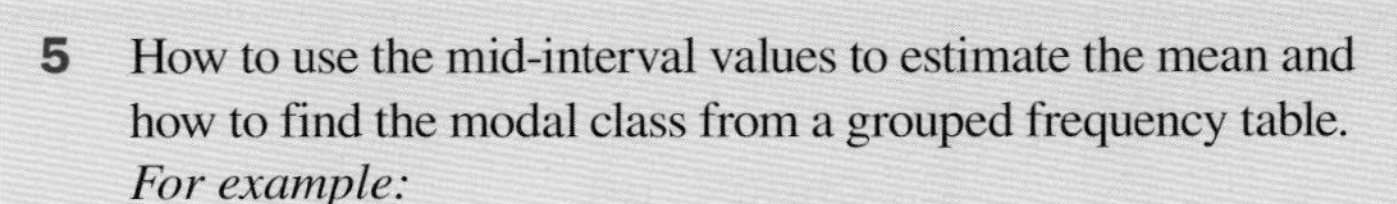

**5** How to use the mid-interval values to estimate the mean and how to find the modal class from a grouped frequency table.
*For example:*

Mid-interval value of the first interval is 7

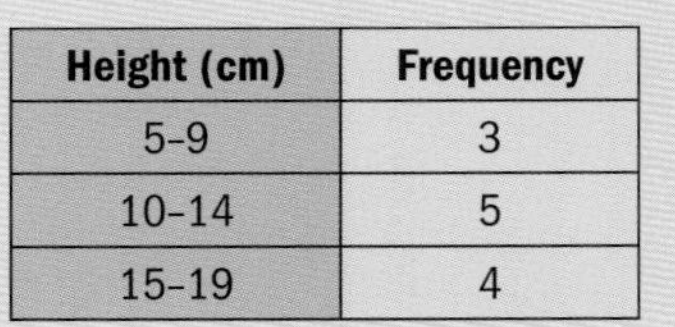

| Height (cm) | Frequency |
|---|---|
| 5-9 | 3 |
| 10-14 | 5 |
| 15-19 | 4 |

$$\text{Mean} = \frac{\Sigma fx}{\Sigma f} = \frac{(3 \times 7) + (5 \times 12) + (4 \times 17)}{12}$$
$$= 12.4 \text{ (to 3 s.f.)}$$

So the mean height is 12.4 cm.
The modal class is the class with the highest frequency: 10–14 cm is the modal height.

**(Note: Estimating the mean from a grouped frequency table is extension work.)**

**5** The marks of 50 students in a test were:

| Mark | Frequency |
|---|---|
| 0-19 | 4 |
| 20-39 | 12 |
| 40-59 | 21 |
| 60-79 | 8 |
| 80-99 | 5 |

**a** Use the mid-interval values to estimate the mean mark.
**b** What is the modal mark?

**(Note: Estimating the mean from a grouped frequency table is extension work.)**

---

**6** How to suggest a question to explore using statistical methods, and how to identify the sets of data needed, how to collect data, what sample sizes to use and what degree of accuracy to use.

**6** Decide what data to collect and how you would collect it if you wanted to find out whether boys or girls are faster at reading.

---

**7** How to identify primary or secondary sources of suitable data.
Primary data is data you have collected yourself (e.g. by carrying out a survey). Secondary data is data that someone else has collected.

**7** Give one advantage and one disadvantage of using
**a** primary data
**b** secondary data.

---

**8** How to design, trial and refine data collection sheets.

**8** Design, trial and refine a data collection sheet to find out which are the favourite fruits eaten by children of different ages.

---

**9** How to select the best statistics to use, depending on the data.

**9** Write down when it is
**a** best
**b** worst to use the
**i** mean
**ii** median
**iii** mode.

# Review A

**1** Write these fractions in their simplest form:

**a** $\frac{25}{60}$ **b** $\frac{16}{100}$

**c** $\frac{14}{49}$ **d** $\frac{32}{68}$

**2** Simplify:

**a** $2m + n - 8n - m + 3m$

**b** $6xy - 2ab + 5xy - 3ab$

**c** $p - q + 3q - 12p - q - 4q$

**3** Use isometric paper to draw the 3D object represented by the plan and elevations.

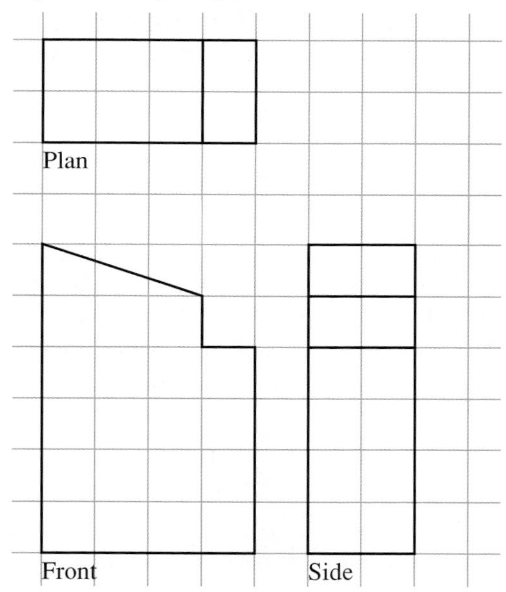


**4** Work out:

**a** $^{-}3 + 4 - 8$ **b** $^{-}6.5 + 8$

**c** $2 - 4.7$ **d** $^{-}3.8 + 5.6$

**e** $^{-}7.2 - 3.9$ **f** $^{-}4 + (^{-}9.3)$

**g** $8 - {}^{-}11.2$ **h** $^{-}3.6 - {}^{-}4.8$

**5** Copy and complete:

**a** 33 300 ml = □ℓ

**b** 0.0008 t = □g

**c** 310 000 cm = □km

**d** 0.08 ℓ = □ml

**e** 0.0045 kg = □g

**f** 40 cl = □ml

**g** 3700 mm = □cm

**h** 17 000 g = □kg

**i** 8.4 t = □kg

**j** 7 ml = □cl

**6** Copy and complete these sentences:

Data that you have collected yourself is called ……… data.

Data collected by someone else is called ……… data.

**7** Work out:

**a** $\frac{7}{8} - \frac{3}{4}$ **b** $\frac{2}{5} + \frac{1}{3}$ **c** $4\frac{2}{5} + 2\frac{1}{4}$

**d** $6\frac{7}{8} - 5\frac{4}{5}$ **e** $4\frac{2}{3} - 1\frac{5}{6}$

**8** Change these statements into algebraic expressions:

**a** Five times the sum of $x$ and 3

**b** 10 more than the product of $p$ and $y$.

**9** How many planes of symmetry do these shapes have?

**a** Regular hexagonal prism

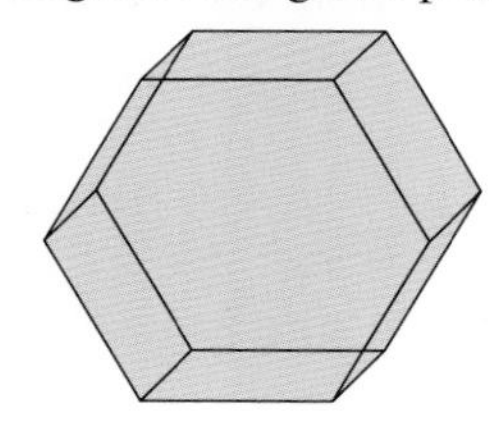

**b** Isosceles triangular prism

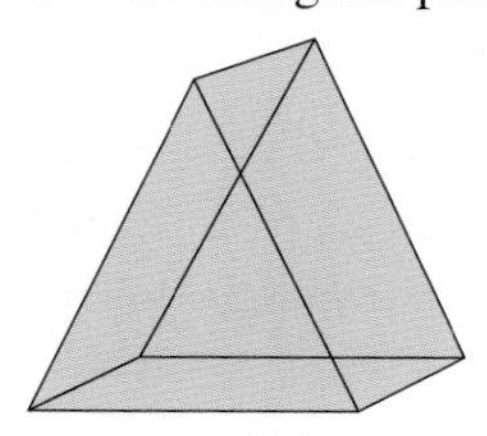

**10** Work out:

**a** $7 \times 10^2$ **b** $1.4 \times 10^3$

**c** $612\,000 \div 10^4$ **d** $4100 \div 10^3$

**11** A piece of fabric is 2.1 m long. A 40-cm length and a 250-mm length are cut from it. How many metres of fabric are left?

**12** The lengths of 30 telephone calls made by Rebekah are, in minutes:

10, 15.5, 17, 21, 29, 36.5, 48, 56.5, 6.5, 1, 14, 23, 44, 3, 2, 18, 26, 41, 12, 8, 16, 5.5, 37, 23.5, 14, 9, 32, 12.5, 7, 18

Construct a grouped frequency table to show this data, choosing suitable class intervals.

**13** A piece of metal is $2\frac{1}{4}$ m long. How long will it be if I cut $1\frac{3}{5}$ m of metal from it?

**14** **a** Derive a formula for $A$, the area of each rectangle.

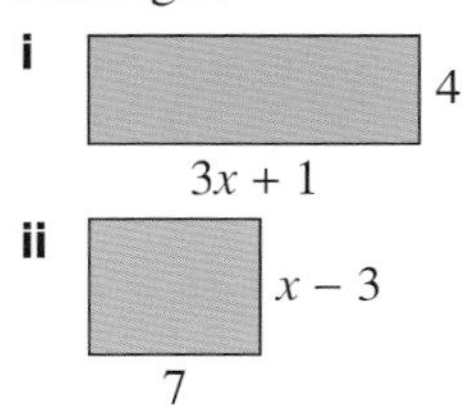


**b** Derive a formula for $P$, the perimeter of each of the rectangles in part **a**.

**15** Copy and complete:

**a** $2^5 = \square$ **b** $49 = 7^{\square}$

**c** $3^{\square} = 27$ **d** $\square^3 = 125$

**16** Draw a line AB and a point C which is 5 cm above the line.
Construct a perpendicular from C to the line AB.

**17** Copy and complete:

**a** $532\,\text{cm}^3 = \square\,\text{ml}$

**b** $1.4\,\text{ha} = \square\,\text{m}^2$

**c** $622\,\text{ml} = \square\,\text{cm}^3$

**d** $8\,800\,000\,\text{cm}^3 = \square\,\text{m}^3$

**e** $0.07\,\ell = \square\,\text{cm}^3$

**f** $380\,\text{mm}^2 = \square\,\text{cm}^2$

**g** $41\,\text{cm}^2 = \square\,\text{mm}^2$

**h** $1.4\,\text{m}^2 = \square\,\text{cm}^2$

**i** $0.0097\,\text{m}^3 = \square\,\text{cm}^3$

**j** $260\,000\,\text{m}^2 = \square\,\text{ha}$

**18** The amounts earned in a month, in dollars, by 7 friends is

\$1200, \$1000, \$800, \$4000, \$800, \$1210, \$1250

Work out the mean, median and mode. Decide which is the best average to represent the data. Explain your choice.

**19** Work out:

**a** $3 \times \frac{1}{4}$ **b** $\frac{2}{5} \times \frac{3}{7}$

**c** $4\frac{1}{4} \times 3$ **d** $6\frac{2}{3} \times 3\frac{3}{5}$

**e** $1\frac{5}{7} \times 3\frac{1}{3}$ **f** $3\frac{1}{5} \times 1\frac{7}{8}$

**20** Simplify:

**a** $m \times m^5$ **b** $p^6 \div p^3$

**c** $5t^2 \times 4t^3$ **d** $y^3 \div y^5$

**21**

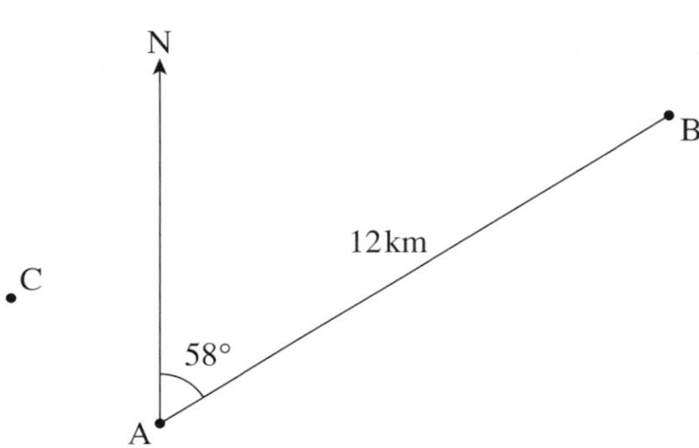


**a** Write down the bearing of B from A.

**b** By measuring the length AB, copy and complete this scale: 1 cm = $\square$ km.

**c** How many kilometres is point C from point A?

**d** What is the bearing of C from A?

**22** Work out:

**a** $7 \div \frac{3}{21}$ **b** $\frac{1}{4} \div \frac{1}{8}$

**c** $\frac{3}{8} \div \frac{4}{5}$ **d** $4\frac{2}{7} \div 1\frac{1}{14}$

**e** $3\frac{3}{5} \div 1\frac{11}{25}$ **f** $2\frac{5}{8} \div 8\frac{3}{4}$

**23** A small bottle of water contains 230 ml. A large bottle of water contains 2.5 ℓ. How many litres of water are there in 3 small bottles and 2 large bottles?

**24** Simplify:

**a** $3^4 \times 3^6$ **b** $2^{15} \div 2^5$

**c** $4^3 \times 4 \times 4^5$ **d** $\frac{7^{10}}{7^2}$

**25** Eva has designed this data collection sheet to find out who gets the most pocket money in her school:

| | |
|---|---|
| **How old are you?** | |
| **Are you female?** | |
| **How much pocket money do you get?** | |
| **What class are you in?** | |
| **Do you earn any money?** | |

**a** What problems are there with this data collection sheet?
How would you improve it? Are there any irrelevant questions?

**b** If Eva was to use the sheet to collect data, what sample size would you suggest? What degree of accuracy should she use for the question about money?

**26** Simplify:

**a** $4x^3 \times 5x^2 \times 2x^4$ **b** $\frac{6x^6}{2x^2}$

**c** $\frac{8t^5y^3}{4ty^2}$

**27** Draw the plan and elevations for this 3D object:

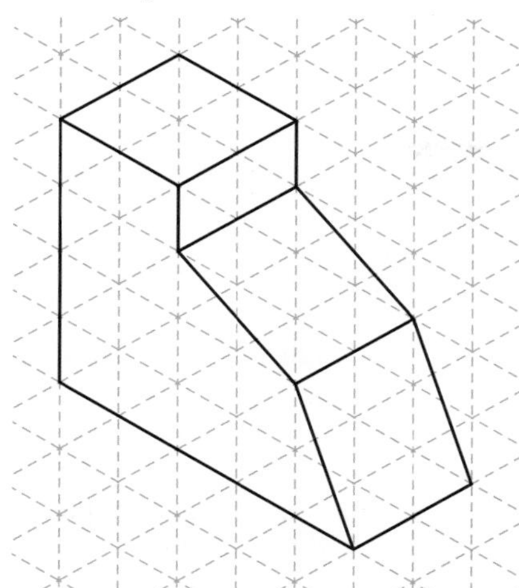

**28** Find the value of each of the following when $t = 2$, $m = {}^-1$ and $c = 3$.

**a** $t(m + c)$ **b** $tc - m^2$

**c** $\dfrac{4m^2 - 2tc}{t + m + c}$

**29** A cake has a mass of $1\frac{1}{4}$ kg. If $\frac{2}{5}$ of it is eaten, what mass of cake is left?

**30** Write down the name of a 3D shape with nine planes of symmetry.

**31** Using the formula $A = 10d - 2w^2$, find $A$ when

**a** $w = {}^-3$ and $d = 0.5$

**b** $w = 4$ and $d = {}^-0.2$.

**32** Draw a circle. Inside the circle construct an inscribed square.

**33** Find the value of

**a** $3r^2 + r$ when $r = \dfrac{1}{2}$

**b** $50 - 200b^2 - 10b$ when $b = {}^-0.4$.

**34** Work out:

**a** $3 \times {}^-7$ **b** ${}^-6 \times {}^-2$

**c** ${}^-4.5 \times 4$ **d** $6 \times {}^-1.2$

**e** $18 \div {}^-6$ **f** ${}^-24 \div {}^-8$

**g** $8 \div {}^-0.2$ **h** ${}^-2.8 \div {}^-1.4$

**35** Cargo is loaded onto an aeroplane. The first item has a mass of 0.04 t, the second item has a mass of 70 kg and the third item has a mass of 54 000 g. What is the mass, in kilograms, of this cargo?

**36** Expand and simplify these expressions:

**a** $(x + 3)(x + 4)$ **b** $(x - 1)(x + 6)$

**c** $(x - 1)(x - 5)$ **d** $(x + 7)^2$

**37** Estimate, to 1 d.p.:

**a** $\pm\sqrt{30}$ **b** $\sqrt{45}$

**c** $\pm\sqrt{5}$ **d** $\sqrt{115}$

**e** ${}^-\sqrt{62}$ **f** $\sqrt{175}$

**38** Write down an expression for the area of each shape.

**a**

$x + 5$

$x - 3$

**b**

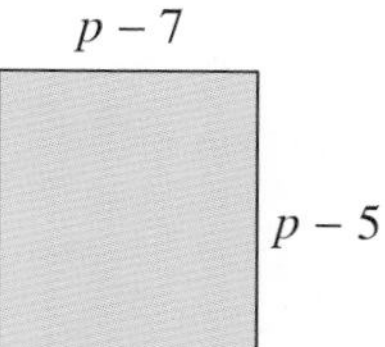


**c**

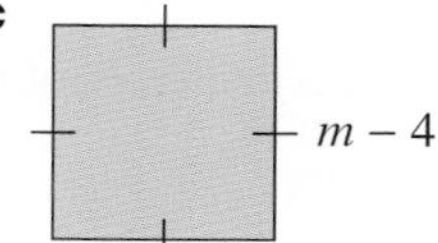


**39** What are the missing numbers?

**a** $2\frac{3}{8} + 1\frac{\square}{3} = 4\frac{1}{24}$

**b** $3\frac{2}{5} - 1\frac{5}{\square} = 1\frac{24}{35}$

**c** $1\frac{\square}{5} \times 4\frac{3}{8} = 6\frac{1}{8}$

**d** $4\frac{1}{\square} \div 1\frac{3}{7} = 2\frac{11}{12}$

**40** Draw a line AB and a point C somewhere on the line. Construct a perpendicular to the line AB from C.

**41** Write as a single fraction:

**a** $\dfrac{w}{3} + \dfrac{w}{4}$ **b** $\dfrac{2e}{5} + \dfrac{3e}{7}$

**c** $\dfrac{2}{5c} + \dfrac{4}{15c}$ **d** $\dfrac{2}{d} - \dfrac{1}{4d}$

**e** $x + \dfrac{3x}{4}$ **f** $5 + \dfrac{4}{h}$

**42** Copy and complete this division grid:

Second number

| First number: ÷ | $^-5$ | 2 | $^-0.5$ |
|---|---|---|---|
| $^-20$ | | | |
| $^-4.5$ | | | |
| 7.5 | | | |

**43** Factorise:

**a** $5g + 10$ **b** $3c + 9v - 18$

**c** $20t + 16$ **d** $xy + xh$

**e** $4rf + 3r + rk$

**44** Copy the text below and then draw lines to match the equivalent scales in the two lists:

| | |
|---|---|
| 1 : 500 | 4 cm : 1 km |
| 1 : 200 | 1 cm : 5 m |
| 1 : 12 500 | 5 cm : 10 m |
| 1 : 25 000 | 8 cm : 1 km |

**45** Estimate, to 1 d.p.:

**a** $\sqrt[3]{70}$ **b** $\sqrt[3]{5}$

**c** $\sqrt[3]{30}$ **d** $\sqrt[3]{160}$

**e** $\sqrt[3]{^{-}12}$ **f** $\sqrt[3]{^{-}500}$

**46** Copy and complete:

**a** $3^0 = \square$ **b** $4^3 \div 4^8 = 4^{\square}$

**c** $7^{-2} = \frac{1}{\square}$ **d** $\frac{1}{8} = 2^{\square}$

**e** $10^{-3} = \square$ **f** $\frac{1}{6} = 6^{\square}$

**47 a** Derive the formula for the perimeter, $P$, of this triangle:

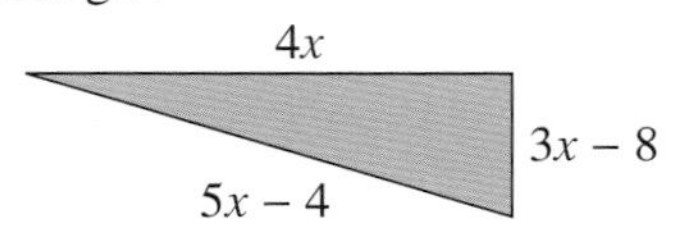


**b** Rearrange the formula to make $x$ the subject.

**c** Use the rearranged formula to find $x$ when $P = 60$.

**d** What are the side lengths of the triangle in part **c**?

**48** Work out:

**a** $\dfrac{3\frac{1}{4}}{2\frac{3}{4} - 1\frac{2}{3}}$ **b** $(1\frac{3}{5})^2$

**49** Work out:

**a** $30\,000 \times 10^{-4}$ **b** $8\,040\,000 \times 10^{-6}$

**c** $0.000027 \div 10^{-5}$ **d** $0.079 \div 10^{-3}$

**50** Suggest a question you want to explore using statistical methods. Identify the sets of data you need. How would you collect the data? What sample size would you use? What degree of accuracy would you use?

**51** B is on a bearing of 132° and 6 km from A. C is 4.5 km from A on a bearing of 200°. Using the scale 1 cm = 1 km, make a scale drawing to show this information. Use your drawing to find the distance of B from C.

**52** Make $x$ the subject of these formulae:

**a** $F = yx - 3$ **b** $m = 4(b + x)$

**c** $V = \dfrac{x(d - p)}{m}$ **d** $R = \dfrac{5}{x} + d$

**53** Copy and complete:

**a** $3.7 \times 10^{\square} = 37\,000$

**b** $\square \div 10^2 = 31.8$

**c** $62\,000 \times 10^{\square} = 6.2$

**d** $0.0008 \div 10^{\square} = 8$

**e** $\square \times 10^4 = 44\,000$

**f** $478\,000 \div 10^{\square} = 478$

**g** $\square \times 10^{-4} = 1.8$

**h** $\square \div 10^{-5} = 900$

**54** This cuboid has side lengths $3\frac{1}{8}$ cm, $1\frac{3}{5}$ cm, and $2\frac{3}{4}$ cm.
What is its volume?

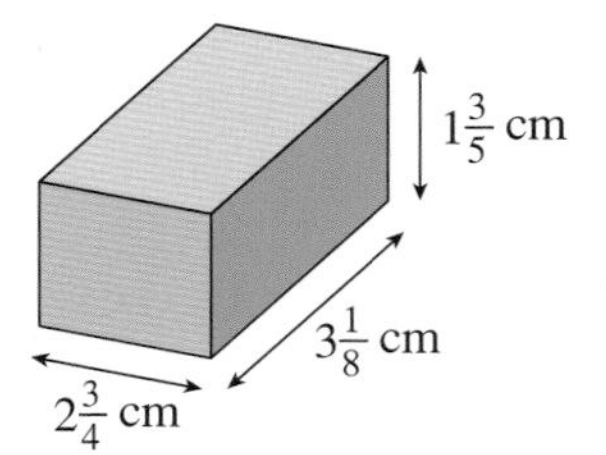


**55** Copy and complete this multiplication grid:

| × | $^{-}1.6$ | $^{-}1.8$ | 4.2 |
|---|---|---|---|
| 3.4 | | | |
| $^{-}1.2$ | | | |
| $^{-}2.3$ | | | |

**56** A farmer has 12 ha of land. 35 000 m² of his land is not suitable for grazing his animals. What is the area of his land, in hectares, that is suitable for grazing his animals?

**57 a** Work out $\sqrt{0.36}$

**b** Work out $\sqrt{0.49}$

**c** Use your answers to **a** and **b** to estimate $\sqrt{0.4}$, to 2 d.p.

**58** Draw a circle. Inside the circle construct an inscribed regular hexagon.

**59** Write as a single fraction:

**a** $\dfrac{3}{gh} + \dfrac{d}{k}$

**b** $\dfrac{ab}{3c} + \dfrac{5c}{4a}$

# 7 Rounding, multiplying and dividing

## Objectives

- Round numbers to a given number of decimal places or significant figures; use rounding to give solutions to problems with an appropriate degree of accuracy.
- Recognise the effects of multiplying and dividing by numbers between 0 and 1.
- Multiply by decimals, understanding where to position the decimal point by considering equivalent calculations; divide by decimals by transforming to division by an integer.

## What's the point?

Giving approximate answers and rounding to a certain number of significant figures is used a lot in the news. A newspaper would not report that there were 23 148 people at a concert. Instead they would write as the headline something like '23 000 attend concert'.

## Before you start

### You should know ...

1 How to round to the nearest ten, hundred or thousand.

*For example*:

246 to the nearest ten is 250 because it is closer to 250 than it is to 240.

8249 to the nearest thousand is 8000 because it is closer to 8000 than it is to 9000.

1750 to the nearest hundred is 1800. It is exactly halfway between 1700 and 1800, which means we round up to 1800.

### Check in

1 Round

**a** 2397 to the nearest hundred

**b** 456 to the nearest ten

**c** 17 159 to the nearest thousand

**d** 650 to the nearest hundred.

# 7.1 Rounding numbers

## Decimal places

You can round decimal numbers to the nearest tenth, hundredth or thousandth. A number line shows how this is done.

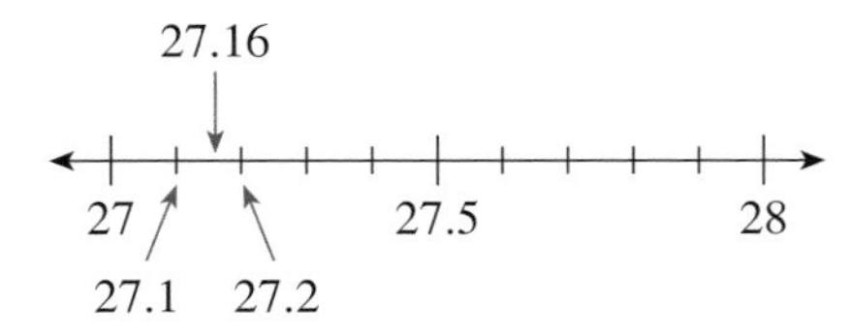


27.16 rounded to the nearest tenth is 27.2 because 27.16 is closer to 27.2 than to 27.1.

You can write

27.16 = 27.2 correct to 1 decimal place (1 d.p.)

On the other hand

27.14 = 27.1 (1 d.p.)

since 27.14 is closer to 27.1 than 27.2.

In general, if the next placed number is 0, 1, 2, 3 or 4, round down. If it is 5, 6, 7, 8 or 9, round up.

**EXAMPLE 1**

Round 4.655 and 0.8723 to
**a** 2 decimal places
**b** 1 decimal place.

**a** The second decimal place is underlined here:

4.65⑤ $\xrightarrow{\text{round up}}$ 4.66 (2 d.p.)

0.87②3 $\xrightarrow{\text{round up}}$ 0.87 (2 d.p.)

**b** The first decimal place is underlined here:

4.6⑤5 $\xrightarrow{\text{round up}}$ 4.7 (1 d.p.)

0.8⑦23 $\xrightarrow{\text{round up}}$ 0.9 (1 d.p.)

### Exercise 7A

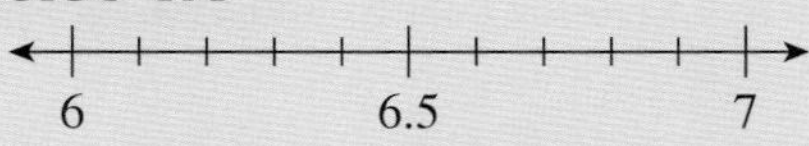


**1** Use the number line to write these numbers correct to 1 decimal place.
**a** 6.14 **b** 6.25 **c** 6.71
**d** 6.38 **e** 6.91 **f** 6.99
**g** 6.41 **h** 6.75 **i** 6.08

**2** Write correct to 1 decimal place:
**a** 3.22 **b** 4.36 **c** 0.82
**d** 0.76 **e** 0.65 **f** 4.98
**g** 6.02 **h** 7.07 **i** 1.19

**3** Write correct to 2 decimal places:
**a** 6.231 **b** 0.782 **c** 0.965
**d** 4.025 **e** 4.204 **f** 6.107
**g** 4.003 **h** 0.028 **i** 0.205

**4** Write these numbers correct to:
**i** 3 decimal places
**i** 2 decimal places
**iii** 1 decimal place.
**a** 6.4183 **b** 0.7062 **c** 0.0215
**d** 9.6008 **e** 7.1808 **f** 0.3144
**g** 6.2999 **h** 4.0818

**5** Seven friends share a meal costing $183. They decide to share the bill equally between them.
**a** How much does each person owe? (Give your answer to 2 d.p.)
**b** If each person pays this amount they will not cover the bill. Why is this? What could the friends do to cover the bill?

**6** Round 0.7459368 to
**a** 4 d.p. **b** 5 d.p.

**7** **a** What is the perimeter of a rectangle with length 6.12 cm and width 4.77 cm, correct to
**i** 2 d.p. **ii** 1 d.p.?
**b** What is the area of the rectangle, correct to
**i** 2 d.p. **ii** 1 d.p.?

**8** You need to be sensible when rounding. Say what the problem is in these situations.
**a** A school hall has 456 seats. The head of music rounded to the nearest ten and sold that number of tickets.
**b** A bridge can safely carry a weight of 15.75 tonnes. A sign shows this amount to the nearest whole number.

**9** This is Ayaan's homework. Write down the correct answers for any he has got wrong.
**a** 7.44999 to 1 d.p. is 7.5
**b** 0.399 to 2 d.p. is 0.4
**c** 0.00493 to 2 d.p. is 0.01

**10** Round
**a** 84.98 to 1 d.p. **b** 60.99999 to 3 d.p.
**c** 7.895 to 2 d.p. **d** 7.9949 to 2 d.p.
**e** 0.0489 to 1 d.p.

**11** A measurement is written to 1 decimal place as 3.4 cm. Which of these could be the actual measurement?

| | | |
|---|---|---|
| 3.4499 cm | 34.7 mm | 341 mm |
| 33.81 mm | 0.342 m | 3.35 cm |

## Significant figures

Rounding to one or two decimal places does not always give a simple approximation of a number.

*For example:*
6 381 278.68
rounded to 1 d.p. is
6 381 278.7
… which is not much simpler!

In such cases it is more useful to round to a given number of **significant figures.**

Significant figures show the relative importance of the digits in a number, the first non-zero digit being the most important.

For example, to one significant figure (1 s.f.), 2914 is 3000.

Notice the place value:

| Th | H | T | U |
|---|---|---|---|
| 2 | 9 | 1 | 4 |

The first non-zero digit is the 2, which is the number of thousands. In this case rounding to 1 s.f. is the same as rounding to the nearest thousand.

**EXAMPLE 2**

Write 83 562 to

**a** 2 s.f.
**b** 1 s.f.

The column heads are

| TTh | Th | H | T | U |
|---|---|---|---|---|
| 8 | 3 | 5 | 6 | 2 |

**a** The second digit head is thousands so round to the nearest thousand:
83 562 = 84 000 (2 s.f.)
**b** The first digit head is tens of thousands so round to the nearest ten thousand:
83 562 = 80 000 (1 s.f.)

Here are some other examples:

| Number | 67 341 | 23.478 | 0.03286 |
|---|---|---|---|
| 3 s.f. | 67 300 | 23.5 | 0.0329 |
| 2 s.f. | 67 000 | 23 | 0.033 |
| 1 s.f. | 70 000 | 20 | 0.03 |

Notice that in the number 0.03286 the first non-zero digit is 3 which is in the hundredths place.
So 0.03286 = 0.03 (1 s.f.).

## Exercise 7B

**1** Write correct to 1 significant figure:

| | | |
|---|---|---|
| **a** 420 | **b** 8314 | **c** 396 |
| **d** 48 | **e** 4.81 | **f** 91 265 |
| **g** 10 034 | **h** 0.0032 | **i** 7.0036 |

**2** Write to 2 significant figures:

| | | |
|---|---|---|
| **a** 962 | **b** 489 | **c** 6183 |
| **d** 17 638 | **e** 17.34 | **f** 26.92 |
| **g** 0.00389 | **h** 133.68 | **i** 8.035 |

**3** Copy and complete:

| Number | 613 752 | 1.6831 | 0.004753 |
|---|---|---|---|
| 3 s.f. | | | |
| 2 s.f. | | | |
| 1 s.f. | | | |

**4** Round
**a** 0.0046572 to 3 s.f.
**b** 0.0189 to 1 s.f.
**c** 0.00022278 to 2 s.f.
**d** 0.399 to 2 s.f.

**5** Ramsingh wrote the following.
**i** 6384 = 6000 (1 s.f.)
**ii** 6384 = 6400 (2 s.f.)
**iii** 816952 = 8 (1 s.f.)
**iv** 0.0356 = 0.03 (1 s.f.)
**v** 0.089 42 = 0.089 (2 s.f.)
**vi** 899 = 800 (2 s.f.)
**a** Which of these are correct?
**b** Write down the correct answers for any Ramsingh got wrong.

**6** Round 82 736 to:
**a** 1 s.f. **b** 2 s.f. **c** 3 s.f. **d** 4 s.f.

**7** Round 0.003565 to:
**a** 1 s.f. **b** 2 s.f. **c** 3 s.f. **d** 4 s.f.

**8** A number rounded to 1 s.f. is 60. Write down five examples of what the number could be.

**9** Imagine your job is to write the headlines for a newspaper. For each fact below, decide whether you would write an *approximate* number or the *exact* number in your headline. If *approximate*, write down the number you would use instead.

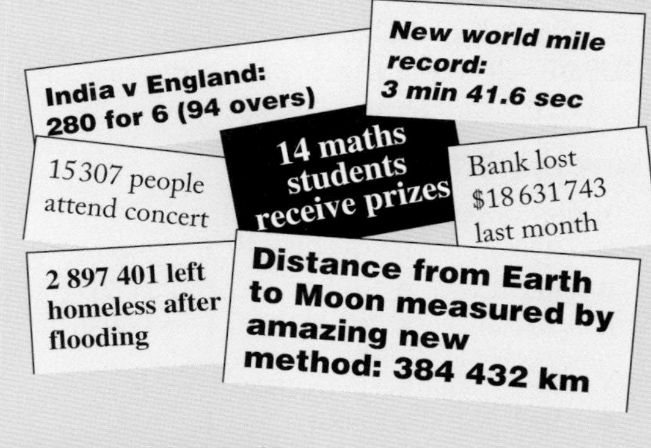


**10** Round 8 134 767 to
- **a** 4 s.f. **b** 3 s.f.
- **c** 2 s.f. **d** 1 s.f.

**11** A whole number rounded to 1 s.f. is 5000. What is
- **a** the largest number it could have been
- **b** the smallest number it could have been?

**12** Calculate $0.053 \times 1.4$, giving your answer
- **a** exactly
- **b** correct to 2 decimal places
- **c** correct to 1 significant figure.

**13** Calculate $0.6234 \div 17.2$ and write your answer
- **a** correct to 2 decimal places
- **b** correct to 3 significant figures.

## Standard form

**Note that standard is not mentioned specifically in the Cambridge Secondary 1 curriculum framework, and is included here as extension work.**

Approximating very large or very small numbers to one or two significant figures is one way of making them easier to work with. However, for huge and tiny numbers, even approximations are troublesome to write down.

For example the mass of the Moon is 73 500 000 000 000 000 000 000 kg to 3 s.f. It takes quite some time to write this out.

Very small numbers are just as awkward; a proton's mass is about

0.000 000 000 000 000 000 000 001 673 g!

Scientists use **standard form** (or **scientific notation**) to write such numbers.

In standard form, the number is written as:

Number between 1 and 10 $\times$ 10 to a power.

*For example:*

$$8000 = 8 \times 1000 = 8 \times 10^3$$
$$7\,000\,000 = 7 \times 1\,000\,000 = 7 \times 10^6$$

The idea is the same whatever the number.

**EXAMPLE 3**

Write in standard form:

**a** 81 432 **b** 91 285 000

**a** $81\,432 = 8.1432 \times 10\,000$

$= 8.1432 \times 10^4$

**b** $91\,285\,000 = 9.1285 \times 10\,000\,000$

$= 9.1285 \times 10^7$

In standard form the mass of the Moon (to 3 s.f.) becomes

$7.35 \times 10^{22}$ kg

which is much easier to write than

73 500 000 000 000 000 000 000 kg.

### Exercise 7C (extension)

**1** Write in standard form:

| | | | | | |
|---|---|---|---|---|---|
| **a** | 2000 | **b** | 400 | **c** | 80 |
| **d** | 90 000 | **e** | 4000 | **f** | 700 000 |
| **g** | 3 000 000 | **h** | 40 000 000 | **i** | 100 000 |

**2** Write in full:

| | | | | | |
|---|---|---|---|---|---|
| **a** | $3 \times 10^2$ | **b** | $5 \times 10^3$ | **c** | $6 \times 10^5$ |
| **d** | $2 \times 10^7$ | **e** | $3 \times 10^6$ | **f** | $4 \times 10^9$ |

**3** Write in standard form:

| | | | | | |
|---|---|---|---|---|---|
| **a** | 420 | **b** | 6300 | **c** | 170 000 |
| **d** | 23 000 | **e** | 61 300 | **f** | 9230 |
| **g** | 416 | **h** | 98 100 | **i** | 6 310 000 |

**4** Write in full:

**a** $1.6 \times 10^3$ **b** $2.8 \times 10^2$
**c** $3.81 \times 10^2$ **d** $4.75 \times 10^5$
**e** $3.01 \times 10^8$ **f** $1.6 \times 10^9$

**5** These are not numbers written in standard form. Explain why not.

**a** $54.3 \times 10^7$
**b** $0.03 \times 10^8$
**c** $7.1 \times 10^3 \times 10^2$

**6** A factory produces 270 000 000 sweets in a week. How many sweets does the factory produce in a year? Give your answer in standard form.

**7** The diameter of the Milky Way galaxy is about $9.46 \times 10^{20}$ m. How many kilometres is this? Give your answer in standard form.

**8** Write the numbers from Question **5** in standard form.

Remember that

$$\begin{aligned} 10^3 \times 10^2 &= (10 \times 10 \times 10) \times (10 \times 10) \\ &= 1000 \times 100 \\ &= 100000 \\ &= 10^5 \end{aligned}$$

In general

$$a^m \times a^n = a^{m+n}$$

Also

$$\begin{aligned} 10^5 \div 10^2 &= \frac{10 \times 10 \times 10 \times 10 \times 10}{10 \times 10} \\ &= \frac{100000}{100} \\ &= 1000 \\ &= 10^3 \end{aligned}$$

In general

$$a^m \div a^n = a^{m-n}$$

So $10^2 \div 10^5 = 10^{2-5} = 10^{-3}$

What is $10^{-3}$?

$$\begin{aligned} 10^2 \div 10^5 &= \frac{10 \times 10}{10 \times 10 \times 10 \times 10 \times 10} = \frac{1}{10 \times 10 \times 10} \\ &= \frac{1}{10^3} \end{aligned}$$

So $10^{-3} = \frac{1}{10^3} = \frac{1}{1000} = 0.001$

In the same way

$$10^{-1} = \frac{1}{10} = 0.1$$
$$10^{-2} = \frac{1}{100} = 0.01$$
$$10^{-3} = \frac{1}{1000} = 0.001$$
$$10^{-4} = \frac{1}{10000} = 0.0001 \text{ etc.}$$

You can use negative powers of 10 to write very small numbers in standard form.

*For example:*

$$\begin{aligned} 0.003 &= 3 \div 1000 = 3 \times \frac{1}{1000} \\ &= 3 \times 10^{-3} \end{aligned}$$

$$\begin{aligned} 0.00005 &= 5 \div 100000 = 5 \times \frac{1}{100000} \\ &= 5 \times 10^{-5} \end{aligned}$$

More complex numbers can be written in standard form in the same way. Just remember that the first number must lie between 1 and 10.

**EXAMPLE 4**

Write in standard form:

**a** 0.0526 **b** 0.00007845

**a** $0.0526 = 5.26 \div 100$

The first number must lie between 1 and 10

$$\begin{aligned} &= 5.26 \times \frac{1}{100} \\ &= 5.26 \times 10^{-2} \end{aligned}$$

**b**
$$\begin{aligned} 0.00007845 &= 7.845 \div 100000 \\ &= 7.845 \times \frac{1}{100000} \\ &= 7.845 \times \frac{1}{10^5} \\ &= 7.845 \times 10^{-5} \end{aligned}$$

In standard form the mass of a proton, to 4 s.f., is

$1.673 \times 10^{-24}$ g

which is much easier to write than
0.000000000000000000000001673 g

## Exercise 7D (extension)

**1** Work out:

**a** $6 \times 10^{-2}$  **b** $4 \times 10^{-1}$
**c** $8 \times 10^{-3}$  **d** $5 \times 10^{-4}$
**e** $2 \times 10^{-6}$  **f** $7 \times 10^{-5}$

**2** Write in standard form:

**a** 0.63  **b** 0.0074
**c** 0.028  **d** 0.00013
**e** 0.02356  **f** 0.00082
**g** 0.00000391  **h** 0.0016
**i** 0.00383

**3** Write in standard form:

**a** 0.0000306  **b** 0.0000004925
**c** 0.0042831  **d** 0.000901
**e** 0.00000000025  **f** 0.0000000846
**g** 0.000000086
**h** 0.000000000003675

**4** The mass of a red blood cell, to 1 s.f., is $9 \times 10^{-14}$ kg. What is this in grams? Give your answer in standard form.

### ACTIVITY

The Eiffel Tower in Paris is about 300 m tall.
That is, $3 \times 10^2$ m.
You can show it on a number line like this:

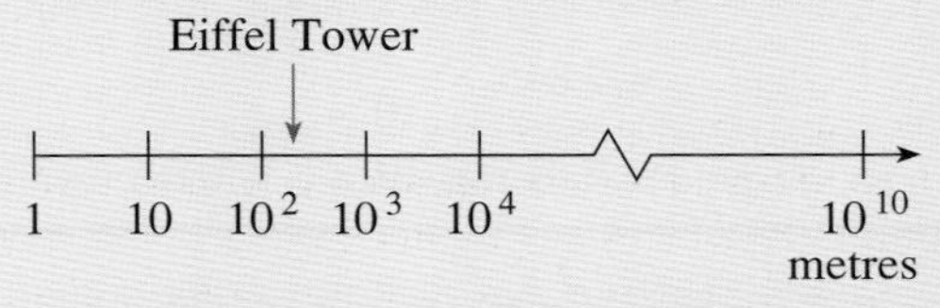


**a** Copy the number line and insert other lengths of objects or distances, up to $10^{10}$ metres.

**b** Repeat for this number line to show lengths or distances which are very small

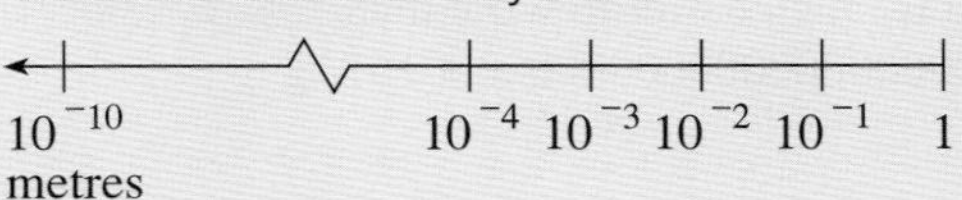


### TECHNOLOGY

You can review how to round a number by visiting the website
www.mathsisfun.com
and following the links to Numbers, Decimals Menu and Rounding Numbers.

Check out the Significant Digit Calculator. It will solve all your significant figure problems!

## 7.2 Estimation

When you have to measure something, it is a good habit to estimate the measurement, as a way of checking your answer.

Similarly, when you have to do a calculation, it is also a good habit to estimate the answer.

Making an estimate is important when you are calculating decimals, because it is easy to put a decimal point in the wrong place during your calculation.

Round numbers to 1 significant figure to estimate.

### EXAMPLE 5

Estimate $3.9 \times 5.1$

3.9 is 4 to 1 s.f.
5.1 is 5 to 1 s.f.
$4 \times 5 = 20$
So the answer for $3.9 \times 5.1$ is about 20.

## Exercise 7E

**1** Estimate the answer:

**a** $1.9 \times 3.2$  **b** $4.8 \times 12.1$
**c** $2.36 \times 11.01$  **d** $10.33 \times 4.98$
**e** $6.87 \times 6.21$  **f** $(4.31)^2$

**2** Estimate the answer:

**a** $(3.84)^3$  **b** $2.7 \times 3.1 \times 6.2$
**c** $3.8 \times 10.1 \times 1.9$  **d** $2.8 \times 0.7 \times 6.9$
**e** $(0.11)^2$  **f** $0.107 \times 0.091$
**g** $0.062 \times 0.048$  **h** $0.067 \times 0.081$

**3** Estimate the answer:

**a** $104.3 - 7.28^2$  **b** $\frac{16.4 + 3.82}{4.91}$

**c** $\frac{3.411}{0.217 + 0.115}$

**d** $5.82 \times 3.14 + 5.24$

**4** Check the accuracy of your estimates in Questions **1–3** using a calculator.

You don't always need to round to 1 significant figure to estimate.

**EXAMPLE 6**

Estimate **a** $47.83 \div 2.99$ **b** $0.78 \div 42$

**a** $47.83 \approx 48$
$2.99 \approx 3$

$$48 \div 3 = \frac{48}{3} = 16$$

So the answer for $47.83 \div 2.99$ is about 16.

**b** $0.78 \approx 0.8$
$42 \approx 40$

$$0.8 \div 40 = \frac{0.8}{40} = \frac{8}{400} = \frac{1}{50} = 0.02$$

So the answer for $0.78 \div 42$ is about 0.02.

## Exercise 7F

**1** Estimate the answer in your head:

**a** $11.7 \div 2.8$ **b** $13.8 \div 6.8$
**c** $4.1 \times 0.9$ **d** $6.32 \div 1.87$
**e** $8.24 \div 0.96$ **f** $15.72 \div 2.13$

**2** Estimate the answer to:

**a** $5.42 \div 2.81$ **b** $1.82 \div 2.6$
**c** $14.08 \div 32.2$ **d** $3.256 \div 3.1$
**e** $123.69 \div 1.9$ **f** $6.72 \div 320$

**3** Check the accuracy of your estimates in Questions **1** and **2** using a calculator.

**4** **a** $\frac{39 \times 71}{54}$ **b** $\frac{28 \times 18}{589}$
**c** $\frac{31 \times 7.9}{3021}$ **d** $\frac{501 \times 81.2}{30.9}$

The correct answers to these calculations (to 2 d.p.) are included in this list: 137.24, 1316.54, 4.72, 0.86, 51.28, 12.46, 0.08. Without doing the calculations, pick out the correct answer for each one.

**5** Using estimation, show which of these calculations are definitely incorrect.

**a** $24.318 \div 19.3 = 21.6$
**b** $0.455 \times 18.4 = 8.372$
**c** $\frac{4.98 + 3.67}{1.8} = 1.015$

## ACTIVITY

**A calculator game for two players**

**Rules**

1. The first player, A, decides on a target range. For example, 850–860.
2. The second player, B, chooses a number and an operation, say 36.4 and ×.
3. Player A tries to estimate a number so that (36.4 × number) lies between 850 and 860.
4. Using a calculator, B checks if A's answer lies within the target range. If yes, A gets one point. If no, B makes an estimate and receives one point if it is within target range.
5. The game continues with players taking turns to choose new ranges and new numbers and operations. The winner is the first player to reach 10 points.

## Rounding appropriately

Often a question does not require an exact answer, particularly when that answer is a decimal with a large or infinite number of decimal places. A question will not always tell you how to round your answer, however, and you need to think carefully about how to do this. The guidelines below should help. Remember, these are not rules but sensible suggestions, for use only if the question does not tell you how to round.

1. It is a good idea to write lots of digits, 5 or 6 significant figures or more, in your working before rounding your final answer. If you over-round or make an error in rounding your answer, your working will still show the correct numbers and you can still gain marks.
2. Do not round part way through a calculation. Where there are two parts to a question, use your unrounded answer to the first part in your calculation for the second part. Only round your final answers.
3. If an answer is a short, terminating decimal do not round it.
4. If an answer is a recurring decimal you do not need to round it. You can write it with recurring dots to show an exact number.
5. Answers involving money, for example dollars, should be given to 2 d.p.. This means they will be to the nearest cent (for example).
6. If you are doing an exam, check the front of the exam paper. It may contain instructions about the

number of decimal places or significant figures to which you should write your answers. If you are preparing for an exam with rules for rounding, get in the habit of using these rules in all your maths lessons.

**7** Use the same accuracy as the question, or slightly better. For example, if all numbers in the question are given to 1 d.p. then answers to 1 or 2 d.p. are sensible.

In Exercise 7G use these guidelines to decide whether to round your answers, and, if so, how to round them. In each case write which guideline you have used.

### Exercise 7G

**You may use a calculator for this exercise.**

**1** Change these fractions to decimals.

**a** $\frac{3}{20}$ **b** $\frac{4}{11}$

**c** $\frac{3}{7}$ **d** $\frac{3}{16}$

**2** Work out:

**a** 2.74 + 3.681 **b** 1.748 × 3.26

**c** 7.4 ÷ 1.7

**3** Work out the missing angle, $x$.

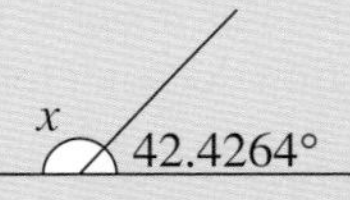


**4 a** What is the area of a rectangle with length 2.73 cm and width 4.68 cm?

**b** What is the width of a rectangle with area 14 cm$^2$ and length 7.6 cm?

**5** Cloth costs $13.17 per metre. What is the cost of

**a** 7 m **b** 2.3 m

**c** 0.41 m?

**6 a** What area of metal is required to make the road sign shown in the diagram?

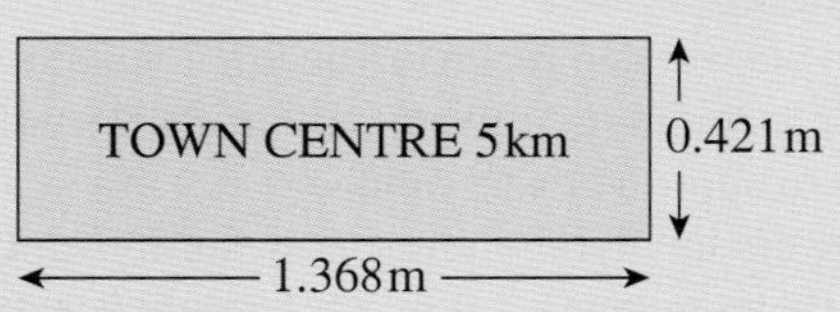


**b** If the metal for the road sign costs $15.19 per square metre, what is the cost of the metal required?

## 7.3 Multiplying and dividing decimals

### Multiplying and dividing by numbers between 0 and 1

It is a common misconception that multiplying a number by something always makes it bigger. Look at these examples where 4 is multiplied by different numbers:

$4 \times 7 = 28$ the answer is bigger than 4
$4 \times 2.5 = 10$ the answer is bigger than 4
$4 \times \frac{1}{2} = 2$ the answer is smaller than 4
$4 \times 10 = 40$ the answer is bigger than 4
$4 \times 0.1 = 0.4$ the answer is smaller than 4

Multiplying by a number between 0 and 1 decreases the value.

Multiplying by 1 leaves the value unchanged.

Multiplying by a number bigger than 1 increases the value.

Similarly it is a common misconception that when you divide a number by something it becomes smaller. Look at these examples where 24 is divided by different numbers:

$24 \div 12 = 2$ the answer is smaller than 24
$24 \div 6 = 4$ the answer is smaller than 24
$24 \div \frac{1}{2} = 24 \times 2 = 48$ the answer is bigger than 24
$24 \div 3 = 8$ the answer is smaller than 24
$24 \div 0.1 = 24 \times 10 = 240$ the answer is bigger than 24

Dividing by a number between 0 and 1 increases the value.

Dividing by 1 leaves the value unchanged.

Dividing by a number bigger than 1 decreases the value.

### Exercise 7H

**1** Which questions will have answers less than 30?

| | | |
|---|---|---|
| 30 × 1.8 | 30 ÷ 20 | 30 × 0.04 |
| 30 ÷ 0.8 | 30 × 0.5 | $30 \times \frac{1}{3}$ |
| 30 × 2.6 | $30 \div 2\frac{2}{5}$ | 30 × 5 |
| 30 ÷ 30 | 30 ÷ 0.1 | $30 \div \frac{4}{7}$ |
| 30 ÷ 2.7 | $30 \div \frac{5}{2}$ | 30 × 2.4 |

**2** Which questions will have answers greater than 50?

| | | |
|---|---|---|
| 50 ÷ 0.5 | 50 × 0.5 | 50 × 1.7 |
| $50 \div \frac{1}{8}$ | 50 ÷ 20 | $50 \times \frac{2}{3}$ |
| 50 ÷ 100 | 50 × 0.15 | 50 ÷ 0.29 |
| $50 \times 4\frac{3}{8}$ | 50 ÷ 3.8 | $50 \times \frac{5}{3}$ |
| 50 ÷ 0.23 | 50 × 0.4 | 50 ÷ 0.8 |

**3** Which of these answers are clearly incorrect?

**a** $47 \times 0.83 = 39.01$
**b** $59 \div 2.5 = 23.6$
**c** $3.8 \times 1.6 = 2.28$
**d** $17 \div 0.32 = 5.44$

## Multiplying decimals

Multiplying decimals is the same as multiplying whole numbers. You need to be careful where you put the decimal point.

One way to do this is to make an estimate of the answer.

**EXAMPLE 7**

Work out $13.4 \times 6.2$

Estimate $13.4 \times 6.2 \approx 13 \times 6 = 78$

Multiplying the numbers without decimal points:

$$\begin{array}{r} 134 \\ \times\ 62 \\ \hline 268 \\ 8040 \\ \hline 8308 \end{array}$$

The estimate shows that the answer to $13.4 \times 6.2$ is around 80, so the decimal point must be placed to give 83.08.
That is, $13.4 \times 6.2 = 83.08$

Another way to multiply decimals is simply to count the number of digits after the decimal point.

**EXAMPLE 8**

Work out $1.73 \times 1.9$

$173 \times 19 = 3287$

so

| | |
|---|---|
| $1.73$ | 2 decimal places |
| $\times\ 1.9$ | 1 decimal place |
| $3.287$ | 3 decimal places |

There are 2 digits after the decimal point in 1.73 and 1 digit after the decimal point in 1.9. Hence there should be $2 + 1 = 3$ digits after the decimal point in the answer.

To show why Example 8 works, consider these equivalent calculations:

$$1.73 \times 1.9 = \frac{173}{100} \div \frac{19}{10} = \frac{173 \times 19}{100 \times 10} = \frac{3287}{1000}$$

$$1.73 \times 1.9 = 173 \div 100 \times 19 \div 10$$
$$= 173 \times 19 \div 100 \div 10 = 3287 \div 1000$$

### Exercise 7I

**1** Given that $56 \times 23 = 1288$, work out:

**a** $5.6 \times 2.3$ **b** $56 \times 2.3$
**c** $0.56 \times 2.3$ **d** $0.56 \times 23$
**e** $0.056 \times 0.23$ **f** $5.6 \times 0.23$
**g** $0.56 \times 0.023$ **h** $560 \times 0.0023$

**2** Calculate:

**a** $0.3 \times 6$ **b** $0.13 \times 6$
**c** $1.41 \times 5$ **d** $0.38 \times 7$
**e** $6.18 \times 9$ **f** $7.3 \times 12$
**g** $13 \times 0.13$ **h** $29 \times 0.814$

**3** What is the area of a rectangle with

**a** length 2.3 m, width 6.8 m
**b** length 19.3 cm, width 8.3 cm
**c** length 14.32 m, width 6.14 m?

**4** Cloth costs $43.15 per metre. What is the cost of

**a** 12 metres
**b** 6.1 metres
**c** 0.45 metres?

**5** Patsy walks 4.2 km in one hour. How far does she travel in

**a** 5 hours
**b** 1.3 hours
**c** 0.6 hours?

**6** Raymond worked out the multiplication $1.6 \times 0.4$ as follows:

$$1.6 \times 0.4 = 1\frac{6}{10} \times \frac{4}{10}$$
$$= \frac{16}{10} \times \frac{4}{10}$$
$$= \frac{64}{100} = 0.64$$

Work out these multiplication using Raymond's method.

**a** $1.5 \times 0.5$ **b** $6.2 \times 0.8$
**c** $0.8 \times 0.3$ **d** $0.84 \times 0.6$
**e** $1.3 \times 2.4$ **f** $0.94 \times 3.1$
**g** $6.45 \times 2.14$ **h** $17.2 \times 0.68$

**7** **a** Do the multiplications in Question **2** using Raymond's method from Question **7**.
**b** Which of the three methods shown in Examples 7 and 8 and Question **7** do you prefer? Explain.
**c** Which method would you use to explain the multiplication of decimals to a new student? Why?

## Dividing decimals

To divide decimals by decimals, it is usually easier to change the division so that the divisor is an integer (a whole number).

**EXAMPLE 9**

Calculate:

**a** $8.4 \div 0.4$ **b** $3 \div 0.08$

**a** $8.4 \div 0.4 = \frac{8.4}{0.4}$

$= \frac{8.4 \times 10}{0.4 \times 10}$

$= \frac{84}{4} = 21$

**b** $3 \div 0.08 = \frac{3}{0.08}$

$= \frac{3 \times 100}{0.08 \times 100}$

$= \frac{300}{8} = 37.5$

### Exercise 7J

**1** Calculate:

**a** $8 \div 0.2$ **b** $6 \div 0.3$
**c** $4.2 \div 2$ **d** $17.4 \div 5$
**e** $12 \div 0.4$ **f** $12 \div 0.04$
**g** $9 \div 0.3$ **h** $16 \div 0.08$

**2** Calculate:

**a** $8.5 \div 0.5$ **b** $6.4 \div 0.4$
**c** $6.4 \div 0.04$ **d** $0.64 \div 0.4$
**e** $13.4 \div 0.5$ **f** $23.68 \div 0.4$
**g** $5.472 \div 1.2$ **h** $12.88 \div 0.23$

**3**

A ball of string contains 21.6 m of string. How many 0.16 m lengths of string can be cut from it?

**4** How many 0.3-litre bottles of juice can be filled from an urn containing 45.9 litres?

**5** The area of a rectangular field is $100.8\text{m}^2$. What is the width of the field if its length is

**a** 14.4 m **b** 11.2 m
**c** 6.3 m **d** 0.35 m?

**6** What is the cost of beef per kilogram, if 6.2 kg of beef sells for \$34.10?

**7** Fatima works out the division $6.24 \div 0.4$ as follows:

$6.24 \div 0.4 = 6\frac{24}{100} \div \frac{4}{10}$

$= \frac{624}{100} \div \frac{4}{10}$

$= \frac{624}{100} \times \frac{10}{4}$

$= \frac{156}{10}$

$= 15\frac{6}{10} = 15.6$

Work out these divisions using Fatima's method.

**a** $8 \div 0.2$ **b** $3.4 \div 0.2$
**c** $8.5 \div 1.7$ **d** $4.5 \div 0.15$
**e** $1.28 \div 3.2$ **f** $0.576 \div 1.2$
**g** $3.61 \div 0.019$ **h** $14.4 \div 0.24$

### Puzzle

Copy and complete this cross-number puzzle.

| | | | | |
|---|---|---|---|---|
| A | | B | ■ | C |
| | ■ | D | E | |
| F | | ■ | G | |
| | ■ | H | | |
| I | | | | ■ |

*Across*

A $0.8 \times 405$

D $542.88 \div 0.58$

F $3.8 + 7.2$

G $403.7 - 390.7$

H $\frac{19.2 \times 1.2}{0.12}$

I $378.856 \div 0.058$

*Down*

A $5.8 \times 5720$

B $0.7^2 \times \frac{5}{0.05}$

C $318 \times 24$

E $1340.64 \div 0.42$

H $16.9 \div 1.3$

# Consolidation

### Example 1

Write

**a** 4.378 to 1 decimal place
$= 4.4$ (1 d.p.)

**b** 7.1436 to 2 decimal places
$= 7.14$ (2 d.p.)

### Example 2

**a** Write to 1 significant figure:

**i** 20 871
$= 20\,000$ (1 s.f.)

**ii** 4175
$= 4000$ (1 s.f.)

**iii** 0.01652
$= 0.02$ (1 s.f.)

**b** Write to 2 significant figures:

**i** 20 871
$= 21\,000$ (2 s.f.)

**ii** 18 381
$= 18\,000$ (2 s.f.)

**iii** 0.01652
$= 0.017$ (2 s.f.)

**iv** 0.0003841
$= 0.00038$ (2 s.f.)

### Example 3

Work out:

**a** $0.68 \times 0.045$

An estimate is $1 \times 0.04 = 0.04$

$$\begin{array}{r} 0.68 \\ \times\, 0.045 \\ \hline 340 \\ 2720 \\ \hline 0.03060 \end{array}$$

So $0.68 \times 0.045 = 0.0306$

**b** $0.035 \div 0.5$

$$\frac{0.035}{0.5} = \frac{0.035 \times 10}{0.5 \times 10} = \frac{0.35}{5} = 0.07$$

### Example 4

**(Note: this is extension work.)**

Write in standard form:

**a** 63 100
$= 6.31 \times 10000 = 6.31 \times 10^4$

**b** 0.00038
$= 3.8 \div 10000 = 3.8 \times 10^{-4}$

### Example 5

Without calculating, which of these answers are incorrect?

**a** 21.2 3 0.43 5 9.116 **b** 16.3 4 5.5 5 89.65
**c** 2.4 3 3.9 5 9.36 **d** $28 \div 0.76 = 25.45$

---

**a** $21.2 \times 0.43 = 9.116$ could be correct as multiplying by a number between 0 and 1 decreases the value and 9.116 is smaller than 21.2.

**b** $16.3 \div 5.5 = 89.65$ is clearly incorrect as dividing by a number bigger than 1 decreases the value but 89.65 is bigger than 16.3.

**c** $2.4 \times 3.9 = 9.36$ could be correct as multiplying by a number bigger than 1 increases the value and 9.36 is bigger than 2.4.

**d** $28 \div 0.76 = 25.45$ is clearly incorrect as dividing by a number between 0 and 1 increases the value but 25.45 is smaller than 28.

### Exercise 7

**1** Round

**a** 0.435 to
**i** 1 d.p. **ii** 2 d.p.

**b** 17.451 to
**i** 1 d.p. **ii** 2 d.p.

**c** 3.89941 to
**i** 1 d.p. **ii** 2 d.p. **iii** 3 d.p.

**d** 78.000849 to
**i** 1 d.p. **ii** 4 d.p.

**2** Write correct to 1 significant figure:

**a** 850 **b** 9317 **c** 41 290
**d** 43 651 **e** 0.763 **f** 0.072
**g** 0.0065 **h** 8.93 **i** 0.000398

**3** Which questions will have answers less than 15?

$15 \times 1.4$ $15 \div 12$ $15 \times 0.32$
$15 \div 0.8$ $15 \times 5$ $15 \times \frac{1}{7}$
$15 \times 1.8$ $15 \div 1\frac{1}{9}$ $15 \div \frac{2}{3}$
$15 \div 30$

**4** Given that $23 \times 56 = 1288$, work out:

**a** $23 \times 5.6$ **b** $2.3 \times 56$
**c** $2.3 \times 5.6$ **d** $2.3 \times 0.56$
**e** $0.23 \times 56$ **f** $0.023 \times 0.56$
**g** $0.23 \times 0.056$ **h** $0.023 \times 0.056$

**5** Write correct to 2 significant figures:

**a** 537 000 **b** 41 562 **c** 0.037 206
**d** 0.000 215 **e** 11.931 **f** 1.076
**g** 0.004 76 **h** 1432 **i** 0.3061

**6** Calculate:

**a** $0.3 \times 7$ **b** $1.2 \times 4$ **c** $6.3 \times 8$
**d** $0.21 \times 3$ **e** $2.76 \times 8$ **f** $17.26 \times 9$

**7** Write correct to three significant figure:

**a** 18.403 **b** 157 683 **c** 10 467
**d** 0.030 084 **e** 15.092 **f** 113.387
**g** 21 547 **h** 13.406 **i** 21.004

**8** Find:

**a** $0.8 \div 4$ **b** $1.6 \div 2$ **c** $2.8 \div 7$
**d** $0.64 \div 4$ **e** $0.08 \div 4$ **f** $2.184 \div 7$
**g** $7 \div 0.1$ **h** $8 \div 0.01$ **i** $23 \div 0.001$
**j** $8 \div 0.2$ **k** $6 \div 0.03$ **l** $24 \div 0.04$

**9** Which questions will have answers greater than 34?

$34 \div 0.5$ $34 \times 0.3$ $34 \times 2.7$
$34 \div \frac{7}{8}$ $34 \div 1.42$ $34 \times \frac{3}{5}$
$34 \div 44$ $34 \times 0.15$ $34 \div 0.2$
$34 \times 1\frac{3}{4}$

**10** Here are some numbers written to 3 significant figures. Some of them are wrong. Pick out the wrong ones and write them correctly.

**a** 539 010 ⟶ 539 000
**b** 0.005 706 ⟶ 0.005 70
**c** 0.050 621 ⟶ 0.0506
**d** 508 716 ⟶ 508 700
**e** 15 480 000 ⟶ 15 500 000
**f** 0.010 06 ⟶ 0.0100

**11** Calculate:

**a** $0.3 \times 0.4$ **b** $0.6 \times 1.2$
**c** $3.1 \times 0.1$ **d** $1.21 \times 3.1$
**e** $6.5 \times 2.65$ **f** $4.2 \times 3.2$
**g** $8.04 \times 0.04$ **h** $7.3 \times 0.06$
**i** $3.62 \times 1.07$

**12** Estimate the answers to these calculations.

**a** $7.41 \times 3.8$ **b** $14.79 \div 4.56$
**c** $\frac{19.6 + 10.43}{5.27}$ **d** $52.1 - 5.3^2$

**13** Find:

**a** $0.4 \div 0.1$ **b** $6.4 \div 0.8$
**c** $8.4 \div 0.7$ **d** $0.4 \div 0.01$
**e** $5.6 \div 0.07$ **f** $8.1 \div 0.03$
**g** $14.4 \div 3.6$ **h** $24.3 \div 2.7$
**i** $0.72 \div 1.8$ **j** $0.76 \div 0.5$
**k** $0.4 \div 0.5$ **l** $1.3 \div 0.04$

**14 a** How many lengths of 0.2 m can be cut from 1.2 m?
**b** How many 0.010-ℓ injections of antibiotic are contained in a 0.1-ℓ bottle?
**c** How many 0.125-kg measures of flour are provided by a 50-kg sack?

**(Note: Questions 15–20 are extension work.)**

**15** Write in standard form:

**a** 4000 **b** 20 000 **c** 70 000 000

**16** Write each number in full.

**a** $3 \times 10^7$ **b** $4 \times 10^6$ **c** $7 \times 10^9$

**17** Write in standard form:

**a** 40 000 **b** 2170 **c** 3401
**d** 90 800 **e** 11 200 **f** 25 000

**18** Write each number in full.

**a** $1.5 \times 10^3$ **b** $6.32 \times 10^2$
**c** $1.628 \times 10^4$ **d** $3.94 \times 10^6$

**19** Write each number in full.

**a** $7 \times 10^{-1}$ **b** $3 \times 10^{-2}$ **c** $4 \times 10^{-3}$
**d** $9 \times 10^{-4}$ **e** $6 \times 10^{-5}$ **f** $5 \times 10^{-6}$

**20** Write in standard form:

**a** 0.16 **b** 0.37 **c** 0.022
**d** 0.094 **e** 0.0064 **f** 0.0037

## Summary

### You should know ...

**1** How to round numbers to a given number of decimal places or significant figures.

*For example:*

0.0384 = 0.04 (2 d.p.)
2348 = 2000 (1 s.f.)
0.005796 = 0.0058 (2 s.f.)
46 891.47 = 46 900 (3 s.f.)
46 891.47 = 47 000 (2 s.f.)
46 891.47 = 50 000 (1 s.f.)

### Check out

**1 a** Write correct to 2 d.p.:
**i** 0.8723 **ii** 0.9452

**b** Write correct to 3 s.f.
**i** 8712 **ii** 0.004 864

**c** Write correct to 2 s.f.
**i** 48 487 **ii** 0.0195

**d** Write correct to 1 s.f.
**i** 58 **ii** 0.000783

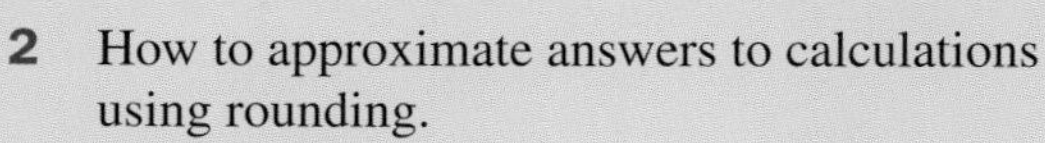

**2** How to approximate answers to calculations using rounding.
*For example:*

$18.29 \div 5.56 \approx 18 \div 6$
so $18.29 \div 5.56 \approx 3$

**2** Estimate:

**a** $6.321 \times 7.8$
**b** $19.79 \div 4.12$
**c** $\dfrac{39.7 - 15.18}{4.87}$
**d** $8.89^2 - 80.657$

---

**3** The rules about multiplying a starting number that is positive:
Multiplying by a number between 0 and 1 decreases the value.
Multiplying by a number bigger than 1 increases the value.
Multiplying by 1 leaves the number unchanged.
Dividing by a number between 0 and 1 increases the value.
Dividing by a number bigger than 1 decreases the value.
Dividing by 1 leaves the number unchanged.

**3** Without calculating, which of these answers are incorrect?

**a** $21.4 \div 2.5 = 8.56$
**b** $3.8 \times 0.77 = 4.926$
**c** $7.81 \times 4.8 = 3.7488$
**d** $9 \div 0.02 = 4.5$

---

**4** How to multiply and divide decimals by decimals.
*For example:*

$6.12 \times 3.4$
$612 \times 34 = 20\,808$ by long multiplication.

| So | | |
|---|---|---|
| | $6.\underline{1}\underline{2}$ | 2 decimal places |
| | $\times\ 3.\underline{4}$ | 1 decimal place |
| | $20.\underline{8}\underline{0}\underline{8}$ | 3 decimal places |

$6.88 \div 1.6$

$\dfrac{6.88}{1.6} = \dfrac{68.8}{16}$ multiply numerator and denominator by 10

$= \dfrac{34.4}{8} = \dfrac{17.2}{4} = \dfrac{8.6}{2}$ cancelling

$2\overline{)8.6}$ = 4.3

The answer is 4.3

**4** Work out:

**a** $2.4 \times 0.37$
**b** $7.21 \times 3.4$
**c** $1.44 \div 0.03$
**d** $0.696 \div 0.12$

---

**5** How to write a number in standard form.
*For example:*

$629\,300 = 6.293 \times 100\,000 = 6.293 \times 10^5$
$0.0049 = 4.9 \div 1000 = 4.9 \times 10^{-3}$

**(Note: this is extension work.)**

**5** Write in standard form:

**a** 8000 **b** 7236
**c** 0.038 **d** 183 000
**e** 0.000062 **f** 0.03

**(Note: this is extension work.)**

# Equations and inequalities

## Objectives

- Construct and solve linear equations with integer coefficients (with and without brackets, negative signs anywhere in the equation, a positive or negative solution); solve a number problem by constructing and solving a linear equation.
- Understand and use inequality signs ($<$, $>$, $\leqslant$, $\geqslant$); construct and solve linear inequalities in one variable; represent the solution set on a number line.
- Solve a simple pair of simultaneous linear equations by eliminating one variable.
- Use systematic trial and improvement methods to find approximate solutions of equations such as $x^2 + 2x = 20$ (1, 2 and 7).

## What's the point?

Inequalities are used in everyday life. For example, a thermostat in a car might cause a valve to open when the engine gets above a certain temperature (95°C, say), allowing water to circulate and cool it down. If the engine's temperature falls below a certain temperature (say 85°C), the thermostat causes the valve to close, reducing the water circulation and raising the engine temperature. We can use the inequalities $T > 95$°C and $T < 85$°C to help show this information.

## Before you start

### You should know ...

**1** What these symbols mean:

$>$ means 'is greater than'
$<$ means 'is less than'
$\geqslant$ means 'is greater than or equal to'
$\leqslant$ means 'is less than or equal to'

### Check in

**1** What are the missing inequality symbols?

**a** A bus can hold at most 50 passengers.
Passengers □ 50

**b** To be a heavyweight boxer, you should have a mass of more than 90 kg.
Mass □ 90

**c** To enter a competition you must be under 12 years old.
Age □ 12

**d** To go through to the next round in a quiz you need at least 50 points.
Points □ 50

**e** A parcel costs $15 to post when its mass is more than 5 kg but not over 10 kg.
5 □ mass □ 10

## 8.1 Solving linear equations

An **equation** is a statement showing the equality of two expressions.

*For example:*

$$4x + 3 = 4$$

or $3x - 7 = 6 - 2x$

are both equations.

To solve linear equations you can use the **balance** method. This means that what you do to one side of an equation you must do to the other to maintain equality.

*For example:*
In the equation

$$x + 3 = 8$$

adding 2 to both sides

$$x + 3 + 2 = 8 + 2$$

maintains equality.

You can use this powerful idea to solve all linear equations.

**EXAMPLE 1**

Solve:

**a** $x - 7 = 8$

**b** $3x + 2 = 14$

**a**

$$x - 7 = 8$$

add 7 to both sides] $x - 7 + 7 = 8 + 7$

$$x = 15$$

**b**

$$3x + 2 = 14$$

subtract 2 from both sides] $3x + 2 - 2 = 14 - 2$

$$3x = 12$$

divide both sides by 3] $\frac{3x}{3} = \frac{12}{3}$

$$x = 4$$

Notice that you add or subtract numbers to simplify the expressions on each side of the equation. The aim is to put the unknown terms (letters) on one side of the equation and the constant terms (numbers) on the other side.

Notice how in Example 1, part **b**, you do the inverse operations to get the $x$ on its own. The inverse of $+2$ is $-2$ and the inverse of $\times 3$ (shown by $3x$) is $\div 3$.

You should remember that in algebra when a term is written as a fraction this means dividing. For example $\frac{x}{7}$ means $x \div 7$. You should also remember that the inverse of, for example, $\div 7$ is $\times 7$.

Sometimes the unknown term (the letter) is not on the left-hand side of the equation. This makes no difference to the method you use to solve it; the aim is still to get the letter on its own.

**EXAMPLE 2**

Solve:

**a** $\frac{x}{4} + 3 = 8$

**b** $22 = 3p - 8$

**a** $\frac{x}{4} + 3 = 8$

subtract 3 from both sides] $\frac{x}{4} + 3 - 3 = 8 - 3$

$\frac{x}{4} = 5$

multiply both sides by 4] $\frac{x}{4} \times 4 = 5 \times 4$

$x = 20$

**b** $22 = 3p - 8$

add 8 to both sides] $22 + 8 = 3p - 8 + 8$

$30 = 3p$

divide both sides by 3] $\frac{30}{3} = \frac{3p}{3}$

$10 = p$

## Exercise 8A

**1** Solve:

**a** $x - 3 = 5$ **b** $m + 3 = 5$

**c** $x + 4 = 11$ **d** $16 = x - 15$

**e** $0 = x + 1$ **f** $y - 8 = 6$

**2** Solve:

**a** $3x = 9$ **b** $5t = 15$

**c** $4x = 12$ **d** $36 = 6x$

**e** $21 = 7x$ **f** $11y = 33$

**3** Solve:

**a** $\frac{x}{3} = 2$ **b** $\frac{p}{4} = 5$ **c** $6 = \frac{x}{4}$

**d** $\frac{x}{9} = {}^{-}3$ **e** $\frac{r}{16} = 1$ **f** ${}^{-}2 = \frac{x}{11}$

**4** Solve:

**a** $3x + 1 = 4$ **b** $1 = 2x - 1$

**c** $3p - 3 = 12$ **d** $4x + 5 = 25$

**e** $47 = 6x - 7$ **f** $3n - 8 = 13$

**5** Solve:

**a** $\frac{d}{2} + 3 = 4$ **b** $5 = \frac{x}{3} + 4$

**c** $\frac{x}{4} - 1 = 1$ **d** $\frac{x}{7} + 2 = 5$

**e** $\frac{n}{5} - 4 = 0$ **f** $5 = \frac{x}{6} - 7$

You can solve more complex equations using the same idea. In each case try to put the like terms together.

**EXAMPLE 3**

Solve:

**a** $5x - 2 = 2x - 11$

**b** $7 - 3x = 3(2x - 5)$

**a** $5x - 2 = 2x - 11$

First, collect the $x$ terms on the left-hand side of the equation by subtracting $2x$ from both sides:

$-2x$] $5x - 2 - 2x = 2x - 11 - 2x$

$3x - 2 = {}^{-}11$

Then, collect the constant terms on the right-hand side by adding 2 to both sides:

$+2$] $3x - 2 + 2 = {}^{-}11 + 2$

$3x = {}^{-}9$

$\div 3$] $\frac{3x}{3} = \frac{{}^{-}9}{3}$

$x = {}^{-}3$

**b** $7 - 3x = 3(2x - 5)$

First, multiply out the brackets:

$7 - 3x = 6x - 15$

To collect $x$-terms, add $3x$ to both sides:

$+3x$] $7 - 3x + 3x = 6x - 15 + 3x$

$7 = 9x - 15$

To collect constant terms, add 15 to both sides:

$+15$] $7 + 15 = 9x - 15 + 15$

$22 = 9x$

$\div 9$] $\frac{22}{9} = \frac{9x}{9}$

$x = 2\frac{4}{9}$

## Exercise 8B

**1** Solve:

**a** $3x + 2 = 2x + 6$

**b** $5y - 2 = y + 6$

**c** $4x + 3 = 3x + 7$

**d** $2p - 3 = p - 5$

**e** $3x - 4 = 2x + 5$

**f** $6x - 9 = 3x + 3$

**2** Solve these equations:

**a** $6 - x = x$
**b** $6 - 2x = x$
**c** $4 - 3x = 2 + x$
**d** $4t - 3 = 6 + 3t$
**e** $6x - 13 = 7 - 4x$
**f** $13 - 5n = {}^{-}11 - 3n$

**3** Solve:

**a** $3(x + 1) = 4x - 2$
**b** $2(x - 2) = 3x - 2$
**c** $7(p - 1) = 2 - 2p$
**d** $3(2x - 1) = 4x - 5$
**e** $2(x + 5) = 3(x + 1)$
**f** $2(y - 2) = 5(2 - y)$

**4** Solve:

**a** $3(x + 1) + 3 = 2x - 4$
**b** $4(2y - 1) - 3 = 3 - 2y$
**c** $7(1 - x) + x = 2 - x$
**d** $18x + 3(1 - 4x) = 3(x + 2)$
**e** $3 + 4(2n - 1) = 6n - 2$
**f** $4 - 2(x + 2) = 3(x - 5)$

**5** 3 is added to a number and the result is multiplied by 4 to give 32.

**a** Denote the number by $x$ and write down an equation.
**b** Solve the equation to find the value of $x$.

**6** A triangle has perimeter $6x$ cm. The lengths of the sides of the triangle are $x$, $x + 2$, $x + 4$ cm.

**a** Write down an equation relating the perimeter to the lengths of the sides.
**b** Solve the equation and hence find the length of each side of the triangle.

**7**

$4x$

$x + 6$

The perimeter of the rectangle above is 48 cm.

**a** Form an equation for the perimeter of the rectangle.
**b** Solve the equation to find the length and width of the rectangle.

**8** The sum of two consecutive even numbers is 226. What are the numbers?

In Exercise 8B, Question **8** is more easily solved by forming an equation first. Did you do this? The next example and exercise will show you how to construct equations to solve.

Use the Equation Calculator at

www.algebrahelp.com

to check your solutions to the equations in Exercise 8B.

How successful were you?

Need more assistance with equations? Look at the sites

www.algebrahelp.com

or

www.coolmath.com

Both sites provide complete courses on solving simple linear equations.

Both sites have lots of practice questions.

Try them!

Remember to check your answers.

## 8.2 Constructing and solving equations

Many problems can be solved by first constructing equations.

**EXAMPLE 4**

The sum of three consecutive numbers is 60. What are the numbers?

Let $x$ be the first number.

So $x + 1$ is the next consecutive number and $x + 2$ is the next consecutive number.

The numbers add up to 60, so:

$$x + (x + 1) + (x + 2) = 60$$

$$3x + 3 = 60$$

$-3]$ $$3x = 57$$

$\div 3]$ $$x = \frac{57}{3} = 19$$

If $x = 19$, then $x + 1 = 20$, and $x + 2 = 21$, so the numbers are 19, 20 and 21

## Exercise 8C

**1** The sum of the angles on a straight line is 180. Write an equation for each diagram, then find the value of the unknown letter.

**a** 140° $x$°

**b** 70° $y$° 60°

**2** **a** Write down an equation in terms of $w$ for the area of this rectangle.

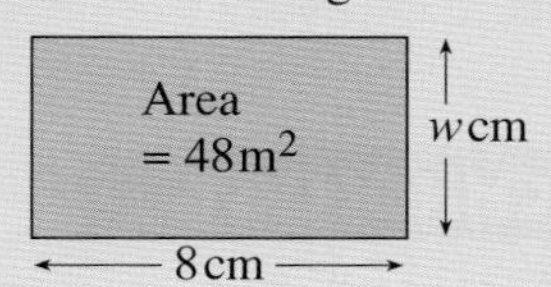


**b** Solve the equation to find $w$ and write down the width of the rectangle.

**3** **a** Write down an expression in terms of $x$ for the perimeter of this isosceles triangle.

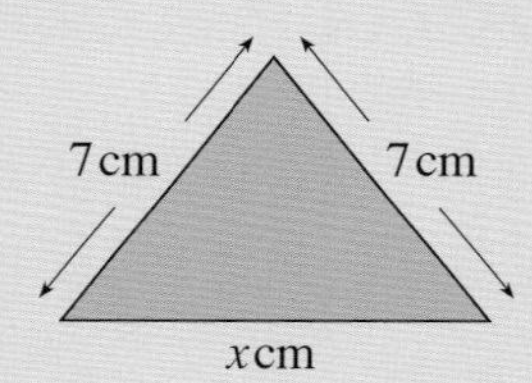


**b** If the perimeter is 20 cm, find the value of $x$.

**4** **a** The perimeter of this isosceles triangle is 32 cm. Write down an equation in terms of $y$ for the perimeter.

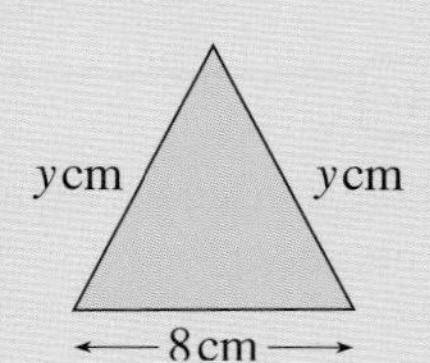


**b** Solve the equation to find $y$ and write down the lengths of the two equal sides of the triangle.

**5** In an exam Curtis scored $x$ marks and Khalid scored 20 marks more than Curtis.

**a** What was Khalid's score?

**b** What was the total score of Curtis and Khalid?

**c** If their total score was 130 what was Curtis's actual score?

**6** **a** Write down an expression for the perimeter of this trapezium.

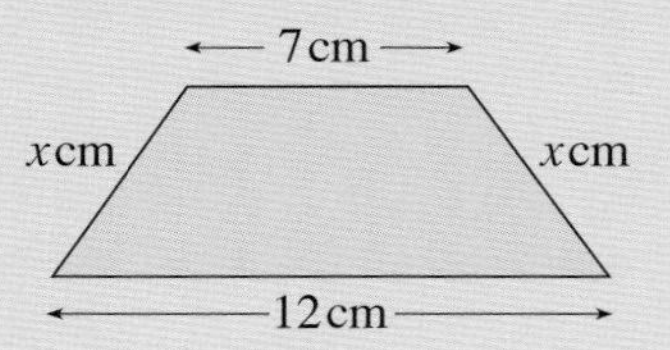


**b** If the perimeter is 30 cm, find the value of $x$.

**7** **a** The width of the rectangle shown is $w$ cm.

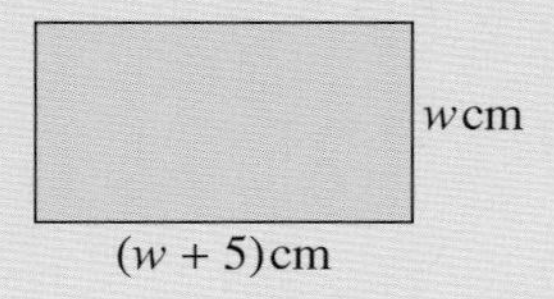


If its length is 5 cm longer than its width write down an expression for the perimeter of the rectangle.

**b** If the perimeter is 50 cm find the dimensions of the rectangle.

**8** One side of a rectangle is 4 cm longer than the other. The perimeter of the rectangle is 28 cm. Find the length of each side.

**9** Peter is three years older than his sister Paula. If their combined age is 21 find their actual ages.

**10** Two consecutive whole numbers add up to 31. What are they?

**11** Three consecutive whole numbers add up to 72. What are they?

**12** The sum of three consecutive odd numbers is 69. Find them.

**13** An adult's ticket to a cricket match costs twice as much as a child's ticket. The cost for Mr and Mrs Brown and their five children is $3.15.

**a** If a child's ticket costs $c$ cents, write an equation to show this information.

**b** Solve your equation to find the cost of each ticket.

**14** Ram bought 5 tickets for the cinema. He paid \$20 and got 75 cents change. Let $t$ cents stand for the cost of a ticket and write an equation to show this information.

Solve your equation to find $t$.

**15** A number is multiplied by 3 and then 7 is added. The result is doubled, giving the answer 80. Using $n$ to represent the original number, write an equation to show this information.

Solve the equation to find $n$.

**16** The area of the shaded L-shape is equal to the area of the white square. What is the missing side length?

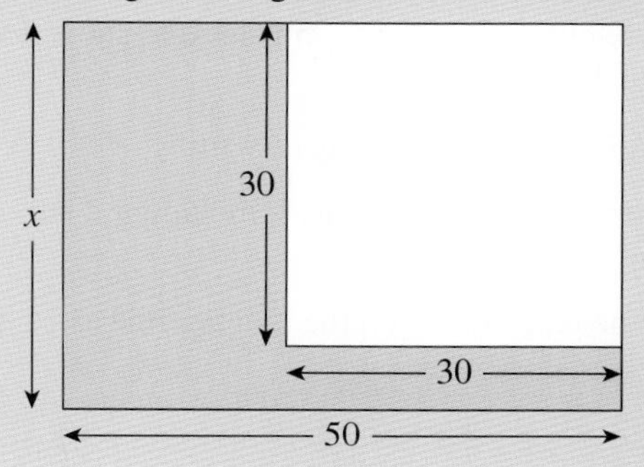


## 8.3 Linear inequalities

Look at the diagram. The red labels read '$x$' and the yellow labels read '1 kg'.

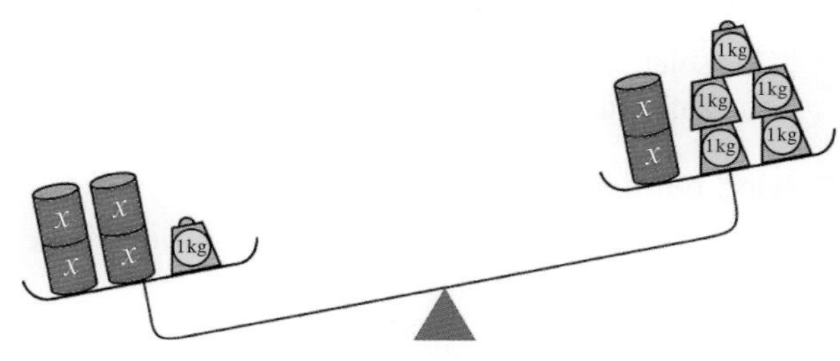


The left-hand side is heavier than the right. The information can be shown by the inequality:

$$4x + 1 > 2x + 5$$

Make a copy of the diagram in pencil. Use your copy and a rubber to explain why:

**a** $2x + 1 > 5$

**b** $2x > 4$

**c** $x > 2$

Example 5 shows how an inequality can be set out and solved in a similar way to an equation.

**EXAMPLE 5**

Solve $\quad 4x - 1 > 2x + 5$

Subtract $2x$ from each side:
$-2x]\quad 2x - 1 > 5$

Add 1 to each side:
$+1]\quad 2x > 6$

Divide each side by 2:
$\div 2]\quad x > 3$

That is, $x$ is greater than 3

Example 5 shows adding or subtracting terms or numbers on both sides of an inequality, and dividing by positive numbers on both sides of an inequality. Multiplying or dividing by a negative number on both sides of an inequality is more problematic.

Consider the inequality $5 > 3$

If you multiply (or divide) both sides by $^-1$ you get $^-5 > {}^-3$, which is not true. In fact, $^-5 < {}^-3$

This brings us to an important rule:

- If you multiply or divide an inequality by a negative number you reverse the inequality sign.

**EXAMPLE 6**

Solve:

**a** $^-x \leqslant 4$ **b** $5 - 3x > {}^-1$

**a** $\quad ^-x \leqslant 4$

$\times {}^-1$, reverse inequality] $\quad x \geqslant {}^-4$

**b** $\quad 5 - 3x > {}^-1$

$-5]\quad {}^-3x > {}^-6$

$\div {}^-3$, reverse inequality] $\quad x < 2$

Example 6, part **b**, could have been done a different way. It requires an extra line of working but avoids dividing by a negative. Use this method if you think it is likely you will forget to reverse the inequality signs:

$5 - 3x > {}^-1$
$(+3x)\quad 5 > 3x - 1$
$(+1)\quad 6 > 3x$
$(\div 3)\quad 2 > x$

This is the same answer as $x < 2$. That is, in both the inequalities $x < 2$ and $2 > x$ the narrow end of the inequality symbol is at the $x$ and the wide end is at the 2, showing that 2 is the larger value.

## Exercise 8D

**1** Use the method shown in Example 6 to solve each inequality.

**a** $5x + 2 > 17$
**b** $7x + 6 > 41$
**c** $4x + 3 \leqslant 15$
**d** $6x + 1 < 19$
**e** $4x + 3 > x + 9$
**f** $5x + 2 \geqslant x + 8$
**g** $x + 7 < 3x + 4$
**h** $7x + 4 < 2x + 19$

**2** Solve:

**a** $3x - 7 < 8$
**b** $2x - 3 \geqslant 15$
**c** $8x - 7 > 3x + 8$
**d** $4x + 10 < 7x - 11$
**e** $4x + 3 \geqslant 18 - x$
**f** $2x + 1 \leqslant 7 - x$
**g** $5x - 17 > 3x - 1$

**3** Solve:

**a** $^-x > 2$
**b** $^-x \leqslant {^-5}$
**c** $^-2x \geqslant 8$
**d** $4 - 5x < 34$
**e** $55 \leqslant 22 - 11x$

**4** Solve each inequality:

**a** $8(5x - 7) - 9 \leqslant 55$
**b** $5(3x + 2) + 6 > 91$
**c** $4(7x + 6) - 8 \geqslant 72$
**d** $\dfrac{2x - 3}{2} \leqslant 5$
**e** $\dfrac{3(x + 2)}{4} < 9$

## Number lines, solution sets and combined inequalities

$x \leqslant 4$ means $x$ is less than or equal to 4.

If $x$ is less than 4, we write $x < 4$.

When the equals symbol is not included with an inequality symbol, this is a **strict** inequality.

If $x$ is a positive whole number and if $x < 4$ then $x$ must be 1, 2 or 3.

If $x$ is any number and $x < 4$ then $x$ could be any number less than 4, such as $^-1$, $1\frac{1}{2}$ or 0.1.

The statement $4 > x$ is equivalent to $x < 4$.

### The number line

A number line is a useful way of showing a result and also of finding which numbers satisfy inequalities. For example,

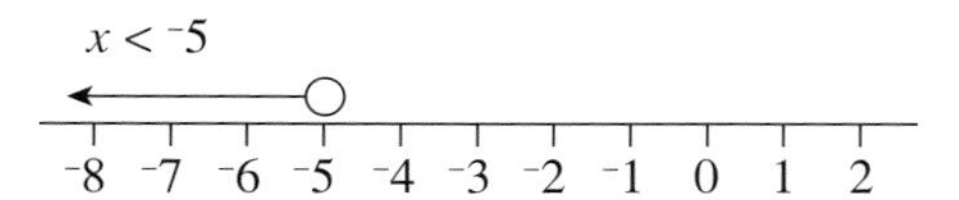


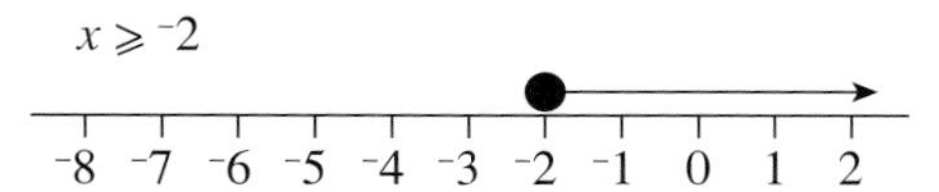


**Note:** The filled-in circle ● for $x \geqslant {^-2}$ is used to show that $x$ can equal $^-2$.

The empty circle ○ for $x < {^-5}$ is used to show that $x$ is not equal to $^-5$.

$2 \leqslant x < 7$ is a shorthand way of writing

'the values of $x$ between 2 and 7, including 2 but excluding 7'.

This can be represented on the number line as:

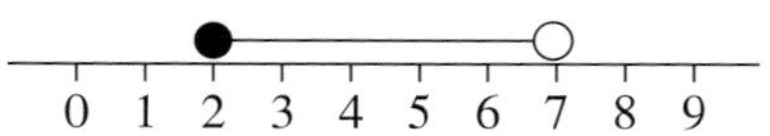


The filled-in circle means *including*; the empty circle means *excluding*.

The whole numbers from 1 to 9 which make $x + 1 > 7$ a true statement are 7, 8 and 9. This set of numbers is called the **solution set** of the inequality.

- The solution set of an inequality is the set of numbers containing the values that make the inequality true.

### Solving combined inequalities

You have seen examples when two inequality symbols are used to describe an inequality. For example $2 < x \leqslant 8$.

These can be solved in a similar way to before. Make sure that you do the same thing to all three sections of the inequality.

**EXAMPLE 7**

Solve $2 \leqslant 2x - 4 < 14$

+4]  $6 \leqslant 2x < 18$

÷2]  $3 \leqslant x < 9$

This can be represented on a number line as

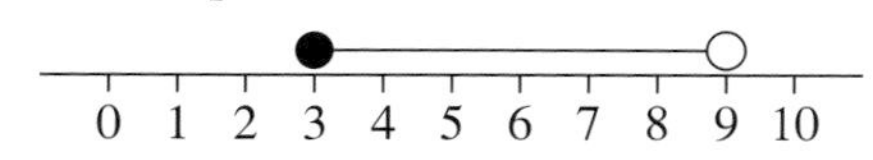

## Exercise 8E

**1** Use a number line to show the solution to the inequality in each part of Questions **1** and **2** of Exercise 8D.

**2** Which *whole* numbers from 1 to 9 make the inequality a true statement?

**a** $x + 3 > 7$ **b** $x + 5 < 8$
**c** $2x > 10$ **d** $3x < 12$
**e** $2x + 3 > 15$ **f** $5x - 1 < 24$

**3** If $x$ can be any *whole* number from 1 to 20, find the solution set for:

**a** $x + 4 > 19$ **b** $x - 7 > 10$
**c** $x + 9 < 13$ **d** $x - 1 < 5$
**e** $2x + 5 < 13$ **f** $3x - 7 > 35$

**4** Write the inequality represented by each of these number lines.

**a** $^{-}1$ 0 1 2 3

**b** $^{-}3$ $^{-}2$ $^{-}1$ 0 1 2

**c** $^{-}2$ $^{-}1$ 0 1 2 3

**d** $^{-}10$ $^{-}5$ 0 5 10 15

**e** $^{-}10$ 0 10 20 30 40

**5** Solve these combined inequalities.

**a** $1 < x + 3 < 9$
**b** $3 < x - 6 \leqslant 10$
**c** $2 \leqslant x + 1 \leqslant 8$
**d** $5 \leqslant 2x - 3 < 17$
**e** $8 < 4(3x - 1) \leqslant 44$

**6** For each part in Question **5**

**i** use a number line to show the solution of the inequality
**ii** if $x$ can be any whole number, write down the numbers in the solution set for the inequality.

**7** If $x$ can be any *whole* number from 21 to 25, find the solution set for:

**a** $7x - 15 > 6x + 8$
**b** $5x + 23 < 3x + 71$
**c** $2x - 3 > 60 - x$

**8** Which *whole* numbers from 1 to 9 make both inequalities true?

**a** $x + 3 > 7$ *and* $x - 2 < 6$
**b** $2x + 1 > 11$ *and* $3x - 1 < 23$

**9** If $x$ can be any *whole* number from 1 to 20, find the solution set for:

**a** $4x + 3 < 55$ *and* $2x - 5 > 13$
**b** $3x + 2 > 11$ *and* $5x + 3 < 38$

**10** A rectangle has area greater than $24\,\text{cm}^2$. The width of the rectangle is 3 cm. What can you say about its length?

**11** A rectangle has length less than 10 cm. What can you say about its width if the rectangle has a perimeter of 50 cm?

For Questions 10, 11 and 12, construct and solve an inequality.

**12** Sam is 6 years older than Amy. If the sum of Sam's and Amy's ages is less than 24, what can you say about Amy's age?

**13** Write down an inequality that has this solution:

**a** $x > 5$
**b** $x \leqslant 4$
**c** $^{-}3 < x \leqslant 10$

**14** Ruth used the method shown to solve the inequality $x + 4 < 3x - 2 \leqslant 2x + 10$.

Split into two separate inequalities:
$x + 4 < 3x - 2$ and $3x - 2 \leqslant 2x + 10$

Solve each separately:

$x + 4 < 3x - 2$
$(-x)$ $4 < 2x - 2$
$(+2)$ $6 < 2x$
$(\div 2)$ $3 < x$

$3x - 2 \leqslant 2x + 10$
$(-2x)$ $x - 2 \leqslant 10$
$(+2)$ $x \leqslant 12$

Then combine these two inequalities again to give the answer:

$3 < x \leqslant 12$

Use Ruth's method to solve these inequalities.

**a** $3x + 10 \leqslant 6x + 4 < 2x + 28$
**b** $3x - 3 \leqslant 5x - 1 < 4x + 2$
**c** $3x - 7 < 6x + 2 \leqslant 5x + 11$

## 8.4 Simultaneous equations

The equations you have looked at so far involve just one unknown. When there are two unknowns you need two equations in order to find the values of both unknowns. Two such equations are known as **simultaneous** equations.

*For example:*

$3x + 2y = 8$
$2x - 3y = {}^{-}1$

are a pair of simultaneous equations.

The solutions to both equations will be the same.

For $3x + 2y = 8$, a possible solution would be $x = 4$ and $y = {}^{-}2$.
This gives

$3 \times 4 + 2 \times {}^{-}2$
$= 12 - 4$
$= 8$

which is what we want.

Trying the same values in $2x - 3y$ gives us

$2 \times 4 - 3 \times {}^{-}2$
$= 8 + 6$
$= 14$

but this is not ${}^{-}1$, as required for the second equation, $2x - 3y = {}^{-}1$.

So while $x = 4$ and $y = {}^{-}2$ is a solution to the first equation it isn't a solution to the second equation, and it is not the solution to the pair of simultaneous equations.

Here is a practical example of simultaneous equations:

Ahmad goes into a café and orders 5 cups of tea and 1 cup of coffee. This costs him $7.
Nadia goes into the same café and orders 2 cups of tea and 1 cup of coffee. This costs her $4.

This can be shown by the simultaneous equations

$5T + C = 7$
$2T + C = 4$

Ahmad and Nadia want to know the cost of 1 cup of tea and 1 cup of coffee.

Ahmad has ordered 3 more teas than Nadia and his bill is $3 more than Nadia's bill.
This can be shown as $3T = 3$.

This equation can be found by subtracting $2T + C = 4$ from $5T + C = 7$. Solving this gives $T = 1$, so the cost of 1 cup of tea is $1.

So when Nadia bought 2 teas and 1 coffee, the 2 teas will have cost $2.

She can use this to find out the cost of 1 coffee. This can be shown as substitution into the equation:

(substitute $T = 1$) $\quad 2T + C = 4$
$(-2)$ $\quad 2 + C = 4$
$C = 2$

So 1 cup of coffee costs $2.

This is the **elimination method** for solving simultaneous equations.

Usually, the easiest way to solve a pair of simultaneous equations is to eliminate either the $x$ or $y$ term and solve the remaining linear equation.

Look again at the first example of simultaneous linear equations:

$3x + 2y = 8$ [1]
$2x + 2y = 6$ [2]

Subtract equation [2] from equation [1]:

$3x + 2y = 8$ [1]
$2x + 2y = 6$ [2]
$x = 2$

Substituting $x = 2$ into equation [1]

$3 \times 2 + 2y = 8$
$6 + 2y = 8$
$(-6)$ $\quad 2y = 2$
$(\div 2)$ $\quad y = 1$

Hence, $x = 2, y = 1$ is the solution to the equations.

Sometimes you need to multiply one or both equations by an appropriate number to make the coefficients of $x$ or $y$ the same before adding or subtracting the equations.

**EXAMPLE 8**

Solve:

$3x + 4y = 13$ [1]
$2x + 3y = 9$ [2]

---

To make the coefficients of $x$ the same, multiply [1] by 2 and [2] by 3.

[1] $\times$ 2: $\quad 6x + 8y = 26$ [3]
[2] $\times$ 3: $\quad 6x + 9y = 27$ [4]

To eliminate the $x$ term, subtract equation [3] from equation [4].

[4] $-$ [3]: $\quad y = 1$

Substituting $y = 1$ into equation [1]

$3x + 4 \times 1 = 13$
$3x + 4 = 13$
$-4]$ $\quad 3x = 9$
$\div 3]$ $\quad x = 3$

Hence, $x = 3, y = 1$ is the solution to the equations.

You can check your answers by substitution. For Example 8, the solution is $x = 3, y = 1$. To check this look at each equation in turn:

[1] is $3x + 4y = 3 \times 3 + 4 \times 1 = 9 + 4 = 13$
[2] is $2x + 3y = 2 \times 3 + 3 \times 1 = 6 + 3 = 9$

## Exercise 8F

**1** Solve these simultaneous equations by subtracting them.

**a** $3x + 2y = 5$, $2x + 2y = 4$  **b** $6x + y = 13$, $4x + y = 9$

**c** $4c - 3d = 1$, $3c - 3d = 0$  **d** $5x - 4y = 11$, $3x - 4y = 5$

**e** $2x + 5y = 31$, $2x - y = 1$  **f** $3a - 7b = 10$, $3a + 2b = 28$

**2** Solve these simultaneous equations by adding them.

**a** $4x - y = 3$, $6x + y = 7$  **b** $3x - 5y = {}^{-}2$, $4x + 5y = 9$

**c** $3x - 2y = {}^{-}3$, $3x + 2y = 9$  **d** $4a - 7b = {}^{-}3$, $3a + 7b = 10$

**e** ${}^{-}x - 3y = {}^{-}7$, $x - 2y = 2$  **f** ${}^{-}4m + g = {}^{-}22$, $4m + 2g = 16$

**3** Solve these simultaneous equations by multiplying one equation by an appropriate number, then adding or subtracting them.

**a** $3x - 2y = 10$, $4x + y = 17$  **b** $6x - 5y = 1$, $3x + 2y = 5$

**c** $x - 3y = {}^{-}1$, $4x - 2y = 12$  **d** $3x + 4y = 14$, $5x - 8y = {}^{-}6$

**4** Solve these simultaneous equations.

**a** $2x + 3y = 5$, $5x + 2y = 7$  **b** $6x - 2y = 4$, $4x + 3y = 7$

**c** $3x - 2y = 11$, $2x + 3y = 16$  **d** $5x - 3y = {}^{-}1$, $2x + 4y = 10$

**e** $7x + 3y = 4$, $3x + 4y = {}^{-}1$  **f** $9x - 7y = 4$, $6x - 5y = 2$

**5** Write down a pair of simultaneous for each solution.

**a** $x = 3, y = 5$
**b** $x = 7, y = 2$
**c** $d = 8, e = {}^{-}1$
**d** $p = 4, m = {}^{-}3$

**6** Give your equations from Question **5** to your neighbour to solve.

**7** At a city farm, some children are feeding sheep.

Altogether the sheep and the children have 16 heads and 40 feet.
How many children and how many sheep are there?

**8** Emily buys 23 bottles of water. Some bottles contain still water, while some contain sparkling water. Still water costs 50 cents per bottle and sparkling water costs 75 cents per bottle. Altogether Emily spends $13.50. How many bottles of still water and how many bottles of sparkling water does Emily buy?

### INVESTIGATION

In the diagram below, the symbols ✱, ○ and ◆ each stand for a different number.

| | | | | |
|---|---|---|---|---|
| ✱ | ○ | ✱ | ○ | 16 |
| ○ | ✱ | ✱ | ◆ | ? |
| ◆ | ◆ | ✱ | ✱ | 24 |
| ◆ | 10 | ○ | 6 | |
| 22 | | 18 | | |

A number at the end of a row gives the total value of the symbols in that row.

A number at the bottom of each column gives the total value of the symbols in that column.

What is the missing total for the second row?

If you liked this Investigation ask your teacher for a copy of the simultaneous equation Sudoku puzzle from the Teacher Book, which is more challenging.

## Challenge

Another way of solving a pair of simultaneous equations is by substitution.

For example:

$$x - 2y = 7 \quad [1]$$
$$3x - 4y = 22 \quad [2]$$

Equation [1] is

$$x - 2y = 7$$

[+ 2y] $\quad x = 7 + 2y \quad [3]$

Substituting [3] into [2] (replace $x$ with $7 + 2y$)

$$3(7 + 2y) - 4y = 22$$
$$21 + 6y - 4y = 22$$
$$21 + 2y = 22$$

[− 21] $\quad 2y = 1$

[÷ 2] $\quad y = \frac{1}{2}$

Substituting in [3]

$$x = 7 + 2 \times \frac{1}{2}$$
$$= 8$$

Hence the solution is $x = 8,\ y = \frac{1}{2}$.

Notice that in this approach the two equations in two unknowns are reduced to one equation in one unknown.

### Challenge questions

**1** Solve these equations using the substitution method.

**a** $x - 3y = 1$, $2x - 5y = 3$ **b** $x + 3y = 4$, $3x - 2y = 1$

**c** $x - 5y = 6$, $2x - 13y = 15$ **d** $x + 3y = 10$, $5x - 2y = {}^-1$

**2** Solve these simultaneous equations using the substitution method.

**a** $2x + y = 5$, $5x - 2y = 8$

**b** $3x - y = 6$, $2x + 3y = 15$

**3 a** Use the substitution method to solve the equations in Exercise 8F Questions **3 a** and **c**.

**b** Which method do you prefer?

**c** Now try Exercise 8F, Questions **3 b** and **d** using the substitution method.

**d** Write a sentence explaining what makes the substitution method easier for Exercise 8F, Questions **3 a** and **c** than for Questions **3 b** and **d**.

## TECHNOLOGY

Do an internet search to find some sites on simultaneous equations.

Which ones help you?

# 8.5 Trial and improvement to solve quadratic equations

So far all the equations you have looked at in this chapter have been linear. Equations are **quadratic** if they have an $x^2$ term in them and no powers of $x$ higher than 2.

Here are some examples of quadratic equations:

$x^2 = 5$

$2x^2 - 3x + 1 = 50$

$x = 7x^2$

These equations are not quadratic, however:

$3x - 4 = 50$

$x^3 = 30$

$2x - 5x^4 + x^2 = 1$

Here, the first equation is linear (it has no $x^2$ term), while the other two equations both have powers of $x$ higher than 2 in them.

It is possible to solve many quadratic equations, but not all – some quadratic equations have no solution. For quadratic equations that can be solved, many have two solutions. (Equations with powers of $x$ higher than 2 can have more than two solutions.)

In this section you will learn how to find an approximate solution to a quadratic equation by using a method called **trial and improvement**. This means that you guess a solution, try it out by substituting it into the equation, then use the result to improve your guess. You will gradually get closer and closer to the correct solution. You may be told a maximum and minimum value for your first guess to help give you a starting point. There are other methods for solving quadratic equations that you will learn about later, but trial and improvement is often a good way of dealing with equations which are hard to solve using other methods.

## EXAMPLE 9

One solution to $x^2 + 2x = 20$ is between $x = 0$ and $x = 4$. Use trial and improvement to find this solution to 1 decimal place.

Using a table:

| Guess | Result for $x^2 + 2x$ | Too big or too small? | The target is 20, so the decision is: |
|---|---|---|---|
| $x = 2$ | $2^2 + 2 \times 2 = 8$ | Too small | Try a value bigger than 2 |
| $x = 4$ | $4^2 + 2 \times 4 = 24$ | Too big | Try a value between 2 and 4 |
| $x = 3$ | $3^2 + 2 \times 3 = 15$ | Too small | Try a value between 3 and 4 |
| $x = 3.5$ | $3.5^2 + 2 \times 3.5 = 19.25$ | Too small | Try a value between 3.5 and 4 |
| $x = 3.6$ | $3.6^2 + 2 \times 3.6 = 20.16$ | Too big | The value is between 3.5 and 3.6 |

We need to know the value to 1 decimal place. $x = 3.6$ gives us an answer closer to 20 than $x = 3.5$, so the solution is $x = 3.6$ to 1 d.p.

To be certain you have the correct value to 1 decimal place, you can continue your table to two decimal places. For Example 9:

| Guess | Result for $x^2 + 2x$ | Too big or too small? |
|---|---|---|
| $x = 3.55$ | $3.55^2 + 2 \times 3.55 = 19.7025$ | Too small |

The table shows the solution is between 3.55 and 3.6. Both of these answers are 3.6 to 1 decimal place so we know for sure this is the correct solution.

## Exercise 8G

**1** One solution to $x^2 + 3x = 30$ is between $x = 2$ and $x = 6$.
Use trial and improvement to find this solution to 1 decimal place.

**2** One solution to $2m^2 + 5m = 45$ is between $m = 1$ and $m = 5$.
Use trial and improvement to find this solution to 2 decimal places.

**3** One solution to $3x^2 - 5x + 3 = 21$ is between $x = {}^{-}5$ and $x = 0$.
Use trial and improvement to find this solution to 1 d.p.

**4** The equation $4x^2 - 8x = 52$ has two solutions between $x = {}^{-}5$ and $x = 5$.
Use trial and improvement to find these solutions to 1 d.p.

**5** The equation $5x^2 + 12x - 8 = 40$ has two solutions between $x = {}^{-}5$ and $x = 5$.
Use trial and improvement to find these solutions to 2 d.p.

**6** Construct an equation to represent the area of this rectangle. Use trial and improvement to find the width of the rectangle correct to 2 d.p.

Length 2 cm more than width

Area = 13.7 cm$^2$

**7** The cubic equation $2x^3 - 5x^2 - 23x = {}^{-}10$ has three solutions between $x = {}^{-}5$ and $x = 5$.
Use trial and improvement to find these solutions to 1 d.p.

**8** Solve each of these equations to 1 d.p.
(**Hint:** all solutions are between 10 and ${}^{-}10$)

**a** $x^4 = 3.81$ **b** $(x - 2)^2 = 30$
**c** $2x + (x + 3)^2 = 25$

**9** The square of a number is 40 times its square root. What is the number, to 1 d.p.?
(**Note:** the number is not 0)

### TECHNOLOGY

Trial and improvement questions can be done on a spreadsheet.

Try using a spreadsheet to work out some of the answers to Exercise 8G again.

Find the solutions to a higher level of accuracy than asked for in the questions.

### TECHNOLOGY

Have a game of trial and improvement golf at:
www.studymaths.co.uk/games/trialandimprovementgolf.html

# Consolidation

### Example 1

One side of a rectangle is 3 cm shorter than the other side.
The perimeter of the rectangle is 26 cm.
Construct and solve an equation for the rectangle.
Find the length of each side.

Let the longer side be $x$. The shorter side is $x - 3$.

The perimeter of the rectangle is 26, so
$26 = x + (x - 3) + x + (x - 3)$

Simplify:

$26 = 4x - 6$

+ 6]   $32 = 4x$

÷ 4]   $8 = x$

So the longer side is 8 cm and the shorter side is 5 cm.

### Example 2

Solve:

**a** $3(2x - 5) = 9$

$3 \times 2x - 3 \times 5 = 9$

$6x - 15 = 9$

+15]   $6x = 24$

÷ 6]   $x = 4$

**b** $\frac{12x + 15}{5} = x - 4$

× 5]   $12x + 15 = 5(x - 4)$

$12x + 15 = 5x - 20$

−15]   $12x = 5x - 35$

−5x]   $7x = {}^{-}35$

÷ 7]   $x = {}^{-}5$

### Example 3

Solve the inequalities and represent the solutions on number lines.

**a** $3x + 2 > 17$

−2]   $3x > 15$

÷ 3]   $x > 5$

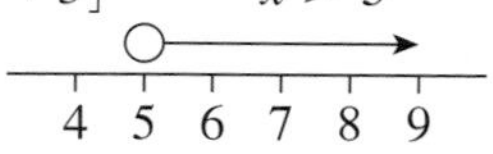


**b** $4x - 2 \leqslant x + 7$

+ 2]   $4x \leqslant x + 9$

−x]   $3x \leqslant 9$

÷ 3]   $x \leqslant 3$

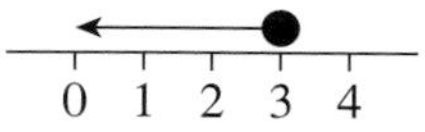


### Example 4

Solve the simultaneous equations.

**a**
$3x + y = 4$   [1]
$5x + 2y = 7$   [2]

2 × [1]:   $6x + 2y = 8$

[2]:   $5x + 2y = 7$

Subtract:   $x = 1$

Substitute in [1]:

$3 \times 1 + y = 4$

$3 + y = 4$

− 3]   $y = 1$

**b**
$2x - 3y = 5$   [1]
$3x + 2y = 14$   [2]

2 × [1]:   $4x - 6y = 10$

2 × [2]:   $9x + 6y = 42$

Add:   $13x = 52$

(÷13)   $x = 4$

Substitute in [1]:

$2 \times 4 - 3y = 5$

$8 - 3y = 5$

−8]   ${}^{-}3y = {}^{-}3$

÷ ⁻3]   $y = 1$

### Example 5

One solution to $x^2 - 5x = 15$ is between $x = 6$ and $x = 8$.
Use trial and improvement to find this solution to 1 decimal place.

| Guess | Result for $x^2 - 5x$ | Too big or too small? |
|---|---|---|
| $x = 7$ | $7^2 - 5 \times 7 = 14$ | Too small |
| $x = 8$ | $8^2 - 5 \times 8 = 24$ | Too big |
| $x = 7.4$ | $7.4^2 - 5 \times 7.4 = 17.76$ | Too big |
| $x = 7.2$ | $7.2^2 - 5 \times 7.2 = 15.84$ | Too big |
| $x = 7.1$ | $7.1^2 - 5 \times 7.1 = 14.91$ | Too small |

$x = 7.1$ gives an answer closer to 15 than $x = 7.2$, so the solution is $x = 7.1$ to 1 d.p.

## Exercise 8

**1** Solve these equations.

**a** $3x + 2 = 8$
**b** $3x - 2 = 10$
**c** $2 - x = 1$
**d** $\frac{x}{3} = 5$
**e** $\frac{x}{3} + 5 = 9$
**f** $4x = 5 - x$
**g** $2x - 3 = x - 4$
**h** $3(x - 7) + 4 = 22$

**2** Solve these inequalities.

**a** $4x < 2$
**b** $x + 3 > 7$
**c** $x - 4 < 8$
**d** $3x - 2 < 19$
**e** $2x + 3 \geqslant 9$
**f** $4x - 3 < 2x + 3$
**g** $5x + 9 > 14 - 5x$
**h** $3(x - 2) \leqslant 12$
**i** $4x - 6 < 4 - x$

**3** Solve:

**a** $3(x + 1) = 2(x + 5)$
**b** $4(x - 1) = 3(x + 2)$
**c** $5(x + 3) = 4(x + 7)$
**d** $5(x - 2) = 3(x + 2)$
**e** $4(x + 3) = 2(x + 7)$
**f** $7(x + 1) = 5(x + 3)$
**g** $5(x - 2) = 2(x + 4)$
**h** $6(x - 3) = 2(x + 3)$

**4** Solve these simultaneous equations.

**a** $3x + 5y = 39$
$3x + 2y = 30$

**b** $6x + 2y = 7$
$5x - 6y = 2$

**c** $7x - 2y = 9$
$3x + 4y = {}^{-}1$

**d** $3x - 2y = 1$
$4x - 3y = 1$

**5** Solve the following inequalities and show the solution on a number line.

**a** $3(x - 3) \geqslant 2(x + 1)$
**b** $3(x - 8) < 12$
**c** $6x - 7 < 2x + 9$
**d** $7(x - 4) > 2(x - 4)$
**e** $8f + 4(f - 6) < 6$
**f** $y + 4(2y - 5) \leqslant {}^{-}2$
**g** $-5 + 3(t + 2) > 13$

**6** The cost of joining a badminton club is \$80 for adults and \$50 for children. Suzette can spend up to \$500 to pay for her friends (who are adults) and her 3 children to join the club.

**a** If Suzette has $w$ friends, write down an inequality to show this information.
**b** Solve your inequality to find out how many of her friends can join.
**c** How much change will she receive from \$500?

**7** Delia has \$945 in 20-dollar and 5-dollar bills. She finds she has 5 times as many 20-dollar bills as 5-dollar bills.

**a** Writing $x$ as the number of 20-dollar bills, form an equation to show the information.
**b** Solve this equation to find the number of bills of each type Delia has.

**8** The cost for 5 adults and 3 children to attend a show is \$210. For 2 adults and 5 children the cost is \$160. Form a pair of simultaneous linear equations and solve them to find the cost of

**a** an adult ticket
**b** a child ticket.

**9** Solve these equations, to 1 decimal place, using a method of trial and improvement.

**a** $x^2 + 5x = 40$, for the solution between $x = 2$ and $x = 6$.
**b** $3t^2 + 8t = 60$, for the solution between $t = 1$ and $t = 5$.
**c** $4x^2 - 8x + 5 = 32$, for the solution between $x = {}^{-}5$ and $x = 0$.
**d** $2x^3 - 6x^2 - 34x = {}^{-}12$, for the solution between $x = 3$ and $x = 9$.

**10** When 2 is added to a certain number and the result is multiplied by 5 the answer is 35. What is that number?

**11** The perimeter of this rectangle is 74 cm.

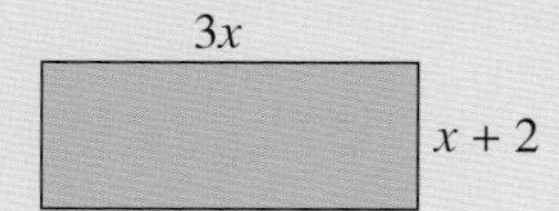


**a** Form an equation for the perimeter of the rectangle.

**b** Solve your equation to find the length and width of the rectangle.

**12** Three times Ann's age four years ago is the same as twice her age in four years' time. What is her present age?

**13**

The sum of three consecutive whole numbers is 273. What are the numbers?

# Summary

### You should know ...

**1** How to construct equations to solve problems involving unknown quantities. You should:

- identify the unknown
- form an equation
- solve the equation
- interpret the solution and answer the question.

**2** How to solve linear equations using the balance method.
*For example:*

$$5(x - 2) = 3x - 7$$
$$5x - 10 = 3x - 7$$
$$+10] \quad 5x - 10 + 10 = 3x - 7 + 10$$
$$5x = 3x + 3$$
$$-3x] \quad 5x - 3x = 3x + 3 - 3x$$
$$2x = 3$$
$$\div 2] \quad x = 1\tfrac{1}{2}$$

**3** How to solve linear inequalities in a similar way to linear equations and show the solution on a number line.
*For example:*

$$2 - 2x < 6(x + 3)$$
$$2 - 2x < 6x + 18$$
$$-18] \quad {}^{-}16 - 2x < 6x$$
$$+2x] \quad {}^{-}16 < 8x$$
$$\text{So} \quad 8x > {}^{-}16$$
$$\div 8] \quad x > {}^{-}2$$

⁻3 ⁻2 1 1 2

### Check out

**1** **a** The perimeter of a square is 28 cm. What is its area?

**b** The length of a rectangular room is 2 m longer than its width. What are the room's dimensions if the perimeter is 16 m?

**2** Solve:

**a** $3x - 1 = x + 5$

**b** $6(x + 2) = 5(x + 4)$

**c** $7 - 2x = 4x - 5$

**3** Find the solutions to:

**a** $4 + 3x < 13$

**b** $4x - 3 > 13$

**c** $3 - 2x \leqslant 5 - 6x$

**d** $\dfrac{5 + 7x}{2} \geqslant 10 + x$

**4** How to solve simultaneous linear equations using the elimination method.

*For example:*

$3x - y = 1$ [1]

$2x + 3y = 8$ [2]

Multiply [1] by 3:

$9x - 3y = 3$ [3]

Add [3] to [2]:

$9x - 3y = 3$

$2x + 3y = 8$

$11x = 11$

so $x = 1$

Substitute $x = 1$ into [1]:

$3 \times 1 - y = 1$

so $y = 2$

**4** Solve using the elimination method:

**a** $2x + 3y = 7$
$x - y = 1$

**b** $3x - 2y = 7$
$2x + y = 7$

**c** $x + 4y = 7$
$3x + 2y = 1$

**d** $4x - 5y = 3$
$3x - 4y = 6$

**5** How to use trial and improvement to find approximate solutions to equations.

*For example:*

One solution to $x^2 - 10x = 18$ is between $x = 9$ and $x = 12$.

Use trial and improvement to find this solution to 1 decimal place.

| Guess | Result for $x^2 - 10x$ | Too big or too small? |
|---|---|---|
| $x = 11$ | $11^2 - 10 \times 11 = 11$ | Too small |
| $x = 12$ | $12^2 - 10 \times 12 = 24$ | Too big |
| $x = 11.4$ | $11.4^2 - 10 \times 11.4 = 15.96$ | Too small |
| $x = 11.5$ | $11.5^2 - 10 \times 11.5 = 17.25$ | Too small |
| $x = 11.6$ | $11.6^2 - 10 \times 11.6 = 18.56$ | Too big |

$x = 11.6$ gives an answer closer to 18 than $x = 11.5$, so the solution is $x = 11.6$ to 1 d.p.

**5** Solve these equations, to 1 decimal place, using trial and improvement.

**a** $x^2 + 9x = 32$, for the solution between $x = 0$ and $x = 4$.

**b** $2y^2 + 3y = 50$, for the solution between $y = 2$ and $y = 6$.

**c** $3x^2 - 5x - 15 = 0$, for the solution between $x = {}^{-}3$ and $x = 1$.

**d** $2x^3 + 3x^2 - 10x = 15$, for the solution between $x = 0$ and $x = 5$.

# 9 Geometry

## Objectives

- Calculate the interior or exterior angle of any regular polygon; prove and use the formula for the sum of the interior angles of any polygon; prove that the sum of the exterior angles of any polygon is 360°.
- Solve problems using properties of angles, of parallel and intersecting lines, and of triangles, other polygons and circles, justifying inferences and explaining reasoning with diagrams and text.
- Know and use Pythagoras' theorem to solve two-dimensional problems involving right-angled triangles.
- Tessellate triangles and quadrilaterals and relate this to angle sums and half-turn rotations; know which regular polygons tessellate, and explain why others will not.
- Find by reasoning the locus of a point that moves at a given distance from a fixed point, or at a given distance from a fixed straight line.

## What's the point?

The study of geometry allows you to answer questions such as: How wide is this river? How tall is this mountain? Tessellations help in the creation of artwork such as the piece shown here.

## Before you start

### You should know ...

**1** The area of a square is given by the formula
Area = length × length

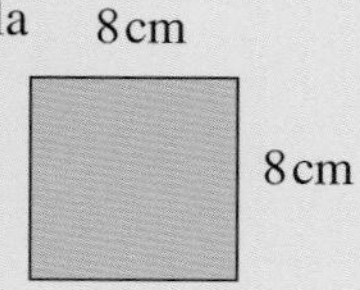


### Check in

**1** **a** What is the area of the square shown on the left?
**b** Find the area of a square with sides $x$ cm.

**2** How to use a protractor to measure angles.
*For example:*

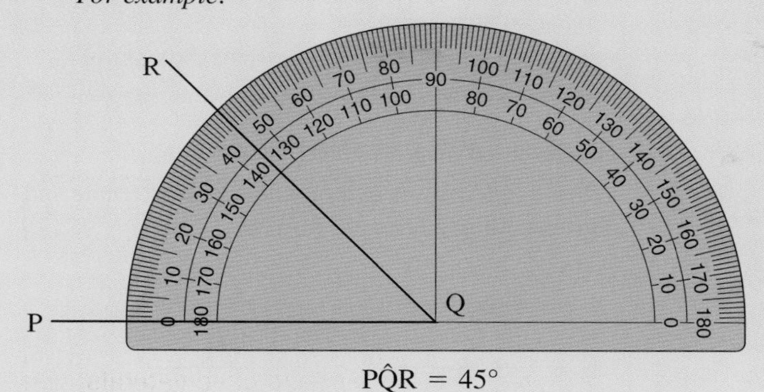


$P\hat{Q}R = 45°$

**2** Measure these angles with a protractor.

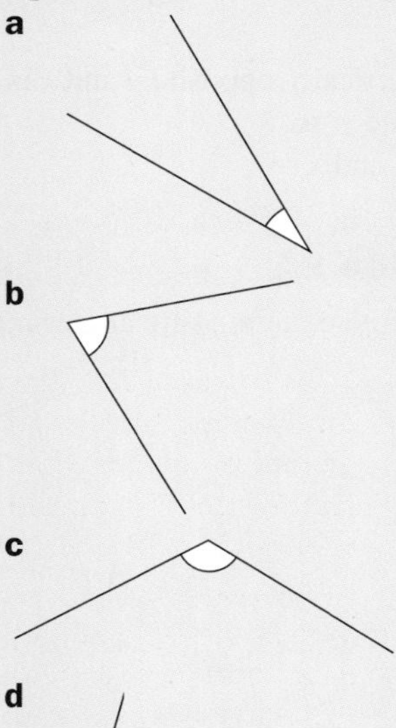


**3** How to solve equations.
*For example:*

$$\begin{aligned} 2x + 50 &= 180 \\ -50] \quad 2x &= 130 \\ \div 2] \quad x &= 65 \end{aligned} \qquad \begin{aligned} x^2 &= 25 \\ x &= \sqrt{25} \\ &= 5 \end{aligned}$$

**3** Solve:

**a** $3x + 10 = 22$
**b** $4x - 5 = 23$
**c** $7(x + 2) = 42$
**d** $8x - 3 = 4x + 17$
**e** $x^2 = 36$
**f** $x^2 = 81$

# 9.1 Properties of angles

## Basic angle facts

**1** Angles at a point add up to 360°.

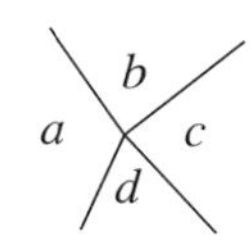


$a + b + c + d = 360°$

**2** Angles on a straight line add up to 180°.

**EXAMPLE 1**

Find angle $x$.

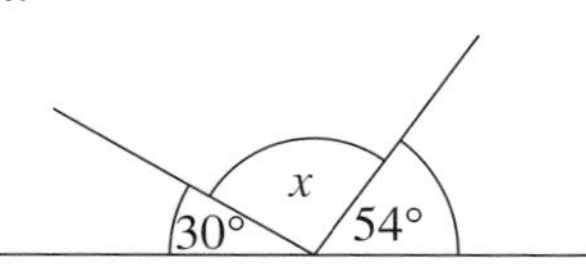


$$\begin{aligned} 30° + x + 54° &= 180° \\ x + 84° &= 180° \\ x &= 96° \end{aligned}$$

**3** Vertically opposite angles are equal.

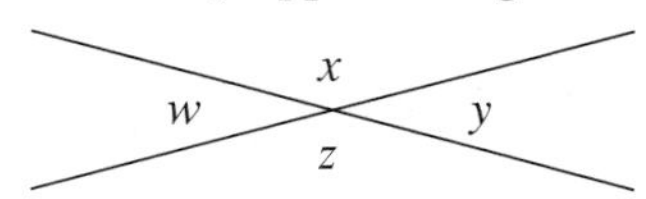


$w$ is vertically opposite $y$ and $x$ is vertically opposite $z$, so
$w = y$ and $x = z$

## Exercise 9A

**1** Calculate the missing angles $a$, $b$, $c$ and $d$.

**a** 245° $a$

**b**

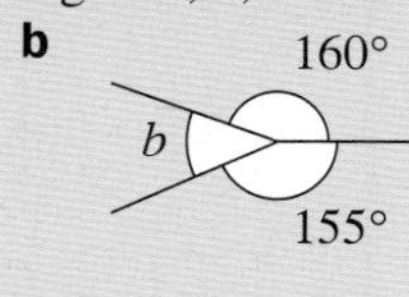


**c** **d**

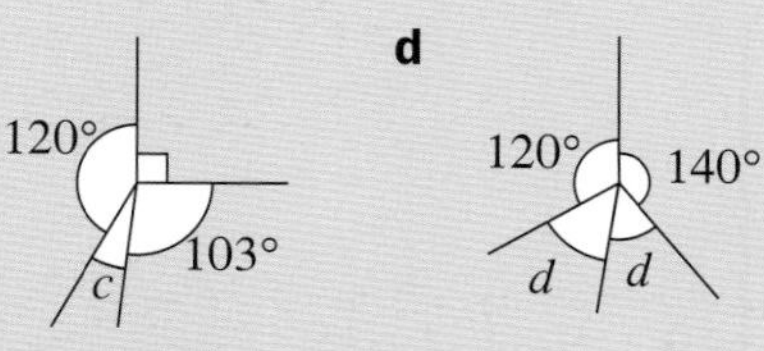


**2** Calculate the missing angles $e$, $f$, $g$ and $h$.

**a**

**b** $f$ 58° 25°

**c**

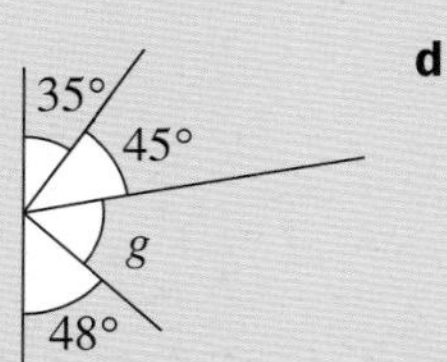


**d**

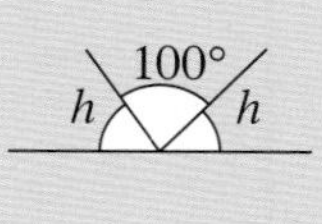


**3** Which angle is vertically opposite $p$?

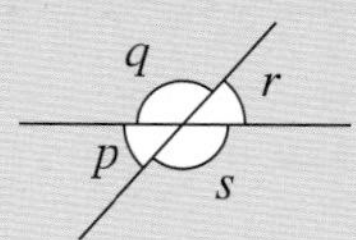


**4** Calculate the missing angles $i$, $j$, $k$ and $l$.

**a**

**b** 143° 37° 37° $j$

**c**

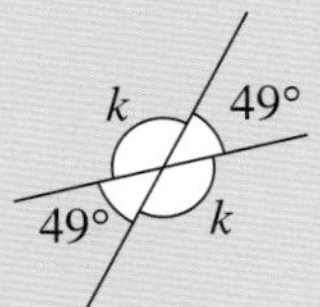


**d**

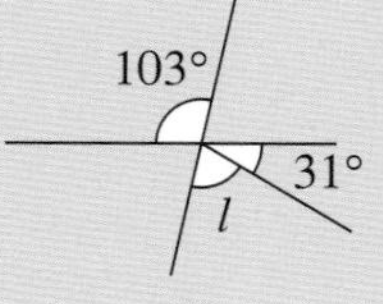


### TECHNOLOGY

Forgotten all about angles? Have a go at Angle Kung Fu at the website
www.bbc.co.uk/keyskills/flash/kfa/kfa.shtml

## Angles and parallel lines

When a line crosses two parallel lines, alternate and corresponding angles are formed.

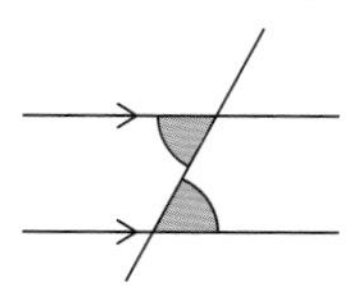

Alternate angles
(Look for a Z-shape)

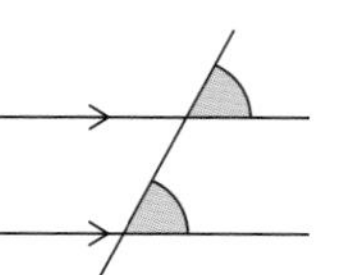

Corresponding angles
(Look for an F-shape)

Alternate angles are equal.

Corresponding angles are equal.

### EXAMPLE 2

Find the angles $x$ and $y$.

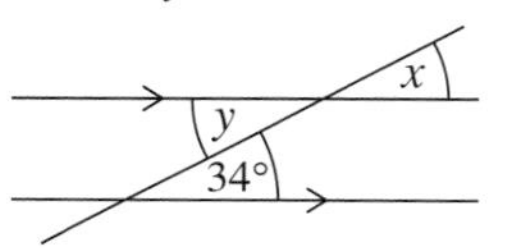


Angle $x$ and 34° are corresponding angles
so $x = 34°$
Angle $y$ and 34° are alternate angles
so $y = 34°$
**Note:** $x = y$ since they are vertically opposite.

## Exercise 9B

**1** Find the lettered angles.

**a**

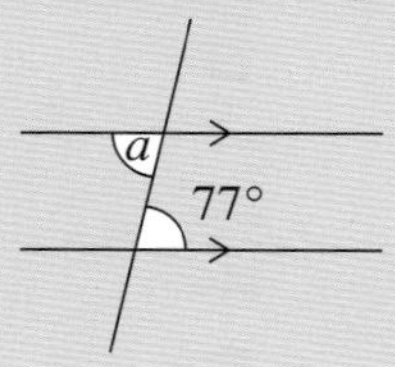

**b**

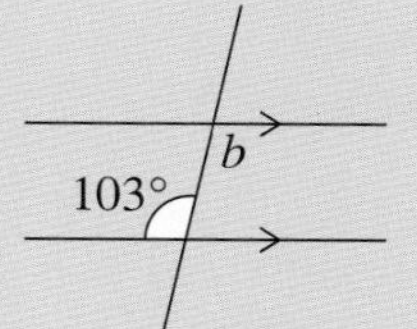


**c**

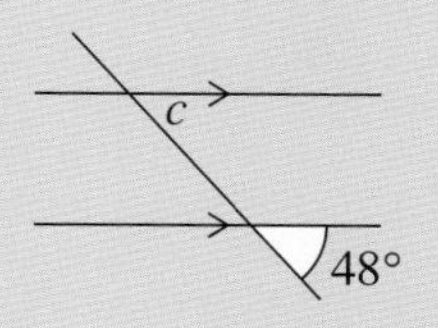


**d**

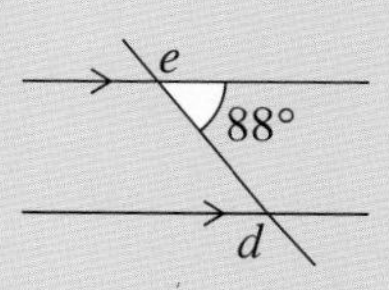


**2** Find the lettered angles. Give reasons for your answers.

**a**

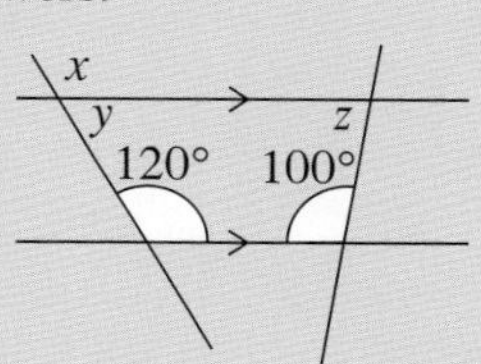


**b**

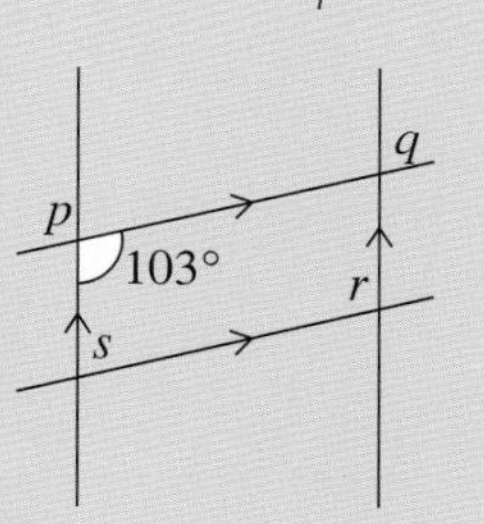


**c**

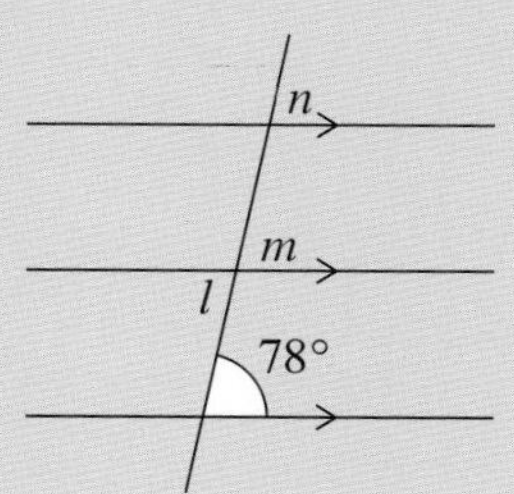


**d**

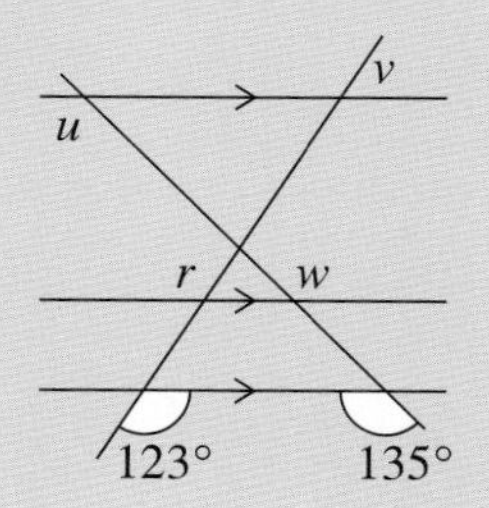


**3** Shape ABCD is a parallelogram. $B\hat{A}D = 72°$.

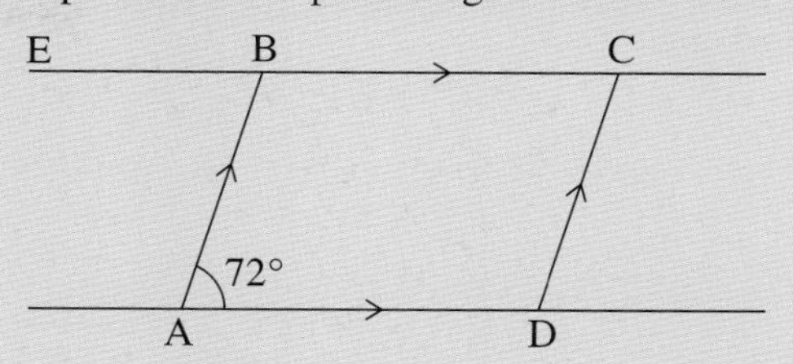


**a** Work out the size of $A\hat{B}E$.

**b** Find $A\hat{D}C$.

**4** Find the angles lettered $a$, $b$, $c$ and $d$.

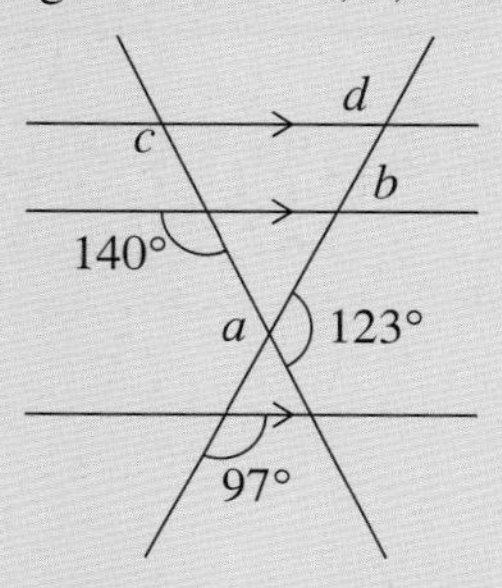


**5** Find the values of $a$ and $b$.

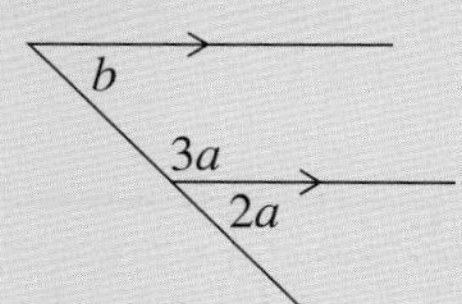


**6** Find the values of $x$, $y$ and $z$.

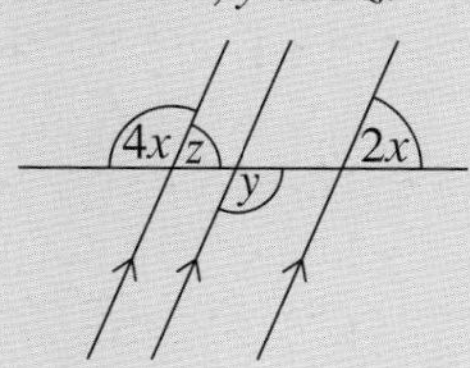


## Angles in triangles and quadrilaterals

- The sum of the interior angles in a triangle is 180°.

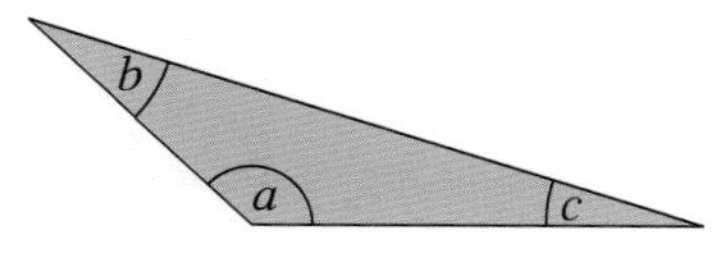


$a + b + c = 180°$

You can see this if you take a paper triangle, tear off the angles and fit them together.

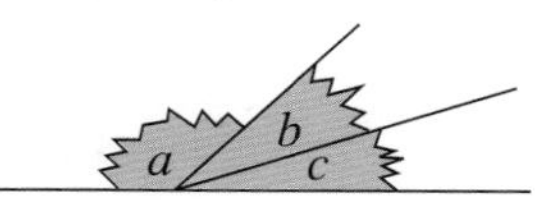

## EXAMPLE 3

Find angle $z$.

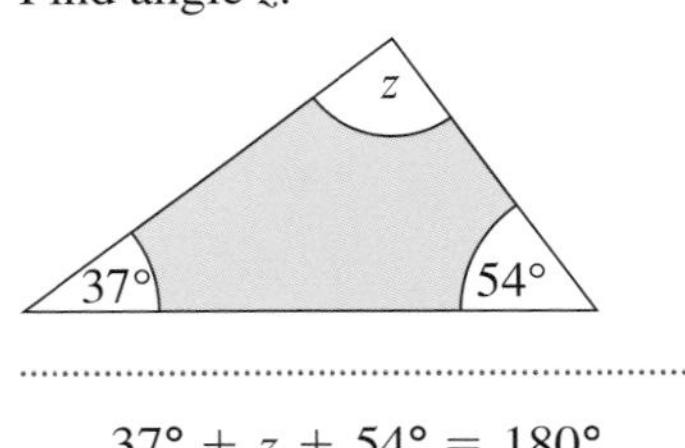


$$37° + z + 54° = 180°$$
$$z + 91° = 180°$$
$$z = 89°$$

Quadrilaterals can always be split into two triangles.

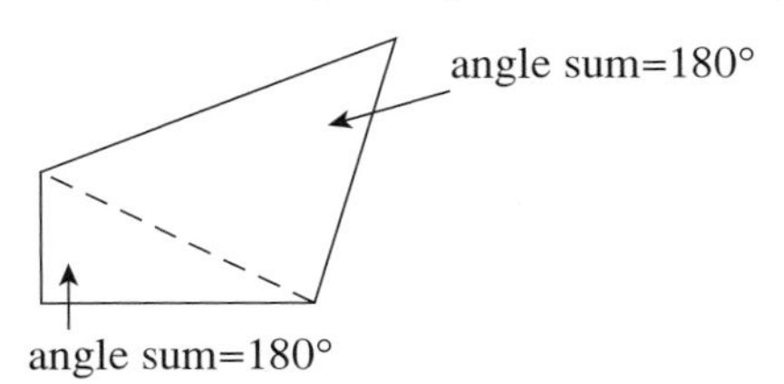


- So the sum of the interior angles in a quadrilateral is 360°.

## EXAMPLE 4

Find the size of angle $p$.

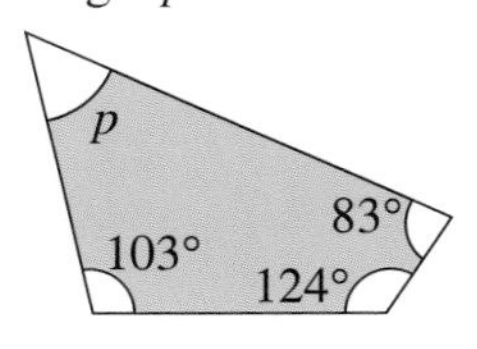


$$p + 83° + 124° + 103° = 360°$$
$$p + 310° = 360°$$
$$p = 50°$$

- The exterior angle of a triangle is equal to the sum of the two opposite interior angles.

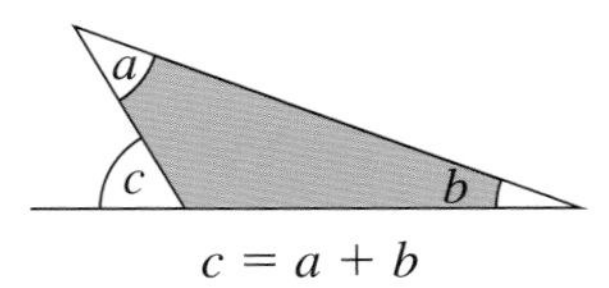


$c = a + b$

## Exercise 9C

**1** Calculate the missing angles.

**a**

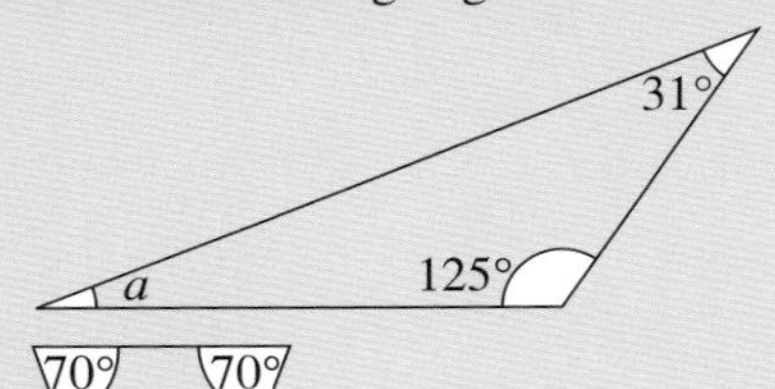


**b**

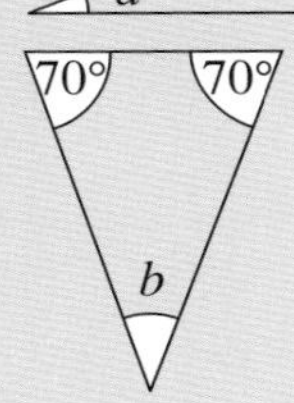


**c**

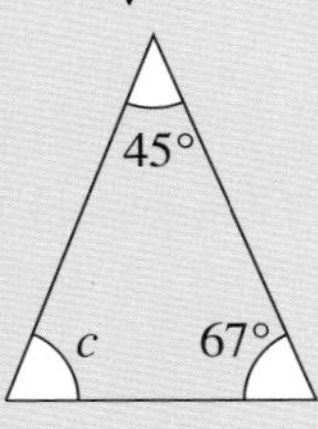


**d**

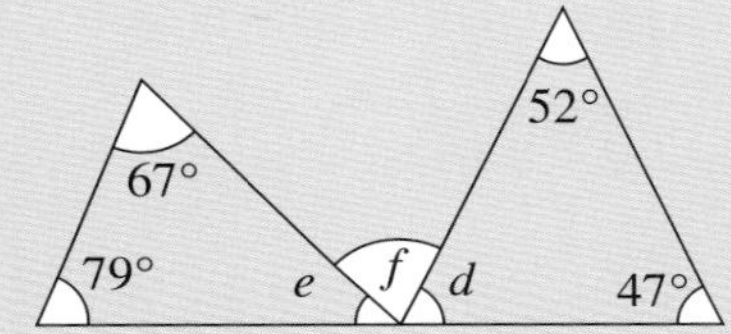


**2** Find the lettered angles.

**a**

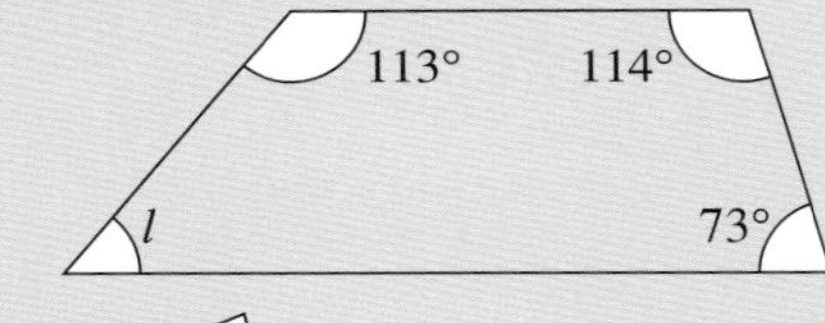


**b**

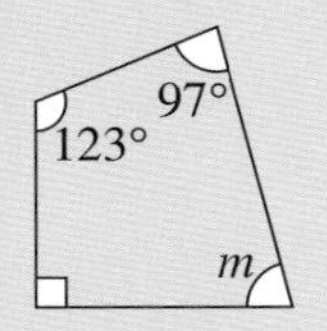


**c**

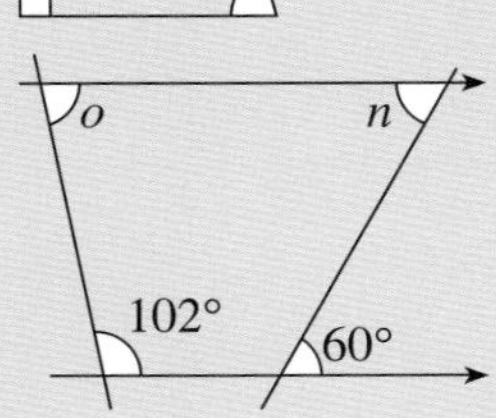


**d**

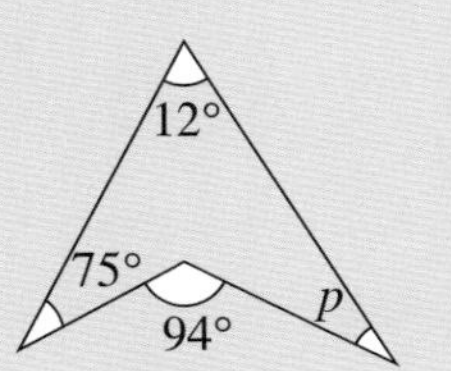

**3** Find the angles $w$, $x$, $y$ and $z$.

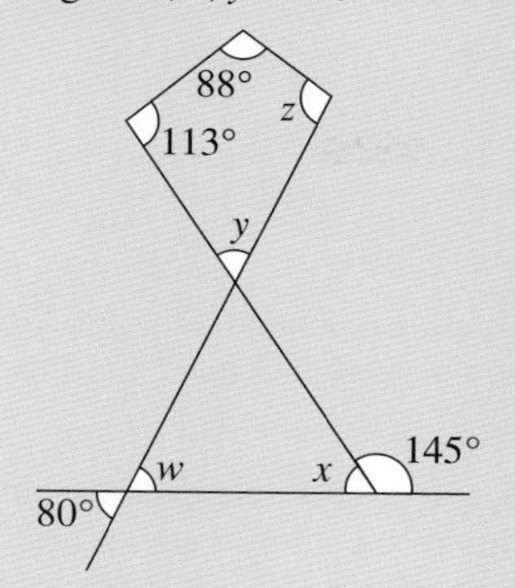


**4** Find angle $g$.

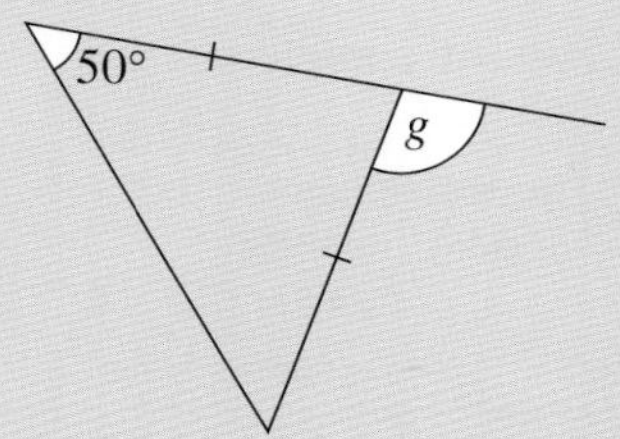


**5** Find the lettered angles.

**a**

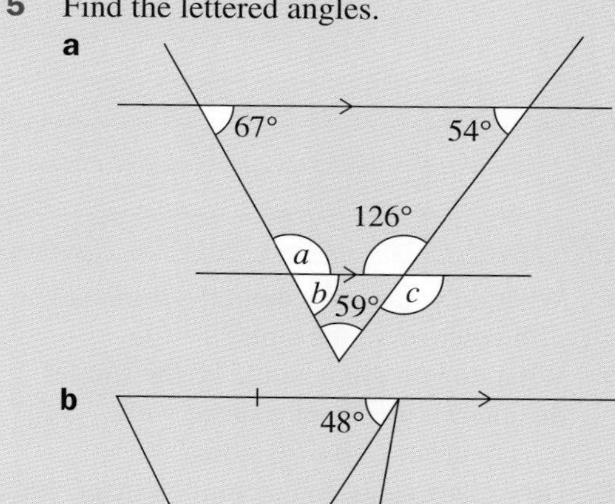


**b**

48°

$d$ $e$ 107° $f$

**6** Use the lettered angles to find all the angles inside these triangles and quadrilaterals.

**a**

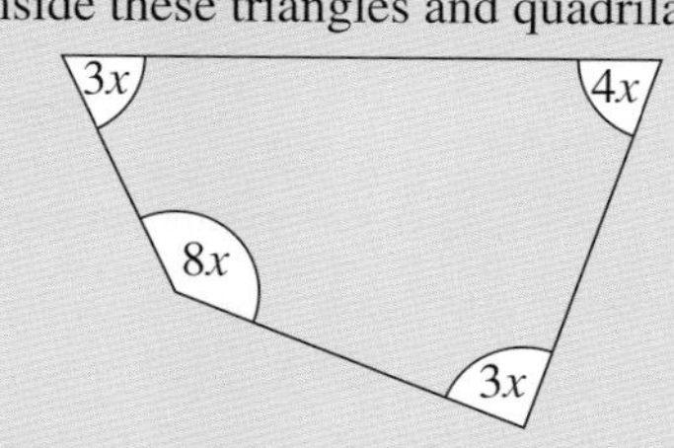


**b**

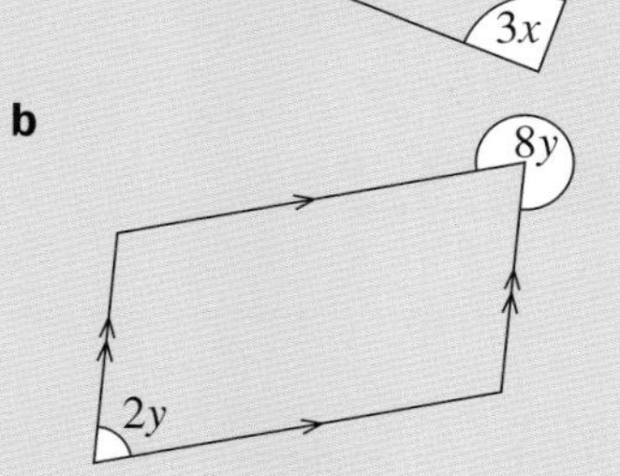


**c**

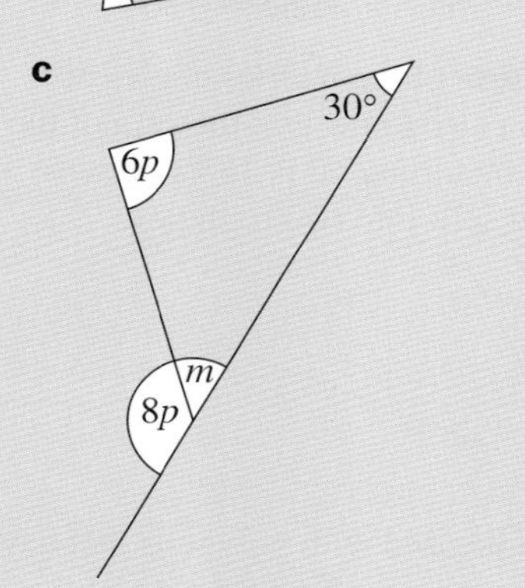


## INVESTIGATION

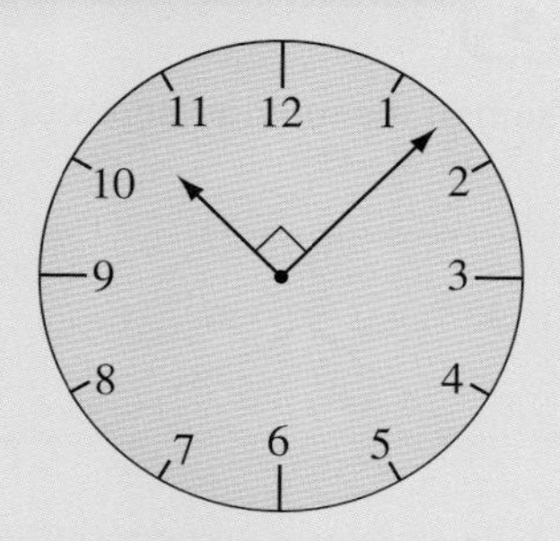


At what time between 10.00 am and 11.00 am are the hour and minute hands at right angles?

At what time between 10.00 am and 11.00 am will the hands form a straight angle?

## 9.2 Angles in polygons

All polygons have **interior** and **exterior** angles.

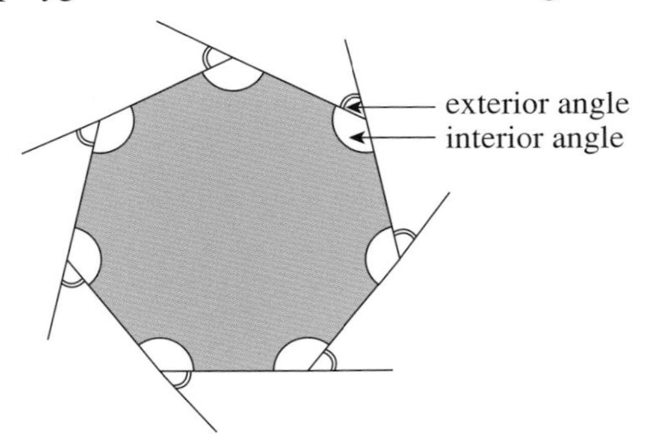


Notice that:

exterior angle + interior angle = 180°

You know that the sum of the interior angles of a triangle is 180°. You also know that the sum of the interior angles of a quadrilateral is 360°, because a quadrilateral can be split into two triangles:

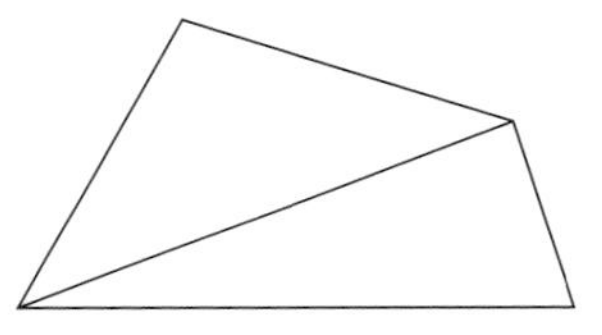

This same idea can be used with polygons that have more than four sides. You can find the sum of the interior angles of a polygon by dividing it up into triangles.

**EXAMPLE 5**

What is the sum of the interior angles of a hexagon?

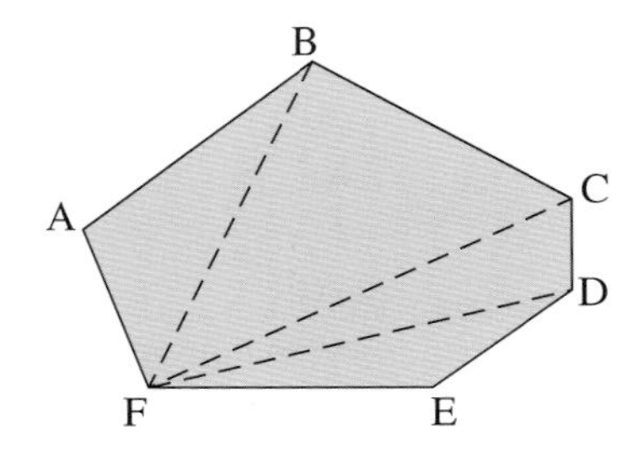


Look at the hexagon ABCDEF. It can be split into four triangles (ABF, BCF, CDF, DEF).

Each triangle has angle sum 180°, so the interior angle sum of the hexagon is 4 × 180° = 720°.

### Exercise 9D

**1**

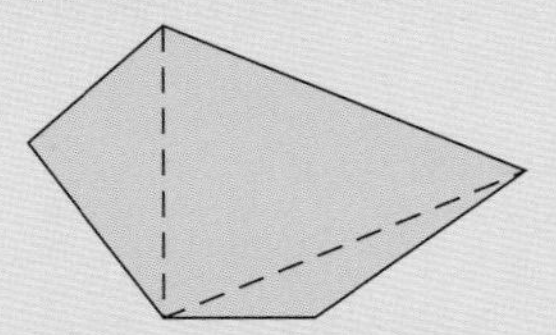

By dividing a pentagon up into three triangles, as shown, show that the sum of the interior angles of a pentagon is 540°.

**2** By dividing these shapes into triangles, find the sum of the interior angles of:

**a** an octagon **b** a heptagon

**c** a decagon **d** a nonagon.

**3 a** Copy and complete:

| Shape | Number of sides, $n$ | Sum of interior angles | $n - 2$ | $(n - 2) \times 180°$ |
|---|---|---|---|---|
| Triangle | | 180° | | |
| Quadrilateral | 4 | | 2 | 360° |
| Pentagon | | | | |
| Hexagon | | | | |
| Heptagon | 7 | | | |
| Octagon | | | | |
| Nonagon | | | | |
| Decagon | 10 | | | |

**b** What do you notice about the third and fifth columns in the table?

**c** What do you think would be the sum of the interior angles of a shape with:

**i** 11 sides **ii** 20 sides?

**d** How many sides do you think a shape with interior angle sum 2700° would have?

**4** Find the lettered angles.

**a** 90°, $a$, 90°, 110°, 130°

**b** 38°, 170°, 118°, 103°, $b$, 165°

**c** 130°, 140°, $c$, 125°, 107°, 111°

**d**

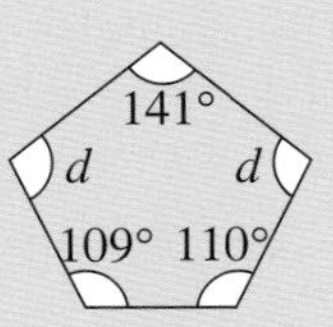


**5** A polygon has $n$ sides. What would be the sum of the interior angles of this $n$-sided polygon?

In Exercise 9D, Question **3**, you should have found that:

- The sum of the interior angles of a polygon with $n$ sides is $180(n - 2)$.

The $n - 2$ is because there are always two fewer triangles than there are sides.

The 180 is because each triangle is worth 180°.

To prove this formula another way, consider any polygon with $n$ sides. In this diagram we show a hexagon.

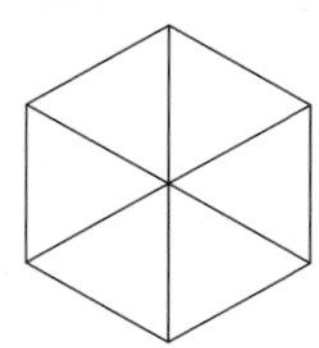

Join the centre of the polygon to all vertices. This makes $n$ triangles. The angles in these $n$ triangles add up to $180n$.

We then need to subtract all the angles at the centre of the polygon which are not interior angles.

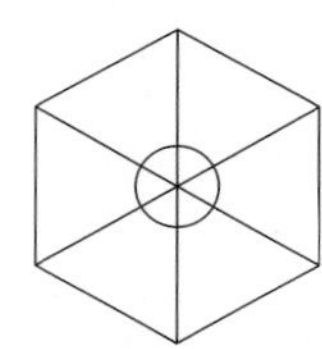

These make a full circle, which is 360°, so the sum of interior angles is $180n - 360°$ which factorises to make $180(n - 2)$.

## Exterior angles

Consider this pentagon.

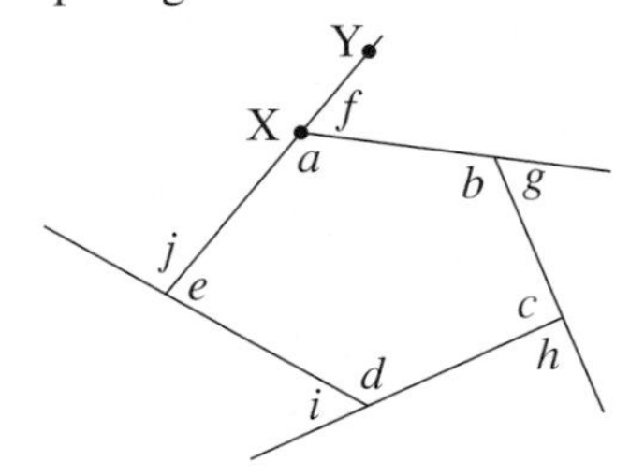


Imagine standing at point X facing in the direction of point Y. Turn round through angle $f$ and imagine walking along the top side of the pentagon. Imagine walking around the entire shape in this way, turning through angles $g$, $h$, $i$ and $j$ in turn until you are back where you started. You have gone through a full turn, which is 360°.

The proof for this follows.

You know that the sum of the interior angles $a + b + c + d + e = 180(5 - 2) = 540°$

You also know that the interior and exterior angle pairs add up to 180°, so $a + f = 180°$, $b + g = 180°$, $c + h = 180°$, $d + i = 180°$ and $e + j = 180°$

The sum of the interior and exterior angles is $a + f + b + g + c + h + d + i + e + j = 180° \times 5$ $= 900°$

The sum of the exterior angles is the sum of the interior and exterior angles minus the sum of just the interior angles.

$$f + g + h + i + j = (a + f + b + g + c + h + d + i + e + j) - (a + b + c + d + e) = 900° - 540° = 360°$$

So the sum of the exterior angles of a pentagon is 360°.

To extend this proof to any polygon we use the same idea:

Sum of interior + exterior angles of an $n$-sided polygon $= 180n$

Sum of interior angles of an $n$-sided polygon $= 180(n - 2)$

Sum of exterior angles = (sum of interior + exterior angles) − (sum of interior angles)

expand brackets] $= 180n - 180(n - 2)$

simplify] $= 180n - 180n + 360$

$= 360$

- The sum of the exterior angles of any polygon is 360°.

## Regular polygons

The interior and exterior angles of this regular hexagon are marked.

You can put all the exterior angles together as you know they add up to 360°:

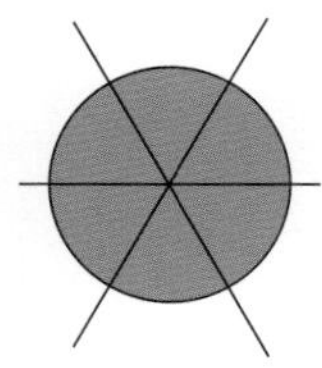

As all the exterior angles are equal:

Exterior angle $= 360° \div 6 = 60°$

So interior angle $= 180° - 60° = 120°$

In general, the **exterior angle** of a regular polygon is given by

- $\dfrac{360°}{n}$, where $n$ = number of sides.

The interior angle = 180° − exterior angle.

## Exercise 9E

1 Find the interior and exterior angles of a regular pentagon.

2 a Calculate the exterior angle of:
   i a regular hexagon
   ii a regular octagon
   iii a regular nonagon
   iv a regular heptagon.
 b Using your answers to part **a**, work out the interior angle of:
   i a regular hexagon
   ii a regular octagon
   iii a regular nonagon
   iv a regular heptagon.

3 Copy and complete this table. You may need to look up the definitions of 'hendecagon' and 'dodecagon'.

| Regular polygon | Equilateral triangle | Square | Hendecagon | Dodecagon |
|---|---|---|---|---|
| Number of sides | | | | |
| Exterior angle | | | | |
| Interior angle | | | | |

4 Calculate the exterior and interior angles of a 36-sided regular polygon.

5 Find the lettered angles.

a
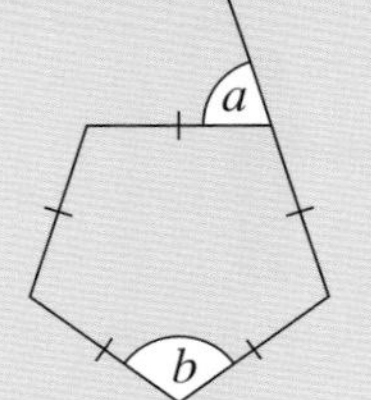


b
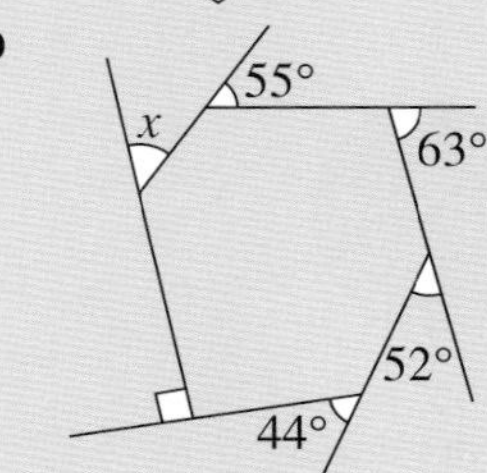


c
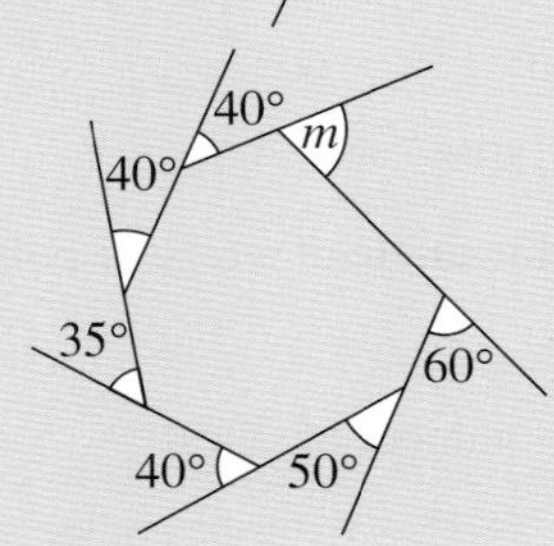


d
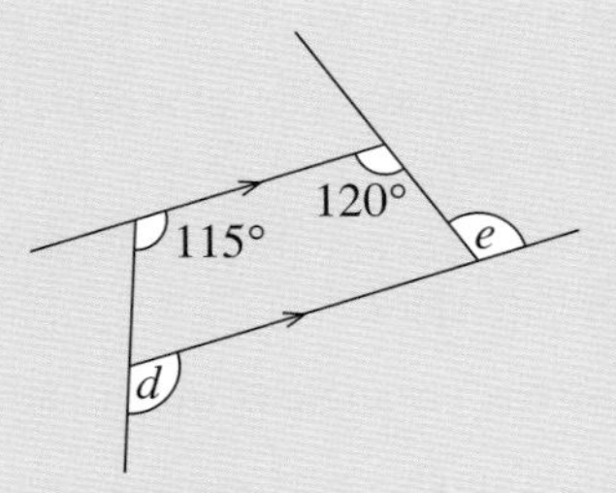


6 A regular polygon has an interior angle of 160°.
 a Work out the size of an exterior angle.
 b How many sides does the polygon have?

7 Can a regular polygon have an interior angle of 110°? Explain your answer.

### INVESTIGATION

A four-sided shape has two diagonals.

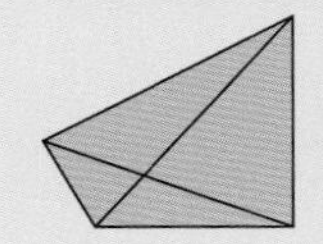

How many diagonals does a five-sided shape have?
A six-sided shape?

Copy and complete the table.

| Number of sides | Number of diagonals |
|---|---|
| 3 | 0 |
| 4 | 2 |
| 5 | |
| 6 | |
| 7 | |
| $n$ | |

How many diagonals would a 23-sided polygon have?

A polygon has 40 sides.

How many diagonals does it have?

## 9.3 Angles in circles

The different parts of a circle have names:

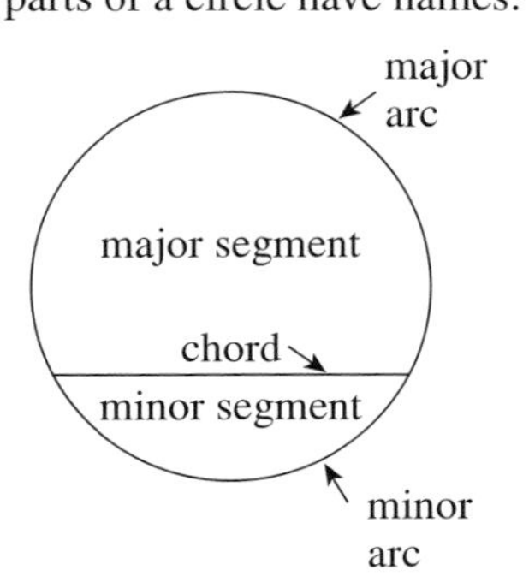

## Angle at the centre of a circle

**ACTIVITY**

Draw a circle with centre O. Mark two points A and B on it, as shown below. Join OA and OB, and measure AÔB.

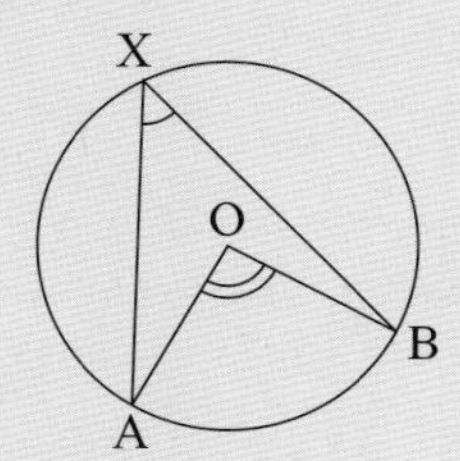


Mark any point X on the major arc. Join AX and BX. Measure AX̂B.

Repeat for other positions of A, B and X.

What do you notice about the sizes of AÔB and AX̂B?

- The angle that an arc of a circle **subtends** at the centre is double the angle which it subtends at any point on the remaining part of the circumference.

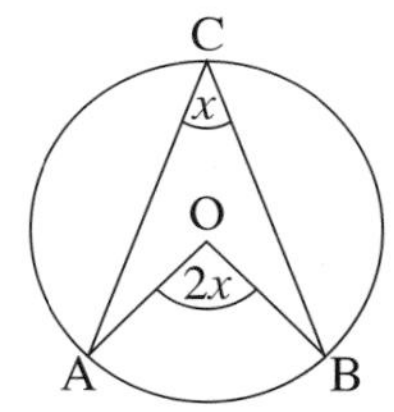


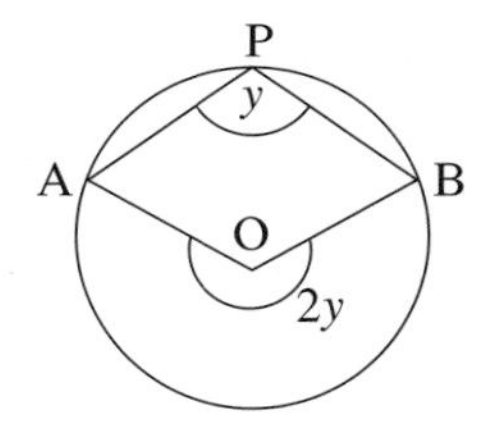


## Angles in the same segment of a circle

**ACTIVITY**

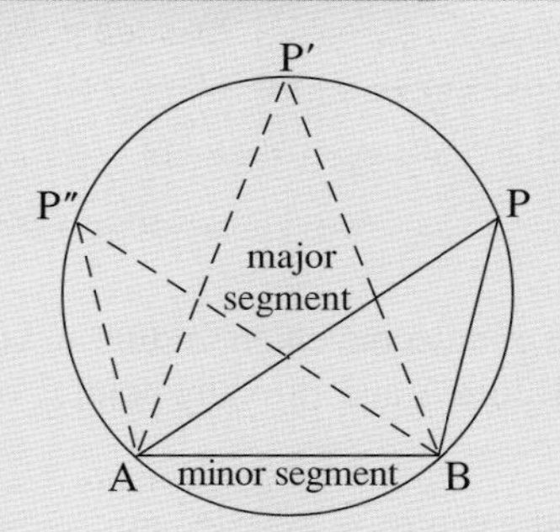


**a** Draw a circle and draw a chord AB on it, as shown above.
**b** Mark a point P on the major arc. Join PA, then PB.
**c** Measure AP̂B.
**d** Repeat for at least four other positions of P.
**e** What do you notice about the sizes of the angles AP̂B, AP̂′B, AP̂″B, etc.?

In the Activity you should have found that:

- Angles in the same segment of a circle are equal.

**EXAMPLE 6**

Find angle POM.

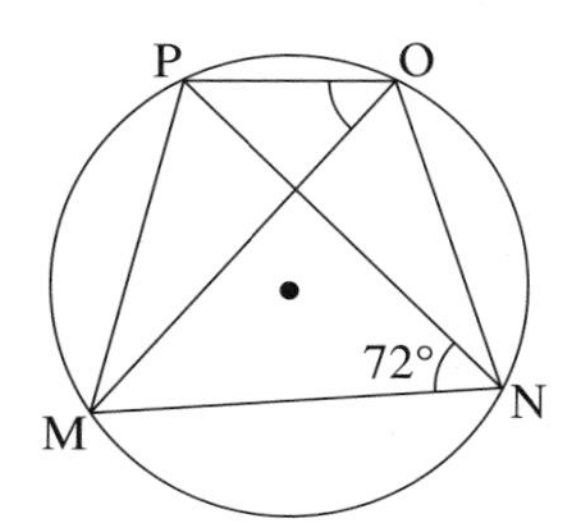


The angles in the major segment from PM are PÔM and PN̂M.

PÔM = PN̂M = 72°

### Exercise 9F

**1** Without measuring, find the value of the unknown angles in each circle.

**a**

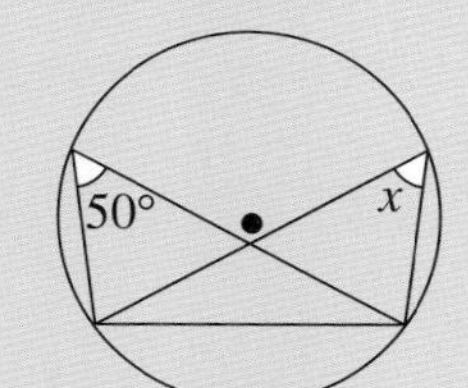


**b**

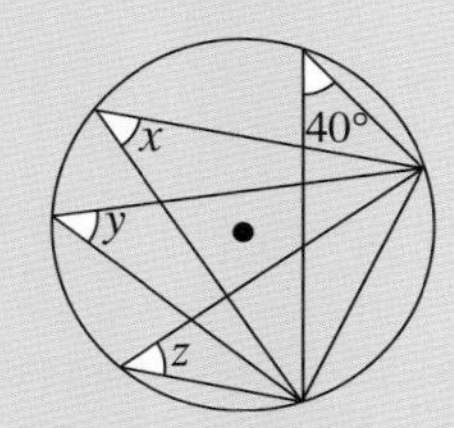


**c**

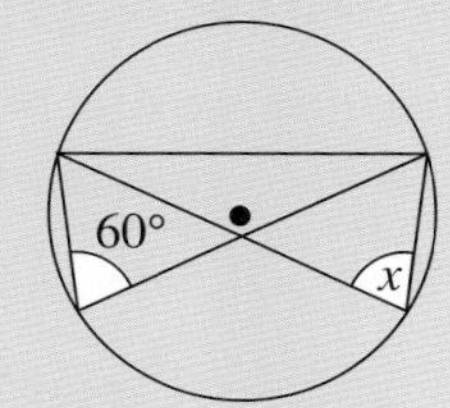


**d**

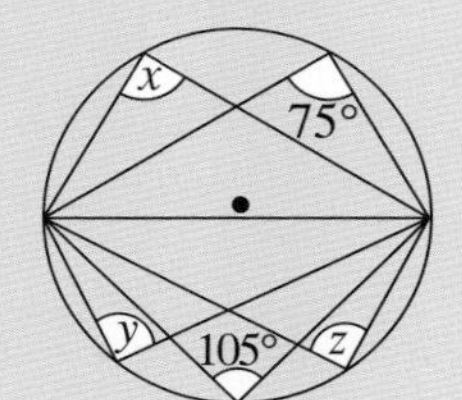

**2**

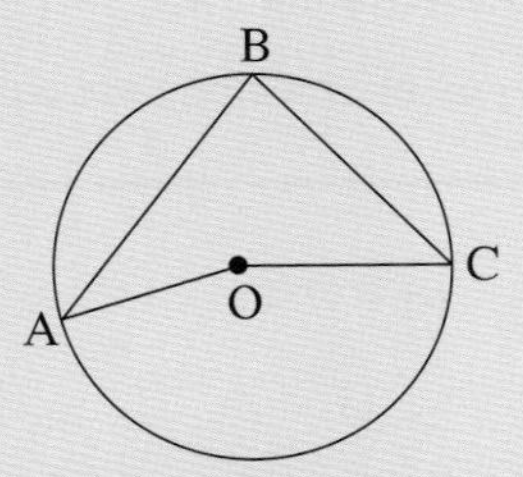


**a** If $A\hat{O}C = 118°$, find $A\hat{B}C$

**b** If $A\hat{B}C = 83°$, find $A\hat{O}C$

**3** AB is a diameter of the circle.

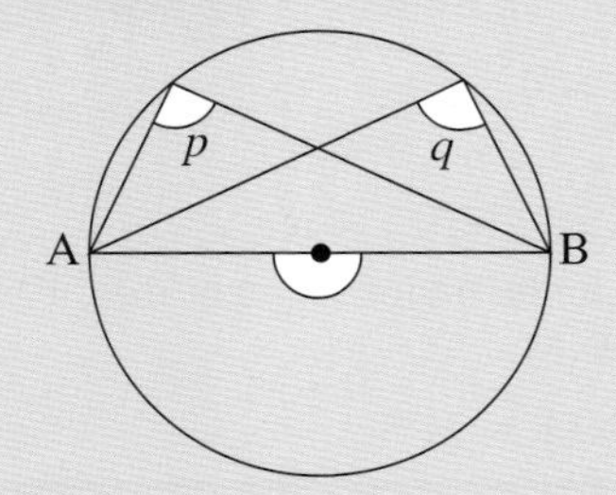


**a** What shape is each segment?

**b** What size is the marked angle at the centre?

**c** What size would you expect angles $p$ and $q$ to be?

**d** Measure $p$ and $q$. Were you correct?

**4**

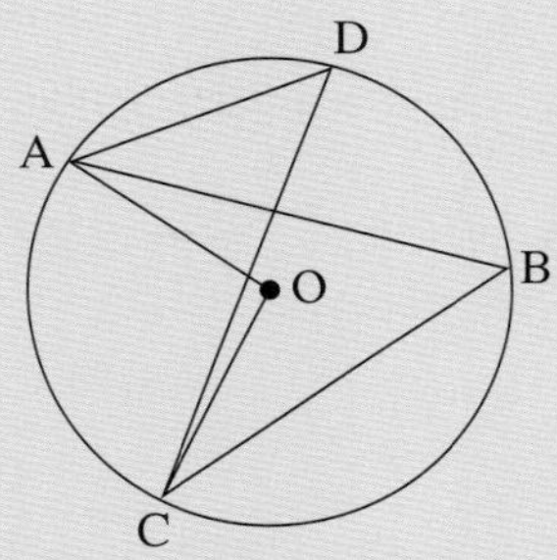


**a** If $A\hat{D}C = 42°$, find $A\hat{B}C$.

**b** If $A\hat{D}C = 56°$, find the reflex angle AOC.

**c** If reflex angle AOC $= 265°$, find $A\hat{D}C$.

**d** If $A\hat{O}C = 88°$ and $D\hat{C}O = 26°$, find $D\hat{A}O$.

**5** Copy these circles and fill in the sizes of as many angles as you can.

**a**

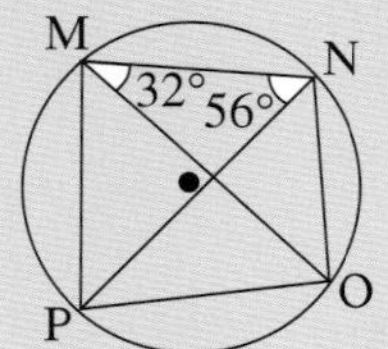


**b**

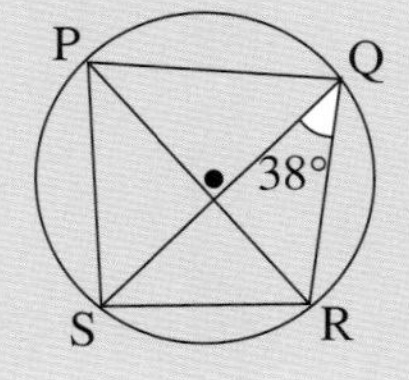


**6** Find the lettered angles

**a**

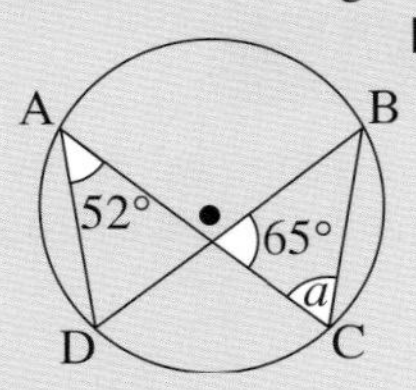


**b**

**c**

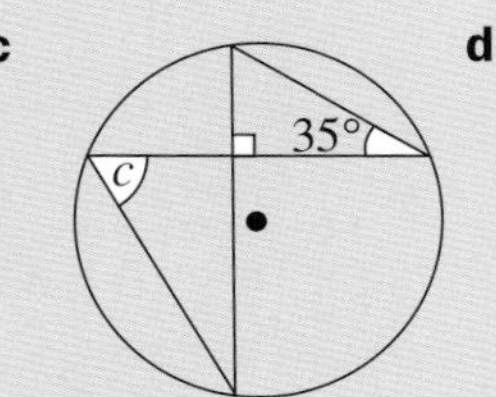


**d**

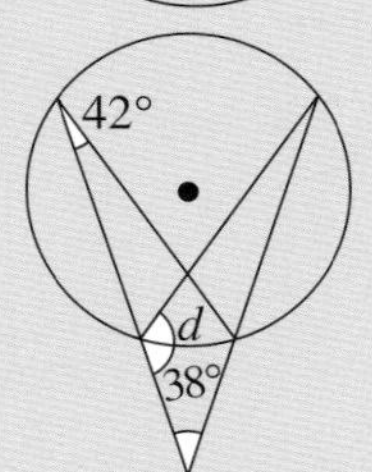


**7**

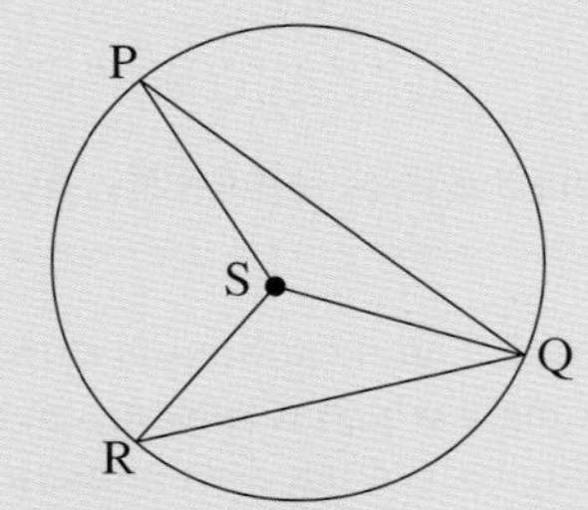


S is the centre of the circle.

**a** If $P\hat{S}Q = 142°$ and $Q\hat{S}R = 92°$, calculate $P\hat{Q}R$.

**b** If $Q\hat{S}R = 115°$ and $P\hat{Q}R = 78°$, calculate $P\hat{S}Q$.

## Angle in a semicircle

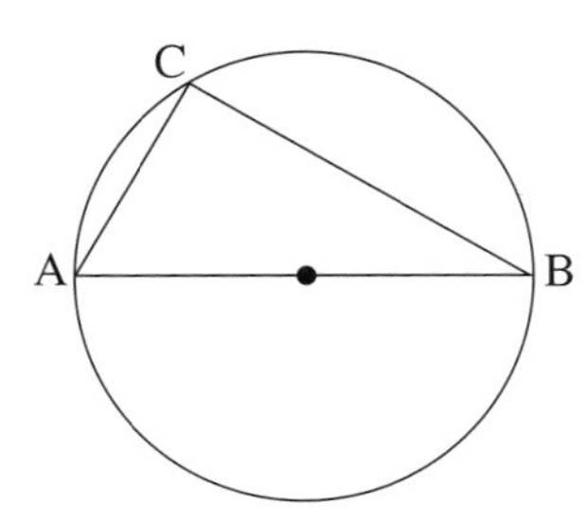


Draw a circle of any convenient radius and draw the diameter, AB. Join A and B to any point C on the circumference.

The angle at the centre $= 180°$

The angle subtended by the chord AB on the circumference is half the angle which the same chord subtends at the centre, so

$A\hat{C}B = 90°$

- The angle in a semicircle is 90°.

Check this by measuring.

## The cyclic quadrilateral

**a** Draw a circle with a dashed chord LM.

**b** Choose a point X on the major arc and a point Y on the minor arc. Join XL, XM, YL and YM as shown in the diagram.

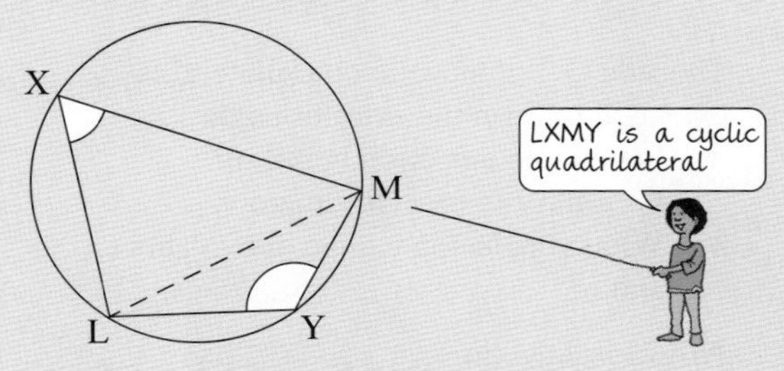


**c** Measure $L\hat{X}M$ and $L\hat{Y}M$. Add the measurements. What do you notice?

**d** On your drawing, join XY with a dashed line to form a new chord.

**e** Do you agree that angles $X\hat{M}Y$ and $X\hat{L}Y$ stand on this chord?

**f** Measure $X\hat{M}Y$ and $X\hat{L}Y$. Add the measurements. What do you notice?

Quadrilaterals inscribed inside circles, like LXMY in the Activity, are called **cyclic quadrilaterals**.

- The sum of opposite interior angles in a cyclic quadrilateral is always 180°.

For example, in the Activity

$$L\hat{X}M + L\hat{Y}M = 180°$$
$$\text{and } X\hat{L}Y + X\hat{M}Y = 180°$$

## Tangents

A tangent is a straight line that touches the circumference of a circle at one point only. This point is called the **point of contact**.

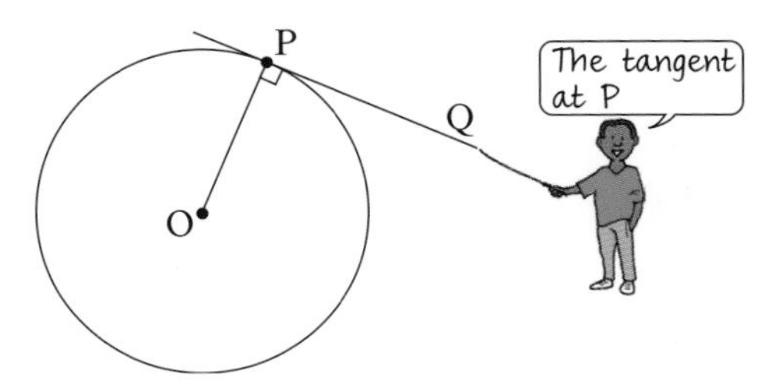


In the diagram $O\hat{P}Q = 90°$.

The tangent is perpendicular to the radius of the circle at the point of contact.

When two tangents are drawn from an external point, the lengths of these tangents are equal.

In the diagram XB = XA.

These two results can help you find missing angles in a variety of situations.

### EXAMPLE 7

In the diagram, find the angle $x$.

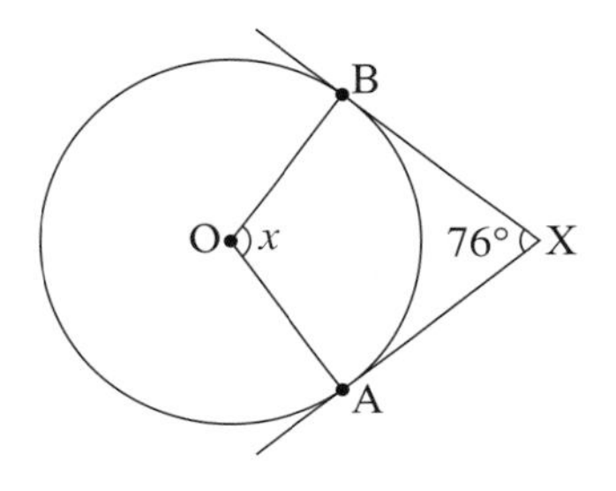


$O\hat{B}X = O\hat{A}X = 90°$ (tangent is perpendicular to radius)

The shape OBXA is a quadrilateral so its angle sum is 360°.

Hence

$$x + 90° + 90° + 76° = 360°$$
$$x = 104°$$

### Exercise 9G

**1**

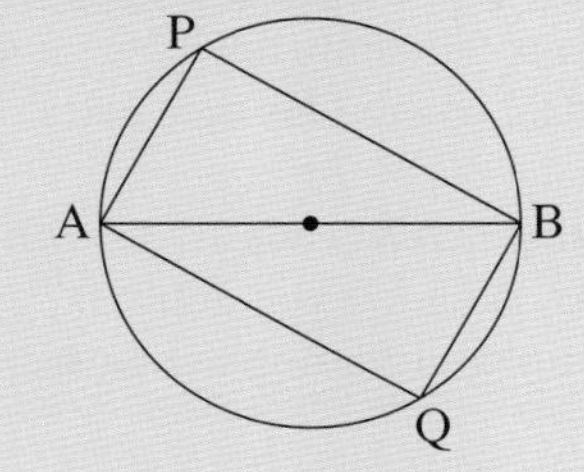


AB is a diameter of this circle.

**a** Find $A\hat{P}B$.

**b** If $Q\hat{A}B = 40°$ what is $A\hat{B}Q$?

**2** Write down the sizes of the angles marked with letters.

**a**

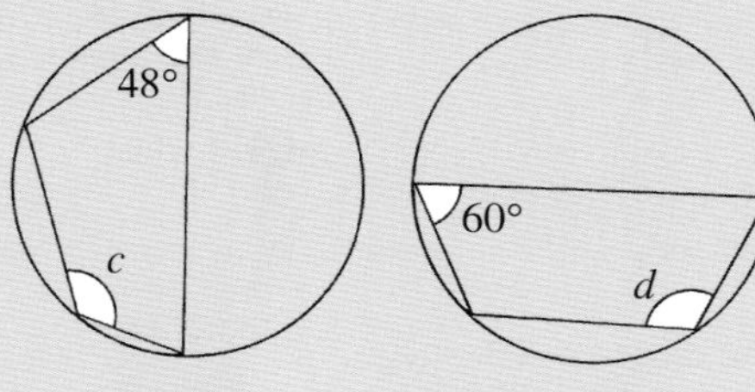


**b**

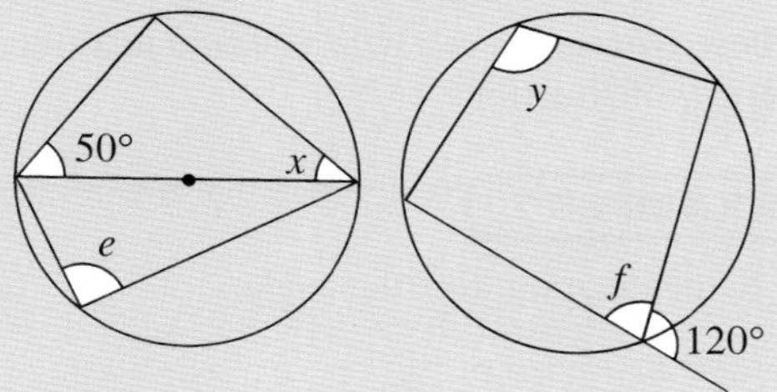


**3** Draw a tangent to three different circles. With a protractor, check that the angle made with the radius at the point of contact is 90°.

**4** Draw copies of this diagram for circles of radius:

**a** 3 cm **b** 4 cm **c** 5 cm

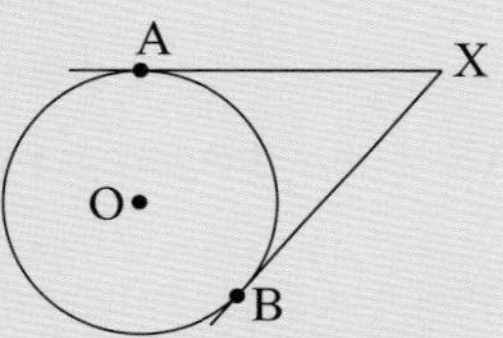


**d** Copy and complete the table:

| Radius (cm) | AX (cm) | BX (cm) |
|---|---|---|
| 3 | | |
| 4 | | |
| 5 | | |

Is AX = BX in each diagram?

**5** **a** For the diagrams you drew in Question **4**, draw the line OX where O is the centre of the circle.

**b** Measure the angles $A\hat{X}O$ and $B\hat{X}O$ with your protractor. What do you notice?

**c** Can you explain your finding in part **b**?

**6**

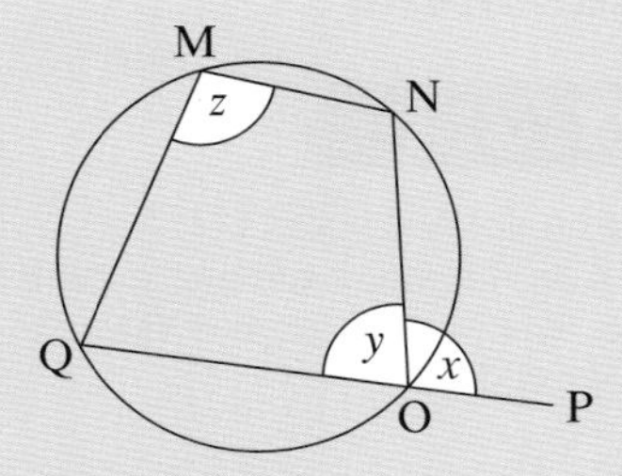


The side QO of a cyclic quadrilateral is extended to form the exterior angle $x$.

**a** State the connection between:

**i** $y$ and $z$ **ii** $x$ and $y$.

**b** What conclusion can be drawn about $x$ and $z$?

**7** ABCD is a cyclic quadrilateral. O is the centre of the circle. $A\hat{O}C = x$ is obtuse. $A\hat{O}C = y$ is reflex.

**a** Write down a relationship between:

**i** $x$ and $q$ **ii** $x$ and $y$ **iii** $y$ and $p$.

**b** What relationship follows for $p$ and $q$, the opposite angles of the cyclic quadrilateral?

**8** In the diagrams, lines drawn from point X are tangents and O is the centre of a circle. Find the lettered angles.

**a**

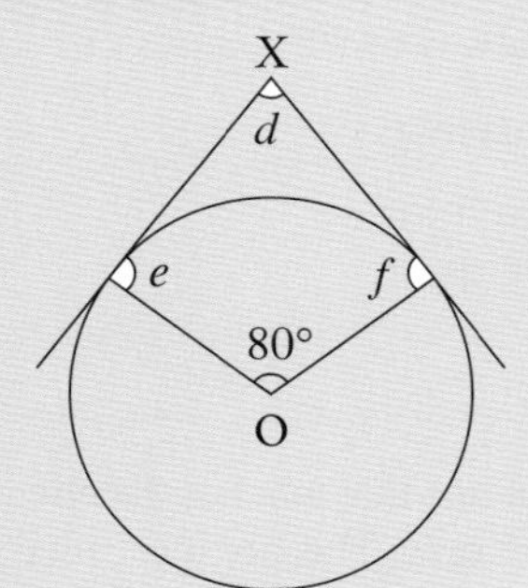


**b**

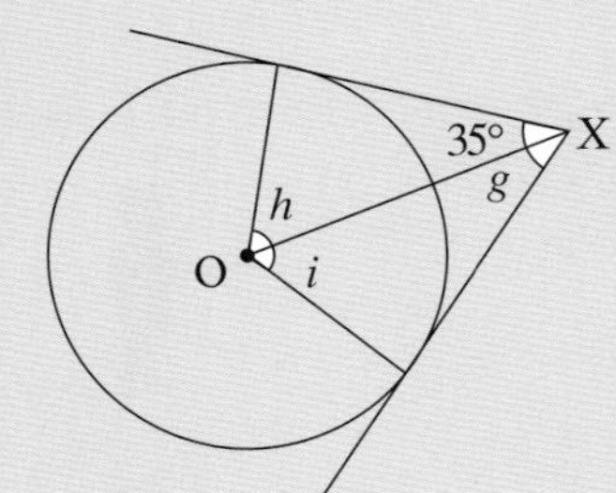

**c**

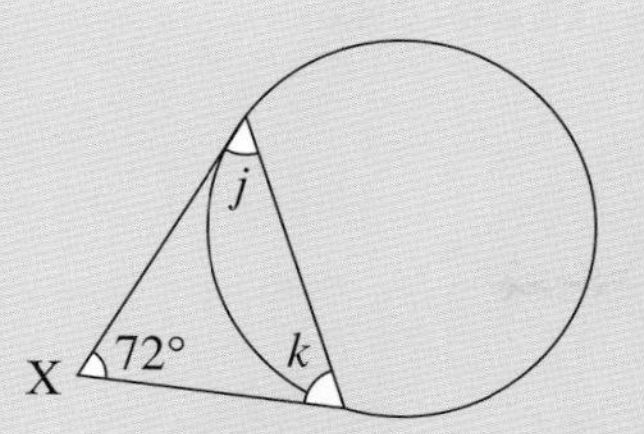


**d**

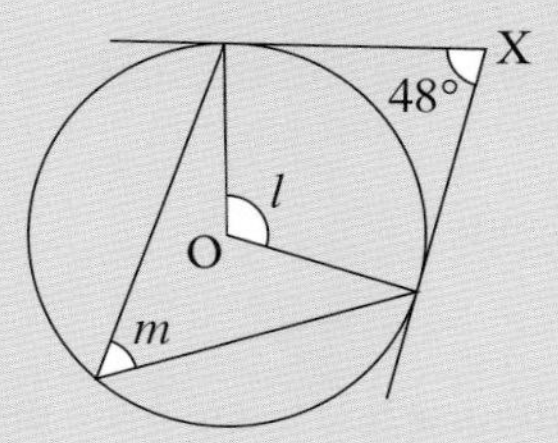


**9** Work out the lettered angles.

**a** **b**

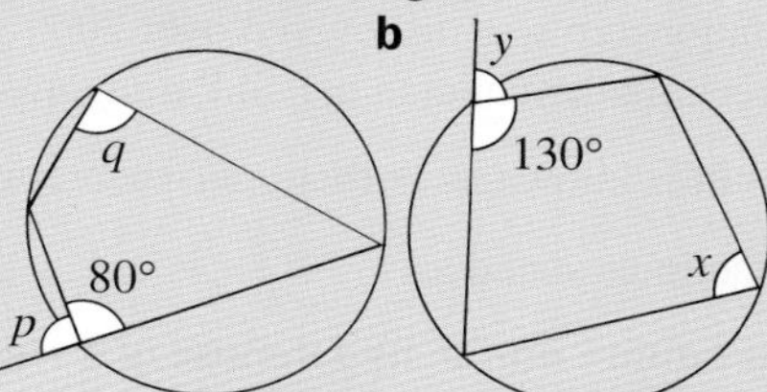


**c** 112° 135° $e$ $f$

**d**

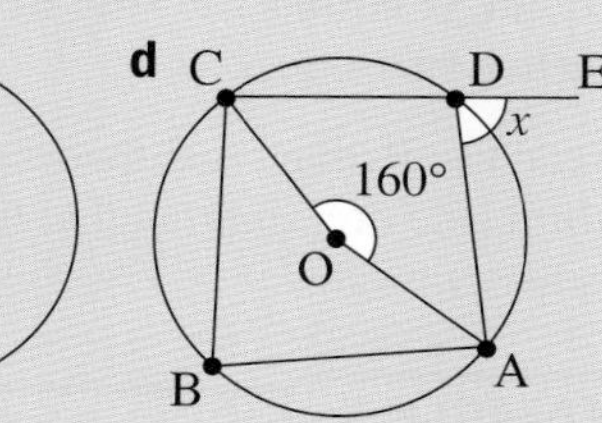


**10** Find angles QP̂S and QÔS. Give reasons for your answers.

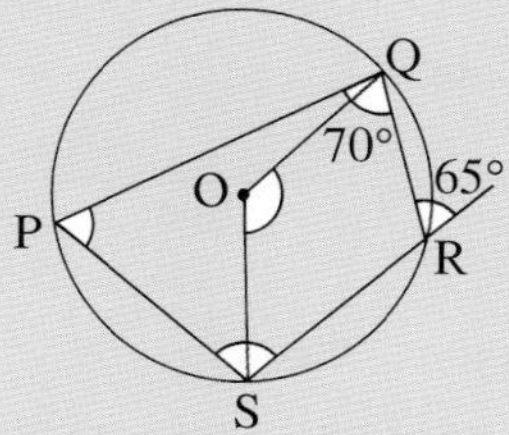


**11** In the figure, O is the centre of the circle. BP is parallel to OQ. If PB̂A = 42°, calculate QP̂A, BP̂Q and OQ̂P.

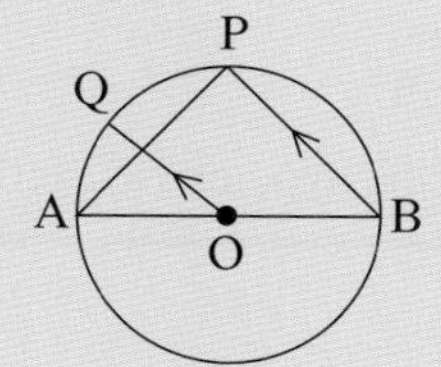


## 9.4 Pythagoras' theorem

### Right-angled triangles

You will need a set square and centimetre-squared paper.

The longest side of a right-angled triangle is called the hypotenuse.

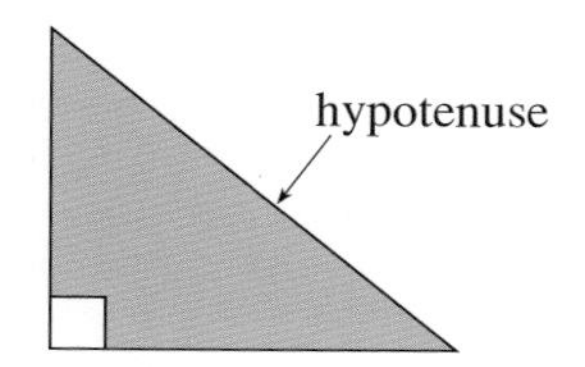


### Exercise 9H

**1** This shape has been drawn on centimetre-squared paper:

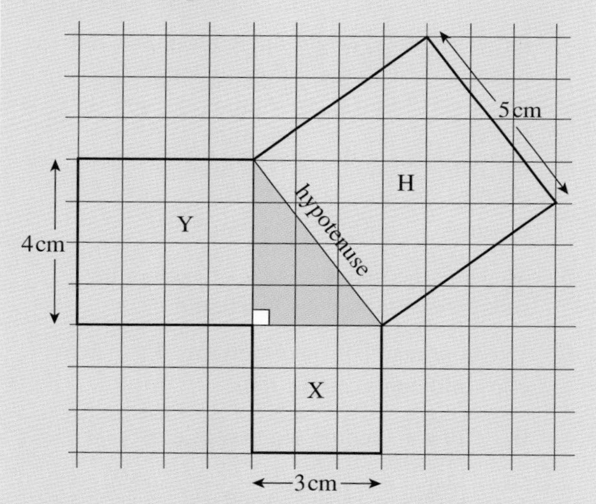


**a** Find the areas of the squares X and Y.

**b** Find the area of the square H.

**c** Can you find a relationship between the areas of squares X and Y and the square H?

**2** Repeat Question **1** for this shape. (Use a counting squares method to find the area of the square H.)

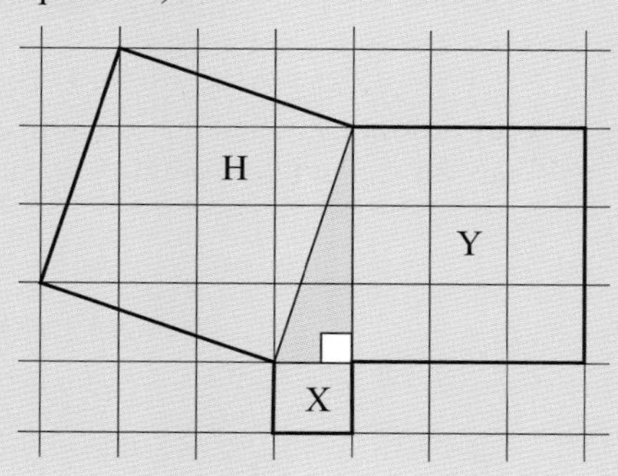

**3** **a** Draw a right-angled triangle on centimetre-squared paper. Make the two shorter sides an exact number of centimetres.
**b** Draw a square on each side. Draw the square on the hypotenuse very carefully, using a set square.
**c** Find the area of each square.
**d** What is the relationship between the three areas?

**4**

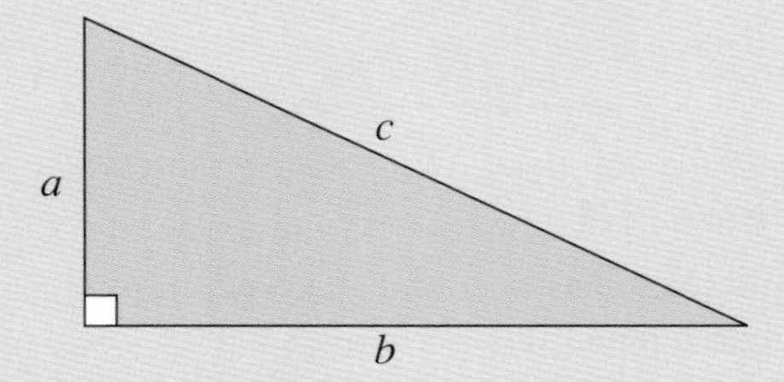


Draw accurately right-angled triangles like the one shown with the following lengths:
**a** $a = 12$ mm, $b = 16$ mm
**b** $a = 30$ mm, $b = 40$ mm
**c** $a = 5$ mm, $b = 12$ mm
**d** $a = 20$ mm, $b = 21$ mm
**e** $a = 12$ mm, $b = 35$ mm

**5** Measure the hypotenuse, $c$ of each of the triangles you drew in Question **4**.

**6** Use your results from Questions **4** and **5** to copy and complete the table:

| | a | b | c | $a^2$ | $b^2$ | $a^2 + b^2$ | $c^2$ |
|---|---|---|---|---|---|---|---|
| **a** | 12 | 16 | | 144 | 256 | 400 | |
| **b** | 30 | 40 | | | | | |
| **c** | 5 | 12 | | | 144 | | |
| **d** | 20 | 21 | | | | | |
| **e** | 12 | 35 | | | | | |

What do you notice about the last two columns in the table?

## Pythagoras' theorem

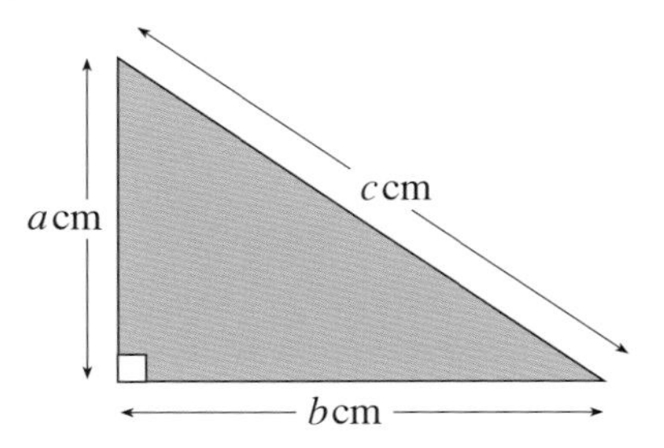


In a right-angled triangle

$$c^2 = a^2 + b^2$$

where **$c$ is the hypotenuse** and $a$ and $b$ are the other two sides.

This relationship was discovered over 2000 years ago by a Greek mathematician called Pythagoras.

This result is known as Pythagoras' theorem.

It is sometimes stated as:

- **The square on the hypotenuse of a right-angled triangle equals the sum of the squares on the other two sides.**

You can use Pythagoras' theorem to find the length of the third side of a right-angled triangle, if you know the other two sides.

**EXAMPLE 8**

Find the length, $c$, of the hypotenuse of the triangle.

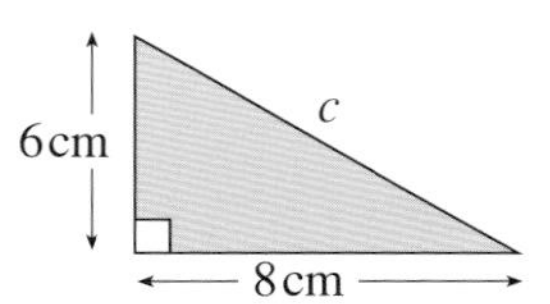


Using Pythagoras' theorem

$c^2 = a^2 + b^2$
so $c^2 = 6^2 + 8^2$
$= 36 + 64$
$= 100$
so $c = \sqrt{100}$
$= 10$ cm

The length of the hypotenuse is 10 cm.

**EXAMPLE 9**

In the triangle, find the length $a$.

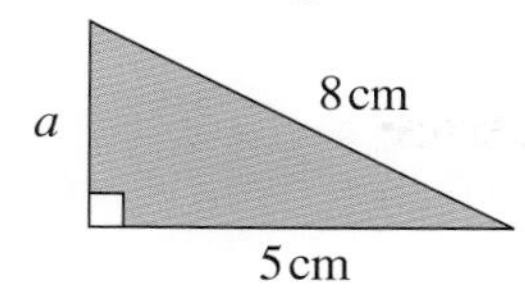


By Pythagoras' theorem:

$$a^2 + 5^2 = 8^2$$
$$a^2 + 25 = 64$$
$$(-25) \quad a^2 = 39$$
$$a = \sqrt{39}$$

so $\quad a = 6.24$ cm (2 d.p.)

## Exercise 9I

**1** Look at the triangle, then copy and complete the statements which follow.

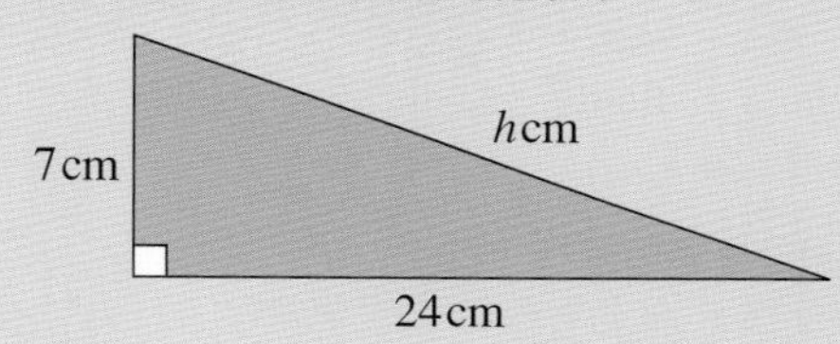


$$h^2 = \square^2 + 24^2$$
$$h^2 = \square + \square$$
$$h^2 = \square$$

so $\quad h = \sqrt{\square} = \square$

**2** Find the length of $c$ in these right-angled triangles.

**a**

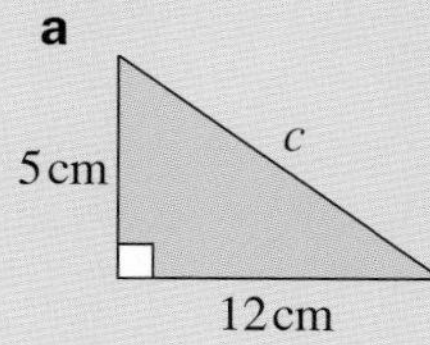


**b**

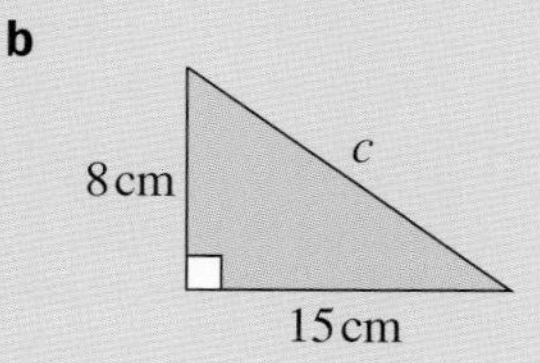


**c**

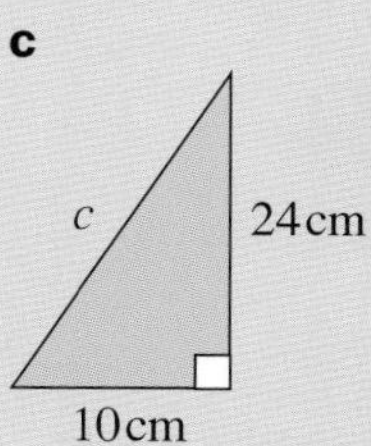


**d**

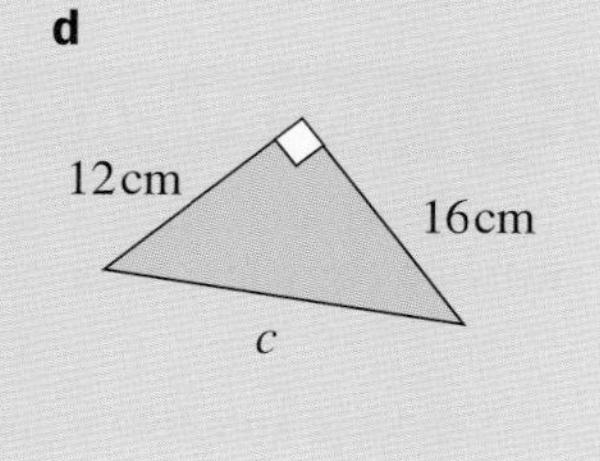


**3** Use your calculator to find the unknown length represented by the letter.

**a**

2 cm, $w$ cm, 3 cm

**b**

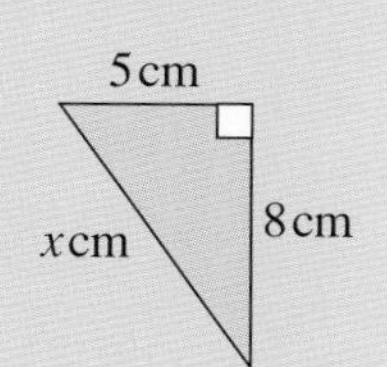


**c**

4 cm, 4 cm, $y$ cm

**d**

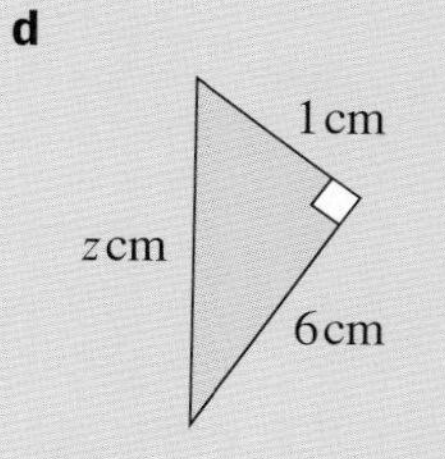


**4** Copy and complete the statements for this triangle.

$$a^2 + \square^2 = 5^2$$
$$a^2 + \square = \square$$
$$a^2 = \square$$

so $\quad a = \sqrt{\square}$

$$a = \square$$

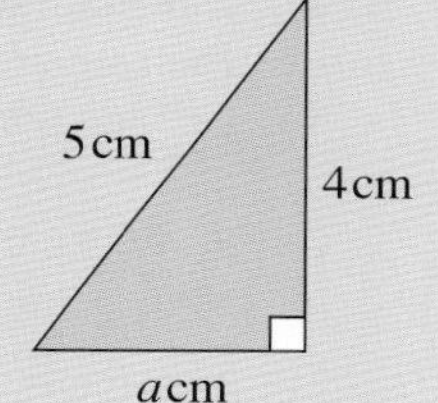


**5** For each triangle, find the unknown length represented by the letter.

**a**

6 cm, 10 cm, $f$ cm

**b**

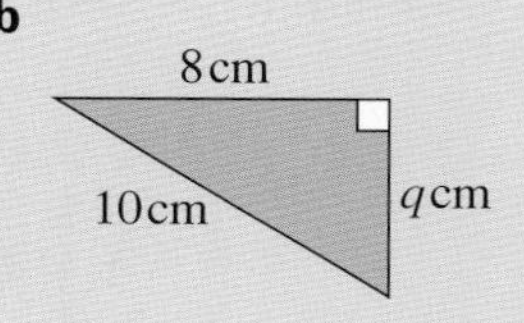


**c**

12 cm, $x$ cm, 15 cm

**d**

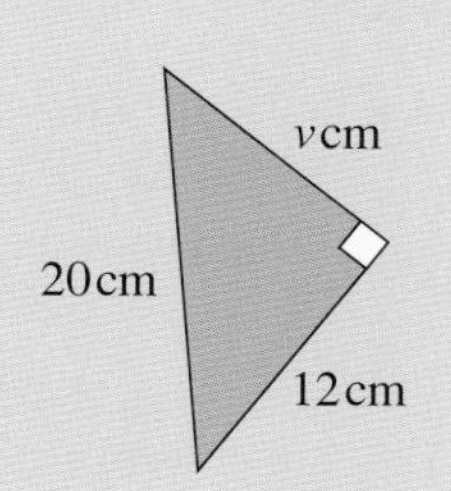

**6** Find the value of $x$.

**a**

$x$ cm, 5 cm, 6 cm

**b**

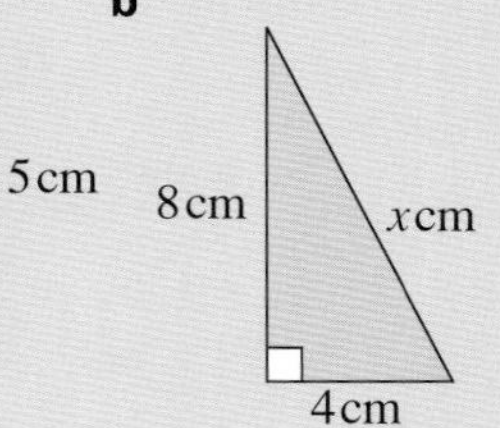


**c**

**d**

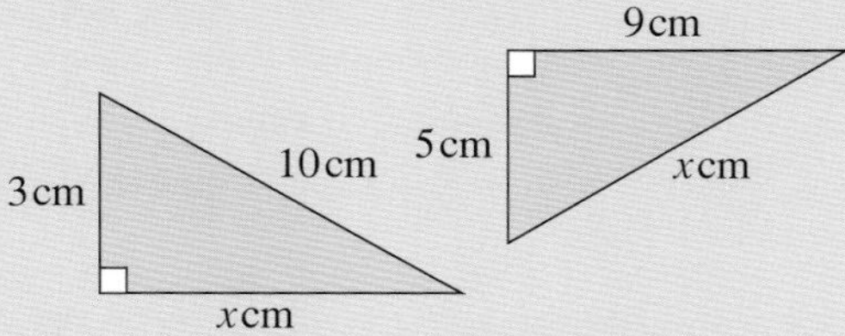


**7 a** Calculate the length of the diagonal in each square.

**i**

5 cm, 5 cm

**ii**

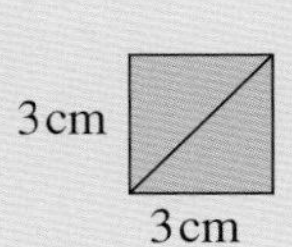


**b** Calculate the length of the diagonal in each rectangle.

**i**

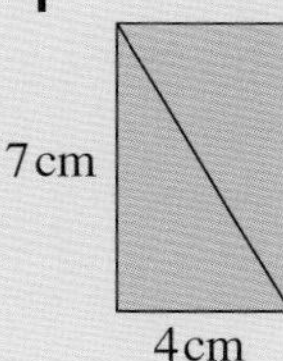

**ii**

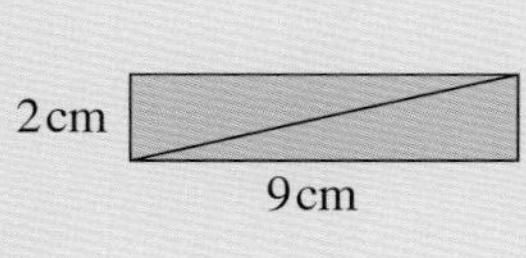


**8** Triangle ABC is isosceles, with AB = AC = 25 cm and BC = 10 cm
Calculate its height AX.

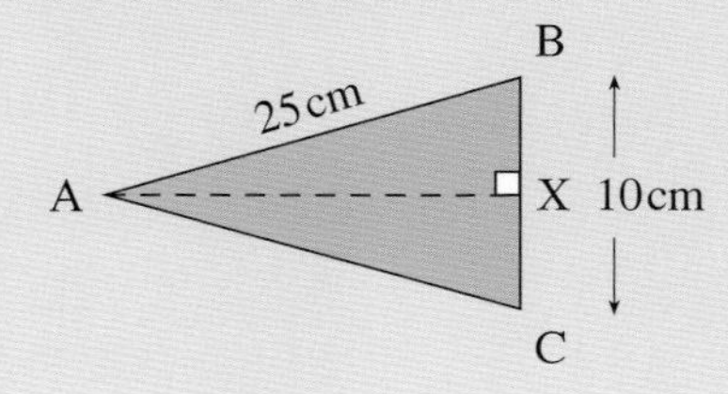


**9** The diagonals of a rhombus intersect at right angles. Find the length of the vertical diagonal in the rhombus shown.

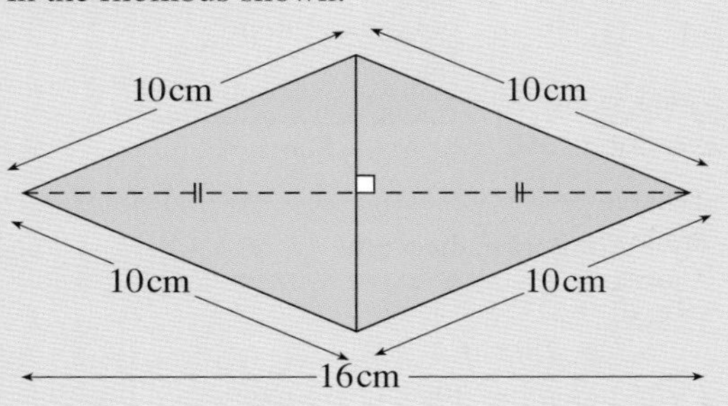


**10** The diagonals of a kite intersect at right angles. Find the length of each diagonal using the measurements given below.

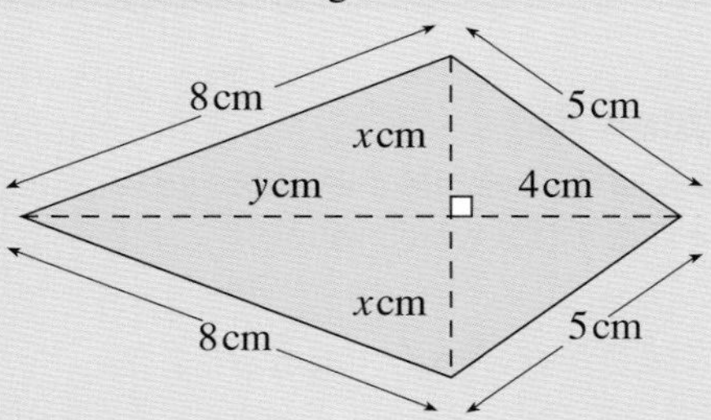


**11** Find the values of $x$ and $y$.

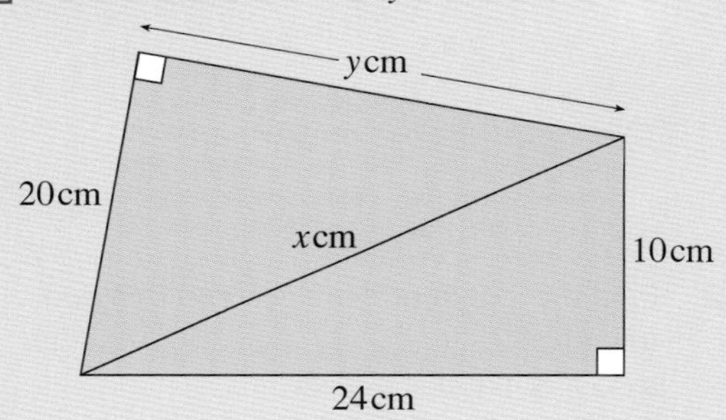


**12** PQRS is a kite and QS is its line of symmetry. Angle P = angle R = 90°, QR = 36 cm and RS = 16 cm. Find the length QS.

## INVESTIGATION

The whole numbers 3, 4, and 5 form a Pythagorean triple because:

$$3^2 + 4^2 = 5^2$$

5, 12, 13 is another Pythagorean triple since

$$5^2 + 12^2 = 13^2$$

What other Pythagorean triples can you find?

## 9.5 Tessellations

In this section you will make patterns from shapes.

You will need card, a pair of scissors, tracing paper and coloured pencils or crayons.

Make an exact copy, on card, of this hexagon. Use tracing paper to help you. (Alternatively, ask your teacher for a copy of the hexagon on card, from the Teacher Book.) Cut out your copy very carefully.

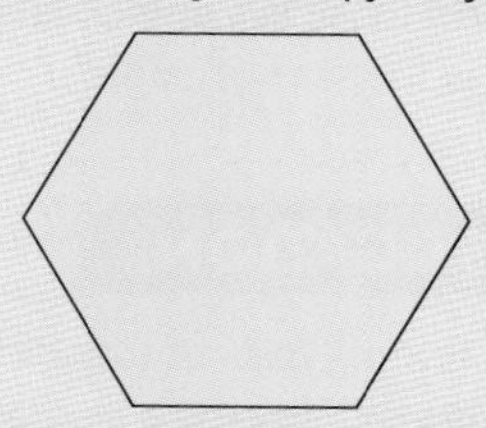

Place your hexagon on a sheet of paper. Draw its outline. Continue drawing outlines to make a larger copy of the pattern below. Colour your pattern with coloured pencils or crayons.

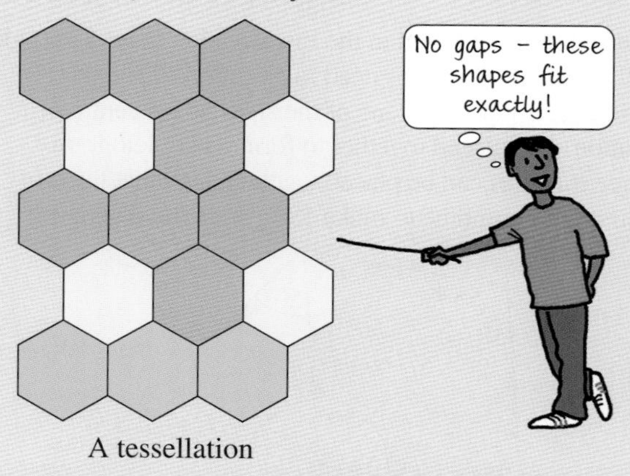


A tessellation

- A pattern made by fitting shapes together, *with no spaces and no overlapping*, is called a **tessellation**.

The pattern you made in the Activity is a tessellation.

You often see tessellations on tiled floors. People have used tessellations for thousands of years.

This photo shows Egyptian floor tiling:

### Exercise 9J

**1** **a** Make an exact copy, on card, of this shape:

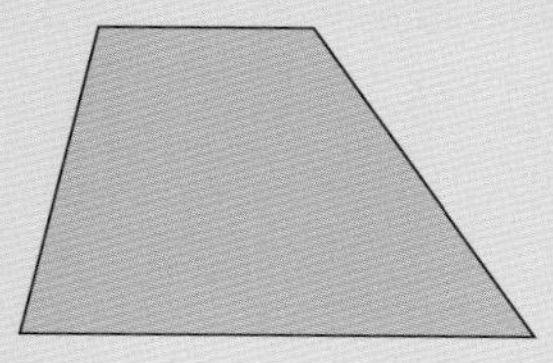

**b** Use your shape to draw a tessellation.

**c** Make sure your tessellation covers at least a quarter of your page.

**d** Colour your pattern.

**2** Repeat Question **1** for each of these triangular shapes.

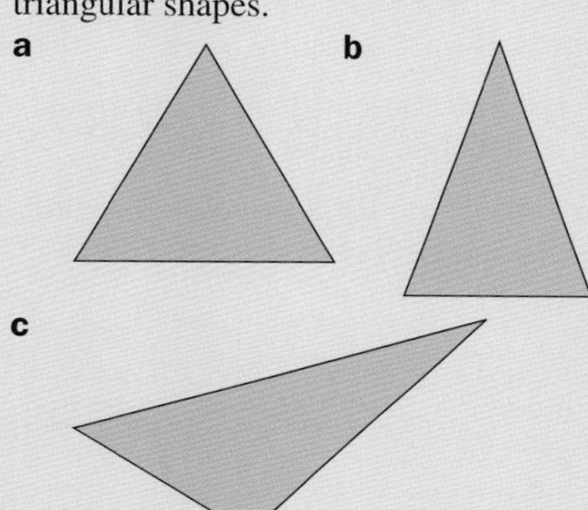


**3** Repeat Question **1** for each four-sided shape:

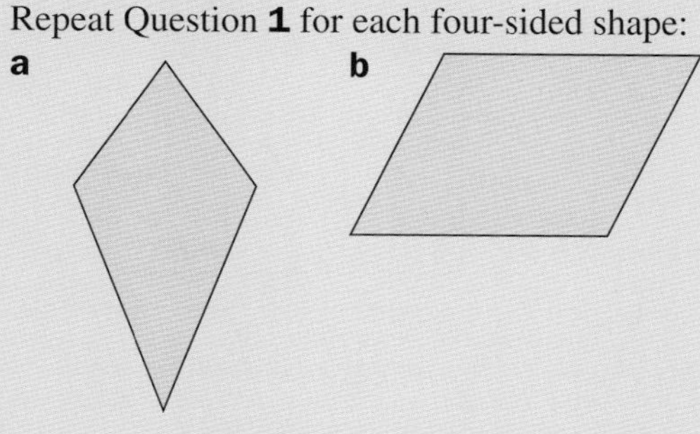


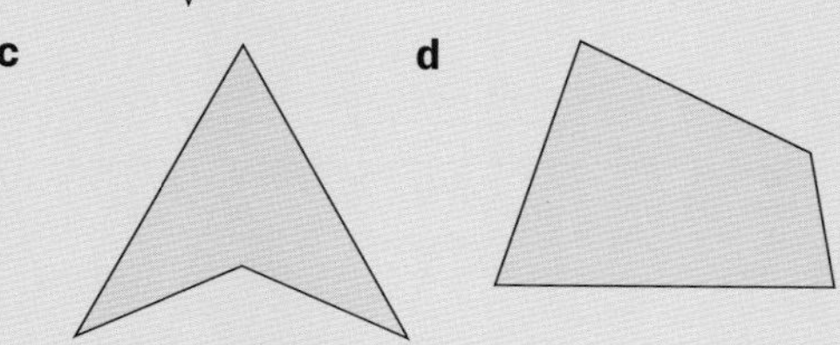


**4** Not all shapes tessellate.

Why is the pattern above not a tessellation?

**5** **a** Make cut-outs of each of these four shapes. (Copies of these shapes can be found in the Teacher Book.)

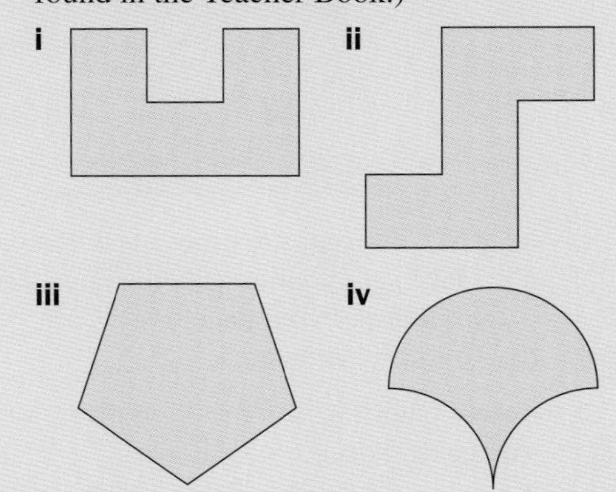


**b** Use your cut-outs to help you decide which of the shapes tessellate.

**6** Use dot paper to see if you can make different tessellations with each of these tiles.

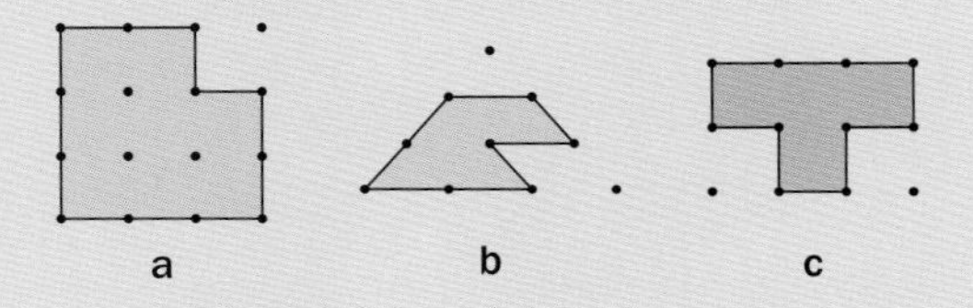


**7** In each of the pictures find the value of the angles marked $a$, $b$, $c$, $d$, $e$.

**a**

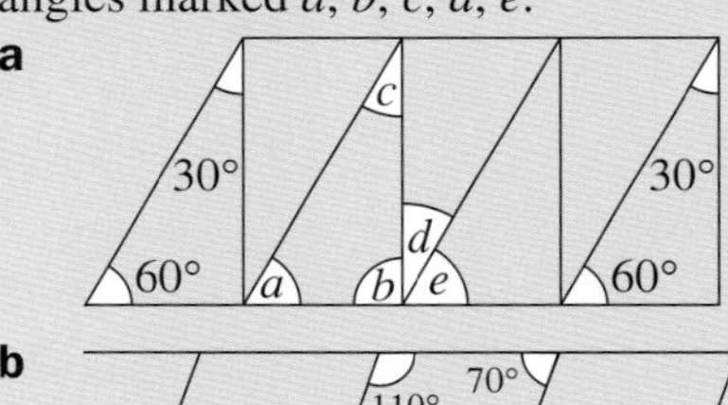


**b**

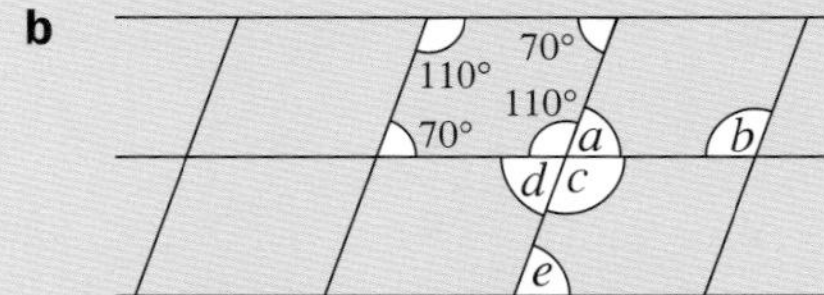


**c**

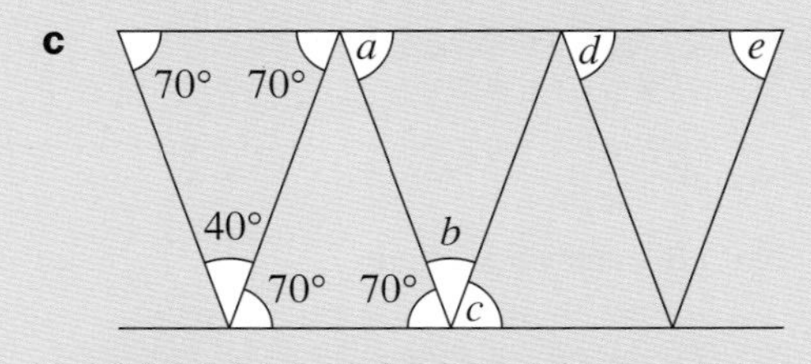


**8** Describe the transformation used to tessellate the triangles and quadrilaterals in Question **7**.

**9** Look at this tessellation of an equilateral triangle.

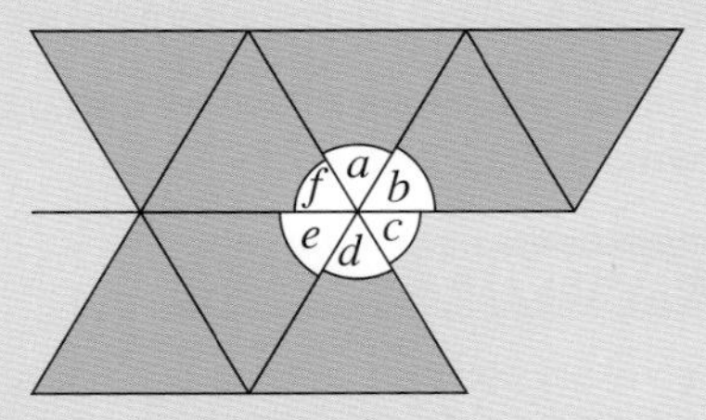


**a** Do you agree that the angles $a$, $b$, $c$, $d$, $e$, $f$ are all equal?

**b** What do these six angles add up to?

**c** What is the size of angle $a$?

## INVESTIGATION

**a** Draw four different types of triangle.
Which ones tessellate?
Can you find a triangle that does not tessellate?

**b** Repeat your Investigation but this time use quadrilaterals.

You should have found in the Investigation that all triangles will tessellate. Triangles need to be rotated through 180° (using the midpoint of one of the sides as the centre of rotation) and placed together to form a parallelogram. Three triangles can be placed together such that three of their angles combine to make 180°, as in the diagram.

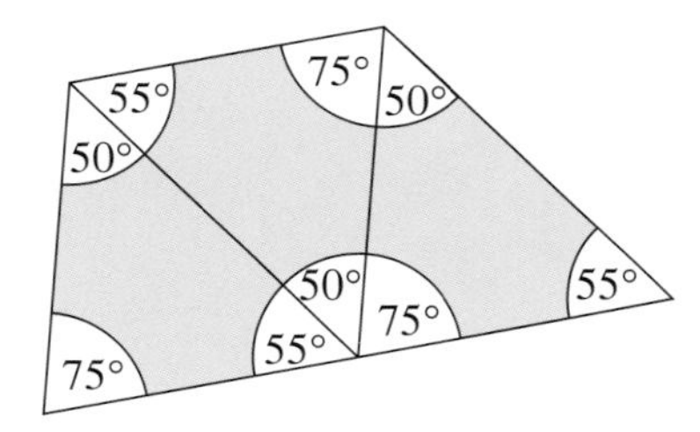


This shape can then be rotated through 180° so that there are six tessellating triangles positioned so that six of their angles join to make a 360° angle. This will work for any triangle.

A similar process can be used for quadrilaterals. The diagram shows quadrilaterals tessellated by rotation through 180° about the midpoint of a side. Four quadrilaterals will join together to make a 360° angle.

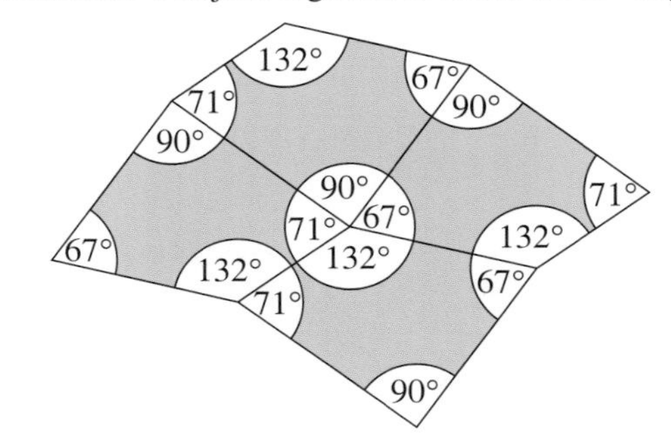

Not all polygons will tessellate like triangles and quadrilaterals. When some polygons are rotated and placed next to each other there are gaps. Look at this diagram of regular pentagons:

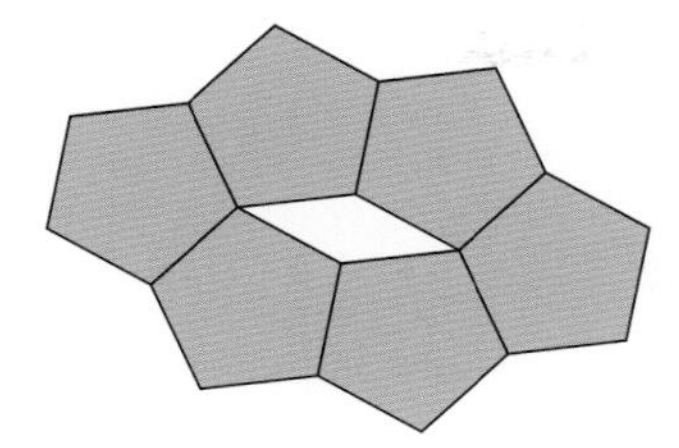

## Exercise 9K

**1** Look at the diagram of repeated regular pentagons.
What is the size of an interior angle of a regular pentagon?
Explain why regular pentagons will not tessellate.

**2** What is the size of the interior angle of a regular octagon?
Will a regular octagon tessellate? Explain your answer.

**3** Repeat Question **2** for a regular heptagon.

**4** Tessellations can be made up of more than one shape. For example, this tessellation uses regular pentagons and quadrilaterals:

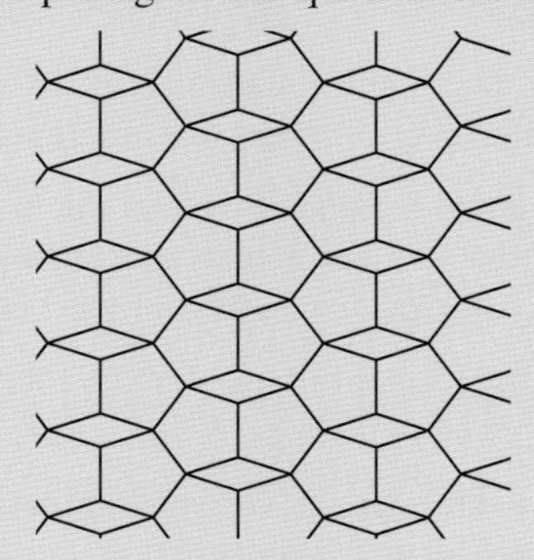

**a** What sort of quadrilateral is used in the tessellation?
**b** What are the angles inside the quadrilateral?

**5** You can tessellate a regular octagon with a quadrilateral. What sort of quadrilateral will it be? Explain your answer by talking about the angles in the tessellation.

**6** A regular hexagon will form a tessellating pattern with two different regular shapes. What are the other two shapes?

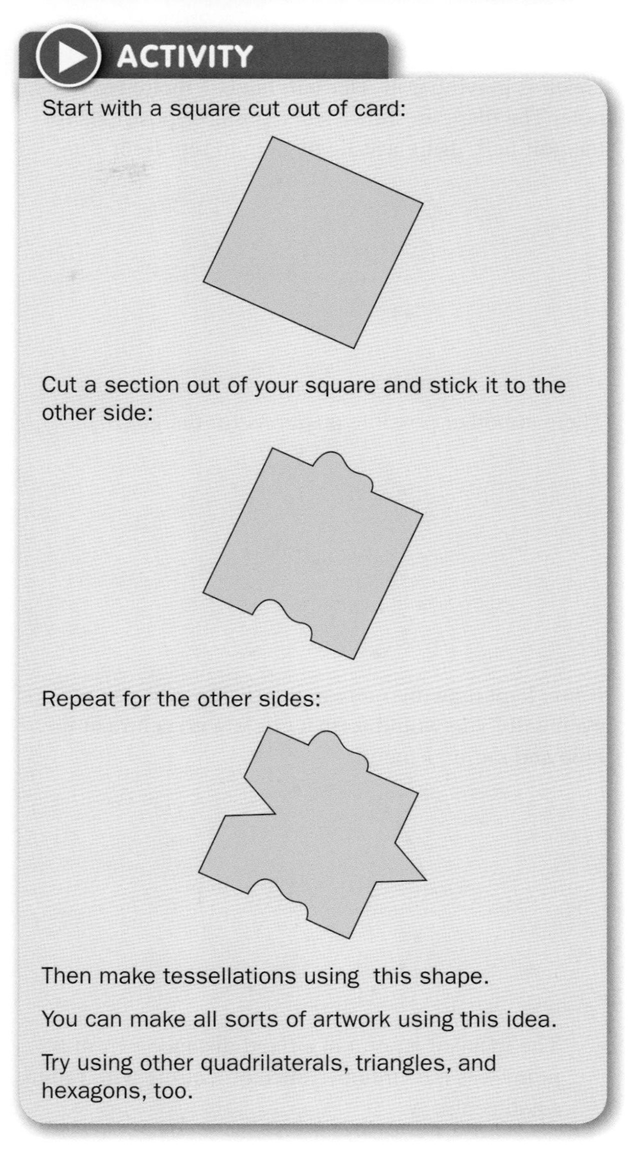

**ACTIVITY**

Start with a square cut out of card:

Cut a section out of your square and stick it to the other side:

Repeat for the other sides:

Then make tessellations using this shape.

You can make all sorts of artwork using this idea.

Try using other quadrilaterals, triangles, and hexagons, too.

## 9.6 Loci

The word *locus* (plural *loci*) means 'place' in Latin. So when we talk about finding the locus of something, we are talking about finding the places where it might be. The locus of a point describes all the possible positions the point can be in, when following a rule. A locus can be a line or a region.

Here is an example.

Helen is playing golf. Her ball lands 1 metre from the hole. The diagram shows some possible positions for the golf ball, using a scale of 2 cm to represent 1 m.

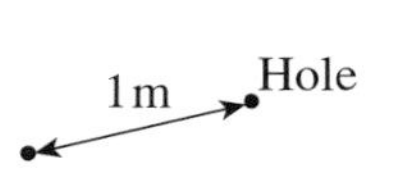


There are other possible positions for the golf ball:

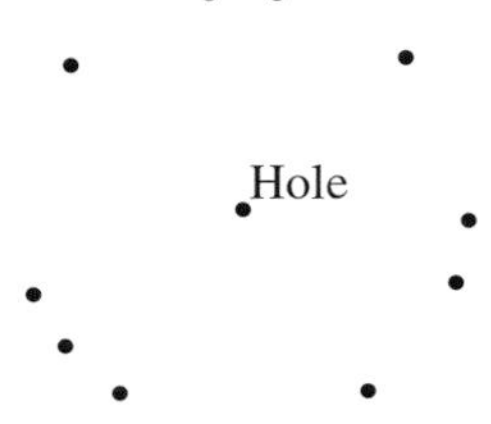


If you kept drawing more and more possible points, eventually you would form a circle with centre at the hole and radius 1 metre:

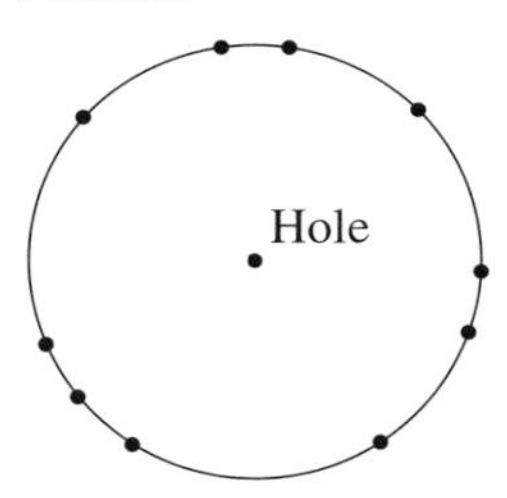


- The locus of a point at a fixed distance $r$ from a point A is the circumference of a circle with radius $r$ and centre A.

On the next hole of the golf course, Helen's ball lands 1 metre from a flagpole which has been left lying on the ground. The diagram shows some possible positions for the golf ball.

There are other possible positions for the golf ball:

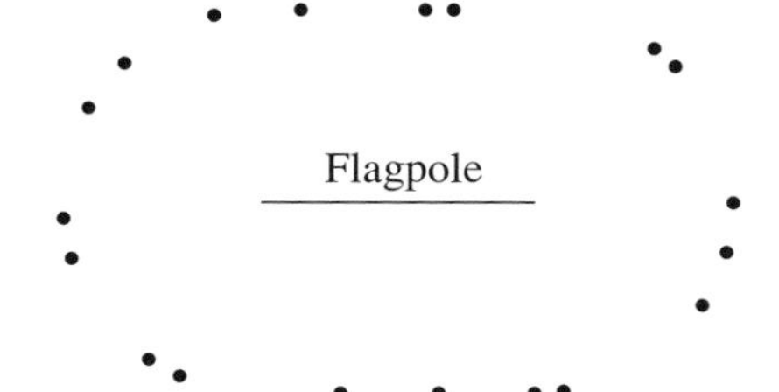


If you kept drawing more and more possible points, eventually you would form two parallel lines the same length as the flagpole and two semicircles at either end of the flagpole:

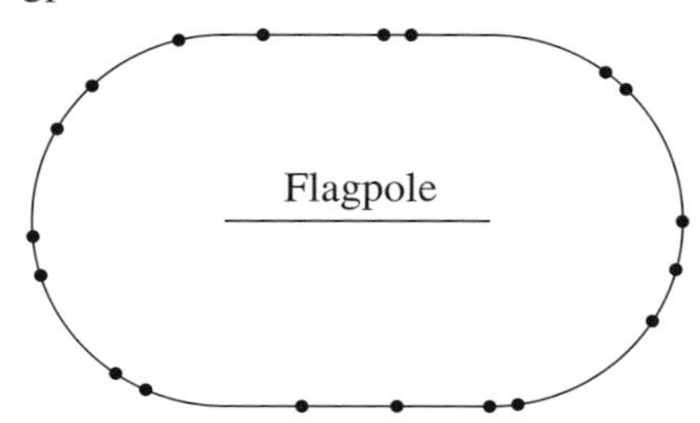


- The locus of a point at a fixed distance $r$ from a line segment AB is two parallel lines $r$ from AB and two semicircles, radius $r$ and centres A and B.

## Exercise 9L

1. Draw the locus of points that are 5 cm from a point B.
2. Draw a line segment CD, making it 4 cm long. Draw the locus of points that are 2 cm from CD.
3. Loci can be regions too. Draw the locus of points that are less than 4 cm from a point E. Check your answer with your teacher.
4. Draw the locus of points that are more than 3 cm from a point F.
5. Draw a line segment GH, making it 5 cm long. Draw the locus of points that are less than 2.5 cm from GH.
6. Draw a line segment JK, making it 4.5 cm long. Draw the locus of points that are less than 2 cm from JK *and* more than 3 cm away from K.

# Consolidation

### Example 1

Find the missing angles.

**a**

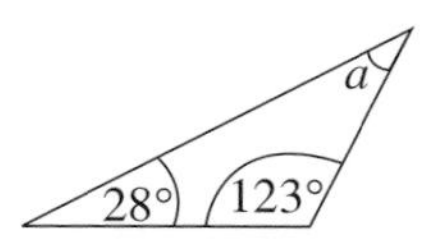


The angle sum in a triangle is 180°

Hence, $28° + 123° + a = 180°$

$151° + a = 180°$

$a = 29°$

**b**

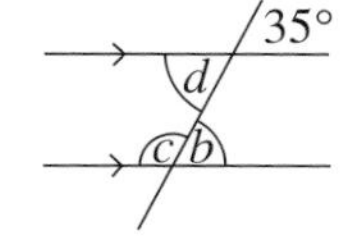


$b = 35°$ corresponding angles

$b + c = 180°$ angles on straight line

$c = 180° - 35° = 145°$

$d = 35°$ alternate angles or vertically opposite angles

**c**

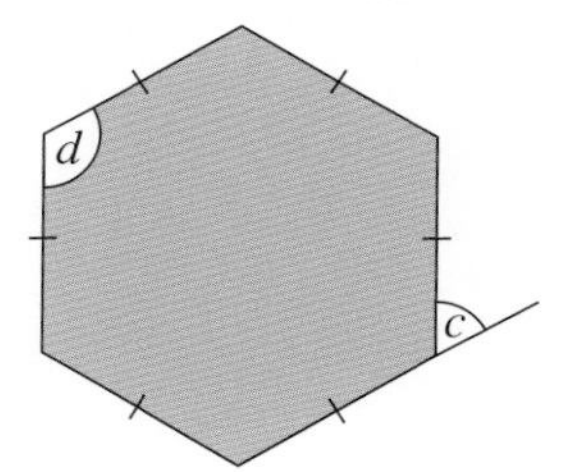


The sum of the exterior angles of a hexagon = 360°

So $6c = 360°$

$c = 60°$

$60° + d = 180°$

$d = 120°$ (interior angle + exterior angle = 180°)

**d**

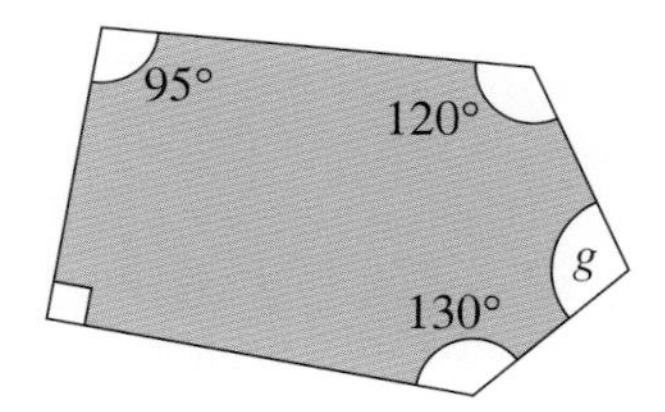


The interior angle sum of a polygon is $180(n - 2)$ where $n$ = number of sides

$180(5 - 2) = 540$

Hence, $95° + 120° + 130° + 90° + g = 540°$

$435° + g = 540°$

$g = 105°$

### Example 2

**a** Find the missing length $x$ in this right-angled triangle.

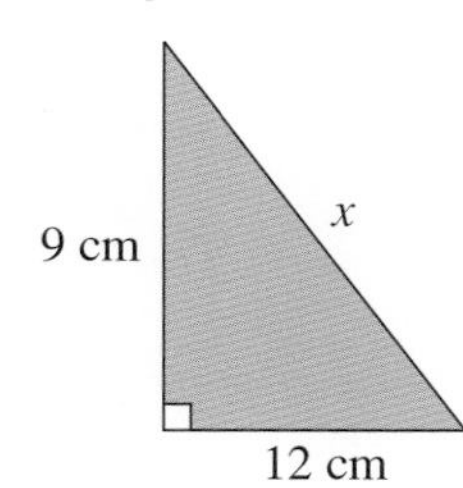


By Pythagoras' theorem:

$9^2 + 12^2 = x^2$

$81 + 144 = x^2$

$x^2 = 225$

$x = \sqrt{225}$

So $x = 15\text{cm}$

**b** Find the length marked $y$ in this right-angled triangle.

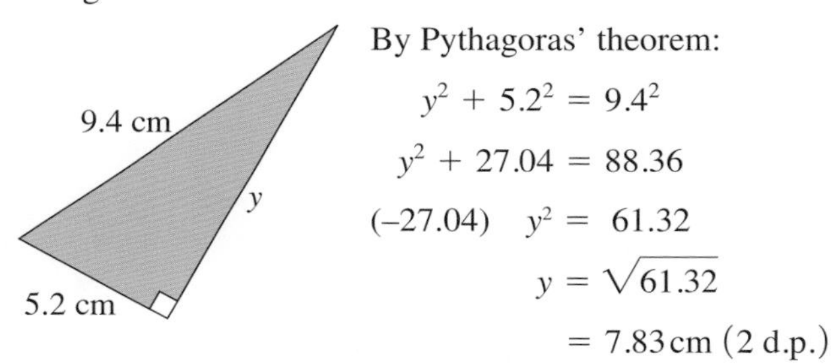


By Pythagoras' theorem:

$y^2 + 5.2^2 = 9.4^2$

$y^2 + 27.04 = 88.36$

$(-27.04)\quad y^2 = 61.32$

$y = \sqrt{61.32}$

$= 7.83\text{cm}$ (2 d.p.)

### Example 3

Will a regular hexagon tessellate with itself?

Yes, because the interior angles of a regular hexagon are all 120°, so when three regular hexagons are put together they make an angle of 360°, meaning there will be no gaps.

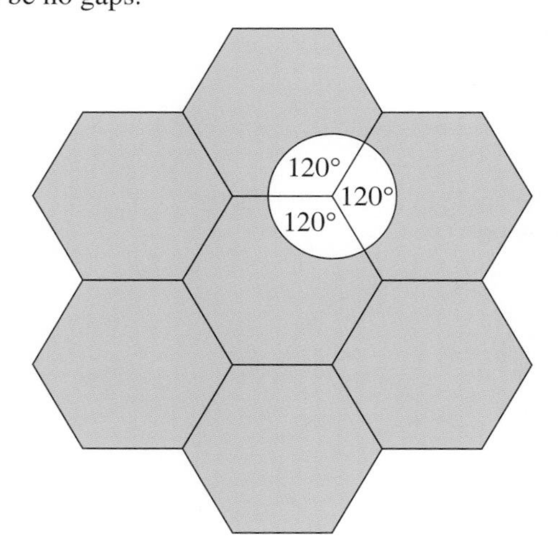

**Example 4**

**a** Draw the locus of a point that is always 2.5 cm from a fixed point A.

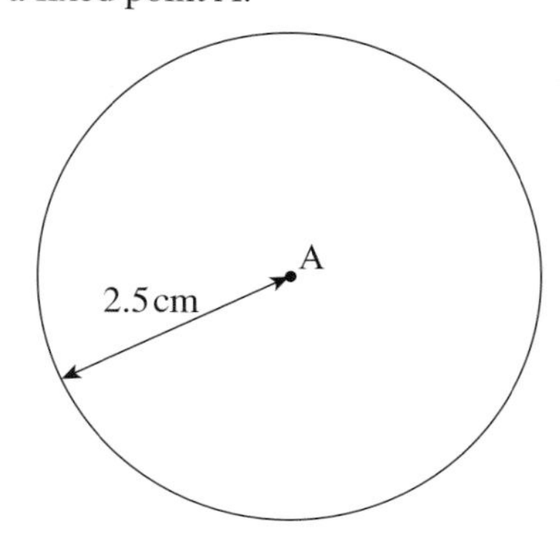


**b** Draw the locus of a point that is always 2 cm from a fixed line segment BC.

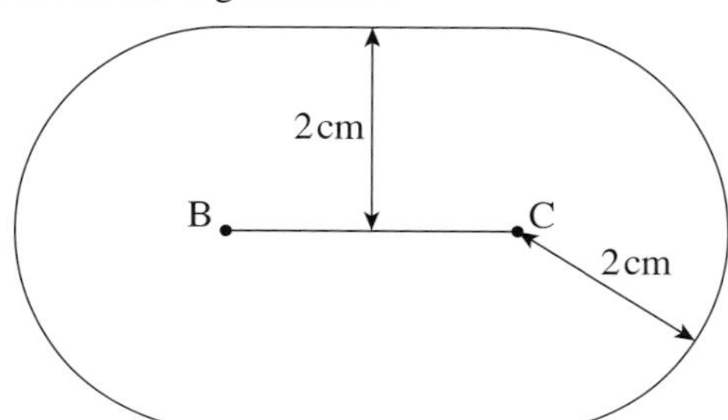


**Example 5**

Find the missing angles.

**a**

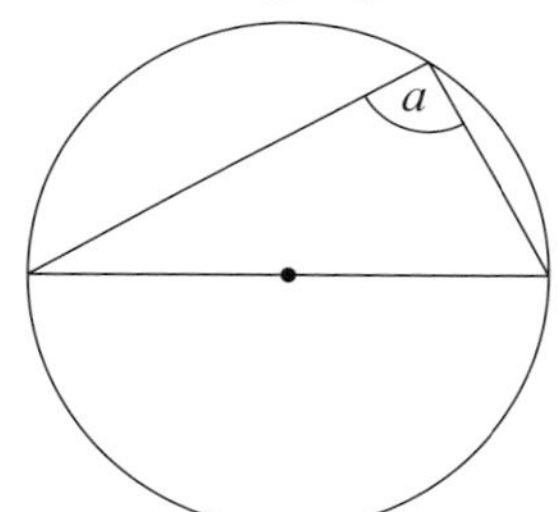


The angle in a semicircle is 90°.
$a = 90°$

**b**

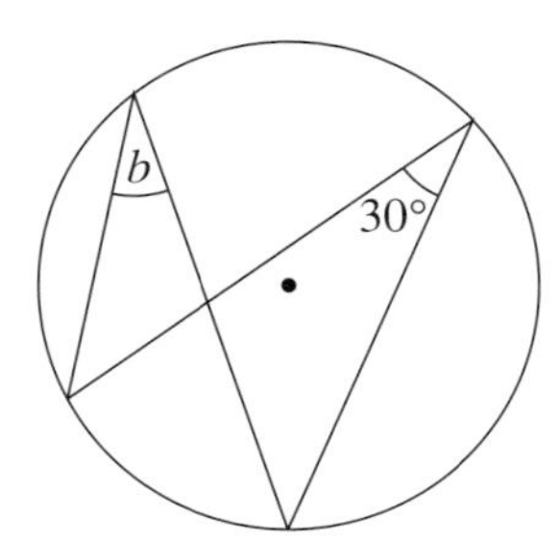


Angles in the same segment of a circle subtended by the same arc are equal.
$b = 30°$

**c**

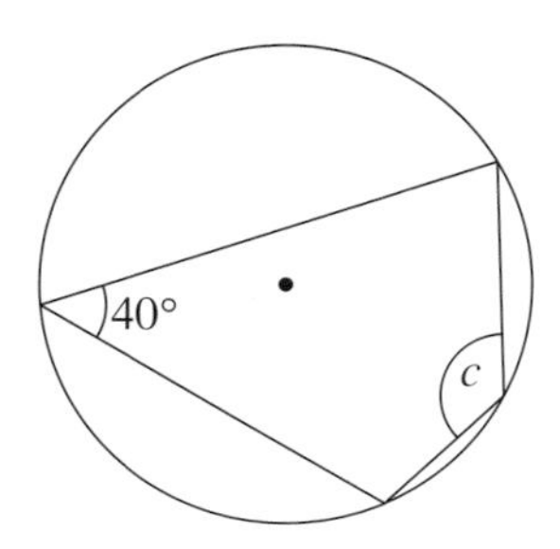


Opposite angles in a cyclic quadrilateral add up to 180°.
$40 + c = 180°$
$c = 140°$

**d**

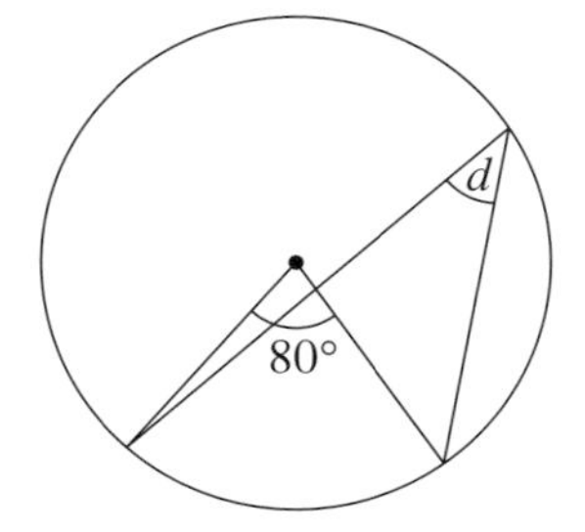


The angle subtended by an arc at the centre is double the angle subtended on the circumference.
$2d = 80°$
$d = 40°$

**e**

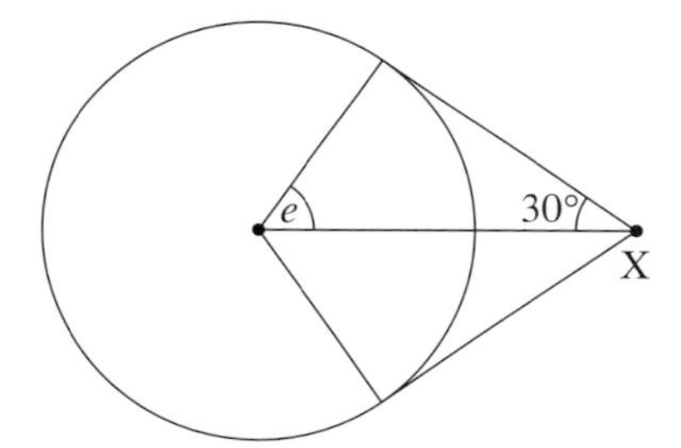


A tangent forms a 90° angle with a radius.
$90° + 30° + e = 180°$
$120° + e = 180°$
$e = 60°$

## Exercise 9

**1** Find the lettered angles:

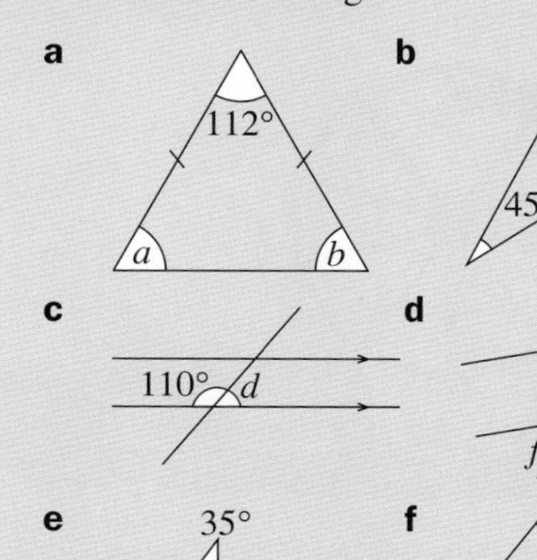


**2** Find the missing lengths in these right-angled triangles.

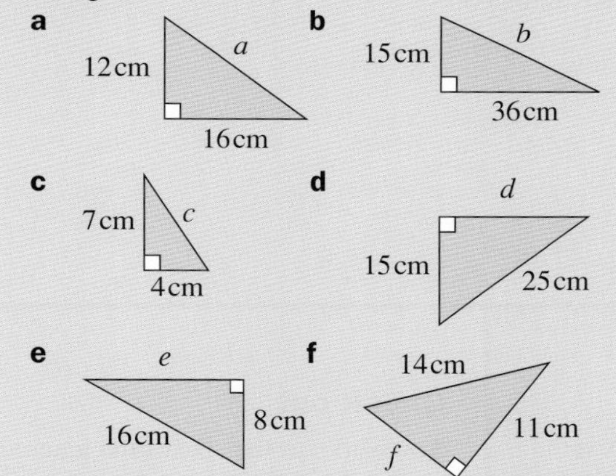


**3** **a** Draw a point. Label it A. Draw the locus of a point that is always 3.5 cm from A.

**b** Draw the line segment BC, where BC is 2.5 cm long. Draw the locus of a point that is always 4 cm from BC.

**4** Draw a tessellating pattern with a scalene triangle.
Explain why any triangle can be used to form a tessellating pattern.

**5** Mr Hassan's gate has a diagonal bracing strut to strengthen it.
What is the length of this bracing strut?

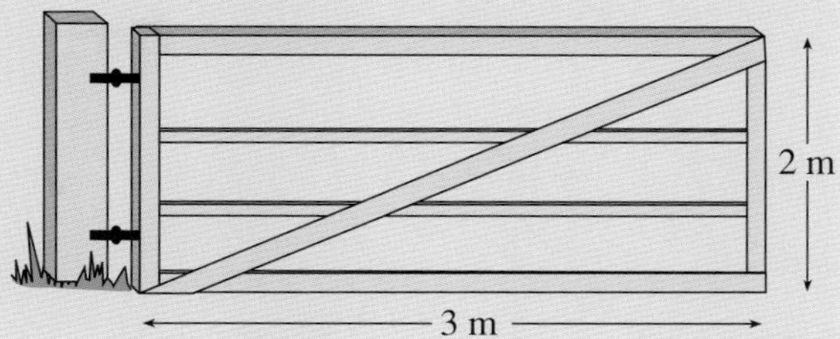


**6** The diagram shows a ladder leaning against Mr Hassan's house. The ladder is 7 metres long. The foot of the ladder is 2 metres from the wall.
How high up the wall does the ladder reach?

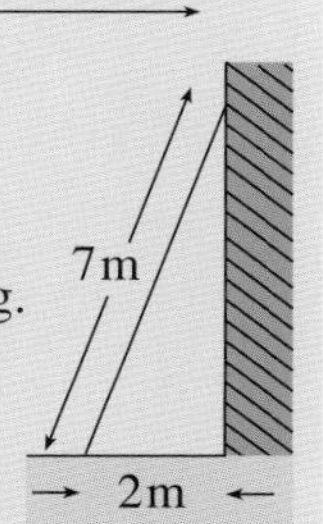


**7** If the ladder in Question **6** was 1 metre longer, and its foot in the same position, how much higher up the wall would it reach?

**8** Suppose in Question **6**, Mr Hassan moves the foot of the ladder 1 m closer to the wall.
How high up the wall does the ladder reach now?

**9** Find the lettered angles.

**a**

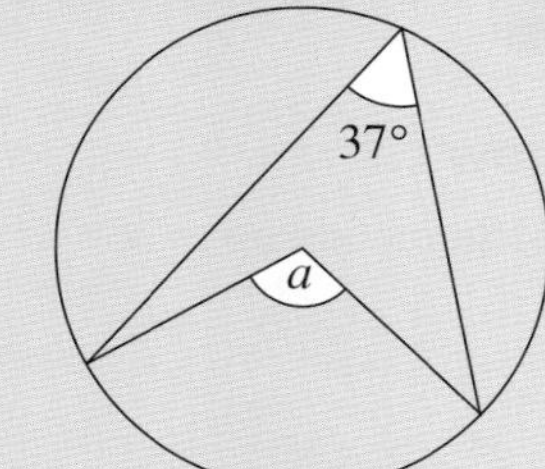


**b**

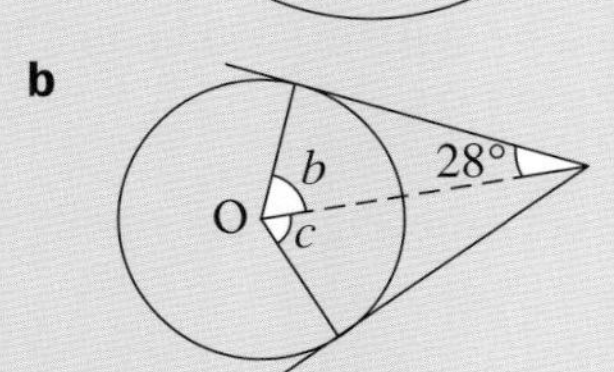


**c**

d
20°

**d**

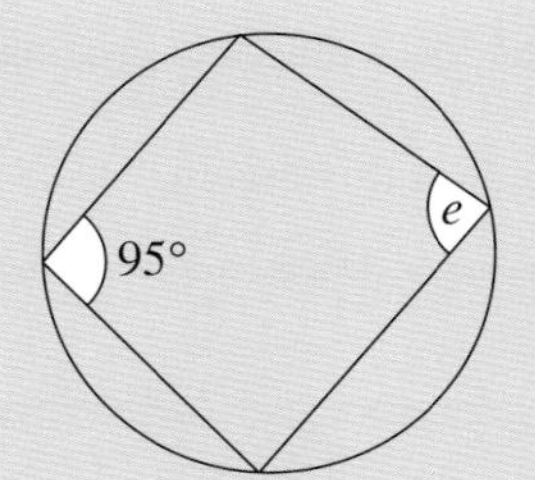


**e**

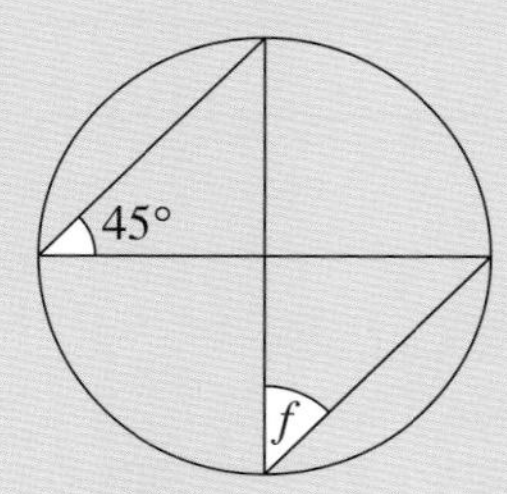


**10 a** Draw a point. Label it D. Draw the locus of a point that is more than 2 cm from D.

**b** Draw the line segment EF, where EF is 4 cm long. Draw the locus of a point that is less than 2 cm from EF.

**11** Draw a tessellating pattern with a quadrilateral of your choice.
Explain why any quadrilateral can be used to form a tessellating pattern.

**12** A big wheel in a fairground is in the shape of a regular decagon. What is the interior angle of the big wheel?

**13** What is **a** the interior angle **b** the exterior angle of a regular 15-sided shape?

**14** Will a regular pentagon tessellate? Explain your answer.

**15** What is the sum of the interior angles of a nonagon?

## Summary

### You should know ...

**1 a** Angles at a point add up to 360°.

$a + b + c + d = 360°$

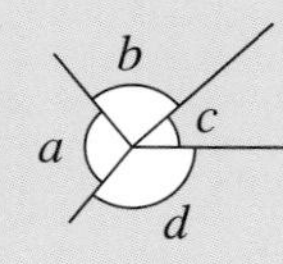


**b** Angles on a straight line add up to 180°.

$a + b + c = 180°$

**c** Vertically opposite angles are equal.

$w = y$

$x = z$

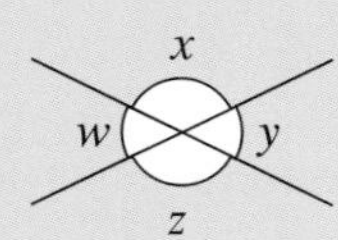


### Check out

**1** Work out the size of the lettered angles.

**a**

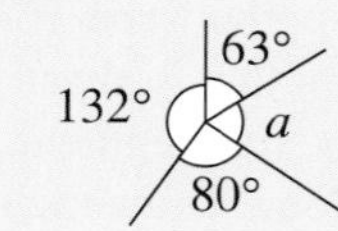


**b**

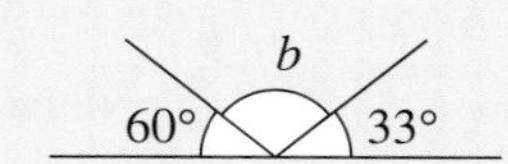


**c**

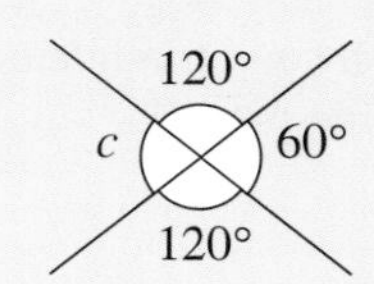

**2** **a** Alternate angles are equal.
$x$ and $y$ are alternate angles,
so $x = y$.

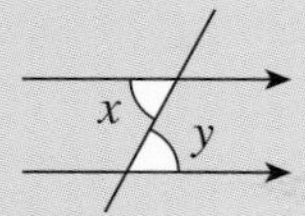


**b** Corresponding angles are equal.
$a$ and $b$ are corresponding angles,
so $a = b$.

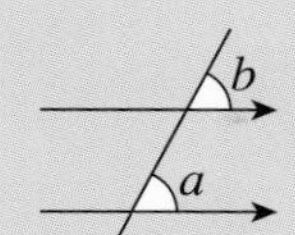


**2** Find the lettered angles.

**a**

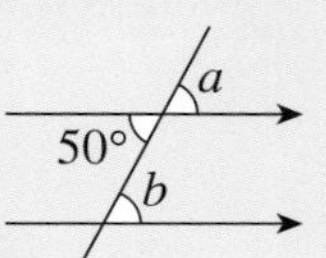


**b**

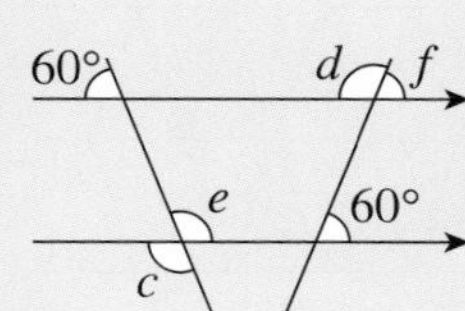


**3** The sum of the interior angles of a polygon is $180(n - 2)$.

The sum of the exterior angles of a polygon is 360°.

**3** **a** Calculate the sum of the interior angles in
- **i** a hexagon
- **ii** an octagon.

**b** What is the sum of the exterior angles of
- **i** a quadrilateral
- **ii** a triangle?

**4** **a** The exterior angle of a regular polygon is given by $\frac{360}{n}$, where $n$ = number of sides.

**b** The interior angle
$= 180° -$ exterior angle.

**4** Calculate the exterior and interior angles of a 12-sided regular polygon.

**5** Pythagoras' theorem.

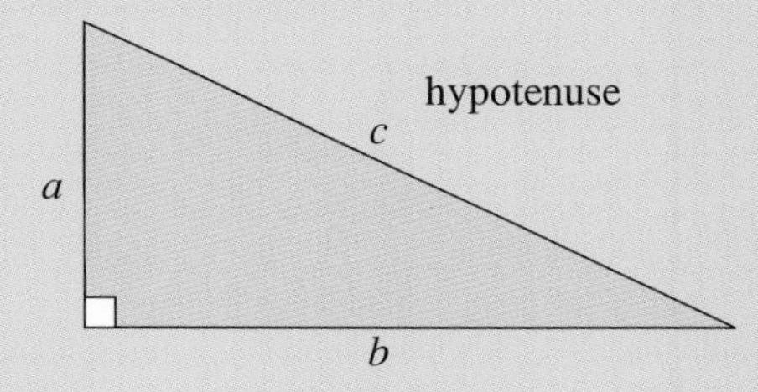


For a right-angled triangle
$a^2 + b^2 = c^2$

**5** Find the missing lengths in these right-angled triangles.

**a** **b**

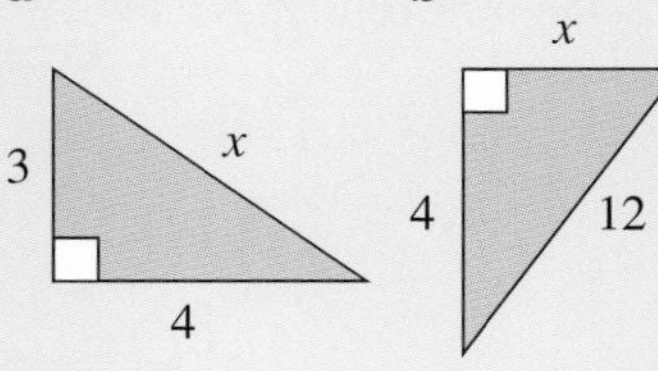


**c**

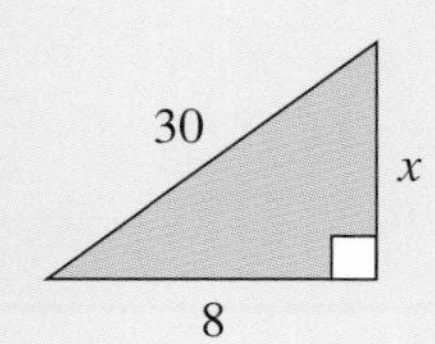

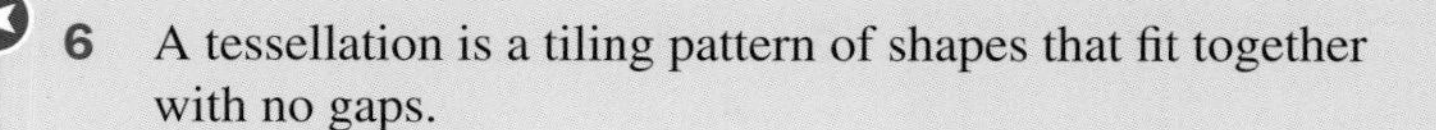

**6** A tessellation is a tiling pattern of shapes that fit together with no gaps.

**7** The locus of point at a fixed distance $r$ from a point A is the circumference of a circle with radius $r$ and centre A.

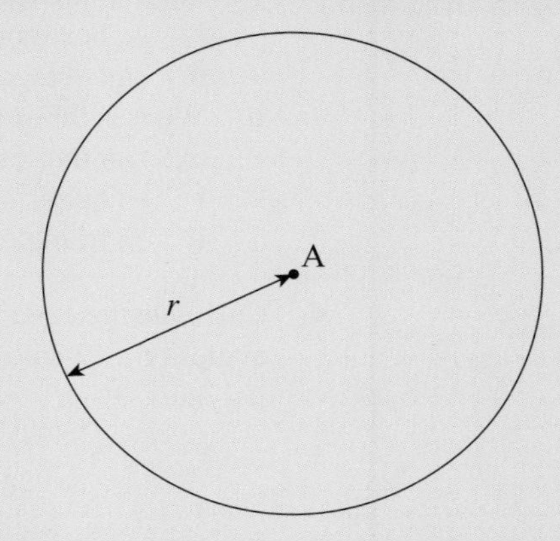


The locus of point at a fixed distance $r$ from a line segment BC is two parallel lines $r$ cm from BC and two semicircles, radius $r$, centres B and C.

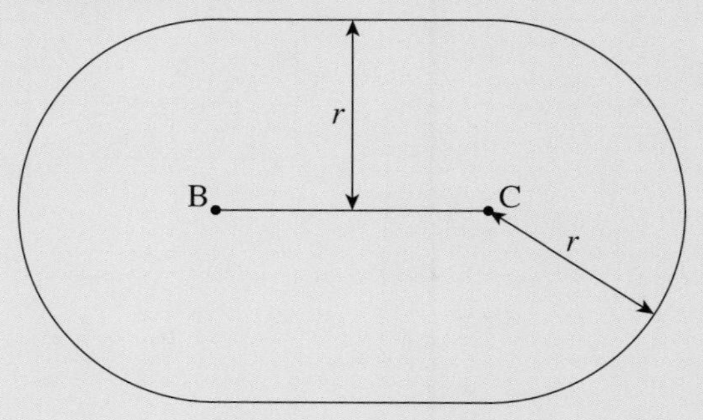


**6 a** Trace this shape.

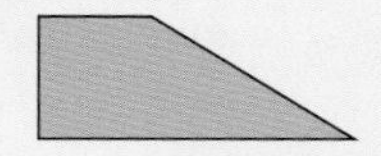

Make a tessellation pattern with it.

**b** Make as many different shapes as you can from four identical equilateral triangles.

**7 a** Draw a point. Label it A. Draw the locus of a point that is always 4.5 cm from A.

**b** Draw a line segment BC where BC is 5 cm long. Draw the locus of a point that is always 1.5 cm from BC.

**8** **a** The angle that an arc of a circle subtends at the centre is double that which it subtends at any point on the remaining part of the circumference.

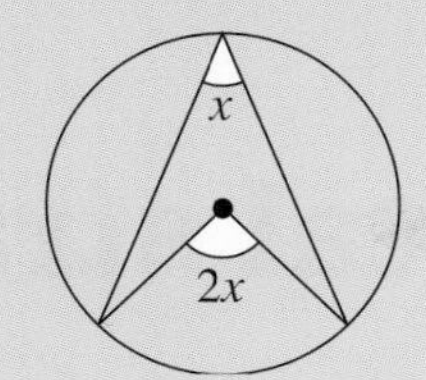


**b** Angles in the same segment of a circle are equal

$\hat{A} = \hat{B}$

**c** The angle in a semicircle is always 90°.

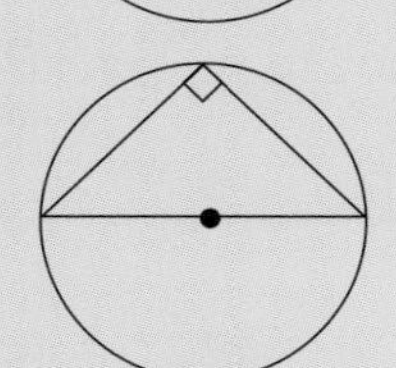

**d** The sum of opposite angles in a cyclic quadrilateral is always 180°.

$\hat{A} + \hat{C} = 180°$
$\hat{B} + \hat{D} = 180°$

**e** The angle between a radius and a tangent is 90°.

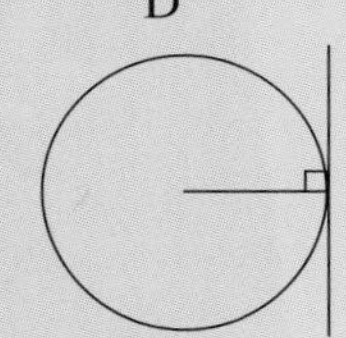

**f** The lengths of two tangents to a circle from a single point outside the circle are equal.

XA = XB

**8** Find the lettered angles.

**a**

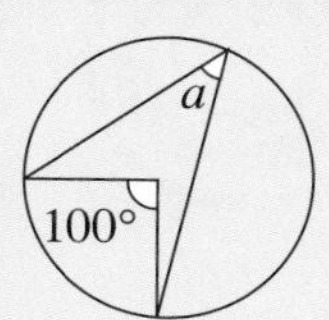


**b**

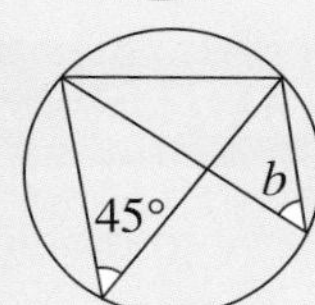


**c**

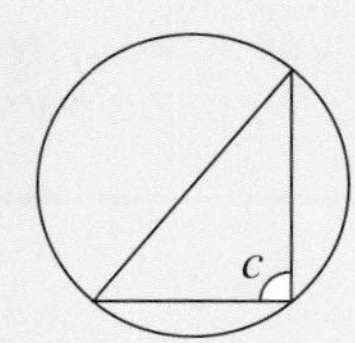


**d**

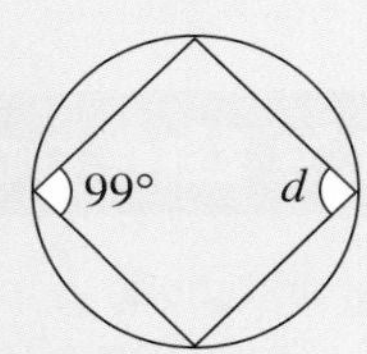


**e**

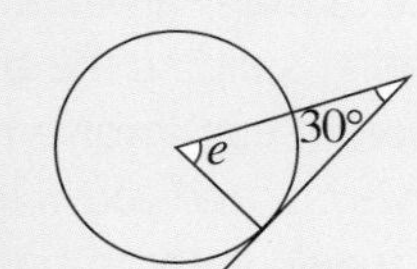


**f**

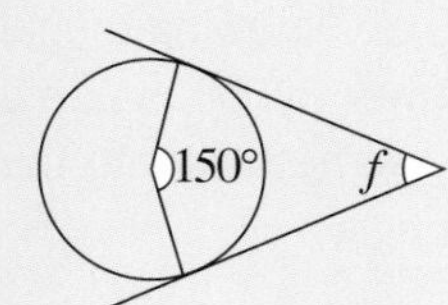

# 10 Mental strategies

## Objectives

- Solve word problems mentally.
- Consolidate use of the rules of arithmetic and inverse operations to simplify calculations.
- Extend mental methods of calculation, working with decimals, fractions, percentages and factors, using jottings where appropriate.

## What's the point?

Mental arithmetic is extremely useful in many areas of life. For example, splitting the bill fairly in a restaurant can involve ratio as well as division, and you will not always have a calculator to hand. Even if you do, knowing roughly what the answer will be is useful, in case you key something into the calculator wrongly.

## Before you start

### You should know ...

1 Correct mathematical language.
*For example:*
*Sum* means 'add up'.
*Difference* means 'subtract'.

### Check in

1 In maths, what do these words mean?
- **a** product
- **b** share
- **c** total

## 10.1 Word problems and strategies

Think about the mental strategies you use to work out calculations. Compare your ideas with others in your class. Does anyone have a quicker or easier method? Do jottings help?

To help you think about different approaches to calculation, consider $54 + 28 + 46$.

You could add the tens: $50 + 20 + 40 = 110$
then add the units: $4 + 8 + 6 = 18$
then combine these: $110 + 18 = 128$

This is like trying to do column addition in your head.

If you find this hard, you can do a running total:

Start with one number: 54
add the tens from the next number: $54 + 20 = 74$
then count on 8 more to get to 82
then add the tens from the next number: $82 + 40 = 122$
then count on 6 more to get to 128.

If you spotted that two of the numbers add up to 100, you could have done:
$54 + 46 = 100$
then $100 + 28 = 128$.

This working is possible because the rules of arithmetic tell us that addition is **commutative**. This means numbers can be added in any order (so, in our example, $54 + 28 + 46$ can be done in any order). Multiplication is also commutative. This idea can help you with many calculations.

Discuss other possible methods for $54 + 28 + 46$ as a class. Which do you prefer? What jottings might be helpful? Using the first method, for example, you might jot down 110 and 18.

Now consider $17 \times 19$. You could do a traditional long multiplication sum. Or you could do $17 \times 20 = 340$ and subtract 19 to get 321, which is much faster and easier to do in your head.

Looking for shortcuts like this in mental strategies can really help with accuracy and speed of calculation.

The first exercise in this chapter includes some word problems, which will enable you to check that you are comfortable with mathematical language. Sometimes students find word problems difficult because they are unsure which mathematical operation they need to do. A glossary is included at the end of the chapter to help you.

Here are some hints for carrying out mental calculations:

- If you are multiplying by a number ending in 9, round the number up, multiply by the rounded number, then subtract the number ending in 9 (as in our example of $17 \times 19$: $17 \times 20 = 340$, $340 - 19 = 21$).
- When adding, begin by looking for pairs of numbers that make 1, 10, 100, etc. Remember, addition is commutative so you can change the order the numbers are added in.
- If you are dividing by 4, you can halve, then halve again.
- If you are dividing by 8, you can halve, then halve again, then halve again.
- To divide by 5 you can double the number then divide by 10.

### Exercise 10A

Work out these sums in your head. For each question, think about the strategy you used. Discuss and compare strategies with your neighbour or with the class. Which is the easiest method for you? What jottings (if any) are helpful?

**1** $73 + 42 + 27 + 58$

**2** $110 - 48$

**3** $15 \times 14$

**4** $2(12 + 4.5)$

**5** $35 ÷ 4

**6** Treble 49

**7** What is the difference between 17.4 and 13.9?

**8** What is the product of 59 and 6?

**9** $70\,500 \times 0.001$

**10** Halve $675

**11** Share $424 equally between 8 people

**12** Find the sum of 27.8 and 18.4

**13** What is the total of all the numbers from 1 to 10?

**14** Double 678

**15** What is the perimeter of a square with side length 89 cm?

**16** What is the area of a square with side length 24 cm?

**17** Find the total of all the factors of 18

**18** It is 8.47 pm. What time will it be in an hour and thirty-four minutes? Give your answer as a 24-hour clock time.

**19** \$23.60 ÷ 5

**20** A 20% discount is given on the cost of a \$135 phone. How much is the discount?

**21** What is 2 to the power of 5?

**22** What number is exactly halfway between 54 and 93?

**23** 82 ÷ 0.01

**24** Work out the cost of a bag of sweets with a mass of 3.2 kg if the sweets cost \$1.30 per kilogram.

**25** A triangle has a base length of 14 cm and an area of 49 cm$^2$. What is the perpendicular height of the triangle?

## 10.2 BIDMAS and inverse operations

You have already learned that the order in which you do operations is BIDMAS, which stands for

Brackets first
then Indices
then Division or Multiplication
then Addition or Subtraction

When following the rules of BIDMAS and calculating mentally, it is useful to jot down each step in the calculation rather than trying to do it all in one go.

**EXAMPLE 1**

Work out $3^2 \times 8 \div (5 - 3)^2$

| | |
|---|---|
| $3^2 \times 8 \div (5 - 3)^2$ | |
| $= 3^2 \times 8 \div 2^2$ | Brackets |
| $= 9 \times 8 \div 4$ | Indices |
| $= 72 \div 4$ | Multiplication |
| $= 18$ | Division |

Notice the order you do multiplication and division in Example 1 doesn't matter. You could also do this:

| | |
|---|---|
| $3^2 \times 8 \div (5 - 3)^2$ | |
| $= 3^2 \times 8 \div 2^2$ | Brackets |
| $= 9 \times 8 \div 4$ | Indices |
| $= 9 \times 2$ | Division |
| $= 18$ | Multiplication |



This is slightly easier, because 8 ÷ 4 and 9 × 2 are easier to do than 9 × 8 and 72 ÷ 4.

Inverse operations can help make calculations with fractions and decimals easier. You have already learned that instead of dividing by a fraction you multiply by the reciprocal of the fraction. Example 2 shows some mental strategies with decimals and fractions combining these with rules of arithmetic.

**EXAMPLE 2**

Work out:

**a** $13.5 \div 0.3 + 5$

**b** $20 - 18 \div 1\frac{4}{5}$

**c** $0.72 + 7 \times 4\%$

**a**

| | |
|---|---|
| $13.5 \div 0.3 + 5$ | |
| $13.5 \div \frac{3}{10} + 5$ | Division first |
| $= 13.5 \times \frac{10}{3} + 5$ | Use inverse operation |
| $= 135 \div 3 + 5$ | Multiplying by 10 first is easier |
| $= 45 + 5$ | Division first |
| $= 50$ | Finally, add |

**b**

| | |
|---|---|
| $20 - 18 \div 1\frac{4}{5}$ | |
| $= 20 - 18 \div \frac{9}{5}$ | Make fraction top heavy |
| $= 20 - 18 \times \frac{5}{9}$ | Use inverse operation |
| $= 20 - 2 \times 5$ | Dividing by 9 first is easier |
| $= 20 - 10$ | Multiplication first |
| $= 10$ | Finally, subtract |

**c**

| | |
|---|---|
| $0.72 + 7 \times 4\%$ | |
| $= 0.72 + 7 \times \frac{4}{100}$ | Change percentage to fraction |
| $= 0.72 + 28 \times \frac{1}{100}$ | Multiply by 4 |
| $= 0.72 + 0.28$ | Divide by 100 |
| $= 1$ | Finally, add |

Note that other methods could be used in Example 2.

In the next exercise you need to decide on the order in which to multiply and divide, and think carefully about the order you add and subtract too. Do not use a calculator.

### Exercise 10B

**1** Andrea wrote this working in her homework:

$7^2 - 4 \times 2 + 3$ Use BIDMAS
$= 49 - 4 \times 2 + 3$ Indices first
$= 49 - 8 + 3$ Then multiply
$= 49 - 11$ Then add
$= 38$ Finally, subtract

What mistake has Andrea made? What is the correct answer?

**2** Work out:

**a** $5^2 - 3 \times 6$
**b** $8 - 1.5 \times 2 + 4$
**c** $2^3 \times 8 \div (3 + 1)^2$
**d** $20 - \frac{15}{3} \times 3$
**e** $10 - 2 \div 4 + 7$
**f** $\frac{4^3}{3 + 5} - 2 \times 3.5$

**3** Work out:

**a** $9 - 15 \div 1\frac{2}{3}$ **b** $14.2 \div 0.2 + 10$
**c** $5 + 1200 \times 2\%$ **d** $10.5 - 7 \times 1\frac{5}{14}$

**4** Work out:

**a** $({}^{-}5 - {}^{-}2) \times 4^2 + 50$
**b** $2.8 + (6 + {}^{-}4) + 1.2^2$
**c** $\frac{4^2 + 2.6}{0.3}$

**5** Write brackets to make these calculations correct.

**a** $3 + 4^2 \times 10 = 190$
**b** $5 + 14 \div 2 - {}^{-}3 = 15$
**c** $10^2 - 10 + 2 \times 8 - 3 = 80$

## 10.3 Factors

A **factor** of a number divides into that number with no remainder.

The best way to make sure that you find all the factors of a number is to find factors in pairs using repeated division.

**EXAMPLE 3**

Find the factors of 36.

First divide by 1:
$36 \div 1 = 36$
1 and 36 are factors because $1 \times 36 = 36$

Then divide by 2:
$36 \div 2 = 18$
2 and 18 are factors because $2 \times 18 = 36$

Then divide by 3:
$36 \div 3 = 12$
3 and 12 are factors because $3 \times 12 = 36$

Then divide by 4:
$36 \div 4 = 9$
4 and 9 are factors because $4 \times 9 = 36$

If you divide by 5 you don't get a whole number so 5 is not a factor.

Then divide by 6:
$36 \div 6 = 6$
6 is a factor because $6 \times 6 = 30$

36 is not divisible by 7 or 8, and the next number to divide by is 9. You already have this as a factor so you know you have found all the factors of 36.

Finally, list the factors in order:
The factors of 36 are 1, 2, 3, 4, 6, 9, 12, 18 and 36.

Factors are useful for many things. The next exercise shows you some of their uses.

### Exercise 10C

**1** Find all the factors of

**a** 15 **b** 24
**c** 30 **d** 25

**2** To find the highest common factor (HCF) of 60 and 84, Harry used this method:
*List all the factors of both numbers and then write down the highest number that is in both lists.*

Here is Harry's working:
*Factors of 60 are 1, 2, 5, 6, 12, 30, 60*
*Factors of 84 are 1, 2, 4, 42, 84*
*So HCF is 2 as it is the highest number in both lists.*

Harry is missing numbers from both his lists of factors and he has not found the highest common factor. Write down all the factors of 60 and 84 and hence find the correct highest common factor.

**3** To find the highest common factor (HCF) of 12 and 30, Rani used this method:

Write both numbers as products of their prime numbers and use common prime factors to find the highest common factor.

Here is part of Rani's working:

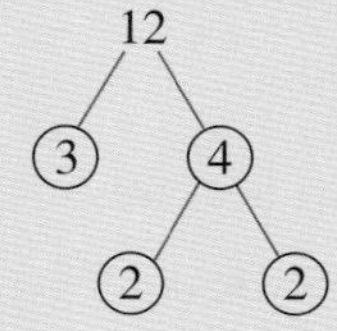


12 = 2 × 2 × 3
30 = 2 × 3 × 5

So HCF of 12 and 30 is 2 × 3 = 6

Draw the factor tree for 30 to show that $2 \times 3 \times 5$ is the correct product of primes. Write down whether you think Rani's working is correct.

**4** Use Rani's method to repeat Question **2**. Which method do you think is easier?

**5** Find the highest common factor of

**a** 20 and 30 **b** 36 and 100
**c** 36, 60 and 42 **d** 126 and 180

**6** Highest common factors are useful when working with fractions. Find the highest common factor of the numerator and denominator to cancel these fractions to their lowest terms.

**a** $\frac{24}{36}$ **b** $\frac{28}{60}$ **c** $\frac{48}{112}$

**7** Calculations with fractions can be made easier by cancelling common factors (any numerator with any denominator) before multiplying. Use this method to work out:

**a** $\frac{24}{49} \times \frac{35}{36}$ **b** $\frac{18}{55} \times \frac{22}{63}$

**c** $\frac{27}{40} \div 1\frac{19}{35}$

**8** A man has a garden measuring 84 m by 112 m. He wants to divide the whole garden into the minimum number of square plots. What is the length of side of each square plot?

**9** I am thinking of a number. My number is a factor of 20. Which of the statements below are true?

My number must be even.
My number must be odd.
My number could be odd or even.

**10** Three strings of different lengths, 120 cm, 180 cm and 420 cm, are to be cut into equal lengths. What is the greatest possible length of each piece?

**11**

| composite | factor | an odd | only two factors |
|---|---|---|---|
| square | 1, 2, 4 and 8 | prime | an even |

Choose from the cards to complete these sentences:

You can tell a number is prime because it has ____________.

8 is not prime because it has ____________ as its factors. 8 is a ____________ number.

16 is a ____________ number because it has ____________ number of factors: 1, 2, 4, 8, 16.

7 is a ____________ number because it has only 1 and 7 as its factors.

12 is not a square number because it has ____________ number of factors.

1 is not prime because its only ____________ is 1.

## INVESTIGATION

### Perfect numbers

Pythagoras was studying perfect numbers over 2500 years ago. The first four perfect numbers were discovered by the ancient Greeks, and to date fewer than fifty perfect numbers are known. The last ten were found during the twenty-first century, and people are still looking for them!

A **perfect** number is a number which is equal to the sum of its proper factors – that is, the sum of its factors excluding the number itself.

For example, the first perfect number is 6. The factors of 6 are 1, 2, 3 and 6. Leaving out 6 itself, the sum of these factors is 6:

$1 + 2 + 3 = 6$

8 is not a perfect number because the sum of its factors (besides 8 itself) is not 8:

$1 + 2 + 4 = 7$

Can you find the second perfect number?

Find out about **abundant** numbers and **deficient** numbers.

Do you think that you could discover the next new perfect number?

# Consolidation

### Example 1

Work out $4^2 \times 15 \div (9 - 4)$.

$4^2 \times 15 \div (9 - 4)$
$= 4^2 \times 15 \div 5$ Brackets
$= 16 \times 15 \div 5$ Indices
$= 16 \times 3$ Division
$= 48$ Multiplication

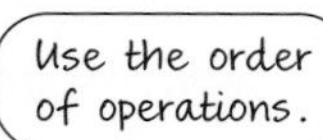


### Example 2

Work out:

**a** $1.28 \div 0.04 - 2$

**b** $0.7 + \frac{27}{35} \div 2\frac{4}{7}$

**c** $(0.3 + 0.6) \times 5\%$

**a** $1.28 \div 0.04 - 2$
$= 1.28 \div \frac{4}{100} - 2$
$= 1.28 \times \frac{100}{4} - 2$
$= 128 \div 4 - 2$
$= 32 - 2$
$= 30$

**b** $0.7 + \frac{27}{35} \div 2\frac{4}{7}$
$= 0.7 + \frac{27}{35} \div \frac{18}{7}$
$= 0.7 + \frac{27}{35} \times \frac{7}{18}$
$= 0.7 + \frac{3}{5} \times \frac{1}{2}$
$= 0.7 + \frac{3}{10}$
$= 0.7 + 0.3$
$= 1$

**c** $(0.3 + 0.6) \times 5\%$
$= 0.9 \times \frac{5}{100}$
$= 0.9 \times 5 \div 100$
$= 4.5 \div 100$
$= 0.045$

## Exercise 10

**1** $82 + 37 + 18$

**2** $3^3 - 2 \times 8$

**3** Write brackets in this expression to make it correct: $3 + 7 \times 10 - 5 = 38$

**4** $140 - 76$

**5** Find the highest common factor of 90 and 225.

**6** $75 ÷ 4

**7** $23 \times 8 \div (3 + 1)^2$

**8** Treble 79.

**9** What is the difference between 28.3 and 19.8?

**10** $\frac{25}{33} \div \frac{15}{44}$

**11** Cancel $\frac{126}{306}$ to its lowest terms.

**12** $3400 \times 0.01$

**13** Find the highest common factor of 72, 120 and 42.

**14** Halve $95.90

**15** $30 - \frac{10 + 5}{3} \times 4$

**16** Share $139.20 equally between 8 people.

**17** $5.8 + 710 \times 2\%$

**18** What is the perimeter of a square with side length 89 cm?

**19** $1 - \frac{24}{55} \times 1\frac{3}{8}$

**20** Find the total of all the factors of 24.

**21** $\frac{3^2 + 2.5}{0.05}$

**22** In a restaurant, a bill for 5 people came to $162.35. They shared the bill equally between them. How much did they each pay?

**23** A 30% discount is given on the cost of a $420 laptop. How much is the discount?

**24** $82 \div 0.01$

**25** Two pieces of wood of different lengths, 150 cm and 210 cm, are to be cut into equal lengths. What is the greatest possible length of each piece?

# Summary

## You should know ...

**1** The language associated with maths.
*For example:*
sum – add
difference – subtract
product – multiply
share – divide
total – add up all the numbers
double – multiply by 2
treble – multiply by 3
halve – divide by 2

**2** Shortcuts you can use to help speed up calculations.
*For example:*
- If you are multiplying by a number ending in 9, round the number up to the nearest 10, multiply by the rounded number, then subtract the unrounded number.
- When adding, begin by looking for pairs of numbers that make 1, 10, 100, etc. Remember, addition is commutative so you can change the order the numbers are added in.
- If you are dividing by 4, you can halve, then halve again.
- If you are dividing by 8, you can halve, then halve again, then halve again.
- To divide by 5 you can double the number then divide by 10.

**3** How to do calculations with decimals, fractions and percentages, including dealing with factors.
*For example:*
Work out: **a** $0.3 \times 30$ **b** $40 \div 6\frac{2}{3}$ **c** 4% of 820

**a** $0.3 \times 30 = \frac{3}{10} \times 30 = 3 \times \frac{30}{10} = 3 \times 3 = 9$

**b** $40 \div 6\frac{2}{3} = 40 \div \frac{20}{3} = 40 \times \frac{3}{20} = 2 \times 3 = 6$

Cancelling the common factor of 20

**c** $4\% \text{ of } 820 = \frac{4}{100} \times 820 = 4 \times \frac{820}{100} = 4 \times 8.2 = 32.8$

**4** How to use the order of operations (BIDMAS).
*For example:*
$2^3 \times 3 - 20 \div (1 + 3)$
$= 2^3 \times 3 - 20 \div 4$ Brackets
$= 8 \times 3 - 20 \div 4$ Indices
$= 24 - 5$ Division and multiplication
$= 19$ Subtraction

## Check out

**1** Without using a calculator
**a** Find the sum of 28.7 and 33.6
**b** Double 79
**c** Find the product of 2.5 and 8
**d** Share \$14.25 between 5 people
**e** Treble 59

**2** Work out:
**a** $\$2.84 \div 4$
**b** $63 \times 19$
**c** $3400 \div 8$
**d** $24.6 + 37.9 + 75.4$
**e** $\$73.25 \div 5$

**3** Work out:
**a** $0.8 \div 200$
**b** $60 \div 4\frac{2}{7}$
**c** 3% of 320

**4** Work out:
**a** $3^3 \times 18 \div (6 + 3)$
**b** $50 - \frac{13}{5} \times 15$
**c** $40 - 40 \div 4 + 7^2$
**d** $\frac{2^3}{2.9 + 1.1} - 2 \times 0.5$

# 11 Compound measures

## Objectives

- Use compound measures to make comparisons in real-life contexts, e.g. travel graphs, and value for money.
- Solve problems involving average speed.

## What's the point?

Speed is an important measure. For example, the faster you travel, the less time you have to spend travelling and the sooner you get to your destination, which is important in many situations. Engineers are always trying to make vehicles that travel faster while being economical and safe, such as high-speed trains.

## Before you start

### You should know ...

1 Metric measures of length, area, mass, capacity and volume, and their abbreviations.

*For example:*

| | |
|---|---|
| cm = centimetres | used for length |
| $km^2$ = square kilometres | used for area |
| kg = kilograms | used for mass |
| ml = millilitres | used for capacity |
| $m^3$ = cubic metres | used for volume |

### Check in

1 Write the units these abbreviations stand for. State whether each is a measure of length, area, mass, volume or capacity

**a** g
**b** $cm^2$
**c** m
**d** ℓ
**e** $cm^3$

## 11.1 Compound measures

A **compound measure** is made up of two or more other measurements. A very commonly used compound measure is speed. Speed is a compound measure because it is calculated from distance and time.

Its units involve distance and time – it is measured in, for example, kilometres per hour, metres per second and (using the imperial system used in the UK, USA and some other countries) miles per hour.

The table shows some examples of compound measures.

| Compound measure | Measurements used | Common units | Formula triangle |
|---|---|---|---|
| Speed | Distance and time | m/s, km/hr (metres per second, kilometres per hour) | Speed = distance ÷ time<br>distance / speed, time |
| Density | Mass and volume | kg/m$^3$ (kilograms per cubic metre) | Density = mass ÷ volume<br>mass / density, volume |
| Pressure | Force and surface area | N/m$^2$ (Newtons per square metre) | Pressure = force ÷ surface area<br>force / pressure, surface area |
| Fuel economy | Distance and capacity | km/ℓ (kilometres per litre) | Fuel economy = distance ÷ capacity<br>distance / fuel economy, capacity |
| Population density | Population and area | people/km$^2$ (number of people per square kilometre) | Population density = population ÷ area<br>population / populati-on density, area |

Notice that the unit for a compound measure tells us what calculation to do. For example, in the units for density of $kg/m^3$, or kilograms per cubic metre, the '/' or 'per' part tells us to divide. So $kg/m^3$ means do the calculation $kg \div m^3$, or mass ÷ volume.

Just as we use centimetres, metres, kilometres and so on to measure length, we can use many different units for compound measures. Speed might be given in metres per second or kilometres per hour; it makes no sense to use the same unit to give the speed of a snail and the speed of a jet plane.

You can use formula triangles like those in the table to work out what calculation to do.

The formula triangle for density is:

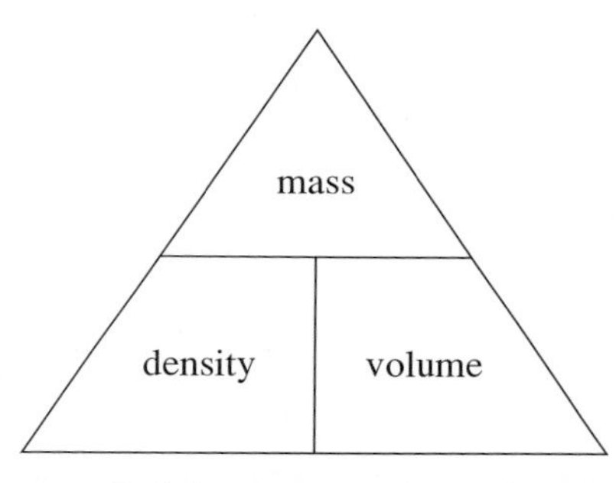


If you want to find density, mass is on the top and volume is below. This means divide.

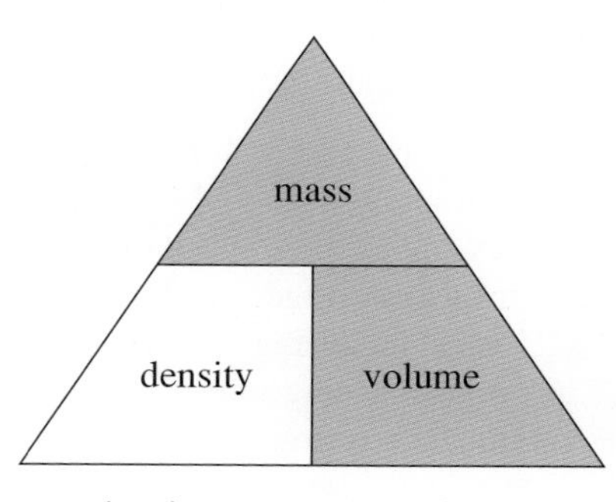


density = mass ÷ volume

If you want to find volume, mass is on the top and density is below. This means divide.

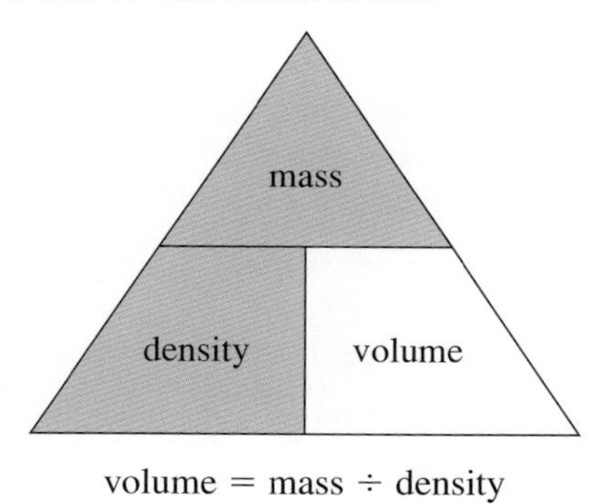


volume = mass ÷ density

If you want to find mass, density and volume are both on the same level. This means multiply.

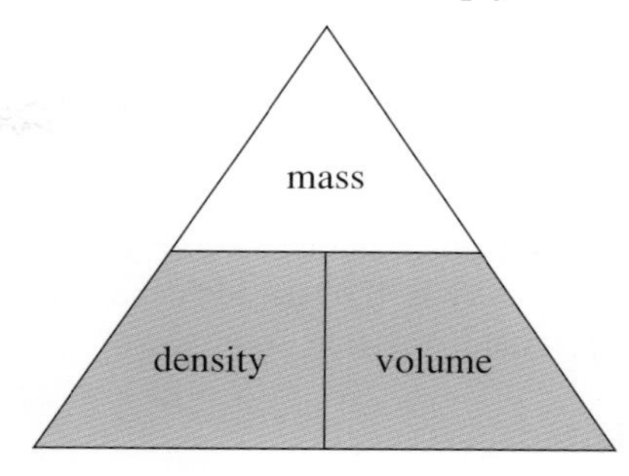


mass = density × volume

Other formula triangles can be used in the same way. If you don't want to use a formula triangle, you can also rearrange the first formula to change its subject, and get the other related formulae.

**EXAMPLE 1**

A car makes a journey of 315 km in $4\frac{1}{2}$ hours. Calculate its average speed.

Using the formula triangle:

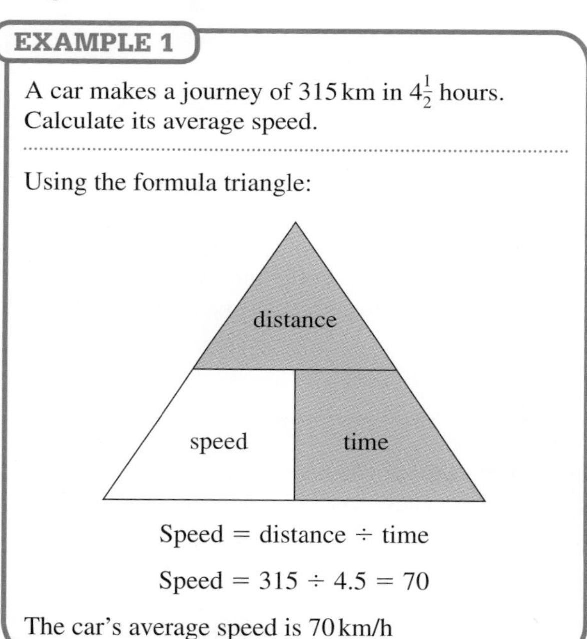


Speed = distance ÷ time

Speed = 315 ÷ 4.5 = 70

The car's average speed is 70 km/h

Note the question in Example 1 talks about **average speed**. The car is unlikely to have maintained a speed of 70 km/h for the entire 315 km. It is more likely that at periods during the journey the car will have been travelling faster or slower than this. Usually questions do not include the word 'average'; instead they just ask for 'speed'. Using common sense, you know that the question means average speed.

Sometimes average speed for a whole journey will be worked out with lots of different speeds (you will see this later in the chapter, in the work on distance–time graphs). The formula for average speed is:

- Average speed = total distance ÷ total time

A common mistake in calculating average speed is to work out the average *of the speeds*, by adding the various speeds together and dividing by the number of those speeds.

**EXAMPLE 2**

A town has a population of 135 000 people. Its population density is 4500 people per square kilometre. What is the area of the town?

Using the formula triangle:

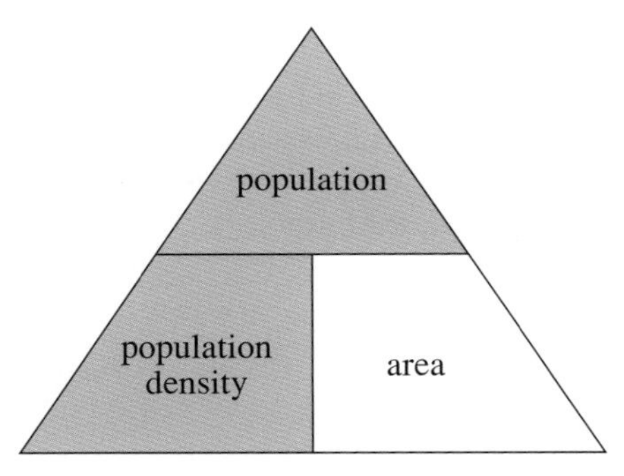


Area = population ÷ population density

Area = 135 000 ÷ 4500 = 30

Since the population density unit was people per square kilometre, the area must be in square kilometres.

Area = $30\,\text{km}^2$

In the next exercise you need to give the units for all your answers. These can be found using the units in the question.

## Exercise 11A

1 Find the average speed, in km/h, of an object that travels
   - **a** 72 km in 2 hours
   - **b** 540 km in 6 hours
   - **c** 8 km in 30 minutes

2 A block of plastic with volume $25\,\text{cm}^3$ has a density of $0.65\text{g/cm}^3$. What is its mass?

3 A town has a population of 80 000 and an area of 12 square kilometres. Calculate its population density (to the nearest whole number).

4 Find the average speed of a cyclist who travels 55 km in $2\frac{1}{2}$ hours.

5 Calculate the density of a piece of metal with mass 540 g and volume $40\,\text{cm}^3$.

6 How long does a train take to make a 715-km journey if it travels at an average speed of 110 km/h?

7 A block of wood exerts a force of 40 N on the surface it stands on. The area of the surface it stands on is $250\,\text{cm}^2$. Calculate the pressure on the surface.

8 A car travels 250 km on 20 litres of petrol.
   - **a** What is the car's fuel economy, in kilometres per litre?
   - **b** The units of fuel consumption are litres per 100 kilometres. What is the fuel consumption for this car?

9 A car is travelling at an average speed of 90 km/h. How far does the car travel in 90 minutes?

Take care with the units as they do not match.

10 John plays 40 matches for his school football team. During those matches he scores 27 goals. Each match lasts 90 minutes. Calculate his scoring rate in
   - **a** goals per match
   - **b** goals per hour
   - **c** goals per minute

11 Work out the time taken, in minutes, for each of these journeys.
   - **a** 20 km at an average speed of 20 km/h
   - **b** 39 km at an average speed of 78 km/h
   - **c** 22.5 km at an average speed of 90 km/h
   - **d** 36 km at an average speed of 120 km/h
   - **e** 8.8 km at an average speed of 44 km/h

12 A cube of aluminium with side length 4 cm has a mass of 172.8 g.
   - **a** Calculate the density of aluminium in $\text{g/cm}^3$.
   - **b** Calculate the mass of an aluminium cube with side length 3 cm.

13 A car travels for 3 hours and 6 minutes at an average speed of 110 km/h. Samia worked out the distance the car travelled like this:

Distance = speed × time
= 110 × 3.6
= 396 km

Is Samia correct? If not, write down the correct working.

**14** Work out the distance travelled for each of these journeys.

**a** 4 hours at an average speed of 75 km/h
**b** 42 minutes at an average speed of 80 km/h
**c** 2 hours and 12 minutes at an average speed of 60 km/h
**d** 54 seconds at an average speed of 8 m/s

## TECHNOLOGY

Have a look at the website
www.bloodhoundssc.com/project/car

Read about the project to develop a supersonic jet- and rocket-powered car designed to travel at speeds of over 1600 kilometres per hour (1000 miles per hour). 'Supersonic' means faster than the speed of sound!

Part of the project's mission is 'To confront and overcome the impossible using science, technology, engineering and mathematics.'

### Value for money

Compound measures are used in value for money comparisons. There are two ways of comparing two items to see which is the best value for money: you can work out the cost per item or unit of mass, capacity, volume, etc., or you can work out the number of items or mass, capacity, volume, etc. per unit of cost.

Have you ever gone into a store and seen two signs like these?

Which is the best buy?

To work this out you need to find the cost of a single pair of socks.

In the first case:
6 pairs cost $43.80
So 1 pair costs $43.80 ÷ 6
= $7.30

In the second case:
4 pairs cost $28.40
So 1 pair costs $28.40 ÷ 4
= $7.10

It is cheaper to buy the socks with the second offer.

In some cases you will need to look at the sizes sold.

**EXAMPLE 3**

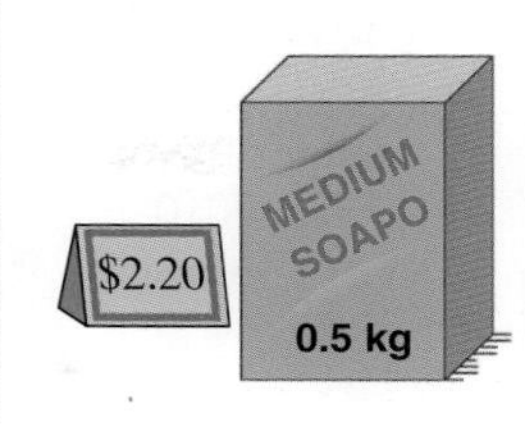


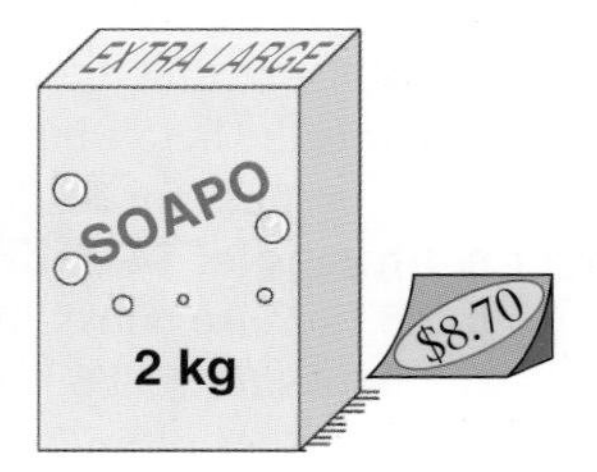


Which soap powder is the better buy?

The extra large size holds 2 kg for $8.70
So 1 kg costs $8.70 ÷ 2 = $4.35

The medium size holds 0.5 kg for $2.20
So 1 kg costs
$2.20 ÷ 0.5 = $4.40

Hence, it's cheaper to purchase the extra large size. (If you can afford it.)

Example 3 shows how to work out the compound measure in dollars per kilogram. You saw that you the cost for each kilogram was lower for the larger box. The same question can be done using kilograms per dollar.

**EXAMPLE 4**

Considering kilograms per dollar, which soap powder from Example 3 is the better buy?

The extra large size costs $8.70 for 2 kg.
So for $1 you get 2 ÷ 8.7 = 0.230 kg (3 d.p.)

The medium size costs £2.20 for 0.5 kg.
So for $1 you get 0.5 ÷ 2.2 = 0.227 kg (3 d.p.)

Hence, the extra large size gives you a greater mass of soap powder for each dollar spent.

For each question you will have to decide which method is the best to use. In this case the method in Example 3 is easier than the method in Example 4, as the decimals involved are easier to work with.

The term **unit cost** is the compound measure 'cost per item'. To find it, divide the cost by the number of items.

## Exercise 11B

1 In each case find the unit cost and then state which is the better buy.
  a 5 oranges for \$2.00 or 4 oranges for \$1.56
  b 6 eggs for \$4.20 or 12 eggs for \$8.10
  c 5 exercise books for \$3.65 or 8 exercise books for \$5.76
  d 7 cartons of milk for \$10.50 or 9 cartons of milk for \$13.32

2 Which would be the best buy?
5 pens for \$10.00 or
7 pens for \$14.70 or
9 pens for \$15.75?
Give reasons for your answer.

3 Which would be the best buy?
10 kg of oranges for \$12.25 or
6 kg of oranges for \$7.29 or
8 kg of oranges for \$9.56?
Give reasons for your answer.

4 Which is the better buy?
1.08 kg of beef for \$7.20 or
1.62 kg of beef for \$10.77?
Give reasons for your answer.

5
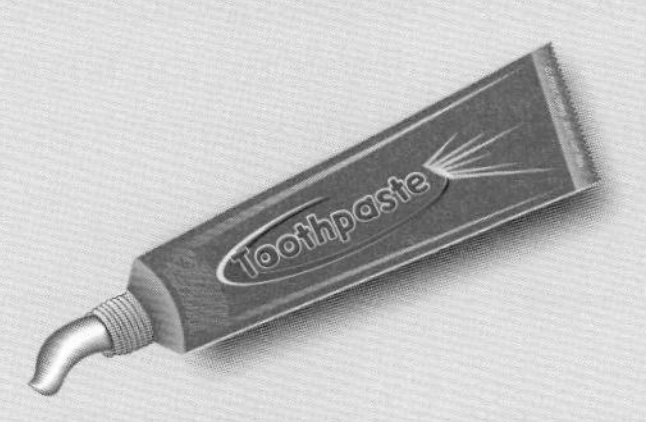


Which is the better buy?
130 g of toothpaste for \$1.17 or
150 g of toothpaste for \$1.31?
Give reasons for your answer.

6 Visit a local supermarket. Find different-sized tubes of toothpaste of a given brand.
  a Write down the sizes and prices.
  b Which is the best buy?
  c Write up your findings and present them to your class.

7 Repeat Question **6** for other items, such as soap powder or bleach.

# 11.2 Real-life graphs

## Travel graphs

Here is a distance–time graph showing a car's journey.

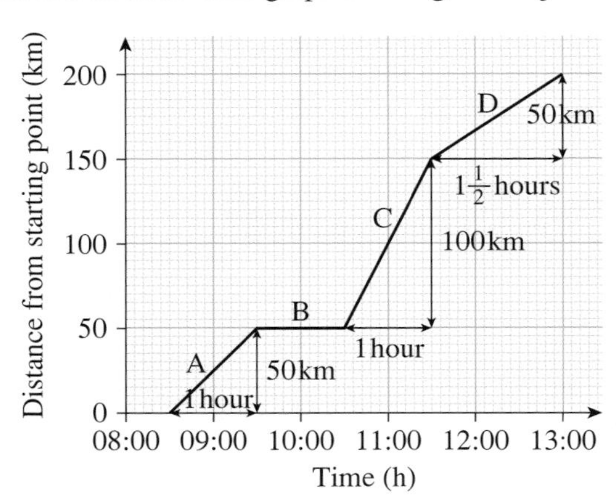


It is possible to tell just by looking at the graph when the car is travelling at its highest speed: it is travelling fastest when the graph is steepest.

To calculate the average speed for each section of the journey you need to find the distance and the time for each section from the graph.

For Section C, you can see the car travels 100 km in 1 hour, so the speed is $100 \div 1 = 100$ km/h. This is the same as working out the **gradient** of the line. You will learn more about this in Chapter 14.

The speed during Section D is $50 \div 1.5 = 33.3$ km/h (1 d.p.).

You can see in Section B that the car does not increase its distance; the graph line is horizontal, showing the car has stopped and is stationary.

In Section A the car is travelling at $50 \div 1 = 50$ km/h.

So the car is travelling fastest in Section C, which is the steepest part of the graph, and its speed during this time is 100 km/h.

To work out the average speed for the entire journey:
Average speed = total distance ÷ total time
$= 200 \div 4.5 = 44.4$ km/h (1 d.p.)

Another type of travel graph is a speed–time graph. These can be used to find the compound measure **acceleration**, which is often measured in km/h$^2$ or m/s$^2$.

The formula triangle for the compound measure of acceleration is:

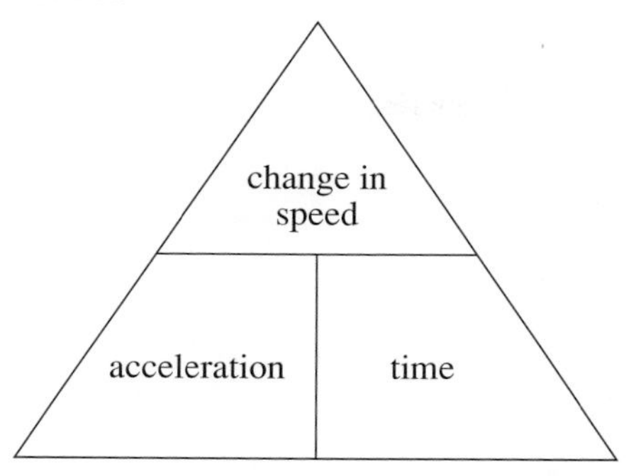


Acceleration is a measure of the rate at which speed is increasing.

**Deceleration** is a measure of the rate at which speed is decreasing. It is calculated in the same way as acceleration: change in speed ÷ time.

Speed–time graphs can also be used to work out distance travelled, as shown in the next example.

**EXAMPLE 5**

The graph represents a car travelling for five hours at a speed which increases steadily from 30 km/h to 70 km/h.

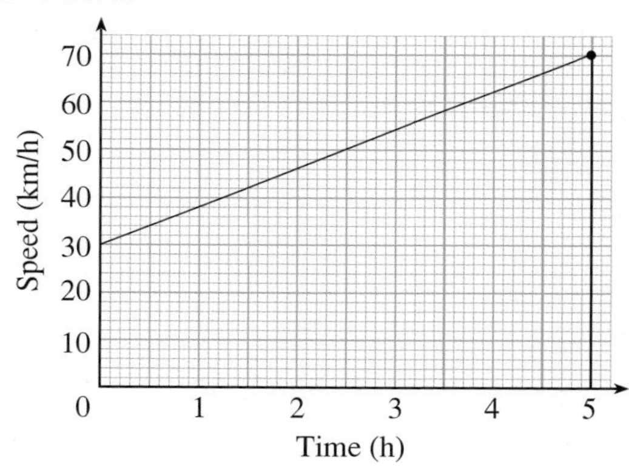


**a** What is the car's average speed?
**b** How far did the car travel in five hours?
**c** What is the acceleration of the car?

**a** Average speed $= \dfrac{30 + 70}{2} = 50$ km/h

**b** Distance travelled = average speed × time
$= 50\text{ km/h} \times 5\text{ h}$
$= 250\text{ km}$

**c** Acceleration = change in speed ÷ time
$= (70 - 30) \div 5$
$= 40 \div 5$
$= 8\text{ km/h}^2$

This is the gradient of the line.

Another way of calculating the distance travelled is to look at the area under the line. That area has the shape of a trapezium. In Example 5:

Area of trapezium $= \frac{1}{2}(30 + 70) \times 5$
$= \frac{1}{2} \times 100 \times 5$
$= 250$

So distance travelled = 250 km.

- The **distance** travelled by an object is given by the **area** under its **speed–time** graph.

## Exercise 11C

**1** Here is the distance–time graph for a car's journey.

**Graph of a car's journey over six hours**

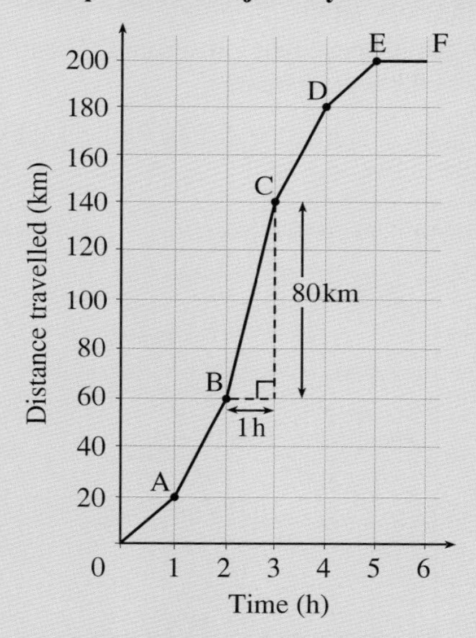


**a** Find the gradients of the line segments OA, AB, CD, DE and EF.

**b** Use these results to write down the average speed of the car during the first, second, fourth, fifth and sixth hours of the journey.

**c** During which of the six hours was the average speed greatest? How can you tell from the graph?

**d** During the last hour the car was stationary. How can you tell this from the gradient of EF?

**e** Work out the car's average speed during the first five hours of the journey. (**Hint:** average speed = total distance ÷ total time.)

**2** The diagram shows the speed–time graph for a car that starts from rest over 5 seconds.

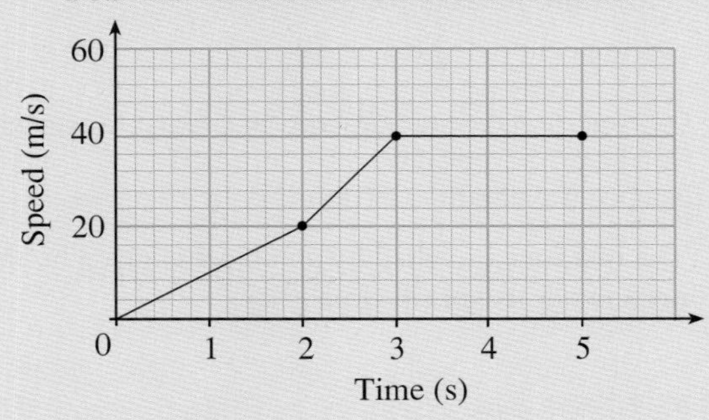


**a** What was the acceleration of the car in the first 2 seconds?

**b** What was the acceleration during the third second?

**c** What happened after 3 seconds? What was the acceleration?

**3** Here is the graph of the distance travelled by a particle in five seconds.

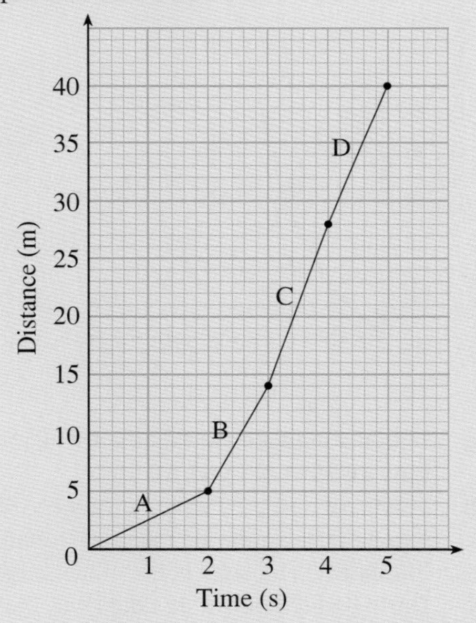


**a** From this graph, find the speed of the particle during

**i** Section A **ii** Section B

**iii** Section C **iv** Section D

**b** What was the average speed of the particle?

**4** The graph shows the speed of a particle over a 15-second time interval.

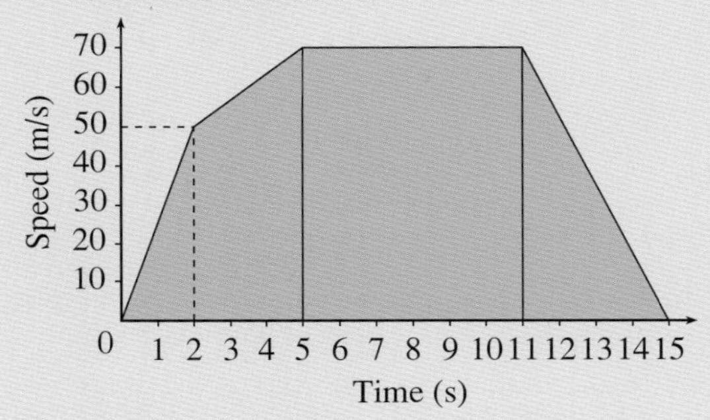


**a** What was the acceleration of the particle

**i** during the first two seconds

**ii** during the next three seconds

**iii** during the six seconds after that?

**b** What was the deceleration of the particle during the last four seconds?

**5** A car starts from rest and reaches a speed of 50 m/s in 6 seconds. It continues at this speed for a further 5 seconds. It then decelerates before coming to a halt 7 seconds later.

**a** Draw a speed–time graph for the car.

**b** What is the car's initial acceleration?

**c** What is the car's final deceleration?

**6** This graph represents a car travelling for five hours at a constant speed of 50 km/h.

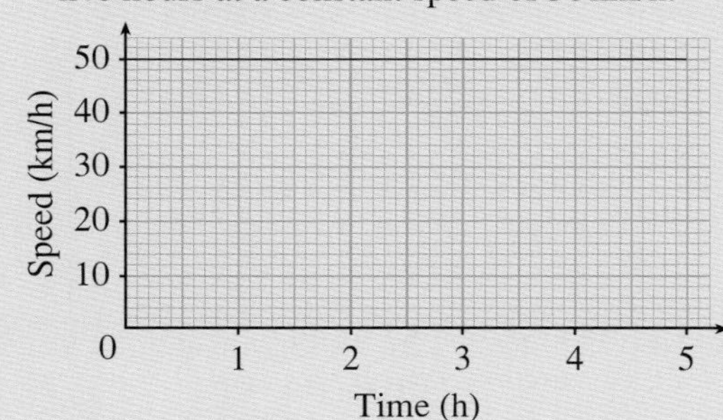


**a** How far did the car travel in the 5 hours?

**b** How can you find this out from the graph?

**7** Dian starts at 9 am and rides his bike 30 km to a football match. He reaches the ground at 11.30 am.

**a** Draw a distance–time graph for Dian's journey.

**b** Joseph leaves at 9.30 am in his car. He gets to the ground at 10.30 am. Draw a line on your graph to show Joseph's journey.

**c** At what time did Joseph overtake Dian?

**8** The graph represents a car travelling for two hours at a speed which increases steadily from 10 km/h and then for three hours at a speed which increases steadily from 50 km/h to 70 km/h.

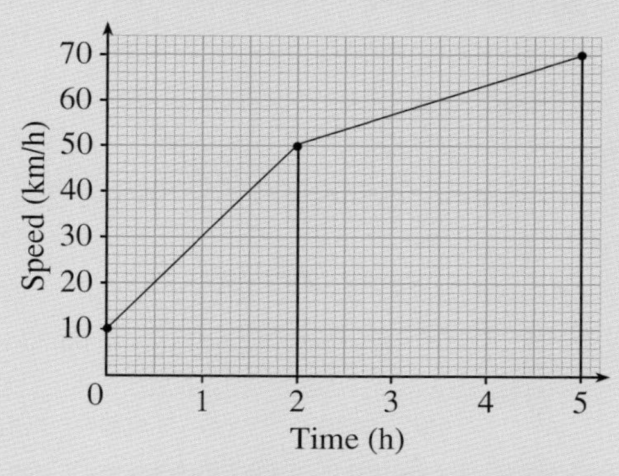


**a** Find how far the car travelled in 5 hours
  **i** by first finding the average speed for each of the two parts of the journey
  **ii** by finding the area of the two trapeziums shown on the graph.

**b** Are your answers to parts **a i** and **ii** the same?

**c** What was the car's acceleration during the first two hours?

**9** A car starts from rest and accelerates for 10 seconds to reach a speed of 30 m/s. Over the next 5 seconds its speed steadily decreases until it is 20 m/s. It continues at this speed for a further 8 seconds before slowing down, then stopping 10 seconds later.

**a** Draw a speed–time graph for the car.

**b** What was the car's initial acceleration?

**c** During what time period was the car's deceleration the greatest?

**10** The graph shows the speed of a particle over a nine-second interval of time.

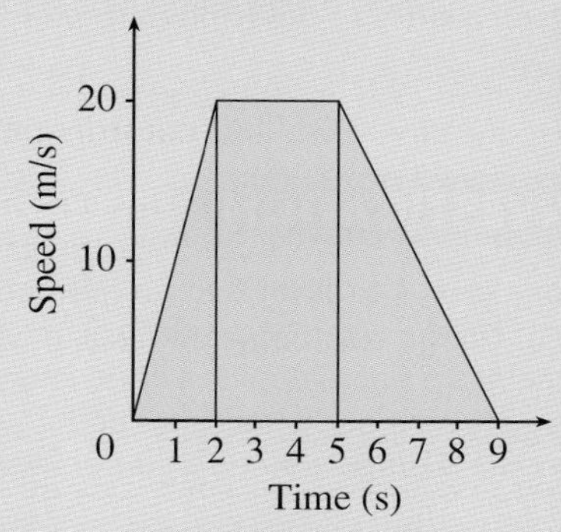


How far did the particle travel

**a** in the first two seconds

**b** between the second and fifth seconds

**c** altogether in the nine seconds?

## Other real-life graphs

As well as travel graphs, there are other real-life graphs that can be used for compound measures.

The next example shows how a value for money graph can be used to make comparisons in a real-life context.

### EXAMPLE 6

The graph shows the text message tariffs of two mobile phone companies, Blue and Green.

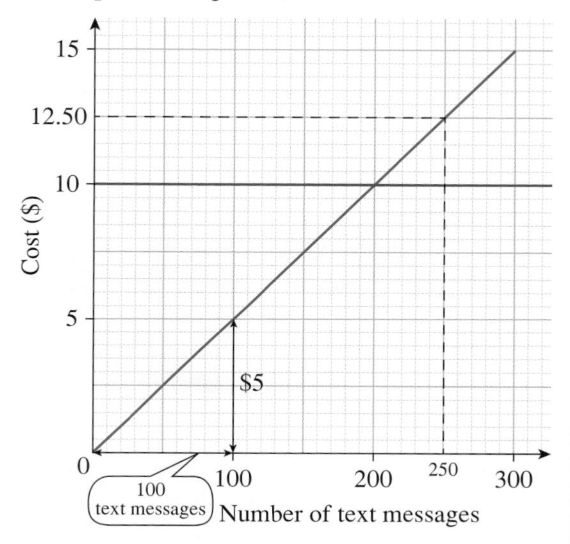


**a** Explain the meaning of the graph in words.

**b** Malik makes very few calls on his mobile phone, so is deciding on which company to choose based on the number of text messages he sends. He usually sends about 250 text messages a month. Which company should he choose? Explain your answer.

**c** When would it make more sense to choose Green rather than Blue?

---

**a** Blue charges a 'flat' rate of $10 per month: the cost doesn't change, regardless of the number of text messages sent. Green charges $5 for every 100 text messages sent (as shown by the arrows on the graph). Using the compound measure unit cost (or cost per text message), Green charge $5 \div 100 = 0.05$, or 5 cents for each text message sent.

**b** Blue is the best value for money for Malik. 250 text messages will cost $10 with Blue and $12.50 with Green (as shown on the graph).

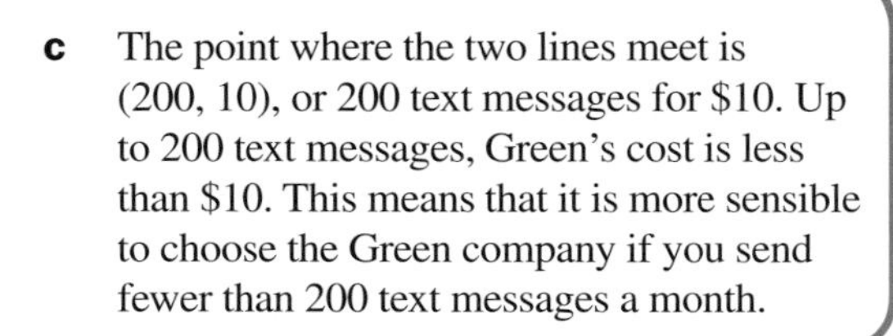

**c** The point where the two lines meet is (200, 10), or 200 text messages for $10. Up to 200 text messages, Green's cost is less than $10. This means that it is more sensible to choose the Green company if you send fewer than 200 text messages a month.

## Exercise 11D

**1** Mr Patel bought a new motor home at the same time as Mrs Templeman bought a second-hand motor home.

The two motor homes have depreciated (gone down in value) as shown in the graph.

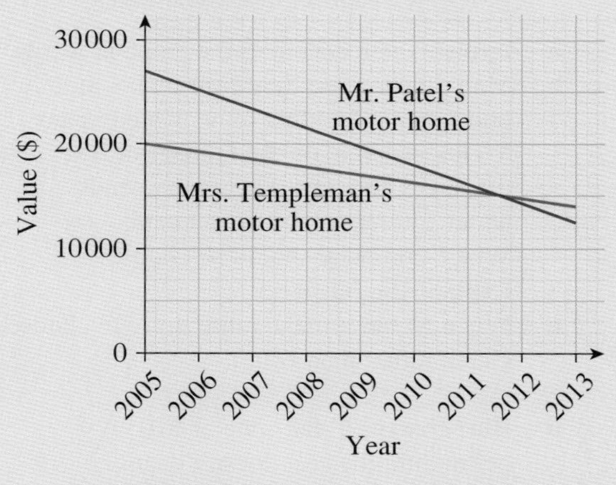


**a** Approximately when were the two motor homes worth the same amount?

**b** Whose motor home has depreciated (gone down in value) more quickly?

**c** What was the approximate difference in value between the two motor homes one year after Mr Patel and Mrs Templeman bought them?

**d** Work out the depreciation rate for each motor home. Use the compound measure loss in dollars per year.

**2** The text tariffs for two mobile phone companies, Red and Yellow, are shown in the graph.

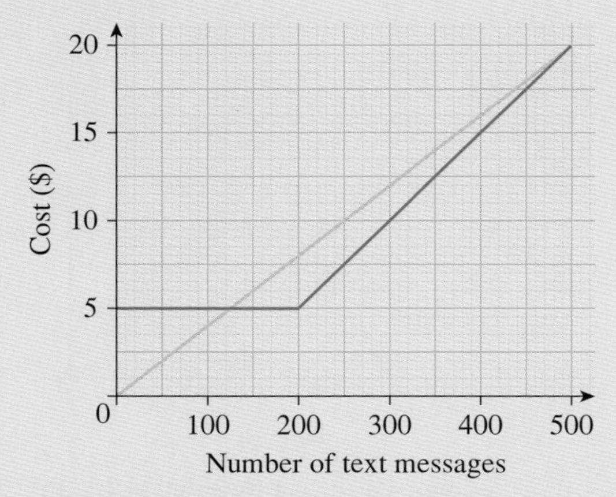


**a** Explain the meaning of the graph in words.

**b** Jane usually sends about 100 text messages a month. Which company should she choose?

**c** When would it make more sense to choose Yellow rather than Red?

**3** Write a 'real-life' story or description to go with this graph.
You may label the axes however you wish.

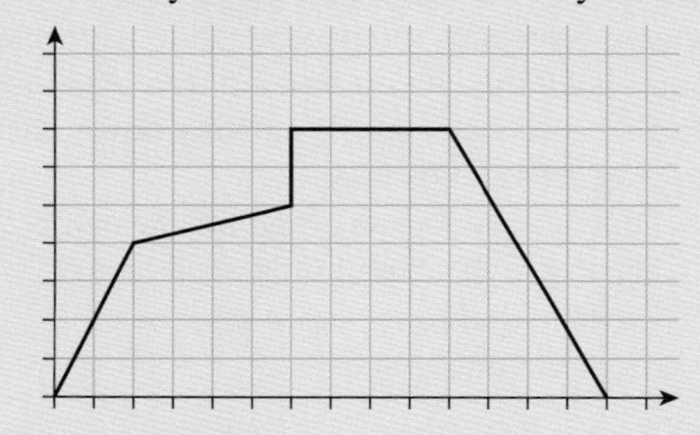

**4** Draw some real-life graphs of your own. Make up a story or description to go with each of them.

### INVESTIGATION

I drove from Town A to Town B at an average speed of 60 km/h. On the return journey I was able to drive faster, at an average speed of 80 km/h. If I travelled exactly the same route for both journeys, why wasn't my average speed for the entire trip 70 km/h?

What average speed for the return journey would make the average speed for the entire trip 70 km/h?

# Consolidation

### Example 1

Work out the average speed of a car that has travelled 260 km in 3 hours 15 minutes.

Using the formula triangle:

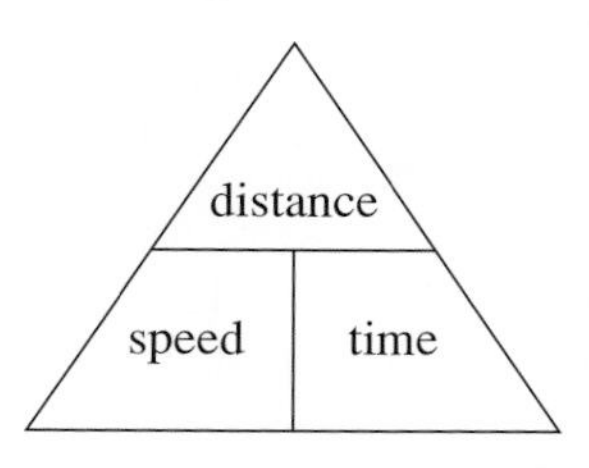


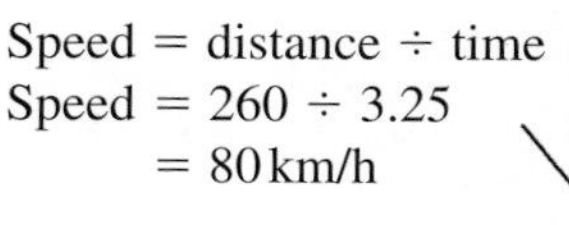

Speed = distance ÷ time
Speed = 260 ÷ 3.25
= 80 km/h

### Example 2

Calculate the mass of a piece of metal with density 11.5 g/cm$^3$ and volume 30 cm$^3$.

Using the formula triangle:

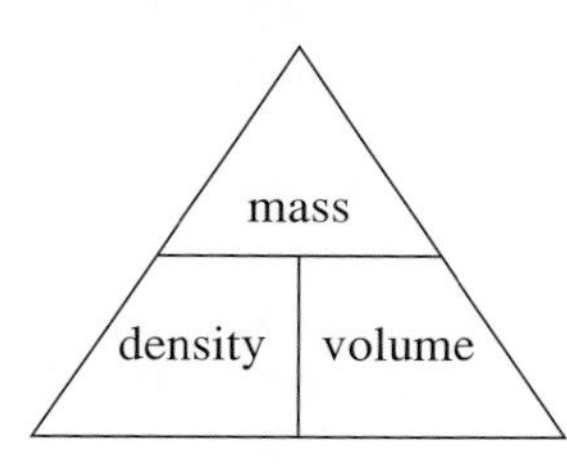


Mass = density × volume
Mass = 11.5 × 30
= 345 g

### Example 3

Which is the best value for money, two medium, 150-g bars of chocolate on special offer, two bars for \$2.64, or one large, 250-g bar for \$2.25?

Using the compound measure price per 100 g:

The medium bars are 300 g for \$2.64
So 100 g cost 2.64 ÷ 3 = \$0.88

The large bar is 250 g for \$2.25
So 100 g cost 2.25 ÷ 2.5 = \$0.90

Hence, the medium bars on special offer are better value.

### Example 4

Draw the distance–time graph that goes with this description:

A car travels from one town to another and then back again. To start with the car travels at a speed of 60 km/h for half an hour. Then it stops for 10 minutes. It then covers the last 20 km of the journey to the town in 15 minutes. The car stops at the town for half an hour. The return journey is non-stop at an average speed of 75 km/h.

The distance travelled for the first part of the journey is: speed × time = $60 \times \frac{1}{2} = 30$ km.
So the car travels 30 km in the first 30 min.

Then the car stops for 10 min.
This is shown by a horizontal line from 30 to 40 min.

The next 20 km takes 15 min.
This is shown by a diagonal line from 30 km at 40 min to 50 km at 55 min.

The car stops for 30 min.
This is shown by a horizontal line from 55 to 85 min.

Finally, the time taken for the return journey is:
distance ÷ speed = $50 \div 75 = \frac{2}{3}$h = 40 min.
This is shown by a diagonal line from 50 to 0 km, between 85 and 125 min.

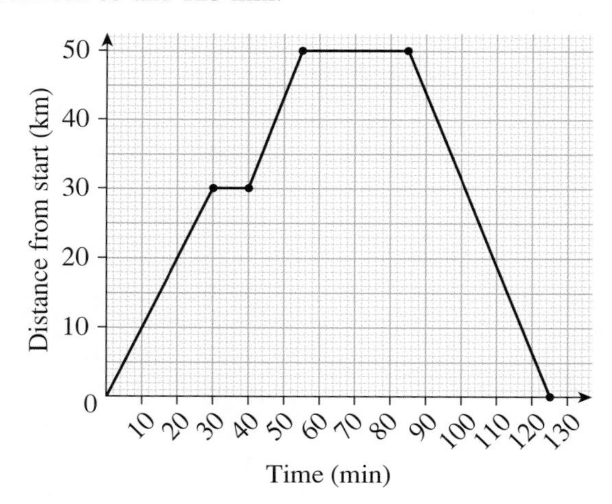

**Example 5**

The diagram shows the speed–time graph for a racing car for ten seconds, starting from rest.

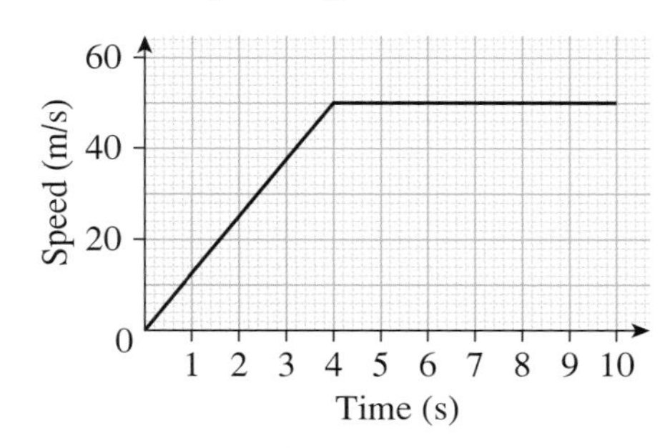


**a** What is the acceleration in the first four seconds?
**b** What does a horizontal line show on a speed–time graph?
**c** How far did the car travel during the first ten seconds of its journey?

**a** Acceleration = change in speed ÷ time (the gradient of the line)
Acceleration $= 50 \div 4 = 12.5\,\text{m/s}^2$
**b** A horizontal line on a speed–time graph shows constant speed.
**c** The distance travelled is given by the area under the graph, which is a trapezium.
Distance $= \frac{1}{2}(6 + 10) \times 50 = 400\,\text{m}$

## Exercise 11

**1** Find the average speed, in km/h, of an object that travels
**a** 180 km in 3 hours
**b** 12 km in 15 minutes

**2** Here is the distance–time graph for a car's journey.

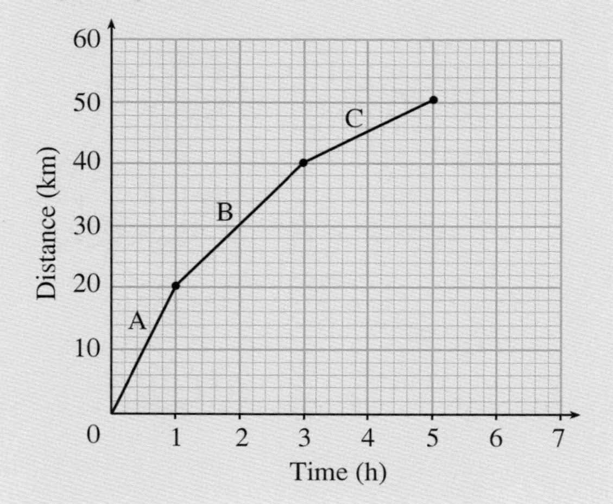


**a** From this graph, find the speed of the car during
**i** Section A
**ii** Section B
**iii** Section C
**b** Work out the average speed for the entire journey.

**3** A block of wood with volume $700\,\text{cm}^3$ has a density of $0.54\,\text{g/cm}^3$. What is its mass?

**4** The distance–time graph shown is for two trains travelling the same route.
Train A is non-stop and Train B has a half-hour stop for staff changeover.

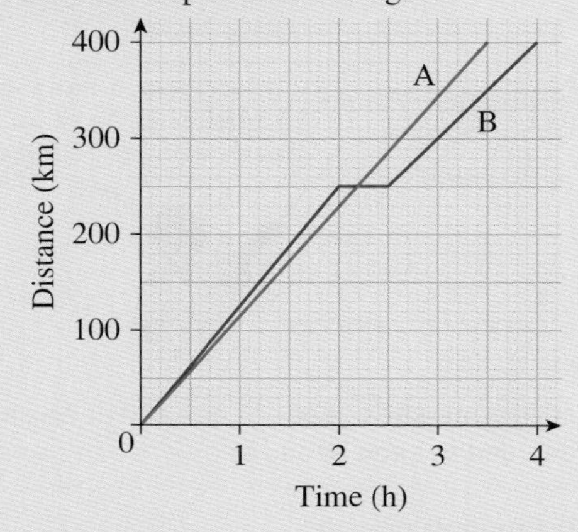


**a** During the first two hours, which train is travelling the fastest?
**b** What is the difference in the speeds of the trains during the first two hours?
**c** Approximately how long into the journey did Train A overtake Train B?
**d** Was Train B travelling faster during the first hour of its journey or the last hour?
**e** How much earlier did Train A arrive at the destination?

**5** Which cheese is the best value for money?

**6** The diagram shows the speed–time graph for a particle.

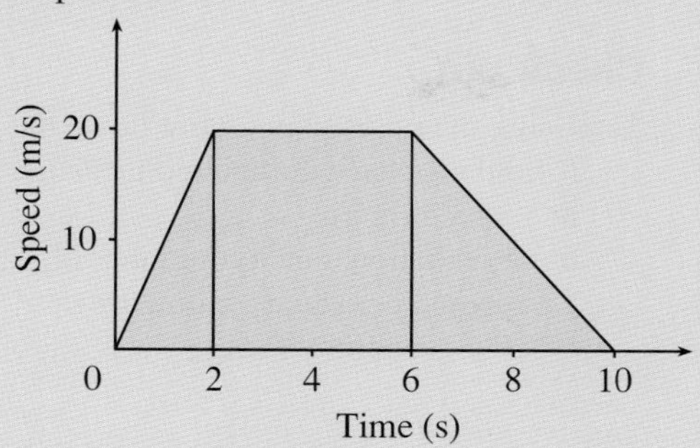


Calculate:

**a** the acceleration of the particle in the first 2 seconds

**b** the deceleration during the last 4 seconds

**c** the distance travelled in the 10 seconds.

**7** How far does a car travel if its average speed is 88 km/h and it travels for $1\frac{1}{2}$ hours?

**8** The diagram shows the speed–time graph for a car during 20 seconds.

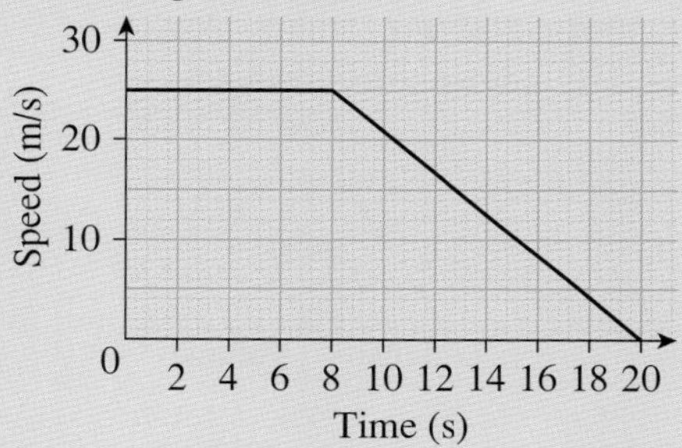


**a** Describe what the graph shows.

**b** Work out the deceleration during the last 12 seconds.

**c** How far has the car travelled in these 20 seconds?

**9** Calculate the density of a piece of metal with mass 540 g and volume 40 cm$^3$.

**10** A car starts from rest and accelerates for 6 seconds to a point A, reaching a velocity of 18 m/s.
It continues at the same speed for a further 12 seconds.

**a** Draw a speed–time graph for this part of the car's journey.

**b** Using your graph, find

**i** the speed of the car after 3 seconds

**ii** the car's acceleration in the first 6 seconds

**iii** the distance travelled in the 18 seconds

**iv** the average speed for the 18 seconds.

**11** Which paint is the best value for money?

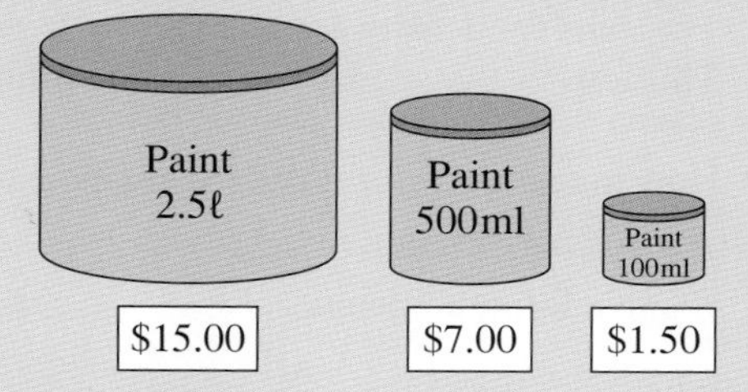


**12** The graph compares the values of two houses, one in the country and one in the city, over time.

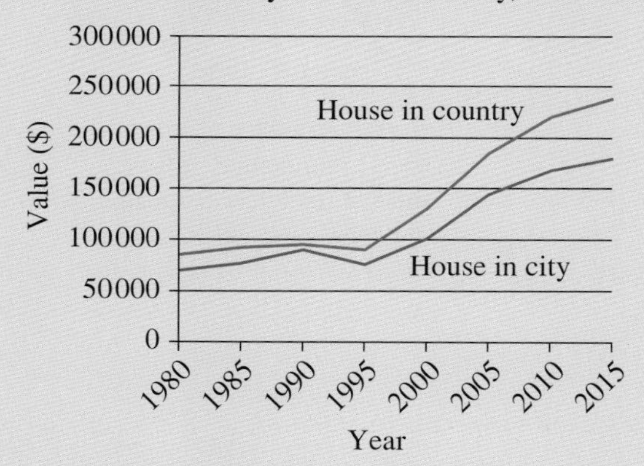


**a** At what point was the gap between the house prices the smallest?

**b** At what point was the gap between the house prices the largest?

**c** Which house has increased in value at a faster rate, overall?

**d** Which house was the best value for money, if bought as an investment in 1980?

**13** A town has a population density of 5500 people/km$^2$ and a population of 797 500. Calculate the area of the town.

**14** Work out the time taken for these journeys, in minutes.

**a** 120 km at an average speed of 48 km/h

**b** 11 km at an average speed of 44 km/h

# Summary

## You should know ...

**1** How to work with compound measures, and how to use the units for compound measures to work out which calculation to do.

*For example:*
The units for density are g/cm$^3$.
This tells us that: density = mass ÷ volume

**2** How to rearrange formulae for compound measures.

*For example:*
Using a formula triangle:

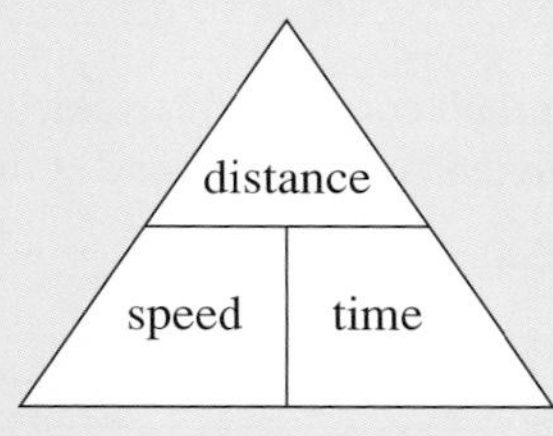


speed = distance ÷ time
time = distance ÷ speed
distance = speed × time

**3** How to work out the best value for money, finding either the unit cost or the quantity for $1.

*For example:*
A 2-ℓ bottle of cola costs $1.24.
Eight 330-ml cans of cola cost $1.98

Cost per litre for the bottle:
$1.24 ÷ 2 = $0.62 per litre

The cans contain 8 × 0.33 ℓ = 2.64 ℓ of cola.

Cost per litre for the cans:
$1.98 ÷ 2.64 = $0.75 per litre

Hence, the bottle of cola is the best value for money.

## Check out

**1** Using the units given, write down the formula for each compound measure.

**a** Speed (m/s)
**b** Population density (number of people per square kilometre)
**c** Fuel economy (km/ℓ)

**2** **a** **i** Find the average speed of an object that travels 205 km in $2\frac{1}{2}$ hours.
**ii** How long does a car take to make a 175-km journey if it travels at an average speed of 100 km/h?
**iii** An aeroplane is travelling at an average speed of 850 km/h. How far does it travel in 4 hours?

**b** Use this formula triangle to write down three formulae.

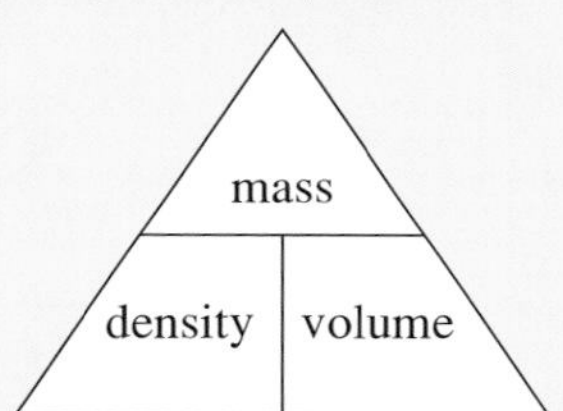


**3** Which is the best buy in each case?

**a** Packs of carrots selling at $1.26 for a 1.2-kg bag or $0.88 for an 800-g bag
**b** A branded pack of 9 muffins for $3.78 or a store's own pack of 4 muffins for $1.40.

**4** A distance–time graph can be used to find the speed an object is travelling at, using the formula
Speed = distance ÷ time.

A horizontal line shows the object has stopped.

Where speed changes during a journey, the average speed can be found using the formula

Average speed = total distance ÷ total time

**4** This is the distance–time graph for an aeroplane journey.

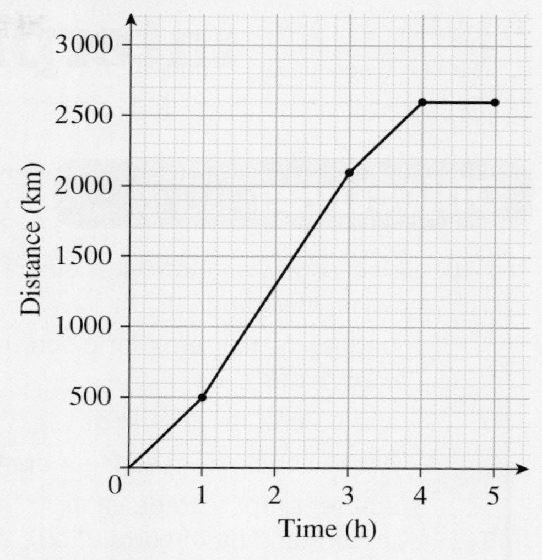


**a** How long was the flight?

**b** What was the speed of the aeroplane during the first hour of the journey?

**c** What was the speed of the aeroplane during the next two hours of the flight?

**d** What was the average speed for the entire flight time (do not include the time when the aeroplane had stopped)?

**5** The area under a speed–time graph gives the distance travelled.

*For example:*

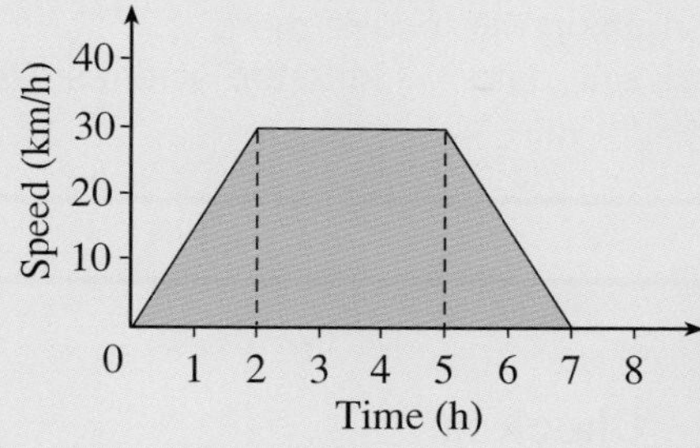


Distance travelled $= \frac{1}{2}(2 \times 30) + 3 \times 30 + \frac{1}{2}(2 \times 30)$
$= 150\,\text{km}$

The acceleration (or deceleration) is given by change in speed ÷ time

During the first two hours of the journey shown in the graph, acceleration $= 30 \div 2 = 15\,\text{km/h}^2$

A horizontal line shows the object is travelling at constant speed.

**5 a** Find the distance travelled for the car journey shown in the graph.

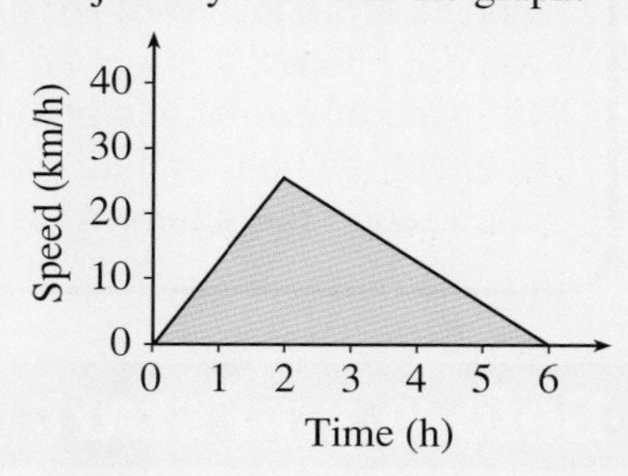


**b** Find the deceleration during the last four hours of the journey.

# 12 Presenting data and interpreting results

## Objectives

- Select, draw, and interpret diagrams and graphs, including:
  - frequency diagrams for discrete and continuous data
  - line graphs for time series
  - back-to-back stem-and-leaf diagrams
  - scatter graphs, to develop understanding of correlation.
- Interpret tables, graphs and diagrams and make inferences to support or cast doubt on initial conjectures; have a basic understanding of correlation.
- Compare two or more distributions; make inferences using the shape of the distributions and appropriate statistics.
- Relate results and conclusions to the original question.

## What's the point?

Graphs are a good way of looking for patterns and relationships in data. For example, scientists may want to investigate any link between the time people spend exercising and the amount of weight they lose. The relationship may not be as strong as you would expect as there are many factors involved. For example, what you eat has a big effect on weight loss, as does your age. The type of exercise is also important; half an hour of running will require more energy than half an hour of walking. Collecting and analysing data will help the scientists understand the different factors and their effects on weight loss.

## Before you start

### You should know ...

**1** How to find the mean, median, mode and range of a set of data.
*For example:*
Find the mean, median, mode and range of
6, 8, 14, 11, 13, 8
To find the mean, add the numbers and divide by how many numbers there are.

### Check in

**1** Find the mean, median, mode and range of
12, 15, 15, 17, 19, 22, 22, 22

Mean = (6 + 8 + 14 + 11 + 13 + 8) ÷ 6
= 60 ÷ 6 = 10
The mean is 10.
To find the median, list the numbers in order then find the middle.
6, 8, 8, 11, 13, 14
The middle two numbers are 8 and 11. The middle of these is (8 + 11) ÷ 2 = 9.5
The median is 9.5.
The mode is the number that occurs most often.
The mode is 8.
The range is the highest number minus the lowest number.
Range = 14 − 6 = 8
The range is 8.

**2** How to draw a stem-and-leaf diagram, and its main features:

- The key – explains what the stem and leaves are worth.
- The stem – often the tens column (but not always, see the key to find out).
- The leaves – often the units column (but not always, see the key to find out). Note: The leaves *must be in numerical order*. It is easier to draw an unordered stem-and-leaf diagram first then draw an ordered one. You are less likely to make a mistake or miss data out that way.
- The spacing – numbers *must be evenly spaced* to maintain the shape clearly.

| Stem | Leaves |
|---|---|
| 1 | 4 |
| 2 | 0 1 |
| 3 | 3 4 6 |
| 4 | 4 5 7 9 9 |
| 5 | 2 6 8 |
| 6 | 3 |

**Key**
1 | 4 means 14 cm

**2** The data below gives the ages of 16 people working in an office.
17, 55, 45, 57, 61, 30, 26, 27, 46, 51, 40, 60, 38, 31, 42, 26

**a** Draw a stem-and-leaf diagram for this data.

**b** Use this stem-and-leaf diagram to find
**i** the median
**ii** the mode
**iii** the range.

**3** The correct statistical language, including the meaning of these terms:

Discrete data
Continuous data
Primary data
Secondary data
Data collection sheet

**3 a** Explain the difference between
**i** discrete and continuous data
**ii** primary and secondary data.

**b** What is a data collection sheet?

# 12.1 Displaying data

## Bar charts and pictograms

One of the simplest ways of displaying discrete data is to use a **bar chart**.

**EXAMPLE 1**

The favourite female singers of a class are:

| Adele | Lady Gaga | Taylor Swift | Rihanna |
|---|---|---|---|
| 5 | 9 | 7 | 3 |

Display this on a bar chart.

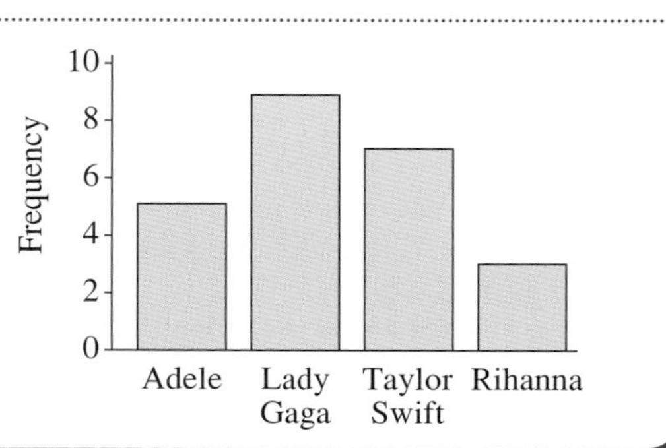


You can also use a **pictogram** to display the information in Example 1. In a pictogram a picture is used to represent the data.

| | |
|---|---|
| Adele | |
| Lady Gaga | |
| Taylor Swift | |
| Rihanna | |

Scale: represents one student

## Exercise 12A

**1** This bar chart shows the favourite colours of a group of students.

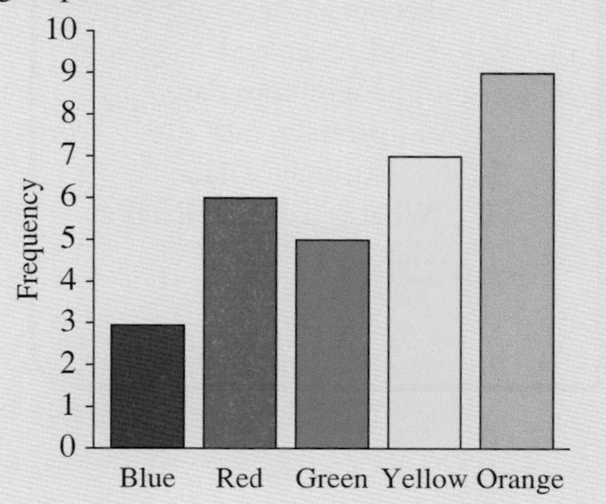


**a** Which colour is the most popular?
**b** Which colour is the least popular?
**c** How many students does the bar chart represent?
**d** Copy and complete this table using the information from the bar chart.

| Favourite colour | Blue | Red | Green | Yellow | Orange |
|---|---|---|---|---|---|
| Number of students | | | | | |

**2** The pictogram shows the number of soft drinks sold by a café last week.

| Soft drinks sold last week | |
|---|---|
| Monday | |
| Tuesday | |
| Wednesday | |
| Thursday | |
| Friday | |
| Saturday | |
| = 2 soft drinks | |

**a** What does each represent?
**b** How many soft drinks did the café sell on Tuesday?
**c** How many were sold on Friday?
**d** On which day did the café sell most soft drinks?
**e** How many soft drinks were sold in total during the week?
**f** Copy and complete the table using information from the pictogram.

| Day | Mon | Tue | Wed | Thur | Fri | Sat |
|---|---|---|---|---|---|---|
| Number of soft drinks sold | | | | | | |

**3** The table shows the favourite sports of some students at Portsmouth Secondary School.

| Volleyball | Football | Hockey | Netball |
|---|---|---|---|
| 40 | 70 | 110 | 60 |

**a** Show the information on a bar chart.
**b** Show the information on a pictogram.

4 The bar chart shows the number of children per family, for the families in one village.

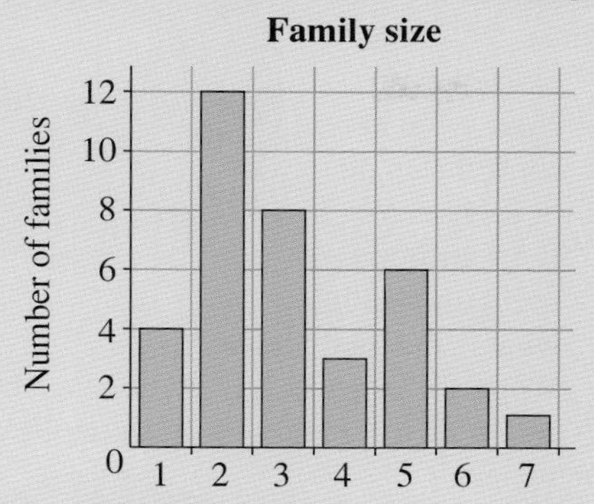


a How many families are there in the village?
b How many children are there in the village?
c Draw a pictogram for the information.

## TECHNOLOGY

You can use a spreadsheet to show a bar chart. For example, type the table from Question **3** of Exercise 12A into a spreadsheet.

- Highlight the data and select the **Column** option from the **Insert** menu.
- Choose a column bar chart from the pop-up window.
- Use the Chart Tools to give your bar chart a title and labelled axes.

## TECHNOLOGY

**Project 1**

a Choose a paragraph in a book. Count the number of times each of the vowels a, e, i, o and u occurs.
b Record your results in a table.

| Vowel | a | e | i | o | u |
|---|---|---|---|---|---|
| Frequency | | | | | |

c Repeat for two more paragraphs.
d Enter your tables into a spreadsheet. Use the spreadsheet to display a bar chart for the number of vowels in each paragraph.
e Write up your results using a word processing program. Be sure to answer questions such as
- Which is the most common vowel?
- Which is the least common vowel?
- Why do you think this may be the case?

Copy and paste your charts into your report.

**Project 2**

a Carry out a survey of your class to find out the number of brothers and sisters each student has.
b Copy and complete the table.

| Number of brothers and sisters | 0 | 1 | 2 | 3 | 4 |
|---|---|---|---|---|---|
| Frequency | | | | | |

c Make two more, separate tables, one for the number of brothers and one for the number of sisters.
d Display charts of your results in a spreadsheet program.
e Using a word processing program, write a report of your findings and illustrate it with your charts.

## Pie charts

A pie chart is another way to display your data. It is useful when you want to show the relative parts of a total.

### EXAMPLE 2

David earned \$100 from his after-school job. He spent it as follows.

| | | | |
|---|---|---|---|
| Shirt | \$30 | Music | \$20 |
| Cap | \$10 | Savings | \$30 |
| Cinema | \$10 | | |

Show this information with a pie chart.

The whole circle, 360°, represents \$100.

Fraction spent on shirt $= \frac{30}{100}$.

So, the angle representing money spent on a shirt

$= \frac{30}{100} \times 360°$

$= 108°$

The pie chart shows David's expenditure.

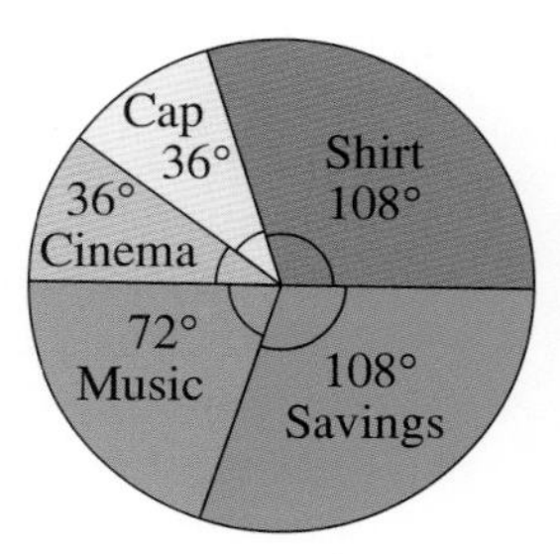

## Exercise 12B

**1** In the School Council election, there are 240 people on the voting list.
At the election for Head of Council, they vote as follows:

150 M. Malik    10 H. James
60 K. Odaro    20 Did not vote

Draw a pie chart to show this.

**2** The 300 students at Marigot High School travel to school as follows:

110 bus    160 walk
20 car    10 cycle

Draw a pie chart to show this information.

**3** The 32 students of Class 1B voted for their favourite subjects. The results were as follows:

English 4    French 2
Maths 12    Social Studies 6
Science 8

**a** Display this data on a pie chart.
**b** What percentage chose English as their favourite subject?

**4** A town council spent its budget as follows:

45% on public services
25% on road maintenance
10% on transportation
10% on sewage improvements
5% on landscaping
5% on investments

**a** Display the data on a pie chart.
**b** If \$3000 was spent on landscaping, what was the total budget?

**5** The pie chart shows the results of a survey of favourite sports among 1080 people.

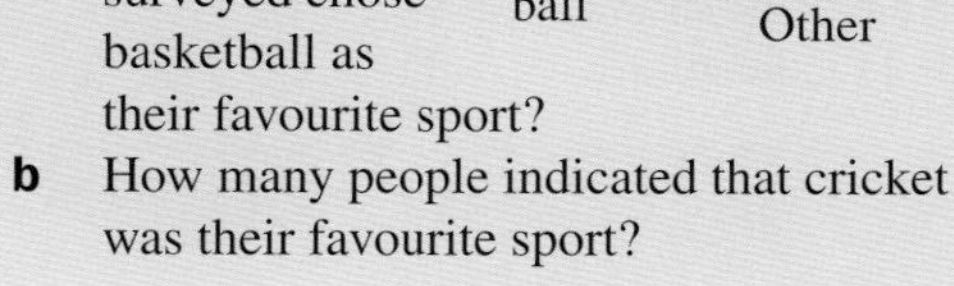


**a** What fraction of the people surveyed chose basketball as their favourite sport?
**b** How many people indicated that cricket was their favourite sport?

**6** The pie chart shows the budget of a town council.

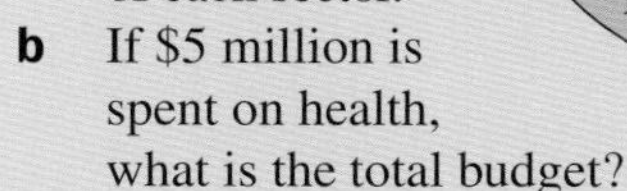


**a** Draw the pie chart again to show the angle of each sector.
**b** If \$5 million is spent on health, what is the total budget?
**c** Work out how much is spent on each area and draw a bar chart to illustrate the data.

**7** The pie chart illustrates the sales of different makes of motor oil.

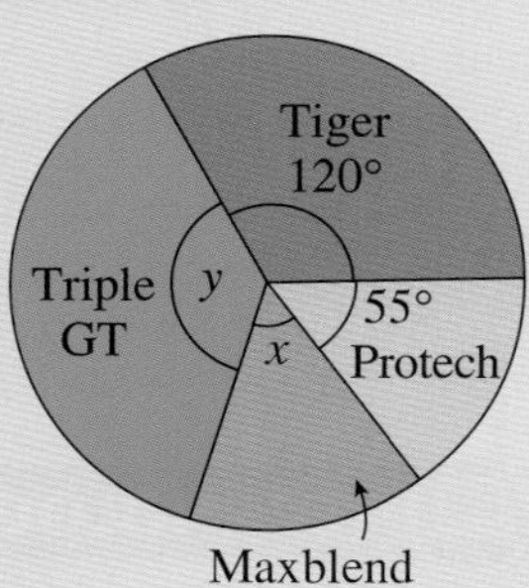


**a** What percentage of the sales does Protech have?
**b** If Maxblend accounts for 15% of the total oil sales, calculate the angles $x$ and $y$.

**8** A fruit importer checks the number of bad oranges in 100 boxes. This is what he found.

| Number of bad oranges | 0 | 1 | 2 | 3 | 4 or more |
|---|---|---|---|---|---|
| Number of boxes | 55 | 32 | 10 | 3 | 0 |

**a** Draw a pie chart to show the data.
**b** How many bad oranges were there altogether?

### TECHNOLOGY

You can use a spreadsheet to show a pie chart. For example, type the data from Question **3** of Exercise 12B into a spreadsheet.

- Highlight the table and select the **Pie** option from the **Insert** menu.
- Select a pie chart from the pop-up window.
- Use the Chart Tools to give your chart a title.

## Histograms

### Grouped continuous data

The heights of 30 plants are shown in the table.

| Height (cm) | Frequency |
|---|---|
| 15-17 | 2 |
| 18-20 | 6 |
| 21-23 | 12 |
| 24-26 | 7 |
| 27-29 | 3 |

Each group of data is a **class** or **interval**.

When a height is given as 21 cm (to the nearest cm) its true value lies between 20.5 cm and 21.5 cm. Therefore, each height in the class 21–23 has a true value between 20.5 and 23.5. These are the **class boundaries** of the 21–23 class.

The **class width** = 23.5 cm − 20.5 cm = 3 cm.
The **class limits** are 21 cm and 23 cm.

You can draw a chart to show this continuous data.

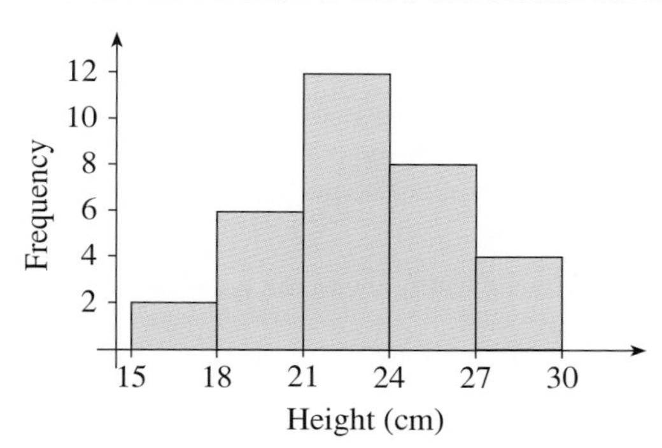


This type of chart is called a **histogram**.

- In a histogram:
  - there are no spaces between the bars
  - the area of each bar represents the frequency.

**EXAMPLE 3**

The time taken by children to travel to school is given in the table.

| Time (minutes) | Frequency |
|---|---|
| 5-9 | 2 |
| 10-14 | 8 |
| 15-19 | 10 |
| 20-24 | 6 |
| 25-29 | 4 |

Draw a histogram to show this information.

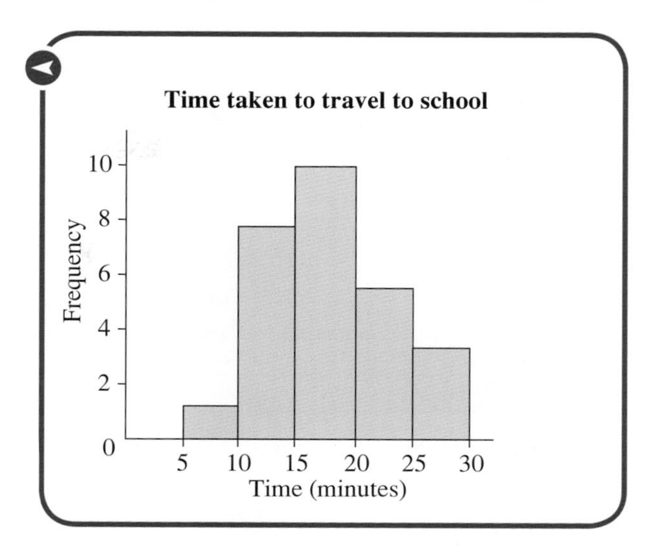


### Exercise 12C

**1** A group of 25 children measure each other's heights and record the answers by putting a tick on a chart:

| Height (cm) | Tick |
|---|---|
| 135-139 | |
| 140-144 | ✓ |
| 145-149 | |
| 150-154 | ✓✓✓ |
| 155-159 | |
| 160-164 | ✓ |
| 165-169 | ✓ |

**a** Suppose the heights are rounded off to the nearest centimetre. In which class would you put a tick for a child whose height is
**i** 140.4 cm **ii** 149.2 cm
**iii** 149.9 cm **iv** 161.3 cm
**v** 164.8 cm **vi** 139.6 cm?

**b** Into which class would you put a height of
**i** 159.5 cm **ii** 139.5 cm
**iii** 144.5 cm?

**c** What is the least height belonging to the class 155–159 cm?

**d** Write down the class boundaries for these classes.
**i** 135–139 cm **ii** 140–144 cm
**iii** 145–149 cm **iv** 150–154 cm

**2** This is the completed frequency table for the 25 children.

| Height (cm) | Frequency |
|---|---|
| 135–139 | 1 |
| 140–144 | 3 |
| 145–149 | 4 |
| 150–154 | 7 |
| 155–159 | 5 |
| 160–164 | 4 |
| 165–169 | 1 |

The histogram for this table has been started. Copy and complete it. Write a title for it.

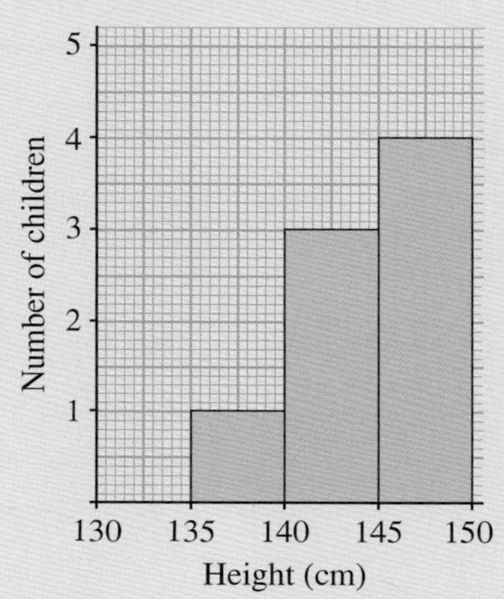


**3** The histogram shows the results of weighing 100 apples to the nearest gram.

**a** How many apples are in the class 110–119 grams?

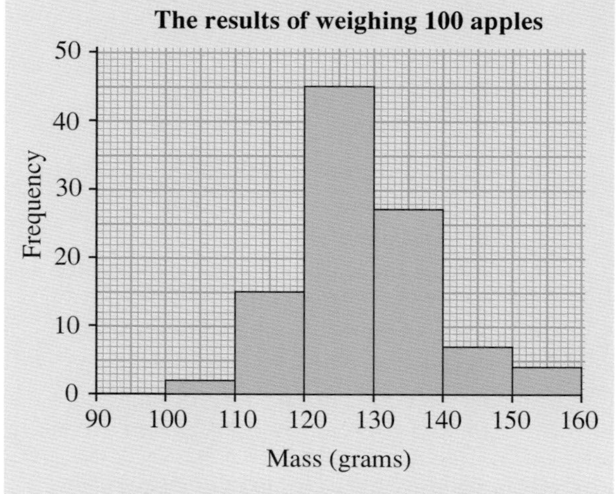


**b** Draw a frequency table to show the information in the graph. Use classes of 100–109, 110–119, . . ., 150–159.

**c** What is
**i** the lower boundary of the fourth class
**ii** the upper boundary of the fifth class?

**d** What is
**i** the minimum possible mass of an apple
**ii** the maximum possible mass of an apple?

**4** In a biology experiment, the lengths of the leaves of a plant are measured and recorded to the nearest millimetre. The results are shown in the table.

| Length (mm) | Frequency |
|---|---|
| 20–24 | 1 |
| 25–29 | 4 |
| 30–34 | 8 |
| 35–39 | 18 |
| 40–44 | 25 |
| 45–49 | 24 |
| 50–54 | 17 |
| 55–59 | 2 |
| 60–64 | 1 |

Draw a histogram of the results. Use one small division to represent one unit, as in the graph in Question **3**.

**5** The heights of 154 boys, to the nearest centimetre, are:

| Height (cm) | 160 | 161 | 162 | 163 | 164 | 165 | 166 |
|---|---|---|---|---|---|---|---|
| Frequency | 4 | 5 | 6 | 9 | 16 | 22 | 27 |

| Height (cm) | 167 | 168 | 169 | 170 | 171 | 172 |
|---|---|---|---|---|---|---|
| Frequency | 25 | 18 | 11 | 6 | 3 | 2 |

**a** Redraw the frequency table using intervals of 160–161, 162–163, . . ., 172–173.

**b** What are the boundaries of the interval 164–165 cm?

**c** What is the boundary between the interval 166–167 cm and 168–169 cm?

**d** Draw a histogram using the intervals in part **a**.

**6** The percentage marks of 100 students in a test were:

| Marks % | No. of students |
|---|---|
| 0-19 | 5 |
| 20-29 | 6 |
| 30-39 | 13 |
| 40-49 | 22 |
| 50-59 | 24 |
| 60-69 | 16 |
| 70-79 | 8 |
| 80-89 | 6 |

**a** Draw up another frequency table using equal intervals of 20 marks.

**b** Illustrate the information using a histogram.

## Frequency polygons

Frequency distributions can also be illustrated by a **frequency polygon**. Frequencies are represented by a single point at the centre of each interval (the mid-interval value). The points are joined by straight lines.

**EXAMPLE 4**

The masses, in kilograms, of 24 children are:

| Mass (kg) | 10-19 | 20-29 | 30-39 | 40-49 | 50-59 |
|---|---|---|---|---|---|
| No. of children | 1 | 2 | 6 | 12 | 3 |

Draw a frequency polygon to show this.

The interval 10–19, for example, has centre

$$\frac{10 + 19}{2} = \frac{29}{2} = 14.5$$

The frequency polygon is:

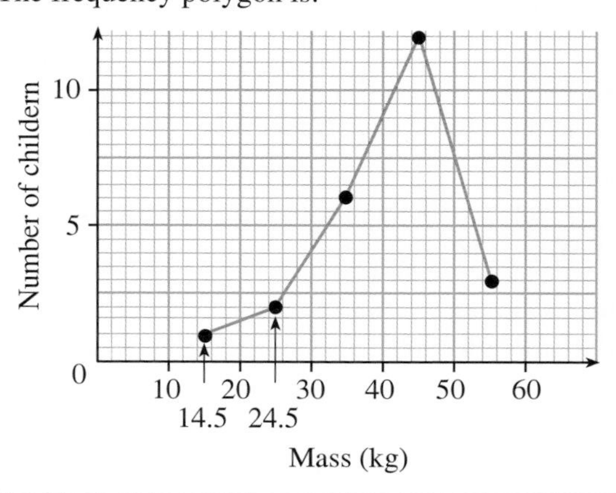


**Note that frequency polygons are not mentioned specifically in the Cambridge Secondary 1 curriculum framework, and are included here as extension work.**

## Time-series graphs

A time series graph is a line graph showing how quantities vary over time. They are often used to compare two quantities visually, as in the next example.

The table shows the average temperatures for each month of the year in Dubai, UAE and in London, UK.

| Month | Average high in Dubai (°C) | Average high in London (°C) |
|---|---|---|
| Jan | 24 | 7 |
| Feb | 25 | 7 |
| Mar | 28 | 10 |
| Apr | 33 | 12 |
| May | 38 | 16 |
| Jun | 40 | 19 |
| Jul | 41 | 21 |
| Aug | 41 | 21 |
| Sep | 39 | 18 |
| Oct | 35 | 14 |
| Nov | 31 | 10 |
| Dec | 26 | 7 |

A time-series graph is useful for looking for trends and for comparing data.

This time-series graph shows the same information as the table.

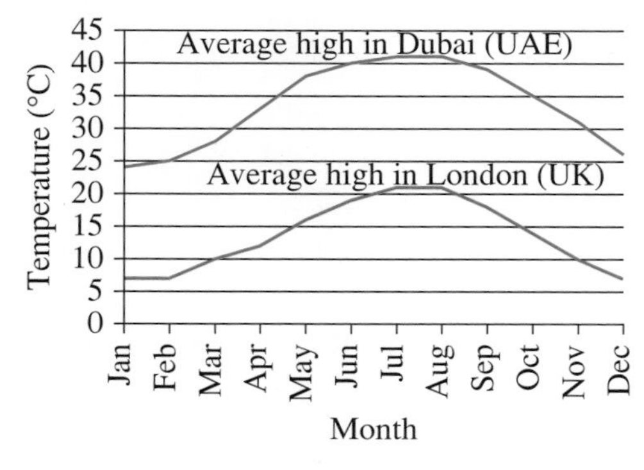


It is much easier to use the graph to compare the data for the two places and to **make inferences**, such as which months are in summer and which are in winter, than it is to use the table, which only shows a list of numbers. To 'infer' or to 'make inferences' means to arrive at a conclusion, a conclusion that can't necessarily be proven, as a result of evidence seen.

This type of graph is often used in holiday brochures to help people decide what time of year to take their holiday, according to the sort of weather they prefer.

## Exercise 12D

**(Note: Questions 1, 2 and 4 are extension work.)**

**1** Draw frequency polygons to illustrate the data in Questions **3** and **4** of Exercise 12C.

**2** Here are the masses of cattle sold at a livestock market:

| Mass (kg) | Frequency |
|---|---|
| 450-499 | 16 |
| 500-549 | 130 |
| 550-599 | 42 |
| 600-649 | 12 |

**a** What is the mid-interval value of the class 500–549 kg?

**b** Draw a frequency polygon to show this information.

**3** Use the data in the table to draw a time-series graph showing the average high temperatures, in degrees Celsius, for Cairo, Egypt and Wellington, New Zealand.

| Month | Average high in Cairo (°C) | Average high in Wellington (°C) |
|---|---|---|
| Jan | 19 | 20 |
| Feb | 20 | 21 |
| Mar | 24 | 19 |
| Apr | 28 | 17 |
| May | 32 | 14 |
| Jun | 34 | 12 |
| Jul | 35 | 11 |
| Aug | 34 | 12 |
| Sep | 33 | 14 |
| Oct | 29 | 15 |
| Nov | 25 | 17 |
| Dec | 20 | 19 |

Describe the trends that you see.
What inferences can you make from the graph?

**4** The lengths of insect larvae are measured to the nearest mm:

| Length (mm) | Frequency |
|---|---|
| 20-24 | 15 |
| 25-29 | 33 |
| 30-34 | 58 |
| 35-39 | 50 |
| 40-44 | 4 |

**a** How many insect larvae were measured?

**b** What is the mid-interval value of the class 20–24 mm?

**c** Draw a frequency polygon for the information.

**5** Here is a time-series graph showing the value of a car over a four-year period:

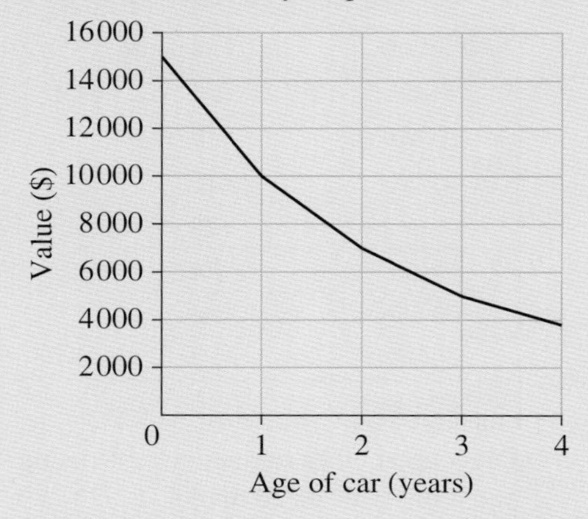


**a** During which year did the car depreciate most (lose the greatest value)?
How does the time-series graph show this?

**b** Describe the trend in the value of the car.

**c** Use the graph to estimate the value of the car after

**i** 6 months

**ii** 18 months.

**d** How much had the car depreciated by after

**i** the first year

**ii** four years?

## Back-to-back stem-and-leaf diagrams

In Book 2 you learned about stem-and-leaf diagrams. You learned that they are similar to histograms, in that the shape of the diagram gives an overall picture of the trend. The advantage of a stem-and-leaf diagram over a histogram is that you can still see all the data values. A **back-to-back stem-and leaf** diagram has all these features plus one more: it allows two different sets of data to be displayed in the same diagram. This means that the two sets of data can be compared with each other.

These two sets of data show the scores of two basketball teams in the same league, from the last 20 games:

**Tigers**
99, 116, 68, 74, 89, 83, 90, 104, 94, 98, 102, 93, 105, 88, 110, 95, 116, 77, 117, 99

**Lions**
112, 61, 92, 76, 88, 63, 71, 93, 75, 76, 80, 82, 95, 100, 63, 64, 79, 84, 111, 76

Bao is a Tigers fan. He thinks the Tigers score higher in their games than the Lions. Here, Bao has made a **conjecture** – an idea or theory yet to be proven.

It is quite hard to draw conclusions from the lists of numbers to support or cast doubt on Bao's conjecture.

However, the data can be displayed on a back-to-back stem-and-leaf diagram, as shown here. Notice the data is in numerical order going outwards, away from the stem.

| Tigers | | Lions |
|---:|:---:|:---|
| 8 | 60 | 1 3 3 4 |
| 7 4 | 70 | 1 5 6 6 6 9 |
| 9 8 3 | 80 | 0 2 4 8 |
| 9 9 8 5 4 3 0 | 90 | 2 3 5 |
| 5 4 2 | 100 | 0 |
| 7 6 6 0 | 110 | 1 2 |

**Key**

60 | 1 means 61 and 8 | 60 means 68

The shape of the distribution of the data on each side of the stem-and-leaf diagram supports Bao's conjecture:

| Tigers | | Lions |
|---:|:---:|:---|
| 8 | 60 | 1 3 3 4 |
| 7 4 | 70 | 1 5 6 6 6 9 |
| 9 8 3 | 80 | 0 2 4 8 |
| 9 9 8 5 4 3 0 | 90 | 2 3 5 |
| 5 4 2 | 100 | 0 |
| 7 6 6 0 | 110 | 1 2 |

**Key**

60 | 1 means 61 and 8 | 60 means 68

More of the Lions' scores are in the lower numbers, and more of the Tigers' are in the higher numbers. The back-to-back plot makes it much easier to see this.

To try to find further support for Bao's conjecture, we could work out the teams' average scores and compare them.

The mode is not a sensible average to pick in this case, as here the modes aren't very representative of the data. The median or mean would be fairer averages to use. The median is easy to find here, since the data has already been ordered. The mean is the best average to use in many cases, as all the data values are used in its calculation. However, we do not need to be so accurate here – we only need to get an impression of the data – so in this case the median is acceptable.

For the Tigers: median $= (95 + 98) \div 2 = 96.5$
For the Lions: median $= (79 + 80) \div 2 = 79.5$

Again, these averages support Bao's conjecture that the Tigers score more than the Lions.

In examples like this we can talk only about supporting or casting doubt on a conjecture. When we only have a sample set of data – such as scores from 20 basketball games – we can't prove an idea is always true, only that it is true for our particular sample.

## Exercise 12E

**1** The back-to-back stem-and-leaf diagram shows the thickness, in millimetres, of two samples of 25 washers, produced by two different machines.

| Machine A | | Machine B |
|---:|:---:|:---|
| 9 8 6 | 1 | 7 7 7 6 8 9 |
| 7 5 5 5 0 | 2 | 1 2 4 5 7 8 8 8 8 9 |
| 9 8 8 7 6 6 5 4 3 2 1 | 3 | 0 1 2 3 5 7 |
| 8 7 5 4 1 0 | 4 | 6 8 |

**Key**

1 | 7 = 1.7 mm and 6 | 1 = 1.6 mm

- **a** Which machine do you think produces thicker washers? Why?
- **b** What is the thickest washer produced by
  - **i** Machine A
  - **ii** Machine B?
- **c** What is the thinnest washer produced by
  - **i** Machine A
  - **ii** Machine B?
- **d** Work out the median washer thickness for
  - **i** Machine A
  - **ii** Machine B.
- **e** Work out the modal washer thickness for
  - **i** Machine A
  - **ii** Machine B.
- **f** Do your answers to parts **b**, **c**, **d** and **e** support or cast doubt on your answer to part **a**?
- **g** Of your answers to parts **b**, **c**, **d** and **e**, which do you think are the most reliable or meaningful?

**2** 15 students sit a Maths exam and a Geography exam. Their results are

**Maths:** 43, 63, 45, 57, 82, 52, 50, 69, 71, 51, 48, 70, 54, 90, 66

**Geography:** 42, 65, 75, 88, 73, 88, 57, 60, 72, 97, 79, 80, 62, 91, 51

- **a** Draw a back-to-back stem-and-leaf diagram to show the students' Maths and Geography results.
- **b** Which exam do you think was easier? Why?
- **c** Work out the mean exam score for each subject. Does this support your answer to part **b**?

**3** Jane manages the finances of several sports clubs. Each club has a membership fee that has to be paid in instalments. 19 people in Club A and 17 people in Club B all paid their membership fees on the same day. The ages of the people who paid are shown in the stem-and-leaf diagram.

| Club A | | Club B |
|---:|:---:|:---|
| 9 9 8 8 7 7 | 0 | |
| 8 5 4 3 2 2 1 1 | 1 | 9 |
| 5 1 0 | 2 | 4 9 |
| 2 1 | 3 | 0 |
| | 4 | 4 5 8 9 |
| | 5 | 4 5 5 6 8 9 |
| | 6 | 3 3 6 |

**Key**

1 | 9 means 19 years old

7 | 0 means 7 years old

**a** How many people aged under 20 paid their fees for
 **i** Club A
 **ii** Club B?

**b** How many people aged over 50 paid their fees for
 **i** Club A
 **ii** Club B?

One of the clubs is a gymnastics club, the other is a golf club. Jane has lost the pieces of paper with the club details on, so she isn't sure which club the two sets of fees were paid for.

**c** Which club do you think is the golf club? Why?

**d** Do you think that Jane can be confident enough to give the golf club money to the club you chose in part **c** without any further checking? Why?

**4** A group of 18 children was randomly selected in each of two different schools. They were asked the distance, in kilometres, that they travelled to school. The results are as follows.

**KMA school:**
2.1, 0.9, 1.0, 4.1, 1.5, 2.2, 1.6, 4.6, 1.8, 0.8, 2.5, 0.8, 2.5, 1.7, 3.0, 1.5, 3.3, 1.8

**Highfield Academy:**
0.6, 2.4, 0.7, 1.3, 4.1, 1.5, 3.8, 1.8, 2.0, 1.9, 2.1, 0.8, 2.4, 3.2, 3.3, 1.3, 4.4, 1.6

**a** Draw a back-to-back stem-and-leaf diagram for the two sets of data.

**b** Comment on what the diagram shows.

Watch the video at

www.youtube.com/watch?v=APlps5sh46s

to discover a really long formula to help you do stem and leaf plots using Microsoft Excel spreadsheets.

Try doing this with the data for the Maths exam from Question **2** in Exercise 12E. If you really want a challenge, try doing the full back-to-back stem-and-leaf diagram for Question **2**.

## 12.2 Scatter graphs and correlation

**Scatter graphs**, also known as 'scatter diagrams' or 'scatter plots', are used to look for a link or relationship between two features (or variables) in a set of paired data. Values for the two features are plotted as points on a graph. If these points tend to lie close to a line on the graph then there is a relationship or link between them. If there is a relationship this is called **correlation**. If points do not lie in any pattern then we say there is no correlation.

You may expect some things to be related to each other. For example, time spent studying for an exam and the exam result.

The amount of time spent revising and the percentage scores in a Maths test for 12 students are shown in this table and scatter graph.

| Time spent studying (hours) | 2 | 4 | 5 | 6 | 3 | 1.5 | 8 | 7 | 9 | 5.5 | 0.5 | 6.5 |
|---|---|---|---|---|---|---|---|---|---|---|---|---|
| Maths test result (%) | 20 | 60 | 46 | 71 | 35 | 31 | 70 | 86 | 94 | 63 | 70 | 52 |

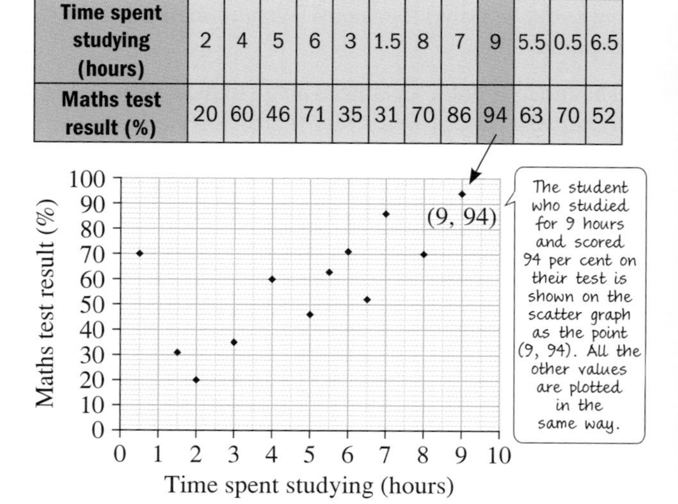

You can see that the points on the graph generally follow an upward trend. This supports our conjecture that the more hours you study the higher your grade tends to be; or the fewer hours of study, the lower the grade. The points don't make a clear line on the graph because there are other factors that affect results, such as ability in a subject, the quality of study done and the ability to work well in an exam.

One point doesn't fit the trend at all. This is the point (0.5, 70) – a student who only studied for half an hour but still managed to get 70 per cent on the test. Sometimes people are able to do well in tests without much revision. Imagine what their result would have been had they done more revision! When there is a result like this one, that doesn't fit the general trend, it is called an **outlier** or **anomaly**.

Sometimes the relationship is not so obvious. Did you know that the size of an ape skull fossil is related to the age of the fossil? Make a conjecture about the relationship you expect.

The table lists the ages (in millions of years) and the volumes (in cubic centimetres), of the fossil skulls of a particular kind of ape.

| Age (million years) | 0.3 | 2.5 | 0.6 | 1.5 | 2.4 | 0.4 | 1.9 | 1.1 | 1.7 | 2.3 | 0.8 |
|---|---|---|---|---|---|---|---|---|---|---|---|
| Volume of skull ($cm^3$) | 206 | 27 | 175 | 93 | 21 | 188 | 56 | 130 | 85 | 32 | 162 |

We can draw a scatter diagram to look for a relationship:

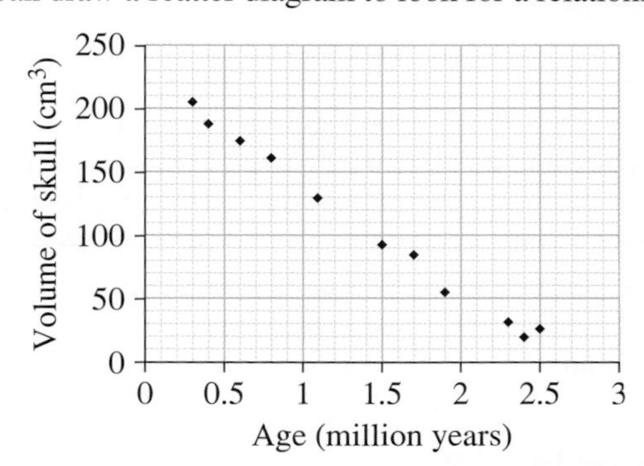


This time we can see a clear downward trend. As the age of the skull increases, the size decreases. Did the scatter graph support or cast doubt on your conjecture? Were you expecting the relationship to be so strong?

If you compare the scatter graphs in our two examples, the first shows an upward trend, but not as strong a relationship as appears in the second graph; the points do not lie as close to a straight line. There is a **weak positive correlation** between the number of hours studied and the result in the Maths test. The second graph shows a downward trend and a stronger relationship, because the points are close to making a straight line. We say there is a **strong negative correlation** between the age and the volume of the ape skull fossils.

Here are some scatter graphs showing different types of correlation. Discuss with your class possible labels for the axes of these graphs.

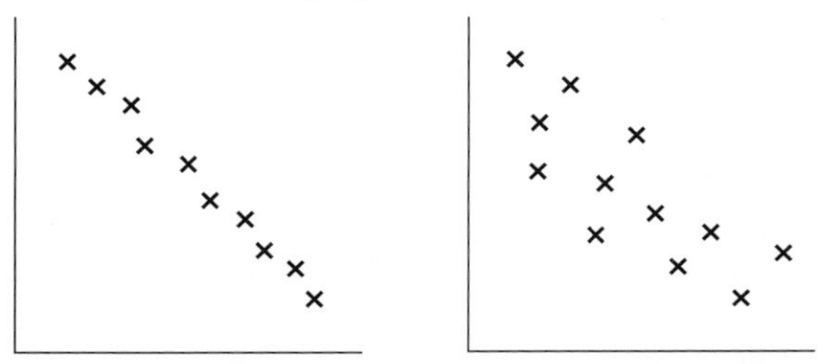

Strong negative correlation | Weak negative correlation

Strong positive correlation | Weak positive correlation

The words 'weak' and 'strong' are subjective (that is, their meaning will differ from person to person) so do not worry too much about these. The more important aspect is whether the correlation is positive or negative.

Sometimes there is no relationship between variables. The points will not show any trend or pattern at all. This is **no correlation**:

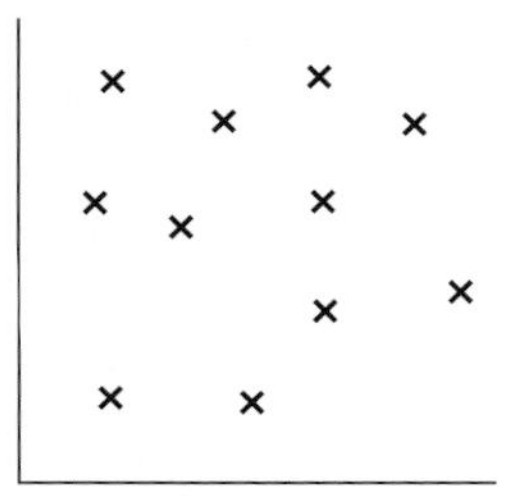

No correlation

You might expect a graph showing no correlation if you plotted Maths test results on the horizontal axis and hair length on the vertical axis.

Look at the next scatter graph. If you are asked to **describe** the relationship it shows, you can say there is a 'strong positive correlation'. If you are asked to **interpret** this, you need to explain what that means, relating the result to the original data. For example, 'as height increases mass increases'.

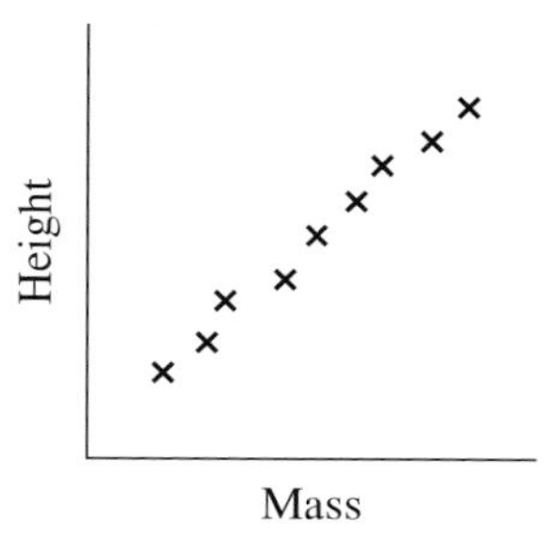


## Line of best fit

When you suspect that there is correlation between two variables you can draw a **line of best fit**. This is a line that best fits the trend of the data values. A line of best fit should:

- follow the trend of the points
- pass through, or very close to, as many points as possible
- have a similar number of points above and below it.

Note that a line of best fit does not have to pass through the origin. This is a common mistake, because many lines of best fit do pass through the origin.

Here are some examples of lines of best fit:

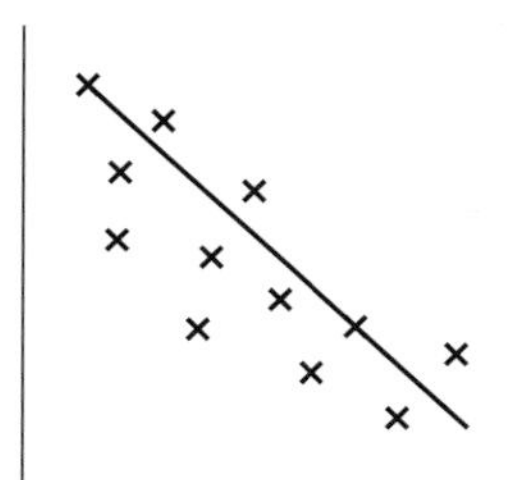

This line follows the downward trend of the points and passes through some of them. It is not good as a line of best fit, though, because there are more points below the line than there are above it.

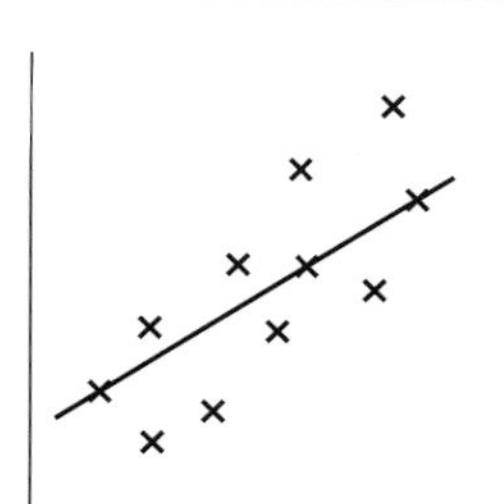

There are the same number of points above and below this line. The line passes through some points. It is not good as a line of best fit, though, because it doesn't follow the upward trend of the points very well; it is not steep enough.

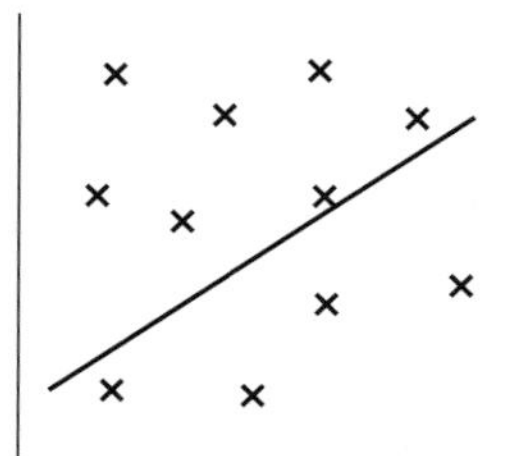

Here, no line of best fit should be drawn, because the points show no correlation.

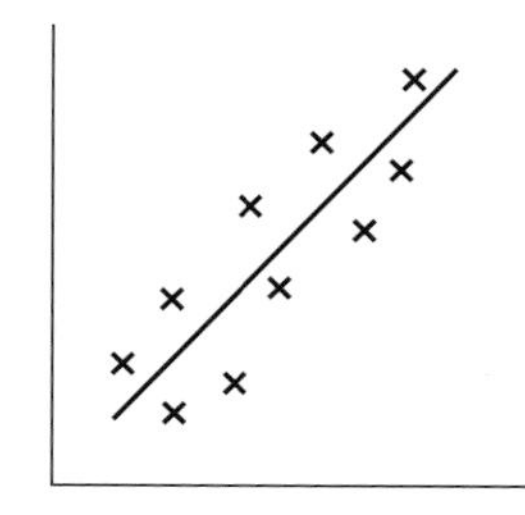

Although this line doesn't pass through any points, it is a good line of best fit because it follows the trend of the points, which lie as close to it as possible.

We can use a line of best fit to help make predictions.

Here is the graph for the ape skull fossils example with the line of best fit:

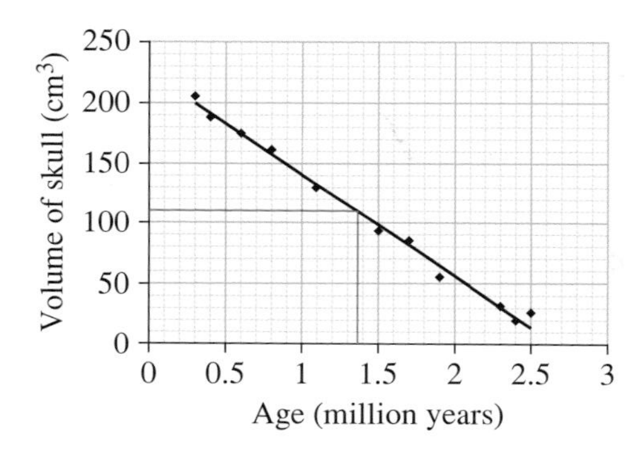


We could use this graph to estimate the age of a skull with a volume of $110\,cm^3$. Drawing a horizontal line from $110\,cm^3$ until it touches the line of best fit, then drawing a vertical line down, the skull is roughly 1.4 million years old. We can be fairly confident about this estimate as the value is right in the middle of our data set and the correlation is quite strong.

### Exercise 12F

**1** Sort these statements into three categories: **always true**, **sometimes true**, and **false**.

- **a** The correlation of a scatter graph is either positive or negative.
- **b** The line of best fit goes through the origin.

- **c** When points do not lie close to a straight line this is weak correlation.
- **d** A line of best fit should be a straight ruled line.
- **e** When points lie in a straight line this is positive correlation.
- **f** A strong correlation is when the points are close to the line of best fit.
- **g** If a graph shows no correlation it is wrong.
- **h** Because there can be lots of lines of best fit the accuracy of the line is not important.
- **i** A line of best fit can be used to estimate the value of one variable when you know the value of the other.

**2** For each of these pairs of variables

- **i** describe the correlation you would expect to see
- **ii** give an interpreation of your answer to part **i**.
  - **a** Height and arm length
  - **b** Leg length and maximum running speed
  - **c** A person's mass and their score in a History test
  - **d** Size of car and fuel economy

**3** The scatter graph shows the ages of people working in an office and their yearly earnings.

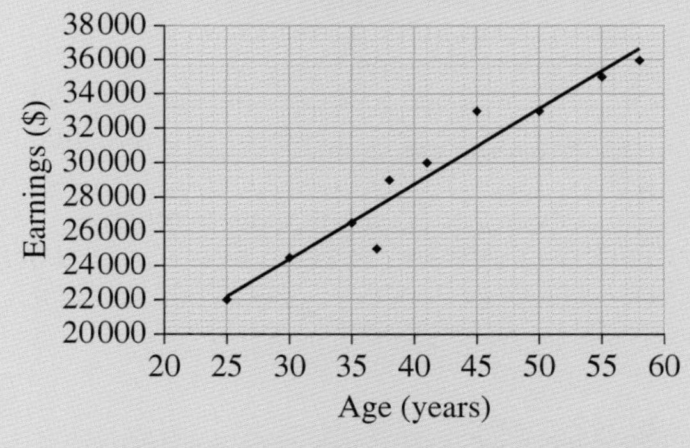


- **a** Use the line of best fit to estimate what a 48-year-old would earn.
- **b** What sort of correlation does the scatter graph show?
- **c** Use the line of best fit to estimate the age of someone earning $23 000 a year.

**4** **a** What sort of relationship do you think there is between the height above sea level and the temperature water boils at?

**b** Use this data to draw a scatter graph. Make the first value on the vertical axis 70°C and the first value on the horizontal axis 1000 m.

| Height above sea level (m) | Temperature of boiling water (°C) |
|---|---|
| 1000 | 96 |
| 1600 | 94 |
| 2200 | 92 |
| 3600 | 88 |
| 4800 | 84 |
| 6400 | 78 |
| 7200 | 76 |

**c** Does the scatter graph support or cast doubt on your conjecture from part **a**?

**d** Draw a line of best fit on your scatter graph.

**e** Use your line of best fit to predict the boiling point of water at 3000 m above sea level.

**f** Use your line of best fit to predict the height above sea level if the boiling point of water is 80°C.

**5** **a** What sort of relationship do you think there is between the number of pages in a book and the mass of the book?

**b** Draw a scatter graph for the data in this table.

| Number of pages | Mass (g) |
|---|---|
| 180 | 360 |
| 65 | 150 |
| 100 | 200 |
| 122 | 255 |
| 112 | 240 |
| 94 | 185 |
| 143 | 300 |
| 68 | 170 |
| 137 | 265 |
| 80 | 180 |
| 75 | 165 |
| 165 | 345 |

- **c** Does the scatter graph support or cast doubt on your conjecture from part **a**?
- **d** Draw a line of best fit on your scatter graph.
- **e** Use your line of best fit to predict the number of pages in a book with a mass of 320 g.

**6** Ten classmates each took a memory test. The test involved looking at a tray with 15 items on it, then seeing how many of the items they could remember. The results of the memory test and the heights of the students are recorded in the table.

| Height (cm) | Number of items remembered |
|---|---|
| 155 | 7 |
| 160 | 8 |
| 173 | 4 |
| 180 | 3 |
| 145 | 3 |
| 165 | 5 |
| 148 | 9 |
| 160 | 2 |
| 177 | 7 |
| 176 | 10 |

- **a** What sort of relationship do you think there is between the number of items remembered and the height of the student?
- **b** Draw a scatter graph for the data in the table.
- **c** Does the scatter graph support or cast doubt on your conjecture from part **a**?

**7** The table lists the values of a particular model of car at various different ages. Yemi wants to find out whether the value of a car will decrease as the car ages, and, if so, whether the correlation between the two variables is strong.

| Age (years) | Value ($) |
|---|---|
| 2 | 22 000 |
| 4 | 10 000 |
| 5 | 13 000 |
| 3 | 16 000 |
| 8 | 2000 |
| 6 | 4400 |
| 7 | 3800 |
| 1 | 21 000 |
| 4 | 16 000 |
| 6 | 9000 |

- **a** Draw a scatter graph for the data in the table. Add a line of best fit.
- **b** Look at your scatter graph. What conclusion can you reach about Yemi's question?
- **c** The correlation shown in the scatter graph is not perfect (a perfect correlation is one in which all points lie on the same straight line). What factors can you think of that affect the value of a car, other than its age?

**8** Students have drawn lines of best fit on these scatter graphs. Some have been drawn well, some have been drawn badly. Write down which lines are drawn badly, and describe what is wrong with them.

**a**

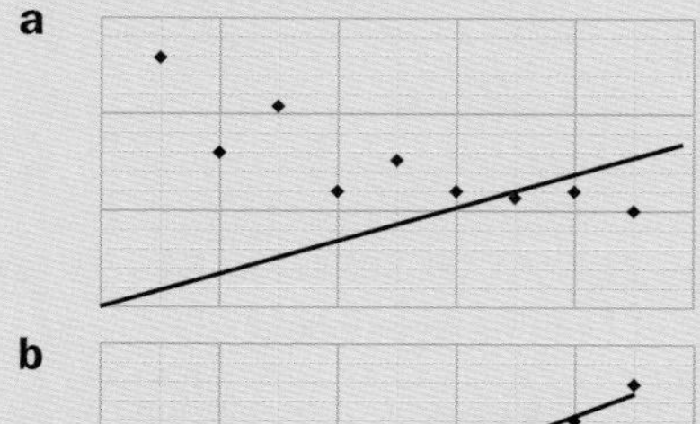

**b**

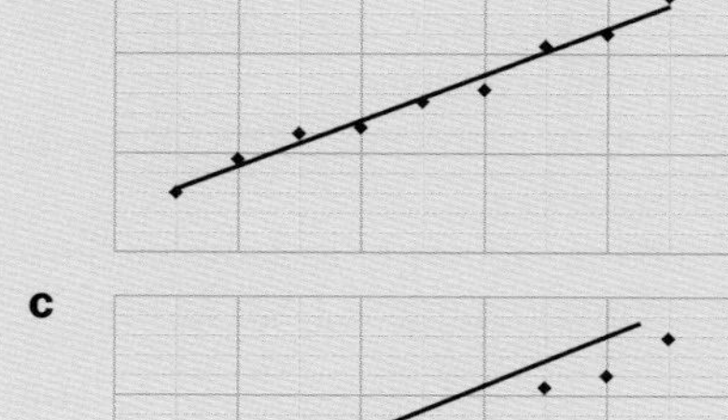

**c**

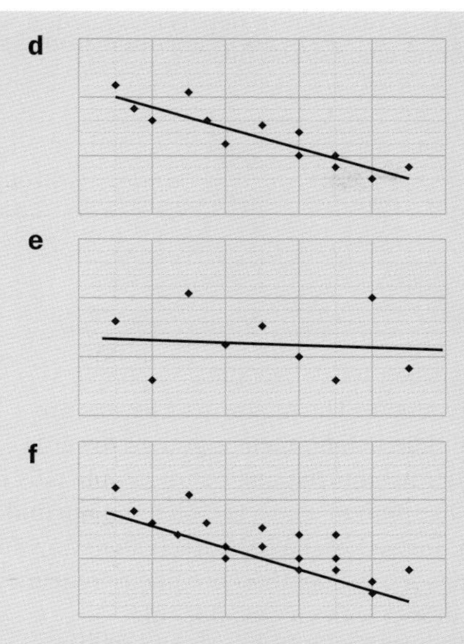


**9** A group of 9 people were asked to write their name backwards. The table shows the time it took them to do this.

| Name | Number of letters | Time (s) |
|---|---|---|
| Sam | 3 | 2.1 |
| Naya | 4 | 3.5 |
| Aisha | 5 | 5.5 |
| Devaj | 5 | 4.8 |
| Hannah | 6 | 3.3 |
| Amanda | 6 | 6.2 |
| Kenisha | 7 | 7.7 |
| Benjamin | 8 | 8 |
| Alexander | 9 | 9.5 |

**a** Do you think there is any correlation between the length of a name and the time it takes to write it backwards? If so, what sort of correlation do you think there is?

**b** Draw a scatter graph for the data. Add the line of best fit, if appropriate.

**c** Benjamin thinks he is really good at writing his name backwards – better than everyone else. Do you think this conjecture is supported by the evidence?

**d** One of the points plotted doesn't fit the results as well as the others. Circle this point on your scatter graph. Write down why you think this result doesn't follow the overall trend.

**e** What name can be given to the point you circled for part **d**?

**10** Habibah drew this scatter graph to show the arm lengths of 5 of her friends and the number of hours of television they each watched at the weekend:

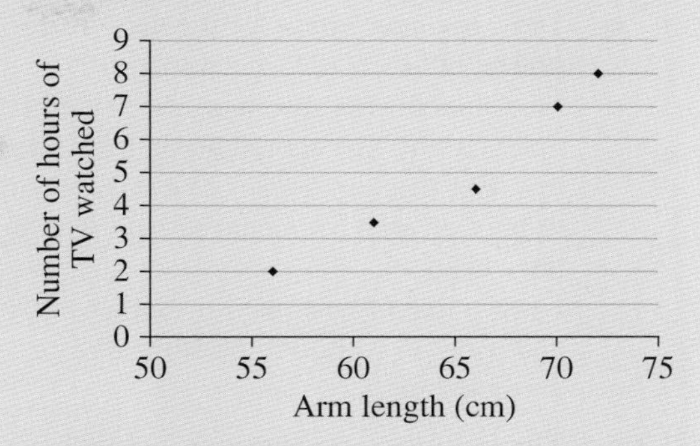


Habibah says, "My graph shows a positive correlation, so the longer your arm length the more hours of television you watch."

Layla says that Habibah is wrong, and that arm length and television watching are not correlated at all.

Who do you agree with? Give a reason for your answer. If you agree with Layla, explain why you think this scatter graph is showing a positive correlation when there isn't one.

**11** The table shows the Maths and Science test results of Class B.

| Maths result | Science result |
|---|---|
| 80 | 92 |
| 43 | 65 |
| 58 | 77 |
| 64 | 61 |
| 90 | 91 |
| 83 | 74 |
| 55 | 38 |
| 22 | 28 |
| 69 | 71 |
| 38 | 44 |
| 29 | 18 |
| 41 | 36 |
| 54 | 25 |
| 40 | 28 |
| 51 | 50 |
| 65 | 43 |
| 75 | 64 |

a Do you expect any correlation between Maths results and Science results? If you do, how strong do you think the correlation will be?

b Draw a scatter graph for the data. Were you right?

c One student was absent on the day of the Science test. She scored 60 on her Maths test. Her teacher decides to use this result to predict what her science result would have been.
Draw a line of best on your scatter graph and use it to work out the student's predicted mark.

12 a Think of two variables you suspect have a positive correlation. Do you think the correlation will be strong or weak?
Design a data collection sheet and collect data on these variables, or find secondary data.
Draw a scatter graph for your data. Relate your results and conclusions to your original question or conjecture.

b Repeat part **a** for two variables you think may have a negative correlation.

c Repeat part **a** for two variables you think may have no correlation.

13 a Find out the meaning of the words 'extrapolation' and 'interpolation' in relation to scatter graphs. Explain which of these is the most reliable and why.

b Find out about 'dependent variables' and 'independent variables'. Write down the meaning of these terms.

## 12.3 Processing and interpreting data

When working with data it is important to know which diagrams to select to present the data in the clearest way. You also need to know which statistics are the most appropriate to calculate – you learned about this in Chapter 6. Usually the purpose of collecting data is to answer a question we are interested in. When you have processed the data you should relate your results and conclusions back to the original question.

Often when a large amount of data is collected, it will form a histogram shaped like this:

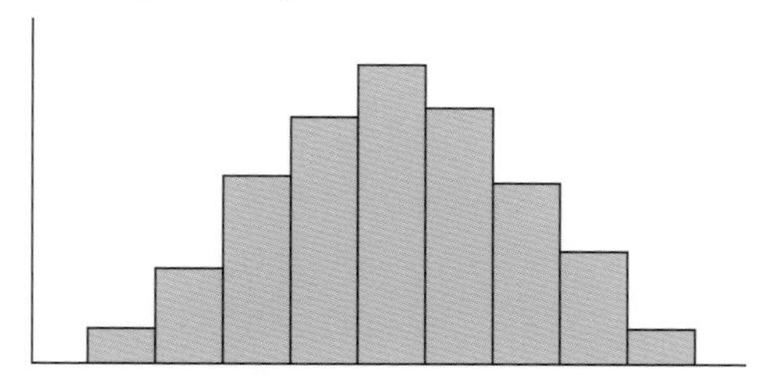

This happens when we collect data on people's heights, for example. The majority of people are around the middle height, then as you go towards shorter or taller heights the number of people tails off. This sort of distribution shape is called the **normal distribution**. Instead of the bars of a histogram you will sometimes see a diagram more like a frequency polygon but drawn with curves, like this:

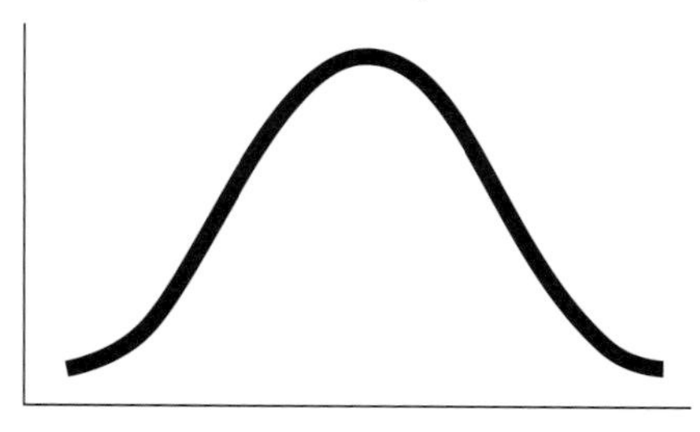

Sometimes the shape is not central but **skewed** to one side, like this:

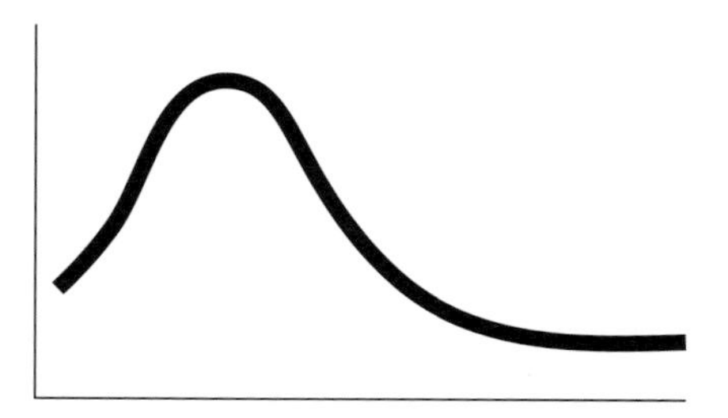

If this was also a diagram showing height data, it shows that there are a lot more people with lower heights than you might expect. One explanation is that the sample group had a large number of children in it. Be careful about the conclusions that you draw, though – you need to be sure that there isn't another explanation. In this case the shape could simply be due to the sample size being small, which could also have this effect.

The next exercise involves everything you have learned in this chapter so far, and the work on averages and range from Chapter 6.

## Exercise 12G

**1** The tables below show how students in Year 9 spend their free time.

**Boys**

| Reading | 12 |
|---|---|
| Sport | 34 |
| Music | 6 |
| Meeting friends | 4 |
| Watching TV | 8 |
| Other hobbies | 16 |
| Total | 80 |

**Girls**

| Reading | 25 |
|---|---|
| Sport | 20 |
| Music | 15 |
| Meeting friends | 25 |
| Watching TV | 10 |
| Other hobbies | 5 |
| Total | 100 |

**a** What are the best diagrams to use to display this data?

**b** Draw one of the diagrams you chose in part **a**.

**c** More girls than boys in Year 9 watch TV. Does your diagram show this or not?

**2** The highest temperature and number of hours of sunshine in Jakarta, Indonesia were recorded on 12 days spread through the year. The results are shown in the table.

| Sunshine (hours) | Highest temperature (°C) |
|---|---|
| 5 | 32 |
| 5.5 | 31 |
| 6.5 | 30 |
| 7.5 | 34 |
| 7 | 31.5 |
| 8 | 29 |
| 9 | 33 |
| 9.5 | 30.5 |
| 10 | 29 |
| 9 | 29.5 |
| 7 | 29.5 |
| 6 | 28 |

**a** What is the best diagram to use to present this data?

**b** What do you expect to see on your diagram?

**c** Draw the diagram you chose in part **a**.

**d** Did you see what you expected to see? Why do you think that was?

**e** If you were to work out the average temperature, which average would you use, mean, median or mode?

**f** Which average would you use for the number of hours of sunshine?

**3** Write down a list of diagrams that are best for displaying

**a** discrete data

**b** continuous data

**c** non-numerical data

**d** paired data.

**4** The graph below shows the distribution of the masses of two different species of monkey.

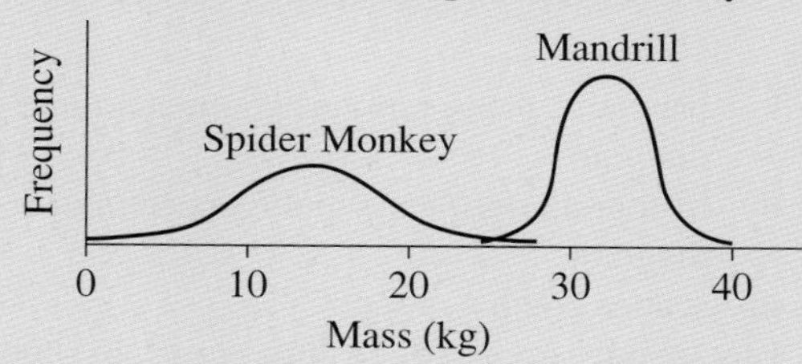


**a** Which species has the lower mean mass?

**b** Which species has the greater range of masses?

**c** Copy the beginning of each sentence and complete it with one of these endings:

is likely to be a spider monkey

is likely to be a mandrill

could be either a spider monkey or a mandrill

**i** A monkey with a mass of 7 kg ............... .

**ii** A monkey with a mass of 26 kg ............... .

**iii** A monkey with a mass of 35 kg ............... .

**d** Old World monkeys are generally bigger and heavier than New World monkeys.

**i** Out of these two species, spider monkey and mandrill, which is more likely to be an Old World species?

**ii** Can we say from the graph whether both these species of monkey are Old World, both are New World, or one is Old World and one New World? Explain your answer.

# Consolidation

### Example 1

**a** Which diagrams are best for presenting
  **i** discrete data
  **ii** continuous data
  **iii** paired data?
**b** When would you use a line graph?

---

**a** **i** bar chart, pictogram, pie chart, stem-and-leaf diagram
  **ii** histogram, frequency polygon, stem-and-leaf diagram
  **iii** scatter graph
**b** Line graphs are useful for time series, to show trends.

### Example 2

Two samples of shells were collected on the seashore. Sample 1 was collected in 2003. Sample 2 was collected from the same place in 2013.

The width of the base of each shell was measured, in millimetres. The results are listed here:

**Sample 1**

| | | | | |
|---|---|---|---|---|
| 21 | 8 | 17 | 26 | 18 |
| 9 | 19 | 25 | 12 | 23 |
| 22 | 24 | 16 | 29 | 31 |

**Sample 2**

| | | | | |
|---|---|---|---|---|
| 32 | 9 | 19 | 29 | 17 |
| 34 | 28 | 32 | 28 | 24 |
| 15 | 23 | 24 | 26 | 27 |

**a** Draw the most appropriate diagram to represent the data.
**b** Use appropriate statistics and the shape of the distribution to make comparisons between the shells from the 2003 and 2013 samples.

---

**a** As the data is continuous, appropriate diagrams to use are a back-to-back stem-and-leaf diagram, histograms or frequency polygons.

A stem-and-leaf diagram retains the original data:

| Sample 1 (2003) | | Sample 2 (2013) |
|---|---|---|
| 9 8 | 0 | 9 |
| 9 8 7 6 2 | 10 | 5 7 9 |
| 9 6 5 4 3 2 1 | 20 | 3 4 4 6 7 8 8 9 |
| 1 | 30 | 2 2 4 |

**Key**
30 | 2 = 32 mm and 1 | 30 = 31 mm

**b** The mode is not an appropriate statistic to use, as there is no mode in 2003. The median and mean are both useful here. The median is easy to work out, while the mean will be the most accurate as all data values are used in the calculation.

2003 median = 21 mm
2013 median = 26 mm

2003 mean = 20 mm
2013 mean = 24.5 mm (1 d.p.)

The averages and the distributions shown in the stem-and-leaf diagram support the conjecture that the shells were bigger in 2013 than in 2003.

### Example 3

The heights of boys aged 15 and the heights of their fathers were measured, in centimetres, and the results listed.

| Height of father (cm) | Height of son (cm) |
|---|---|
| 168 | 153 |
| 173 | 162 |
| 178 | 170 |
| 160 | 155 |
| 175 | 174 |
| 183 | 174 |
| 165 | 159 |
| 171 | 163 |
| 179 | 174 |
| 189 | 176 |
| 162 | 152 |
| 180 | 175 |

**a** Draw an appropriate diagram to represent the data and show the relationship between the two variables.
**b** Predict the approximate height of a boy at the age of 15 if his father is 174 cm tall.

**c** Describe the correlation between the height of a father and the height of his son at age 15. Suggest possible reasons for the strength or weakness of the relationship.

---

**a** The best diagram to use to look for a relationship between paired data is a scatter graph:

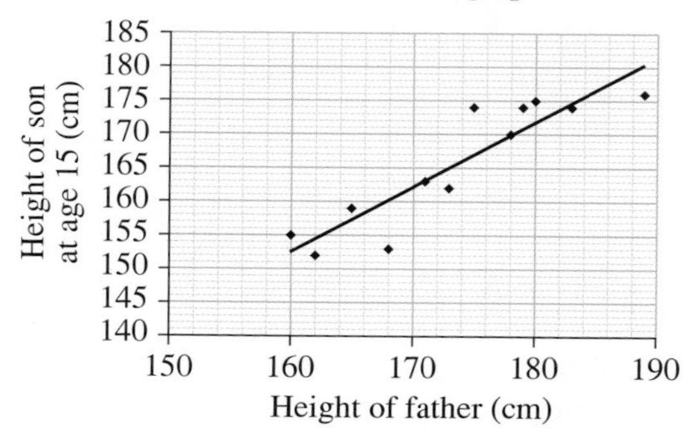


**b** We can use the line of best fit to predict that the height of a 15-year-old son of a 174-cm tall father will be approximately 166 cm:

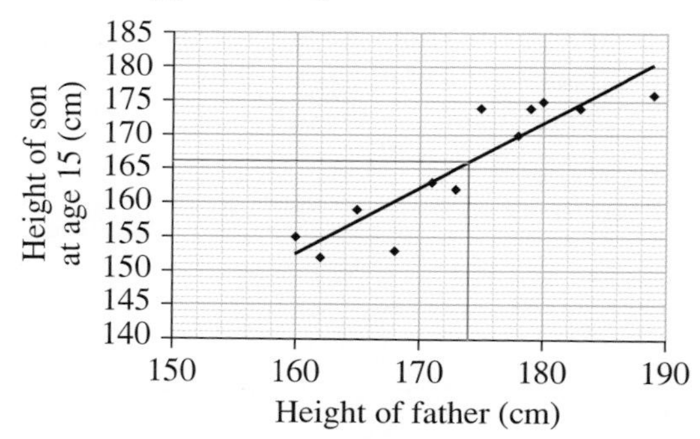


**c** There is a positive correlation between the height of a father and the height of his son at age 15. The correlation is not strong, which could be because the mother will affect a child's height as well as the father. Generally, the taller the father, the taller the son.

## Exercise 12

**(Note: Questions 6 and 8 are extension work.)**

**1** In a memory test, 20 objects are placed on a table and the person taking the test is given a certain amount of time to look at them. The objects are then removed from view and the person has to remember as many items as they can.
Ron thinks that the longer you have to look at the items the more you will remember.

**a** Draw a scatter diagram for the data in the table.

| Time spent memorising items (s) | Number of items remembered |
|---|---|
| 10 | 7 |
| 15 | 5 |
| 20 | 8 |
| 25 | 7 |
| 30 | 12 |
| 35 | 6 |
| 40 | 12 |
| 45 | 16 |
| 50 | 13 |
| 55 | 15 |
| 60 | 17 |

**b** Describe the correlation shown in your scatter graph.
**c** Draw a line of best fit.
**d** Use your line of best fit to predict the number of items that you would remember if you were given 32 seconds to view them.
**e** Do you think this prediction is reliable? Explain your answer.

**2** The heights of 50 plants of a certain species were measured to the nearest centimetre and grouped to give this table:

| Height (cm) | No. of plants |
|---|---|
| 15–17 | 3 |
| 18–20 | 9 |
| 21–23 | 15 |
| 24–26 | 14 |
| 27–29 | 7 |
| 30–32 | 2 |

**a** What is the maximum possible height of the plants?
**b** Draw a histogram to illustrate the information.

**3** The table shows the marks obtained by 100 students in a spelling test.

| Score | *f* |
|---|---|
| 5 | 6 |
| 6 | 9 |
| 7 | 10 |
| 8 | 18 |
| 9 | 32 |
| 10 | 25 |

**a** Represent this data on
 **i** a pie chart
 **ii** a bar chart.

**b** Which of the charts from part **a** do you think shows the trend better?

**4** An organisation gives an aptitude test to all applicants for employment. The results of 100 tests are shown in the table.

| Score | Frequency |
|---|---|
| 1-10 | 5 |
| 11-20 | 8 |
| 21-30 | 11 |
| 31-40 | 12 |
| 41-50 | 20 |
| 51-60 | 16 |
| 61-70 | 13 |
| 71-80 | 7 |
| 81-90 | 5 |
| 91-100 | 3 |

**a** Draw an appropriate diagram to illustrate this information.

**b** What percentage of the applicants scored less than 60?

**c** What percentage of the applicants scored between 50 and 80?

**5** Neema has been given a computer by her parents, to help her with her homework. The number of emails she sends each month for the first year she has her computer is shown in the list.

| Score | Frequency |
|---|---|
| Jan | 4 |
| Feb | 11 |
| Mar | 12 |
| Apr | 19 |
| May | 17 |
| Jun | 22 |
| Jul | 22 |
| Aug | 28 |
| Sep | 25 |
| Oct | 31 |
| Nov | 32 |
| Dec | 35 |

**a** Plot a time-series graph for this data.

**b** Describe the trend in the number of emails Neema sends.

**6** Draw the frequency polygon for the heights of 50 children recorded in the table.

| Height (cm) | *f* |
|---|---|
| 130-134 | 1 |
| 135-139 | 7 |
| 140-144 | 16 |
| 145-149 | 15 |
| 150-154 | 5 |
| 155-159 | 4 |
| 160-164 | 2 |

**7** Danika is a nurse in a hospital. As part of her ward round, she takes the temperatures of all the patients. The doctors have categorised the patients into those who are recovering well and those who are still unwell.
These are the temperatures of the patients who are recovering well:
37.7 37.5 37.6 36.8 37.1 37.5 37.6 38.2
37.2 37.5 37.9 37.5 38.1 37.8 37.6

**a** Copy and complete this back-to-back stem-and leaf diagram, filling in the temperatures of the patients who are recovering well.

| Recovering well | | Still unwell |
|---|---|---|
| | 36 | 2 |
| | 37 | 5 7 |
| | 38 | 3 4 4 6 7 8 8 9 |
| | 39 | 2 2 4 5 |

**Key**
36 | 2 = 36.2 °C and 8 | 36 = 36.8 °C

**b** Use your stem-and-leaf diagram to write down the range of temperatures of those that are
 **i** recovering well
 **ii** still unwell.

**c** Use your stem-and-leaf diagram to write down the median temperatures of those who are
 **i** recovering well
 **ii** still unwell.

**d** Comment on the shape of the distribution and what the answers to parts **b** and **c** tell you.

**8** Here is a frequency polygon.
Draw the corresponding frequency table.

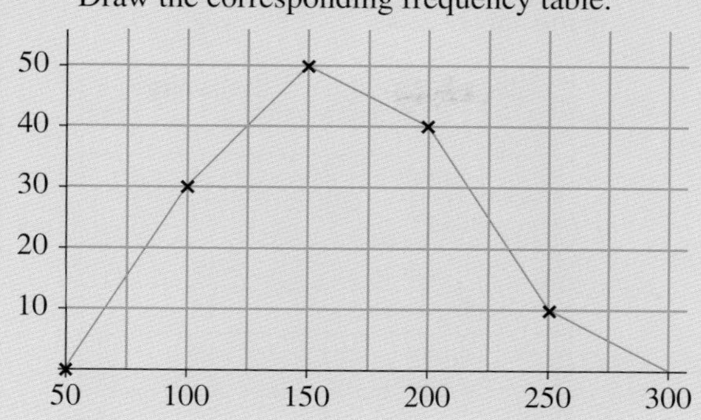


**9** The histogram shows the scores of 50 students in a Biology test.

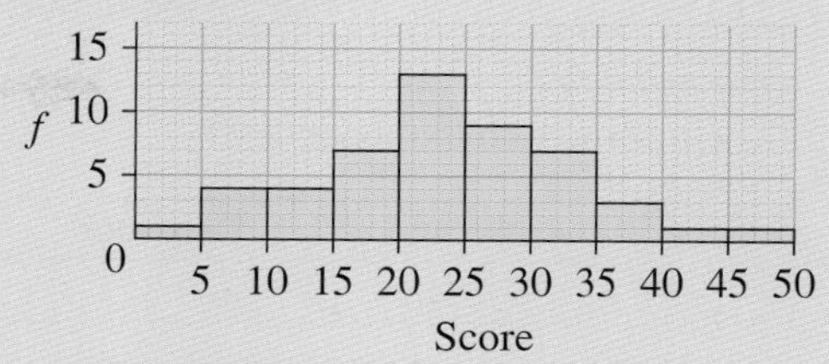


**a** What is the modal group?
**b** Can you tell what the lowest score was on the test? Explain your answer.
**c** What percentage of students scored below 20?
**d** What percentage of students scored above 40?
**e** Describe the approximate distribution the shape of the histogram shows.

## Summary

### You should know ...

**1** How to display data in bar charts, pictograms and pie charts.

*For example:*
The data on this frequency table:

| Score | 0 | 1 | 2 | 3 | 4 | 5 |
|---|---|---|---|---|---|---|
| Frequency | 2 | 4 | 2 | 1 | 3 | 3 |

can be shown as:

Bar chart

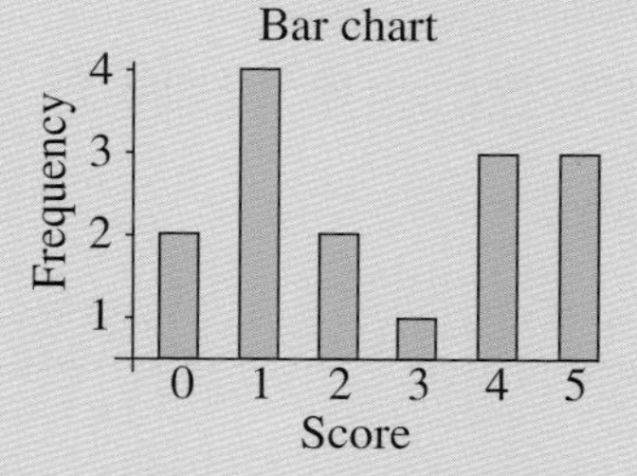


Pictogram

| 0 | ⊗⊗ |
|---|---|
| 1 | ⊗⊗⊗⊗ |
| 2 | ⊗⊗ |
| 3 | ⊗ |
| 4 | ⊗⊗⊗ |
| 5 | ⊗⊗⊗ |

Pie chart

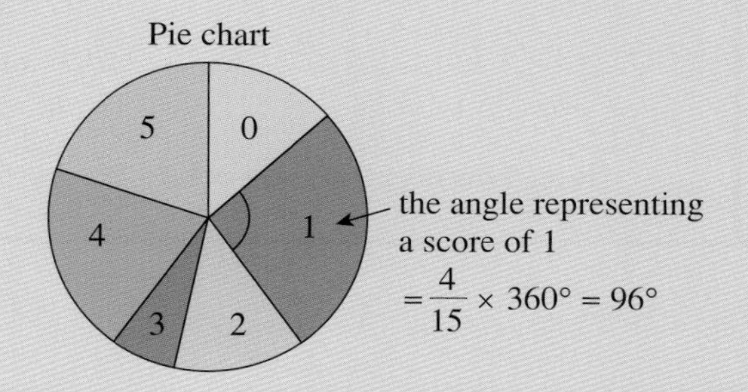


the angle representing a score of 1
$= \frac{4}{15} \times 360° = 96°$

### Check out

**1** Here are the numbers of catches taken by 20 cricketers during a season:

| Catches | 0 | 1 | 2 | 3 | 4 | 5 | 6 |
|---|---|---|---|---|---|---|---|
| Frequency | 1 | 6 | 3 | 4 | 1 | 2 | 3 |

Use the data in the frequency table to draw a:
**a** bar chart
**b** pictogram
**c** pie chart.

2 A histogram has no spaces between the bars. The area of each bar represents the frequency.

*For example:*
The information in this table:

| Height (cm) | Frequency |
|---|---|
| 5-9 | 3 |
| 10-14 | 5 |
| 15-19 | 6 |
| 20-24 | 4 |

can be shown in a histogram:

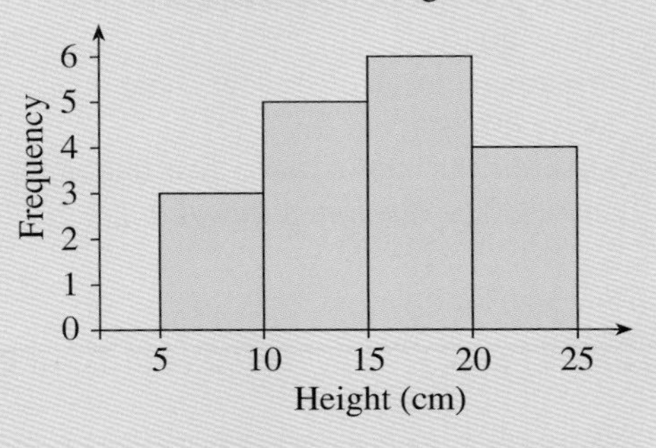


2 Draw a histogram to show the information in this table:

| Weight (kg) | Frequency |
|---|---|
| 10-19 | 1 |
| 20-29 | 4 |
| 30-39 | 9 |
| 40-49 | 11 |
| 50-59 | 15 |
| 60-69 | 27 |
| 70-79 | 21 |
| 80-89 | 16 |
| 90-99 | 7 |

3 How to draw a frequency polygon using mid-interval values.

*For example:*

| Height (cm) | No. of children |
|---|---|
| 120-129 | 3 |
| 130-139 | 1 |
| 140-149 | 7 |
| 150-159 | 4 |

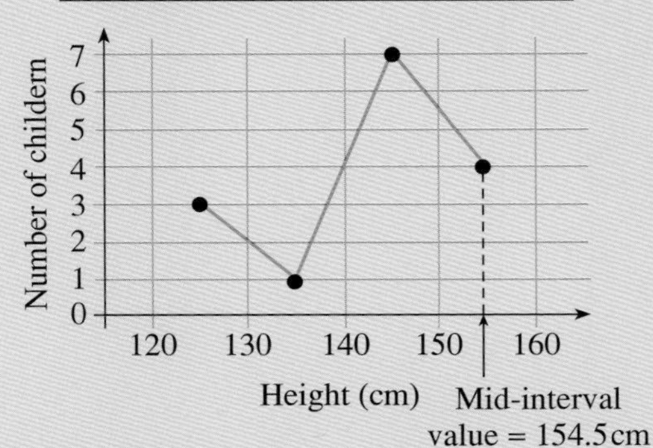


**(Note: This is extension work.)**

3 The number of spectators attending a football match are shown in the table.

| Age (years) | *f* |
|---|---|
| 11-20 | 30 |
| 21-30 | 50 |
| 31-40 | 70 |
| 41-50 | 60 |
| 51-60 | 40 |

Draw a frequency polygon for the data.

**(Note: This is extension work.)**

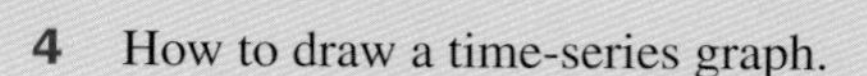

**4** How to draw a time-series graph.

*For example:*
This graph shows the number of visitors, in thousands, to a seaside resort in the UK last year.

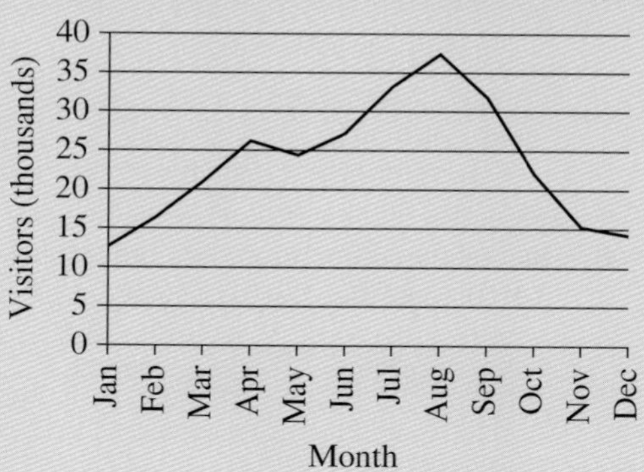


**4** Using the time-series graph on the left, describe the trend in the number of visitors to the seaside resort over the course of last year.

**5** How to draw a back-to-back stem-and-leaf diagram.

*For example:*
Draw a stem-and-leaf diagram for the prices, in cents, of snacks sold in a school canteen in 2010 and 2013.

**2010 prices:** 40, 62, 20, 52, 33, 81, 42, 67, 83, 71, 20, 31, 30, 56, 32

**2013 prices:** 55, 89, 35, 62, 24, 33, 74, 43, 29, 65, 38, 91, 43, 34, 83

Don't forget to write the leaves in numerical order, and to include a key.

| Prices in 2010 | | Prices in 2013 |
|---:|:---:|:---|
| 0 0 | 2 | 4 9 |
| 3 2 1 0 | 3 | 3 4 5 8 |
| 2 0 | 4 | 3 3 |
| 6 2 | 5 | 5 |
| 7 2 | 6 | 2 5 |
| 1 | 7 | 4 |
| 3 1 | 8 | 3 9 |
| | 9 | 1 |

**Key**
2 | 4 means 24 cents and 0 | 2 means 20 cents

It is easier to see the mode, median and range when values are ordered in a stem-and-leaf-diagram. All original data is kept, so it is also possible to work out the mean.

**5** The back-to-back stem-and-leaf diagram shows the distance of people's homes from two different towns, A and B.

| Distance to Town A | | Distance to Town B |
|---:|:---:|:---|
| 7 4 0 | 1 | 9 |
| 8 6 5 1 0 | 2 | 6 8 9 |
| 2 2 0 | 3 | 2 5 7 7 8 |
| 3 2 | 4 | 0 1 3 5 |

**Key**
1 | 9 means 1.9km and 0 | 1 means 1.0km

**a** What is the median distance to
  **i** Town A
  **ii** Town B?

**b** What is the modal distance to
  **i** Town A
  **ii** Town B?

**c** **i** What is the shortest distance to Town A?
  **ii** What is the longest distance to Town B?

| | Statistic | 2010 | 2013 |
|---|---|---|---|
| Average | Mode | 20 | 43 |
| | Median | 42 | 43 |
| | Mean | 48 | 53 |
| Spread | Range | 63 | 67 |

All evidence shows that prices seem to have gone up in 2013. (Note the mode supports this but isn't a very meaningful measure in this example.)

**6** How to draw a scatter graph and interpret it in terms of correlation.

*For example:*

Strong negative correlation

Weak positive correlation

No correlation

**7** How to draw a line of best fit on a scatter graph and use it to make predictions.

A line of best fit should

- follow the trend of the points
- pass through, or very close to, as many points as possible
- have a similar number of points above and below it.

**6** In a talent competition, two judges gave the 10 contestants these scores out of 20:

| Judge 1 | Judge 2 |
|---|---|
| 6 | 5 |
| 5 | 6 |
| 8 | 7 |
| 7 | 7 |
| 9 | 8 |
| 11 | 12 |
| 14 | 12 |
| 12 | 13 |
| 16 | 15 |
| 19 | 17 |

Draw a scatter graph using the data in the table.

**7** **a** Draw the line of best fit on your scatter graph from Question **6**.

**b** Judge 1 had to leave the room temporarily and Judge 2 gave the 11th contestant a mark of 10. Use the line of best fit to predict the mark that Judge 1 would give the 11th contestant.

# Review B

**1** Round

**a** 0.754 to **i** 1 d.p. **ii** 2 d.p.
**b** 39.2481 to **i** 1 d.p. **ii** 2 d.p.
**c** 6.79928 to **i** 1 d.p. **ii** 2 d.p. **iii** 3 d.p.
**d** 1.000349 to **i** 1 d.p. **ii** 4 d.p.

**2** Solve:

**a** $5(x - 2) = 3x - 4$
**b** $2x - 3 = 9 + 4x$
**c** $15 - 4x = 2(3x + 1)$

**3** Prove that the sum of the exterior angles of any polygon is 360°.

**4** Calculate the length of the hypotenuse of a right-angled triangle whose other two sides are of length:

**a** 6 cm, 8 cm **b** 10 cm, 24 cm
**c** 7 cm, 24 cm

**5** A block of wood with volume $600\,\text{cm}^3$ has a density of $0.53\,\text{g/cm}^3$. What is its mass?

**6** In a survey, the masses of students were recorded. The results are shown in the table.

| Mass (kg) | f |
|---|---|
| 20–29 | 16 |
| 30–39 | 26 |
| 40–49 | 34 |
| 50–59 | 44 |
| 60–69 | 40 |
| 70–79 | 24 |
| 80–89 | 12 |
| 90–99 | 4 |

Using this data, draw

**a** a histogram
**b** a frequency polygon.

**(Note: frequency polygons are extension work.)**

**7** Round

**a** 27.48 to **i** 1 s.f. **ii** 2 s.f.
**b** 0.0637 to **i** 1 s.f. **ii** 2 s.f.
**c** 46 587 to **i** 1 s.f. **ii** 2 s.f. **iii** 3 s.f.
**d** 5999 to **i** 1 s.f. **ii** 2 s.f.

**8** Find the sides marked by the letters.

**a**

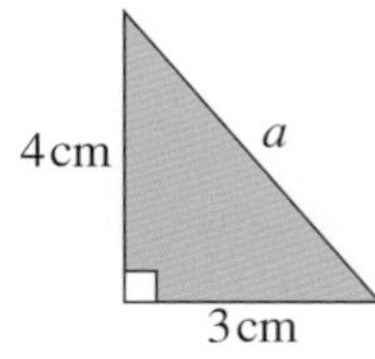


**b**

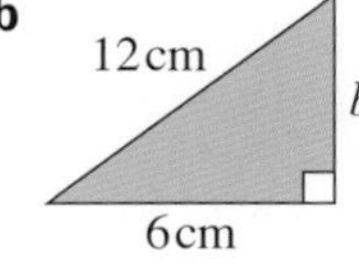


**c**

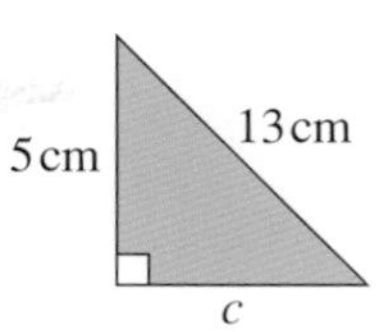


**d**

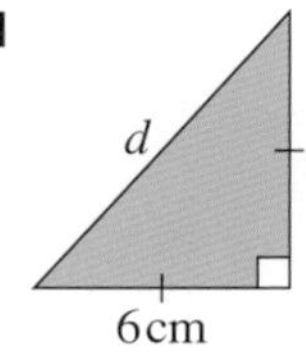


**9** The sum of two consecutive even numbers is 38. Write and solve an equation to find the two numbers.

**10** Calculate the lettered angles.

**a** **b**

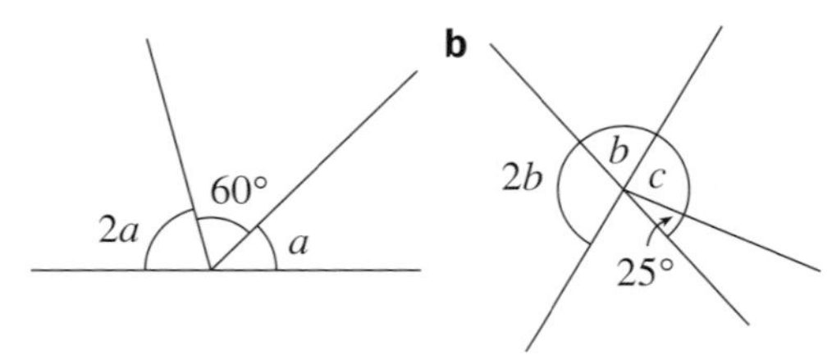


**c**

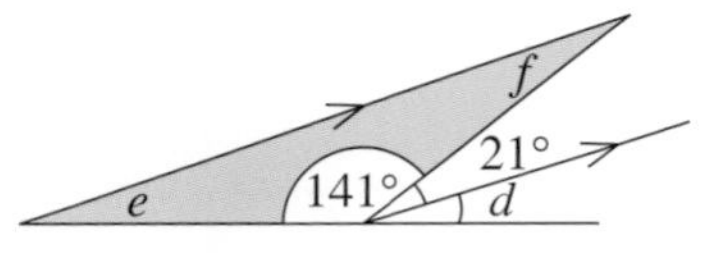


**d**

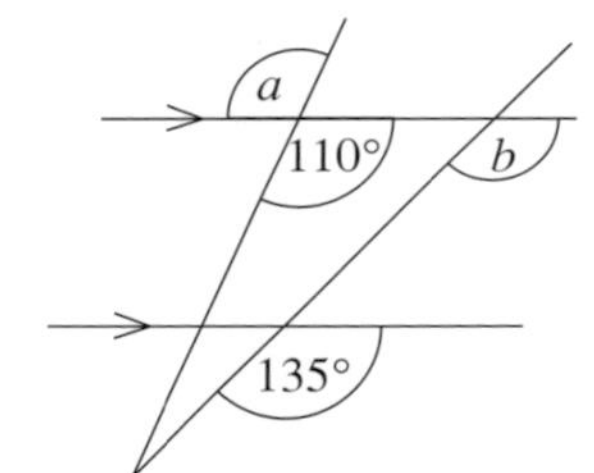


**11** Solve these equations, to 1 decimal place, using trial and improvement.

**a** $x^2 + 3x = 22$, for the solution between $x = 2$ and $x = 6$.
**b** $2t^2 + 7t = 90$, for the solution between $t = 3$ and $t = 7$.
**c** $3x^2 - 2x + 5 = 15$, for the solution between $x = {}^-5$ and $x = 0$.
**d** $3x^3 - 2x^2 - 25x = {}^-10$, for the solution between $x = 1$ and $x = 6$.

**12** Calculate the length of the shortest side of a right-angled triangle whose other two sides are of length:

**a** 60 cm, 61 cm **b** 4 cm, 4.1 cm

**13** The diagram shows a cyclist's journey from home to his friend's house and back again.

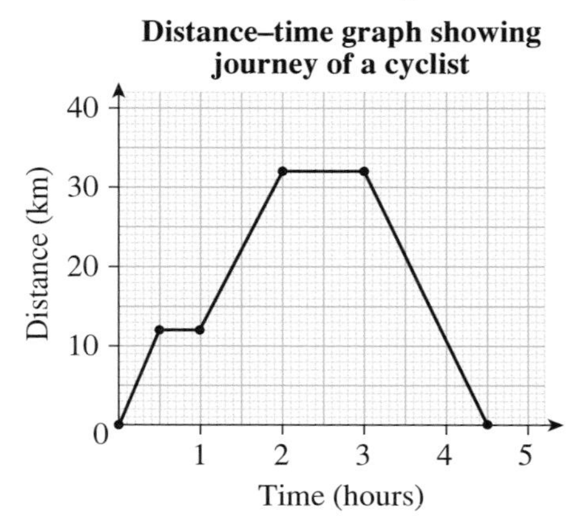


**a** During which part of the journey was the cyclist travelling at the highest speed? What was that speed?

**b** How long did the cyclist stop for in total during the $4\frac{1}{2}$ hours of the journey?

**c** How far does the cyclist live from his friend's house?

**14** Describe the inequalities represented by these number lines. The first has been done for you.

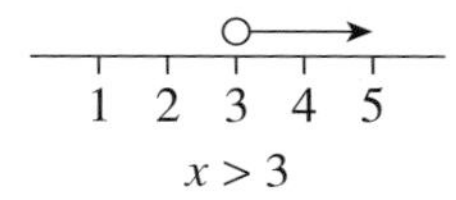


**a**

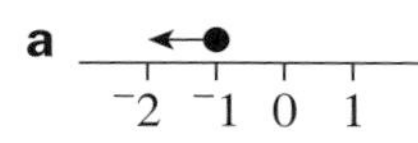


**b**

10 20 30 40 50

**c**

3 4 5 6 7 8 9

**15** The table shows the eye colour of students at a school.

| Eye colour | Number of students |
|---|---|
| Blue | 90 |
| Brown | 45 |
| Green | 25 |
| Grey | 20 |

Represent this data using

**a** a bar chart

**b** a pictograph

**c** a pie chart.

**16** Copy this diagram and draw the pattern formed by tessellations of the triangle.

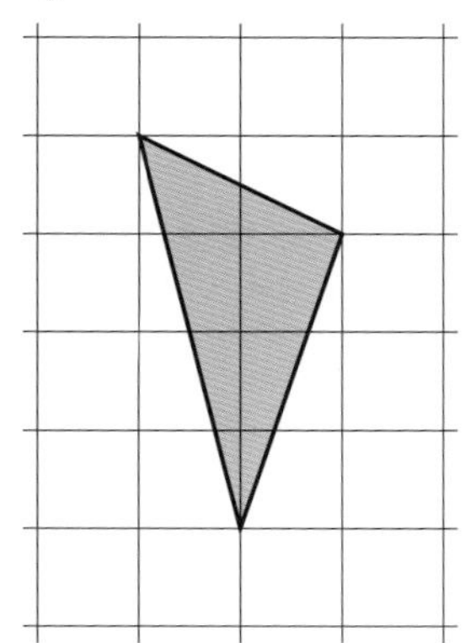

**17** Work out $\frac{35}{77} \div \frac{25}{33}$

**18** Calculate the sizes of the lettered angles.

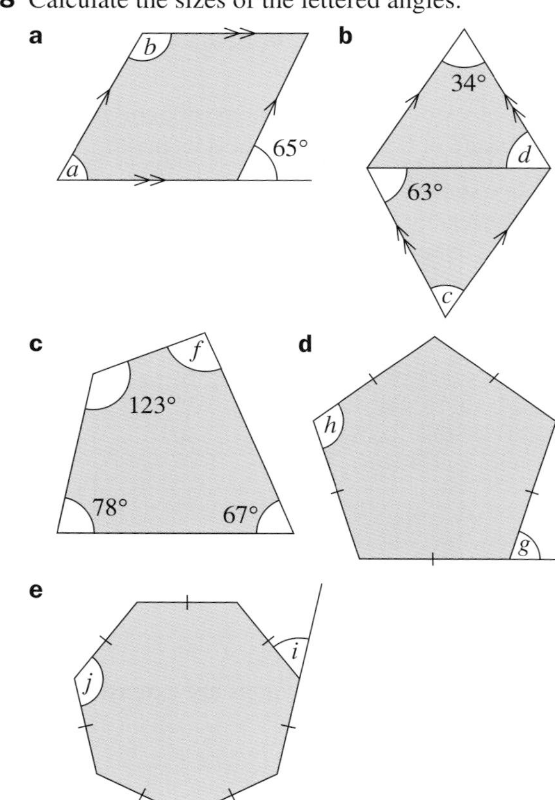


**19** Solve:

**a** $7x + 5 = 11x + 1$
**b** $5x + 14 = 7x + 4$
**c** $4x - 3 = 9x - 23$
**d** $2x - 15 = 11x + 12$
**e** $4x + 3 = 38 - x$
**f** $7x - 5 = 43 - 3x$
**g** $19 - 3x = 41 - 4x$
**h** $49 - 2x = 7 - 9x$

**20** Find, to three significant figures, the length of the diagonal of a square of sides:

**a** 1 cm **b** 10 cm **c** 100 cm

**21** Without working them out, state which questions which will have answers less than 12.

$12 \times 1.3$ $\quad$ $12 \div 7$ $\quad$ $12 \times 0.41$

$12 \div 0.65$ $\quad$ $12 \times 3.4$ $\quad$ $12 \times \frac{1}{9}$

$12 \times 1.01$ $\quad$ $12 \div 1\frac{5}{7}$ $\quad$ $12 \div \frac{4}{5}$

$12 \div 48$

**22** The diagram shows the speed–time graph for an object that starts from rest.

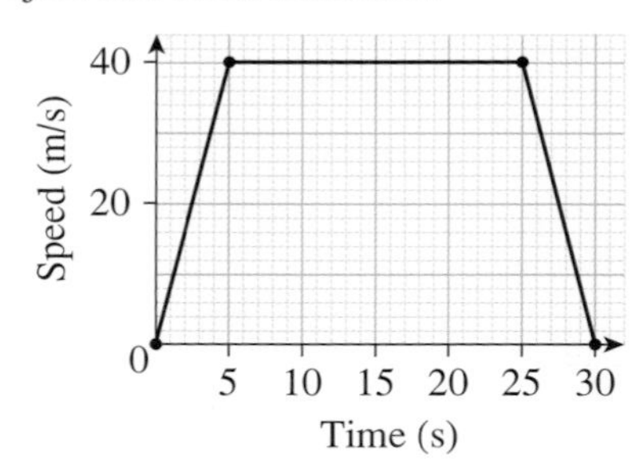


**a** What is the object's acceleration in the first 5 seconds?

**b** What does a horizontal line show on a speed–time graph?

**c** How far did the object travel during the 30-second journey?

**23** Two classes both sat a Geography exam. The results are shown below.

**Class A**

31 50 24 24 41 36 42 35 19 43 36 33 31 49
40 36 32 25 37 40 36

**Class B**

14 28 19 22 45 20 16 18 19 31 29 32 16
27 29 33 27 33 43 7 22

**a** Draw a back-to-back stem-and-leaf diagram to compare the two sets of results.

**b** Work out any relevant statistics to see if they support your ideas about the shape of the distributions.

**24** Draw the locus of points that are 4 cm from a point B.

**25** Given that $27 \times 36 = 972$, work out:

**a** $2.7 \times 3.6$ $\quad$ **b** $0.27 \times 0.36$

**c** $2.7 \times 3600$ $\quad$ **d** $2.7 \times 0.36$

**e** $0.027 \times 3.6$ $\quad$ **f** $0.027 \times 0.036$

**26** Which regular polygons will tessellate?

**27** Assuming that you can use any whole number from 0 to 10, write down the solution set for

**a** $x < 5$ $\quad$ **b** $x > 3$

**c** $x + 5 < 9$ $\quad$ **d** $x - 3 \geqslant 4$

**28** Find the length of the diagonal of a rectangle with sides 7 cm and 8 cm.

**29** Calculate the lettered angles.

**a**

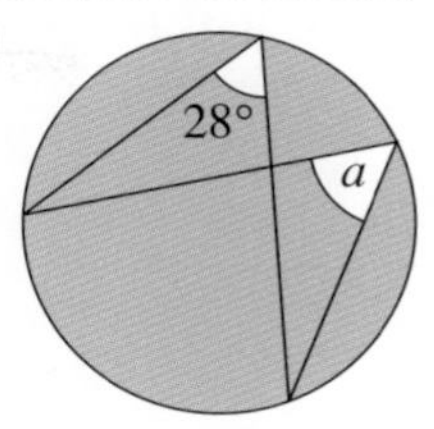


**b**

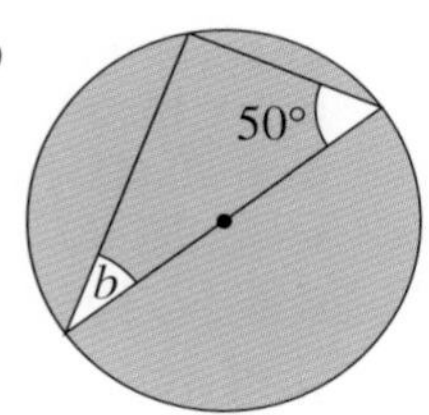


**c**

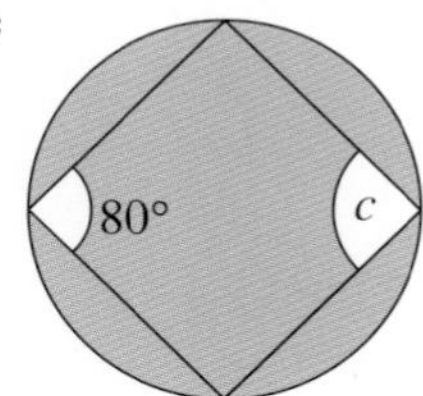


**d**

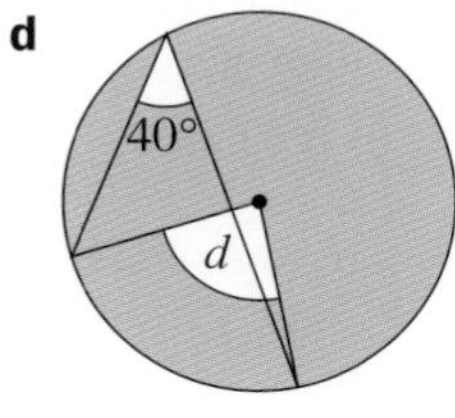


**30** **a** Find the average speed, in km/h, of an object that travels 259 km in 3.5 hours.

**b** How far does a car travel if its average speed is 84 km/h and it travels for 90 minutes?

**c** Work out the time taken, in minutes, for a 68-km journey at an average speed of 85 km/h.

**31** 12 years from now, Ali will be three times older than he was 18 years ago. If Ali is $a$ years old today, write down:

**a** his age 12 years from now

**b** his age 18 years ago

**c** an equation which shows the above information.

**d** Solve the equation to find Ali's age now.

**32** Estimate:

**a** $8.205 \times 4.8$

**b** $19.79 \div 4.86$

**c** $\frac{21.13 + 10.24}{4.97}$

**d** $6 - 3.01^2$

**33** What is **a** the interior **b** the exterior angle in a regular decagon?

**34** Work out $10 - \frac{12 + 6}{3} \times 4$

**35** Represent the solution set for each inequality on a number line.

**a** $x + 8 < 19$ $\quad$ **b** $x - 12 \geqslant 3$

**c** $2x + 5 \leqslant 17$ $\quad$ **d** $3x - 5 > 16$

**36** A triangle with sides of length 3, 4 and 5 is right-angled because $3^2 + 4^2 = 5^2$.

The lengths of the sides of some triangles are given. Which of them are right-angled?

**a** 10, 15, 20 **b** 5, 12, 13
**c** 9, 40, 41 **d** 9, 12, 15
**e** 12, 16, 20 **f** 20, 21, 29
**g** 8, 12, 16 **h** 7, 8, 10

**37** In the diagram, O is the centre of a circle with diameter PQ.

Calculate:
**a** angle PSR
**b** angle POR

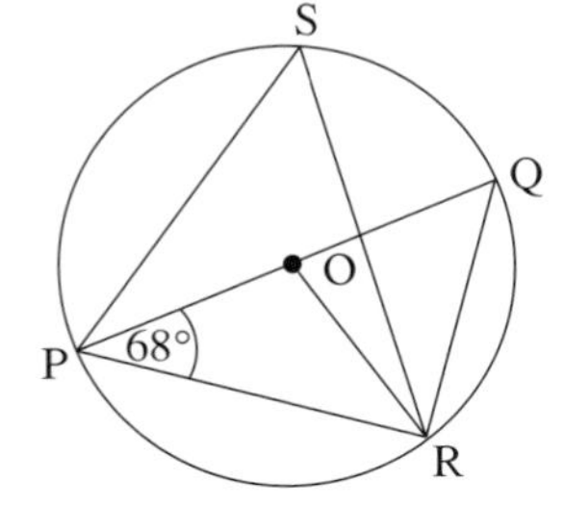


**38** Solve these simultaneous linear equations.

**a** $3x - 4y = {}^{-}1$
$6x + 7y = 13$

**b** $4x - 2y = 3$
$5x + 6y = 8$

**c** $7x - 3y = 1$
$2x - 3y = {}^{-}4$

**39** Use the data in the table to draw a time-series graph showing the average monthly rainfall, in millimetres, for Auckland, New Zealand and Kano, Nigeria.

| Month | Average rainfall in Auckland, New Zealand (mm) | Average rainfall in Kano, Nigeria (mm) |
|---|---|---|
| Jan | 75 | 0 |
| Feb | 80 | 5 |
| Mar | 85 | 10 |
| Apr | 95 | 15 |
| May | 110 | 50 |
| Jun | 130 | 110 |
| Jul | 140 | 175 |
| Aug | 115 | 225 |
| Sep | 90 | 100 |
| Oct | 80 | 15 |
| Nov | 85 | 0 |
| Dec | 90 | 0 |

Describe the trends that you see on the graph. What inferences can you make?

**40** Two boys have \$33 in cash between them. One gives \$6 to the other and finds he now has twice as much money as his friend. Form an equation by letting \$$x$ be the amount one boy had at the start. Hence find how much each boy had to start with.

**41** Without using a calculator, work out:

**a** $0.4 \times 6$ **b** $6.2 \times 4$ **c** $2.38 \times 7$
**d** $0.6 \div 3$ **e** $1.2 \div 4$ **f** $0.09 \div 3$
**g** $1.715 \div 7$ **h** $8 \div 0.01$ **i** $22 \div 0.02$
**j** $9.6 \div 0.04$

**42** Draw a line segment XY 3 cm long. Draw the locus of points that are 2 cm from XY.

**43** Two pieces of metal, 120 cm and 180 cm long, are to be cut into equal lengths. What is the greatest possible length of each piece?

**44** **a** A is the point (1, 4) and B the point (5, 1). Calculate the distance AB.

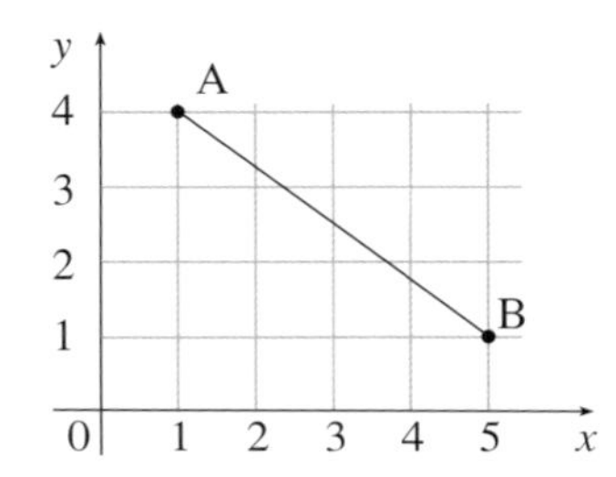


**b** Calculate the distance between the points P(2, 3) and Q(5, 4).

**45** Calculate the density of a metal with mass 483 g and volume 35 cm$^3$.

**46** Mr Johnson needs 320 m of fencing to enclose his rectangular field. If the width of his field is $w$ m and its length is 90 m, find the value of $w$.

**47** The stem-and-leaf diagram shows the number of baskets scored in each match by two basketball teams.

| Team A | | Team B |
|---|---|---|
| 6 5 4 2 0 | 60 | 5 5 |
| 9 8 7 6 5 4 3 1 | 70 | 6 8 9 |
| 9 8 3 | 80 | 0 1 2 4 |
| 4 4 | 90 | 1 2 3 4 7 8 9 |
| 1 | 100 | 0 1 3 |

**Key**
60 | 5 = 65 baskets and 0 | 60 = 60 baskets

**a** Work out the modal score for each team.
**b** Work out the median score for each team.
**c** What do these results tell you?
**d** What conclusions can you make, based on the statistics you calculated in parts **a** and **b** and the shape of the distributions shown in the diagram?

**48** Solve these simultaneous equations. Check your results by substitution.

**a** $x + 5y = 26$
$x + 2y = 14$

**b** $5x + 7y = 18$
$3y - 5x = 22$

**c** $4x - 7y = 41$
$4x - 3y = 29$

**49**

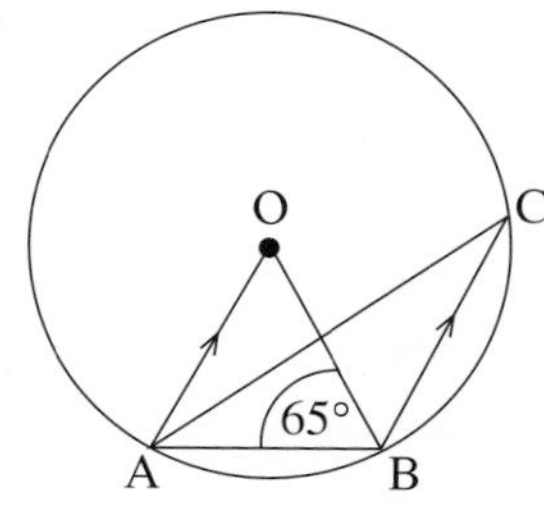


In the diagram, O is the centre of a circle and OA is parallel to BC.

**a** Show that $A\hat{O}B = 50°$

**b** Calculate **i** $A\hat{C}B$ **ii** $A\hat{B}C$

**50** Without using a calculator, work out:

**a** $0.4 \times 0.6$ **b** $5.2 \times 0.03$

**c** $2.3 \times 7.4$ **d** $8.3 \times 2.6$

**e** $14.4 \div 2.4$ **f** $0.3 \div 0.5$

**g** $1.64 \div 0.04$ **h** $0.48 \div 0.012$

**51** Find the formula for the sum of the interior angles of any $n$-sided polygon.

**52** Write brackets in this expression to make it correct: $2 + 7 \times 9 - 5 = 36$.

**53** A group of runners recorded their resting pulse rates and the distance they run per week. The results are shown in the scatter graph.

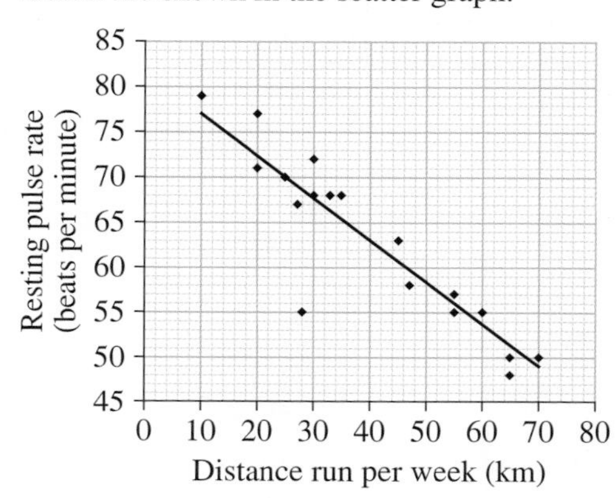


**a** How many runners had a resting pulse rate of 63 beats per minute?

**b** How many runners ran 55 kilometres per week?

**c** Describe and interpret the correlation shown in the graph.

**d** One point on the graph does not fit the trend. Which point is this? Can you suggest a possible reason why this result doesn't fit the trend?

**e** If someone usually runs 40 kilometres in a week, what would you expect their resting pulse rate to be?

**f** If someone has a resting pulse rate of 75 how many kilometres per week do you think they might run?

**54** The diagram shows part of a roof.

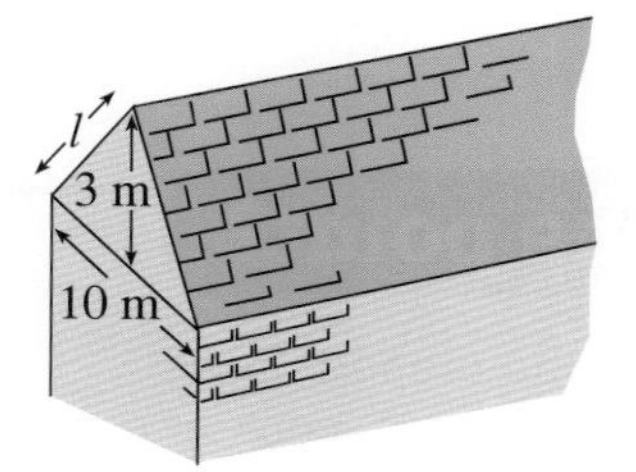


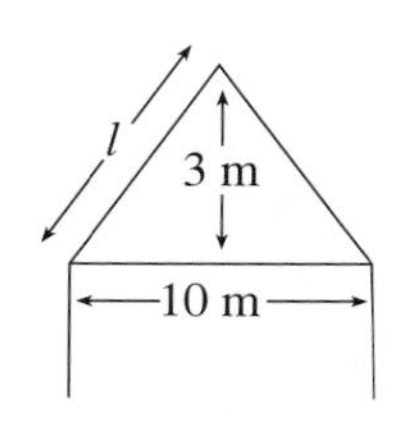


The width of the roof is 10 metres. The highest part of the roof is 3 metres above the top of the front wall.

Find the length, $l$, of the sloping part of the roof.

**55** Explain why a regular pentagon will not tessellate with itself.

**56** Work out $43 - 2 \times 8 - 3$.

**57** Without working them out, state which questions which will have answers greater than 57.

| | | |
|---|---|---|
| $57 \div 0.9$ | $57 \times 0.2$ | $57 \times 3.8$ |
| $57 \div \frac{3}{8}$ | $57 \div 1.84$ | $57 \times \frac{3}{8}$ |
| $57 \div 84$ | $57 \times 0.21$ | $57 \div 0.8$ |
| $57 \times 1\frac{2}{5}$ | | |

**58** I started with a number, $n$. If I subtract 10 from this number and then multiply the result by 7 I get the same number as I do when I add 18 to $n$ and multiply the result by 3. What is $n$?

**59** Find the highest common factor of 126 and 420.

# 13 Ratio and proportion

## Objectives

- Compare two ratios; interpret and use ratio in a range of contexts.
- Recognise when two quantities are directly proportional; solve problems involving proportionality, e.g. converting between different currencies.

## What's the point?

Without ratio and proportion we would struggle to build, cook or run international businesses. Ratio is used in building work, for example to make cement or concrete by mixing quantities in a given ratio. It allows recipes to be scaled up or down to cater for more or fewer people. And businesses use proportion to work with currency exchange rates.

## Before you start

### You should know ...

1 Metric measures of length, mass and capacity, and their abbreviations.
m = metres
mm = millimetres
kg = kilograms
cm = centimetres
g = grams
km = kilometres
ℓ = litres
t = tonnes
ml = millilitres

### Check in

1 From the column on the left, list the units of
**a** length
**b** mass
**c** capacity

2 Equivalent ratios are different ways of showing the same ratio.
*For example:*
6 : 2 = 3 : 1 = 18 : 6

3 Ratios can be simplified by dividing the numbers by their highest common factor (HCF).
*For example:*

24 : 40
÷8 ÷8
3 : 5

2 Write down two ratios that are equivalent to
**a** 5 : 3
**b** 4 : 7

3 Simplify:
**a** 10 : 16
**b** 48 : 72
**c** 28 : 35
**d** 60 : 15 : 45

## 13.1 Simplifying and comparing ratios

A ratio compares the size of two quantities.

In the diagram the ratio of triangles to circles is 3 : 7

The ratio of circles to triangles is 7 : 3

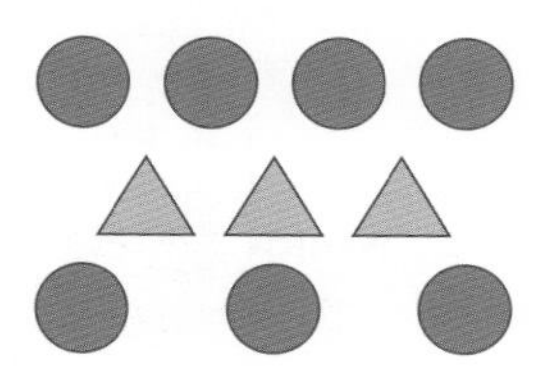

Notice we can also write this fraction and percentage for the diagram:

$\frac{3}{10}$ of the shapes are triangles

30% of the shapes are triangles

A ratio does not have any units; it is simply a way of comparing relative sizes. So when writing a ratio, make sure both numbers have the same units.

To compare two ratios, first simplify them.

**EXAMPLE 1**

Alex takes 1 hour to walk to school.

If he cycles, it only takes him 18 minutes.

Lalita takes 1 hour 28 minutes to walk to school. If she cycles it takes her 24 minutes.

Who is faster at walking compared to cycling, Alex or Lalita?

**For Alex:**

| | |
|---|---|
| 1 hour : 18 minutes | Write the times for walking and cycling as a ratio |
| 60 minutes : 18 minutes | Make the units the same for both times |
| 60 : 18 | Remove the units, as a ratio doesn't have units |
| 10 : 3 | Simplify by dividing by 6 (the HCF of 60 and 18) |

**For Lalita:**

| | |
|---|---|
| 1 hour 28 minutes : 24 minutes | Write the times for walking and cycling as a ratio |
| 88 minutes : 24 minutes | Make the units the same for both times |
| 88 : 24 | Remove the units, as a ratio doesn't have units |
| 11 : 3 | Simplify by dividing by 8 (the HCF of 88 and 24) |

Compare 10 : 3 with 11 : 3

Since the second number (representing cycling time) in both ratios is the same, we only need to look at the first number in each ratio.

10 is less than 11, so Alex has the faster walking time compared to cycling time.

If simplified ratios are still not easy to compare, it is sometimes easier to use percentages.

**EXAMPLE 2**

In basketball practice, two players take shots at the basket.

| | Shots on target | : | Shots missed |
|---|---|---|---|
| **Ajay** | 78 | : | 42 |
| **Ben** | 80 | : | 45 |

Ben had the most shots on target, with 80, but he had more shots in total so he may not have had the highest success rate.

Which player had the higher success rate?

The simplified ratios are 13 : 7 and 16 : 9, which have no numbers in common.

Percentage of shots on target:

**Ajay** $\frac{78}{78 + 42} \times 100 = \frac{78}{120} \times 100 = 65\%$

**Ben** $\frac{80}{80 + 45} \times 100 = \frac{80}{125} \times 100 = 64\%$

So Ajay had the higher success rate.

To be in its simplest form, a ratio shouldn't include decimals or fractions. It should contain only whole numbers.

**EXAMPLE 3**

Write this ratio in its simplest whole-number form:

$0.18 : 27\% : \frac{9}{20}$

$0.18 : 27\% : \frac{9}{20}$ To get rid of the decimal, percentage and fraction, multiply by 100

$18 : 27 : 45$ Divide by the HCF, 9

$2 : 3 : 5$

Sometimes the **unitary method** is a good way to compare ratios. This is where you make one of the numbers 1 in each ratio. The ratio will not necessarily be in its simplest form, but this is acceptable when comparing.

**EXAMPLE 4**

Two different colours of pink paint are made by mixing red and white paint in different ratios.

| | Red | : | White |
|---|---|---|---|
| **Paint A** | 7 | : | 16 |
| **Paint B** | 2 | : | 5 |

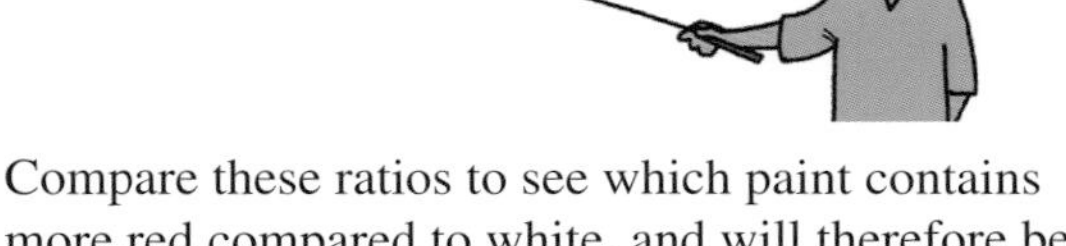

The ratios are already in their simplest form and it isn't easy to compare them.

Compare these ratios to see which paint contains more red compared to white, and will therefore be the darker shade of pink.

Using the unitary method:

**Paint A**

Red : White

7 : 16

÷ 16 ÷ 16

0.4375 : 1

**Paint B**

Red : White

2 : 5

÷ 5 ÷ 5

0.4 : 1

Since 0.4375 is greater than 0.4, Paint A has more red paint compared to white, and will be the darker shade of pink.

## Exercise 13A

**1** Simplify these ratios.

**a** 24 : 42

**b** 78 : 13

**c** 36 : 126 : 162

**2** Write each ratio in its simplest whole-number form.

**a** 0.5 : 3

**b** 4.2 : 3.5

**c** 5 : 2.5 : 4.5

**d** 12 : 8.4

**3** John is 1.6 m tall. Nasim is 156 cm tall. Write down the ratio of John's height to Nasim's height.

**4** Amir takes $\frac{3}{4}$ hour to walk to school. If he runs it only takes him 20 minutes.

Katy takes $\frac{1}{2}$ hour to walk to school. If she runs it takes her 15 minutes.

Who is faster at walking, compared to running?

**5** Write each of these as a ratio in its simplest whole-number form.

**a** $\frac{1}{3} : 4$ **b** $2.3 : \frac{1}{5}$

**c** 45% : 0.65 **d** $0.4 : 75\% : \frac{2}{3}$

**6** In football practice two players took shots at goal.

| | Shots on target | : | Shots missed |
|---|---|---|---|
| **Jenny** | 34 | : | 6 |
| **Aisha** | 52 | : | 13 |

Compare these ratios by working out the percentage of shots taken that were on target. Which player had the highest success rate?

**7** In what ratio are the side lengths of these triangles? Start with the smallest and finish with the largest. Don't forget to simplify the ratio.

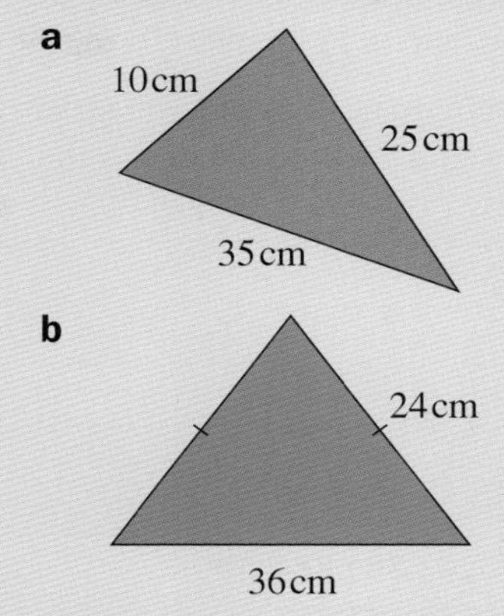


**8** Which pairs of ratios are equivalent?

**a** 14 : 21, 10 : 15

**b** 2 : 5, 10 : 4

**c** 88 : 33, 800 : 300

**d** 34 : 85, 0.4 : 1

**e** 133% : 0.38, 21 : 6

**f** 0.6 : 1, 5 : 3

**9** Two different types of grey paint are being made by mixing black and white paint.

| | Black | : | White |
|---|---|---|---|
| **Paint A** | 3 | : | 8 |
| **Paint B** | 4 | : | 9 |

Use the unitary method to compare these ratios to see which paint contains a higher proportion of black than white, and will therefore be the darker shade of grey.

**10** What is the ratio of days in September to days in February (for a year which is not a leap year)?

**11** This was Ohene's homework on using ratios to compare quantities:

| Question | Working | Answer |
|---|---|---|
| a 5 m : 5 km | Divide by 5 | 1 : 1 |
| b 240 ml : 20 l | Divide by 20 | 12 : 1 |

Ohene has made some mistakes. Correct his homework.

**12** Compare these quantities using ratio.

**a** 5 cm : 350 mm

**b** 4.4 kg : 220 g

**c** 2.4 m : 1600 cm

**d** 2100 ml : 4.9 ℓ

**e** 8400 mm$^2$ : 72 cm$^2$

**f** 0.34 m$^3$ : 380 000 cm$^3$

**13** In 2012 a 22-cm tall tree was planted. One year later it had grown by 30%. Write down the ratio of the tree's height in 2012 to its height in 2013. Write the ratio in its simplest form.

**14 a** Copy and complete this spider diagram with ratios equivalent to the one in the central box.

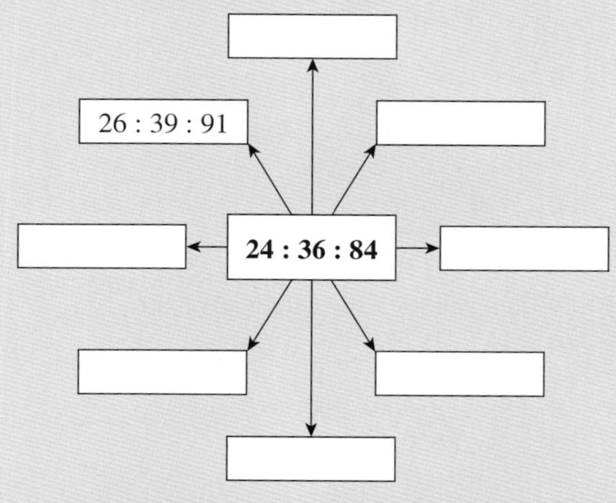


**b** Compare your answers with a friend's answers. Do you agree that all the ratios are equivalent to the one in the central box? Did you find any ratios that are the same?

Try out the ratio Sudoku puzzle at www.nrich.maths.org/4827

## 13.2 Solving ratio problems

### Dividing a quantity in a given ratio

When a quantity is divided into two or more amounts according to a ratio, you can use the ratio to work out how much is in each amount.

**EXAMPLE 5**

\$350 is divided between Emily, Samirah and Lola in the ratio 3 : 2 : 5.

How much do they get each?

For every 3 parts Emily gets, Samirah gets 2 parts and Lola gets 5 parts.
$3 + 2 + 5 = 10$ parts in total

Divide \$350 by 10 to find the value of 1 part:

$350 \div 10 = 35$, so 1 part is \$35

Emily gets $3 \times 35 = \$105$
Samirah gets $2 \times 35 = \$70$
Lola gets $5 \times 35 = \$175$

It's a good idea to check by adding: \$105 + \$70 + \$175 = \$350

You may want to change units before dividing in a ratio, to make calculations easier and final answers more manageable.

**EXAMPLE 6**

Divide 1 hour in the ratio 3 : 7

Dividing 60 minutes in the ratio 3 : 7:

$60 \div (3 + 7) = 60 \div 10 = 6$
$3 \times 6 = 18$
$7 \times 6 = 42$

1 hour divided in the ratio 3 : 7 is 18 minutes and 42 minutes.

In ratio problems you may not be given the total amount. You can still work out the answer by finding the value of one part.

**EXAMPLE 7**

Students at a school were asked to vote for a trip to the museum or a trip to the zoo.
There was a ratio of 10 : 3 in favour of going to the zoo.
130 children voted for going to the zoo.
How many children voted for going to the museum?

10 parts voted for the zoo. This was 130 children.
$130 \div 10 = 13$
So each part is worth 13.

3 parts voted for the museum.
$3 \times 13 = 39$
39 children voted for going to the museum.

Example 7 can be set out in a different way, using equivalent ratios:

Zoo : Museum

10 : 3

$\times 13$ ↓ ↓ $\times 13$

130 : 39

Another type of ratio problem does not provide a total quantity or the quantity in one share. Instead, you are simply told how much bigger (or smaller) one share is than another. You still need to work out the value of one part.

**EXAMPLE 8**

Guntur and Eva share an amount of money in the ratio 8 : 5
Guntur gets \$21 more than Eva.
How much do they get each?

---

Guntur gets 3 more parts than Eva because $8 - 5 = 3$.
These 3 parts are worth \$21.
1 part is worth $\$21 \div 3 = \$7$

So Guntur gets $8 \times \$7 = \$56$
and Eva gets $5 \times \$7 = \$35$

## Exercise 13B

**1** Divide:

- **a** 405 cm in the ratio 7 : 8
- **b** \$228 in the ratio 5 : 1 : 6
- **c** 0.24 kg in the ratio 3 : 5
- **d** 0.8 m in the ratio 2 : 3
- **e** 0.91 ℓ in the ratio 3 : 4
- **f** 2 hours in the ratio 3 : 7 : 5

**2** The angles A, B and C in this triangle are in the ratio 3 : 5 : 4.

Work out the size of each angle.

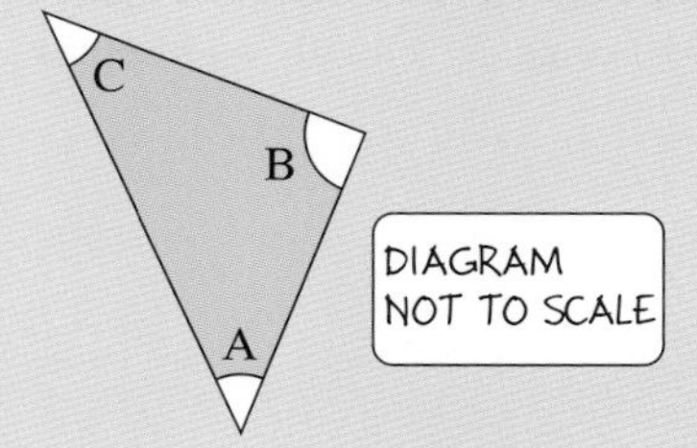


**3** Kamil and Sean share some money in the ratio 3 : 5
Sean gets \$56 more than Kamil.
How much do they get each?

This is Jadee's working for this question:

$3 + 5 = 8$
$56 \div 8 = 7$
Kamil gets $3 \times \$7 = \$21$
Sean gets $5 \times \$7 = \$35$
(Check: $21 + 35 = 56$)

Jadee has made a mistake.

- **a** What mistake has she made?
- **b** Write down the correct working and answer.

**4** Look at the diagram below. How many more rectangles need to be shaded so that the ratio of shaded to unshaded is 2 : 3?

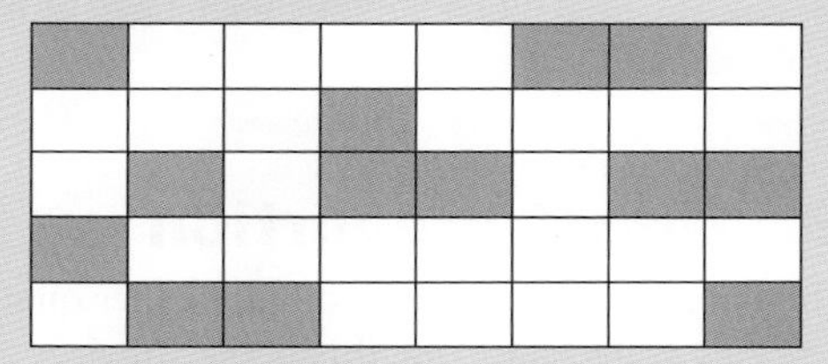

**5** A recipe requires flour, margarine and butter to be mixed in the ratio 4 : 2 : 1 by mass. If the mass of margarine used is 400 g, what is the total mass of all the ingredients?

**6** Find two numbers with the ratio 9 : 4 and whose sum is 169.

**7** Students voted on which of the fruits sold in the school canteen was their favourite.
The votes were in the ratio 5 : 9 : 4 for peach to orange to mango.
153 students chose orange. How many students voted altogether?

**8** The angles in a triangle are in the ratio 2 : 3 : 2.
What sort of triangle is it?

**9** A box contains 80 coloured pens.
The ratio of red to blue to black pens is 3 : 8 : 5.

- **a** A pen is chosen at random. What is the probability that the pen is
  - **i** blue
  - **ii** not blue?
- **b** How many pens are there of each colour?

**10** Jamie and Nashwa share some money in the ratio 10 : 7.
Jamie gets \$36 more than Nashwa.
How much do they get each?

**11** The angles of a quadrilateral are in the ratio 2 : 11 : 7 : 4. What are the angles?

**12** In a survey, the ratio of students who preferred Maths to students who preferred Science was 7 : 4. If 45 fewer students preferred Science, how many students were in the survey?

**INVESTIGATION**

Amira, Beth and Cassie went shopping. At the start of the day, they had money in the ratio 6 : 5 : 4. At the end of the day, their money was in the ratio 5 : 4 : 3. During the day one of them was given \$20 by her brother. How much did they each have at the start of the day?

## 13.3 Direct proportion

- When one quantity increases and another quantity increases at the same rate this is known as **direct proportion**.

There is another type of proportion called 'inverse proportion'. If quantities are inversely proportional, when one quantity decreases the other increases. You will learn more about inverse proportion when you are older. For this chapter, assume that the words 'proportion' and 'proportional' refer only to direct proportion.

An example of two things that are in direct proportion is the number of apples you buy and the amount you pay for them. If you buy three times as many apples as your friend, you will pay three times as much.

When two quantities in direct proportion are put into a table, multiplication patterns can be seen. Using these patterns is often the easiest way to find missing values.

×5

×2 ×3

| Number of apples | 2 | 4 | 12 | 10 |
|---|---|---|---|---|
| Cost of apples (in cents) | 60 | 120 | 360 | 300 |

×30

×2 ×3

×5

You can see from the table that doubling the number of apples doubles the cost. Similarly, if you multiply the number of apples by 5 you multiply the cost by 5.

You can also see the pattern going down the table. The number of apples multiplied by 30 gives the cost. So each apple costs 30 cents.

You can also see dividing patterns in the table, since division is the inverse of multiplication:

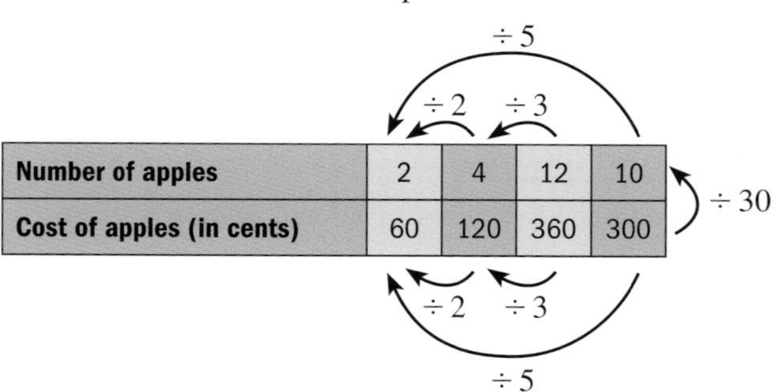

| Number of apples | 2 | 4 | 12 | 10 |
|---|---|---|---|---|
| Cost of apples (in cents) | 60 | 120 | 360 | 300 |



You can see from the table that if you halve the number of apples you halve the cost.

Going up the table, you can see that if you divide the cost by 30 you get the number of apples.

You can also plot these values on a graph:

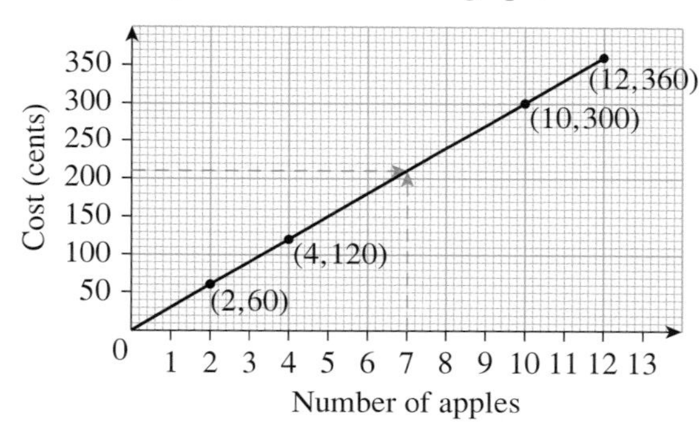


- Plotting two quantities which are directly proportional on a graph will give a straight line passing through the origin $(0, 0)$.

You can use a table or graph to work out missing values. From the graph, you can work out that the cost of 7 apples is 210 cents. From the table, you can use the multipliers or divisors to work out how many apples you can buy for 510 cents: $510 \div 30 = 17$ apples.

## EXAMPLE 9

3 oranges cost 81 cents.
How much do 7 oranges cost?

| Number of oranges | 3 |
|---|---|
| Cost of oranges (if cents) | 81 |

× 27

The multiplier is $81 \div 3 = 27$

| Number of oranges | 3 | 7 |
|---|---|---|
| Cost of oranges (if cents) | 81 | 189 |

× 27

Another way to set out the answer for Example 9 makes it clear you are using the unitary method:

÷ 3 × 7

| Number of oranges | 3 | 1 | 7 |
|---|---|---|---|
| Cost of apples (in cents) | 81 | 27 | 189 |

÷ 3 × 7

Sometimes the multipliers aren't whole numbers.

## EXAMPLE 10

Fill in the missing numbers if $x$ and $y$ are directly proportional to each other.

| $x$ | 4 | 7 | |
|---|---|---|---|
| $y$ | 10 | | 35 |

$10 \div 4 = 2.5$, so the multiplier is 2.5
$7 \times 2.5 = 17.5$
$35 \div 2.5 = 14$

| $x$ | 4 (× 2.5) | 7 (× 2.5) | 14 (÷ 2.5) |
|---|---|---|---|
| $y$ | 10 | 17.5 | 35 |

## EXAMPLE 11

A green paint is a mixture of yellow and blue paint in the ratio 5 : 2.

If 28 litres of yellow paint are used, how many litres of blue paint are needed to make the green paint?

In a table:

| Yellow paint | 5 | 28 |
|---|---|---|
| Blue paint | 2 | |

$2 \div 5 = 0.4$, so the multiplier is 0.4

| Yellow paint | 5 (× 0.4) | 28 (× 0.4) |
|---|---|---|
| Blue paint | 2 | 11.2 |

So 11.2 litres of blue paint are needed.

There are many ways to approach Example 11.

For instance, you could look for the multiplier going across the table.

Since $28 \div 5 = 5.6$, the multiplier going across the table is 5.6.

× 5.6

| Yellow paint | 5 | 28 |
|---|---|---|
| Blue paint | 2 | 11.2 |

× 5.6

Alternatively, you can use equivalent ratios:

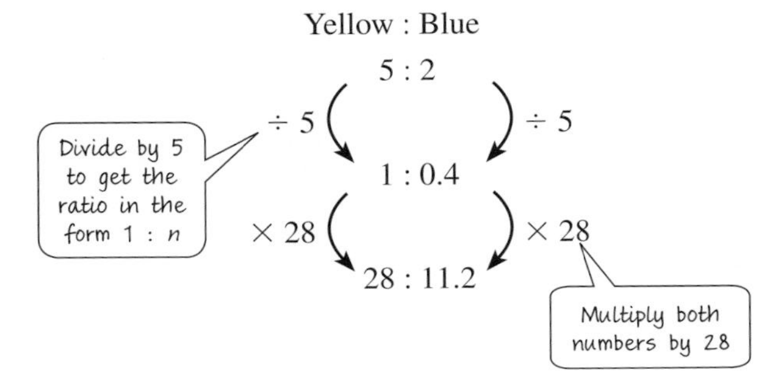

## Exercise 13C

**1** The values of $x$ and $y$ are directly proportional. Copy and complete these tables.

**a**

| $x$ | 6 | 30 | |
|---|---|---|---|
| $y$ | 24 | | 4 |

**b**

| $x$ | 2 | 9 | |
|---|---|---|---|
| $y$ | 14 | | 56 |

**c**

| $x$ | 5 | 8 | |
|---|---|---|---|
| $y$ | | 2.4 | 3.3 |

**d**

| $x$ | 12 | | $^-2$ |
|---|---|---|---|
| $y$ | 18 | 5.4 | $^-3$ |

**2** **a** If 16 tickets cost $280, what do 7 tickets cost?
**b** If 5 pens cost $6.45, what do 7 pens cost?
**c** If 8 boxes of chocolates have a mass of 1.8 kg, what is the mass of 15 boxes?
**d** 2.5 litres of paint cover 37.5 $m^2$ of wall. What area would 7 litres of paint cover?

**3** I change 80 US dollars ($) into 57.60 euros (€).
**a** How many euros would I get for $30?
**b** How many dollars would I get for €288?
**c** Plot these values on a graph, showing dollars on the horizontal axis and euros on the vertical axis.
**d** Draw a line through the three points you have plotted.
**e** If you extend the line in part **d**, does it pass through the origin?
**f** Use the line to estimate how many euros you would get for $73.

**4** A market stall sells fruit at these prices:

| Pineapples | Mangoes | Bananas |
|---|---|---|
| $2.28 for 3 kg | $2.48 for 4 kg | $2.10 for 5 kg |

What is the total cost of 2 kg of bananas, 3 kg of mangoes and 2 kg of pineapples?

**5** Look at the list of ingredients needed to make cupcakes.

To make 12 cupcakes you will need:
110 g flour
2 eggs
250 g butter
110 g caster sugar
4 tablespoons milk
280 g icing sugar

**a** How many cupcakes can you make with a box of 6 eggs (assuming you have plenty of the other ingredients)?
**b** How much flour do you need to make 18 cupcakes?
**c** How much butter do you need for 30 cupcakes?

**6** **a** Use the table to convert between US dollars ($) and British pounds (£).

| $ | 70.00 | | 104.00 |
|---|---|---|---|
| £ | 46.20 | 54.12 | |

**b** I exchange 400 US dollars ($) for 3688 South African rand (R).
At the same rate of exchange
**i** how many dollars would I get for R3227
**ii** how many rand would I get for $450?

**7** A grey paint is a mixture of white and black paint in the ratio 13 : 5.
If 32 litres of black paint are used, how many litres of white paint are needed to make the grey paint?

**8** Which is the best value for a lemonade drink, 2-litre bottles on offer, 3 for $5.16 or 12 cans, each containing 330 ml, for $3.48?

**9** A river which is 3.2 cm long on a map is 14.4 km long in real life.
The same map shows a road which is 10.8 km long in real life.
What is the length of the road on the map?

# Consolidation

### Example 1

A suitcase has a mass of 22 kg. A shoulder bag has a mass of 1500 g.
Use ratio to compare the mass of the suitcase to the mass of the shoulder bag.

| | | | |
|---|---|---|---|
| 22 kg | : | 1500 g | |
| 22 000 g | : | 1500 g | Make the units the same for both masses |
| 22 000 | : | 1500 | Remove the units |
| 44 | : | 3 | Simplify |

### Example 2

Share 0.45 kg in the ratio 7 : 5 : 3

0.45 kg = 450 g — Work in grams to avoid the decimals.

7 + 5 + 3 = 15 parts in total

450 ÷ 15 = 30, so 1 part is 30 g

7 × 30 = 210
5 × 30 = 150
3 × 30 = 90

(Check: 210 + 150 + 90 = 450)

0.45 kg shared in the ratio 7 : 5 : 3 gives shares of 210 g, 150 g and 90 g.

### Example 3

Jamil and Sue share an amount of money in the ratio 9 : 2.
Sue gets $42 less than Jamil.
How much do they get each?

Jamil has 9 − 2 = 7 parts more than Sue.
These 7 parts are worth $42.
42 ÷ 7 = 6, so 1 part is worth $6

Jamil gets 9 × $6 = $54
Sue gets 2 × $6 = $12

(Check: 54 − 12 = 42)

### Example 4

$x$ and $y$ are directly proportional to each other.
Copy and complete the table.

| $x$ | 5 | | 7 |
|---|---|---|---|
| $y$ | 8 | 12 | |

8 ÷ 5 = 1.6, so the multiplier from 5 to 8 is 1.6

12 ÷ 1.6 = 7.5
7 × 1.6 = 11.2

| $x$ | 5 (× 1.6) | 7.5 (÷ 1.6) | 7 (× 1.6) |
|---|---|---|---|
| $y$ | 8 | 12 | 11.2 |

### Example 5

I change 80 British pounds (£) into 6584 Indian rupees (₹).

**a** How many rupees would I get for £75?
**b** How many pounds would I get for ₹16 460?

The multiplier from pounds to rupees is 6584 ÷ 80 = 82.3

| £ | 80 (× 82.3) | 75 (× 82.3) | 200 (÷ 82.3) |
|---|---|---|---|
| ₹ | 6584 | 6172.50 | 16 460 |

## Exercise 13

**1** Simplify:
**a** 120 : 210
**b** 91 : 78
**c** 72 : 180 : 252

**2** Share
**a** $240 in the ratio 3 : 5
**b** 75 g in the ratio 7 : 8
**c** 0.7 m in the ratio 1 : 4
**d** 6.3 km in the ratio 1 : 5 : 3
**e** 3 hours in the ratio 4 : 3 : 8

**3** Write each of these as a ratio in its simplest whole-number form.
**a** 2.8 : 7
**b** $\frac{2}{3} : \frac{4}{5}$
**c** 2 : 3.5 : 0.5
**d** 80% : 1.2 : $2\frac{2}{5}$

**4** I change 450 US dollars ($) into 351 euros (€). At the same exchange rate

**a** how many euros would I get for $240?

**b** how many dollars would I get for €468?

**5** Which is the best value for money, a 2.4-kg bag of potatoes for $2 or a 400-g bag of potatoes for $0.35?

**6** If 3 boxes of chocolates contain 78 chocolates in total, calculate how many chocolates 8 boxes will contain.

**7** To make a fruit drink, pineapple juice, orange juice and mango juice are mixed in the ratio 5 : 3 : 4. How much of each juice do you need to make 0.6 litres of the fruit drink?

**8** Two different types of grey paint are mixed from black and white paint in the ratios shown.

| | Black | : | White |
|---|---|---|---|
| **Paint A** | 5 | : | 8 |
| **Paint B** | 2 | : | 3 |

Which paint is darker?

**9** Compare these quantities using ratio.

**a** 245 mm and 35 cm

**b** 5.5 ℓ and 8800 ml

**c** 6.3 kg and 180 g

**d** 96 $cm^2$ and 4800 $mm^2$

**10** Students in a primary school voted for whether they wanted a blue school uniform or a red school uniform. The result was a ratio of 12 : 5 in favour of blue.
180 children voted for a blue uniform.
How many children voted altogether?

**11** A metal alloy is made from copper and lead in the ratio 3 : 5.

**a** If a block of the alloy contains 3.6 kg of copper, how much lead does it contain?

**b** Another block of the alloy has a mass of 6.24 kg. What is the mass of

**i** copper

**ii** lead in this block?

**12** Maduka plays tennis. During one practice, the ratio of the serves he hit in to the serves he hit out was 23 : 2.

**a** Which of these statements are definitely true or could be true?

**i** He hit 23 times as many serves in as he hit out.

**ii** 92% of his serves were in.

**iii** For every 2 of his serves that were in, 23 serves were out.

**iv** $\frac{2}{23}$ of his serves were out.

**v** He hit 69 serves in and 6 serves out.

**b** For any incorrect statement in part **a**, write the correct statement.

**13** A purple paint is a mix of blue and red paint in the ratio 8 : 5.
If 3.8 litres of red paint are used, will 6 litres of blue paint be enough to make the purple paint?

**14** Amir and Rita share some sweets in the ratio 3 : 7.
Rita gets 52 more sweets than Amir.
How many sweets do they get each?

**15** 7 calculators cost $41.58.

**a** How much do 3 calculators cost?

**b** How many calculators can you buy for $58?

**16** In a chemistry lab, acid and water are mixed in the ratio 2 : 11.
How much acid and how much water are needed to make 585 ml of the mixture?

**17** $x$ and $y$ are directly proportional to each other. Copy and complete the tables by filling in the missing numbers.

**a**

| $x$ | 3 | | 20 |
|---|---|---|---|
| $y$ | 81 | 108 | |

**b**

| $x$ | 8 | 10 | |
|---|---|---|---|
| $y$ | 28 | | 77 |

# Summary

## You should know ...

**1** How to compare ratios.
*For example:*
The ratios show the shots on target to shots missed for two students at archery practice.

| | Shots on target | : | Shots missed |
|---|---|---|---|
| Henrik | 15 | : | 10 |
| William | 28 | : | 22 |

Which player had the higher success rate?

Percentage of shots on target:

Henrik $\frac{15}{15+10} \times 100 = \frac{15}{25} \times 100 = 60\%$

William $\frac{28}{28+22} \times 100 = \frac{28}{50} \times 100 = 56\%$

So Henrik had the higher success rate.

**2** How to solve ratio problems by working out the value of one part.
*For example:*
In a survey, the ratio of students who preferred History to Science was 3 : 7
If 48 more students preferred Science, how many students were surveyed?

$7 - 3 = 4$, so 4 more parts prefer Science
48 students represent 4 parts
$48 \div 4 = 12$, so 1 part is represented by 12 students

$3 + 7 = 10$ parts altogether
$10 \times 12 = 120$, so 120 students were surveyed

**3** How to solve problems involving proportionality, including converting between different currencies.
*For example:*
I change 90 euros (€) into 117 US dollars ($).
Convert between euros and US dollars to complete the table.

| € | 90 | | 234 |
|---|---|---|---|
| $ | 117 | 110.50 | |

$117 \div 90 = 1.3$, which gives the exchange rate: every €1 is worth $1.30

| € | 90 (×1.3) | 85 (÷1.3) | 234 (×1.3) |
|---|---|---|---|
| $ | 117 | 110.50 | 304.20 |

## Check out

**1** Jane and Toyin were at the same archery practice as Henrik and William.

| | Shots on target | : | Shots missed |
|---|---|---|---|
| **Jane** | 58 | : | 42 |
| **Toyin** | 48 | : | 27 |

Write the four students' names in order, starting with the student with the highest success rate.

**2** **a** The angles in a triangle are in the ratio 2 : 3 : 4
**i** What size is the smallest angle?
**ii** What size is the largest angle?

**b** Gill and Nat share some money in the ratio 8 : 3
Gill gets $60 more than Nat.
How much do they get each?

**3** I change 650 US dollars ($) into 35 295 Indian rupees (₹).
**a** How many rupees would I get for $220?
**b** How many US dollars would I get for ₹48 870?

# 14 Sequences, functions and graphs

## Objectives

- Construct tables of values and plot the graphs of linear functions, where $y$ is given implicitly in terms of $x$, rearranging the equation into the form $y = mx + c$; know the significance of $m$ and find the gradient of a straight-line graph.
- Find the approximate solutions of a simple pair of simultaneous linear equations by finding the point of intersection of their graphs.
- Construct functions arising from real-life problems; draw and interpret their graphs.
- Use algebraic methods to solve problems involving direct proportion, relating solutions to graphs of the equations.
- Find the inverse of a linear function.
- Generate terms of a sequence using term-to-term and position-to-term rules.
- Derive an expression to describe the $n$th term of an arithmetic sequence.

## What's the point?

Graphs are pictorial ways of showing relationships and functions. A graph can help you see at a glance how something is changing or what may happen next.

## Before you start

### You should know ...

**1** How to read a scale.

*For example:*

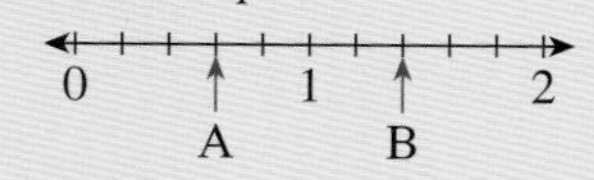


Each small interval represents 0.2, so A = 0.6 and B = 1.4

### Check in

**1** Write down the values of X, Y and Z.

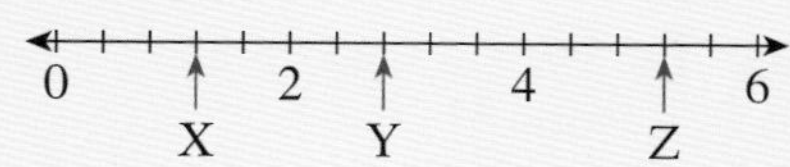

**2** How to read and plot points with positive, negative and decimal coordinates.
*For example:*

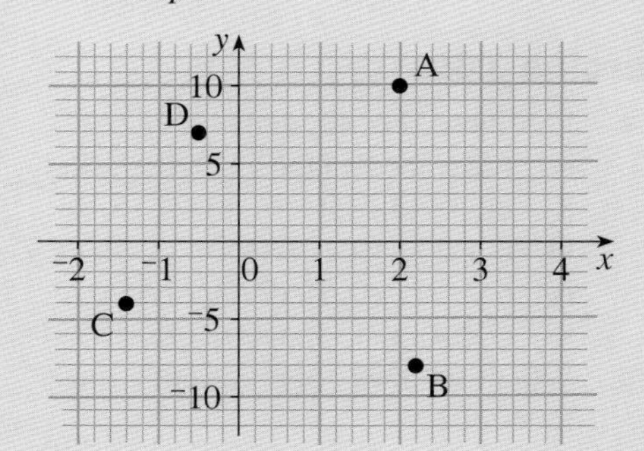


A is the point $(2, 10)$ B is the point $(2.2, {}^-8)$
C is the point $({}^-1.4, {}^-4)$ D is the point $({}^-0.5, 7)$

**3** How to solve linear equations.
*For example:*

$$3x + 12 = 2x + 17$$

$[-12]$ $3x = 2x + 5$

$[-2x]$ $x = 5$

**4** Parallel lines are always the same distance apart.

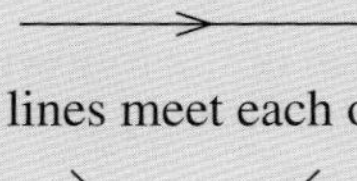

Perpendicular lines meet each other at right angles.

**5** How to change the subject of a formula.
*For example:*
Make $a$ the subject of the formula $v = u + at$.

$v - u = at$

$a = \dfrac{v - u}{t}$

**2 a** Write down the coordinates of the points X, Y and Z.

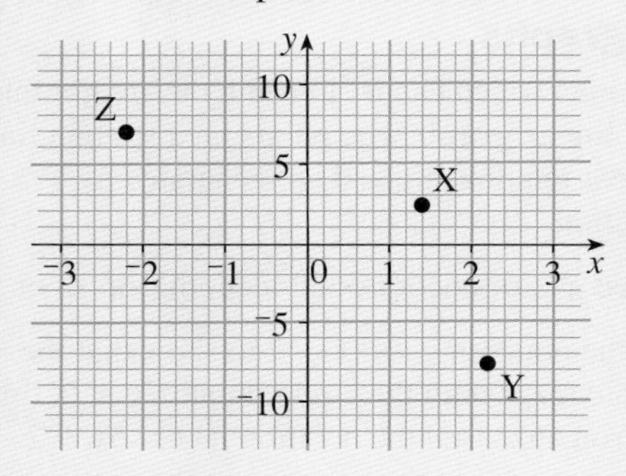


**b** Plot these points.

**i** $(1.8, 6)$ **ii** $({}^-2.5, 4)$

**iii** $({}^-1, {}^-7.5)$ **iv** $(2.5, {}^-9)$

**3** Solve:

**a** $3x + 15 = 21$

**b** $6x - 12 = 18$

**c** $9x = 14x - 25$

**d** $11x + 16 = 14x - 23$

**4** ABCD is a square.

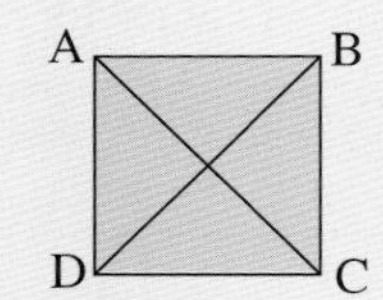


State whether each pair of lines is parallel, perpendicular or neither.

**a** AB and DC

**b** AC and BD

**c** AB and AD

**d** AD and BD

**5** Make $x$ the subject of these formulae:

**a** $y = 2x + 1$

**b** $2y - 3x = 5$

## 14.1 Linear functions

You will need graph paper.

### Linear graphs

A set of points which can be joined by a straight line is called a **linear graph**. The points on a straight line form a linear relationship and are generated by a linear function.

You learned about functions and mappings in Book 2.

You used the notation $x \rightarrow 3x + 2$ to show that the $x$-coordinate maps onto the $y$-coordinate $3x + 2$, for example. You also used the notation $f(x) = 3x + 2$ and $y = 3x + 2$. All of these represent the same linear function.

When the function is in the form $y = mx + c$ (where $m$ and $c$ are numbers) it is often referred to as an equation.

**EXAMPLE 1**

Plot the graph of $y = 2x - 1$ for $x = {}^-2$ to $x = 4$.

First complete the table of values for $x$.

| x | $^-2$ | $^-1$ | 0 | 1 | 2 | 3 | 4 |
|---|---|---|---|---|---|---|---|
| 2x | $^-4$ | $^-2$ | 0 | 2 | 4 | 6 | 8 |
| −1 | $^-1$ | $^-1$ | $^-1$ | $^-1$ | $^-1$ | $^-1$ | $^-1$ |
| y | $^-5$ | $^-3$ | $^-1$ | 1 | 3 | 5 | 7 |

The points which satisfy the relation $y = 2x - 1$ are $(^-2, ^-5)$, $(^-1, ^-3)$, $(0, ^-1)$, $(1, 1)$, $(2, 3)$, $(3, 5)$, $(4, 7)$.

The graph is:

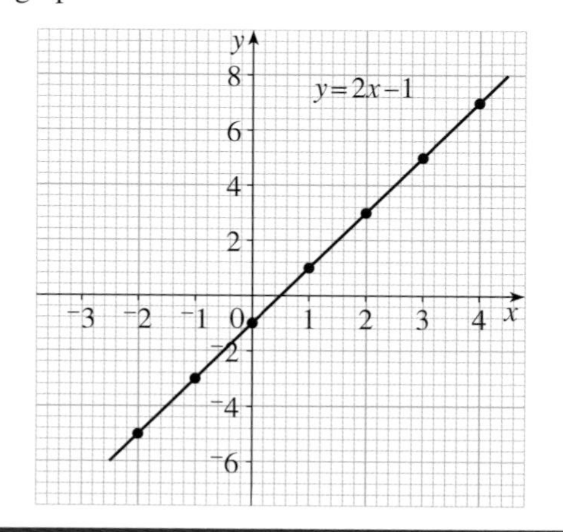


The function $y = 2x - 1$ can be written in words.

'$y = 2x - 1$' means: to find the $y$-coordinate you double the $x$-coordinate and then subtract 1.

### Exercise 14A

**1** **a** Plot these sets of points:

- **i** $(^-2, 1), (^-1, 1), (0, 1), (1, 1), (2, 1)$
- **ii** $(1, 4), (2, 5), (3, 6), (4, 7), (5, 8)$
- **iii** $(^-1, 11), (0, 10), (1, 9), (2, 8), (3, 7)$
- **iv** $(^-2, ^-1), (0, 3), (2, 7), (4, 11), (6, 15)$
- **v** $(^-2, 4), (^-1, 1), (0, 0), (1, 1), (2, 4)$

**b** Which set does not represent a linear function?

**2** **a** Copy and complete the table for $y = 3x + 2$.

| x | $^-2$ | $^-1$ | 0 | 1 | 2 | 3 | 4 |
|---|---|---|---|---|---|---|---|
| 3x | $^-6$ | | 0 | | | 9 | |
| +2 | +2 | | +2 | | +2 | | |
| y | $^-4$ | | | | | | 14 |

**b** Draw the graph of $y = 3x + 2$ using a scale of 1 cm to represent 1 unit on the $x$-axis and 1 cm to represent 2 units on the $y$-axis.

**3** **a** For each of the following, draw tables of values for $x = {}^-2$ to $x = 4$.

- **i** $y = 2x$ **ii** $y = x + 4$
- **iii** $y = 2x + 4$ **iv** $y = 3x - 4$

**b** Using a scale of 1 cm to represent 1 unit on the $x$-axis and 1 cm to represent 2 units on the $y$-axis, plot the graphs of the equations in part **a**.

**4** Using suitable scales, plot the graphs of the following equations for $x = {}^-3$ to $x = 3$.

- **a** $y = 5x - 9$ **b** $y = \frac{1}{2}x + 2$
- **c** $y = 6 - x$ **d** $y = 12 - 2x$

**5** **a** If $x + y = 8$, complete
$y = 8 - \square$
to make $y$ the subject.

**b** Hence plot the graph of $x + y = 8$ for $x = 0$ to $x = 8$.

**6** **a** If $x + 2y = 8$, complete
$2y = 8 - \square$
therefore $y = 4 - \square$
to make $y$ the subject.

**b** Copy and complete the table for this relation.

| $x$ | $^{-}2$ | $^{-}1$ | 0 | 1 | 2 | 3 | 4 |
|---|---|---|---|---|---|---|---|
| 4 | 4 | 4 | | | | | |
| $-\frac{1}{2}x$ | 1 | | | $-\frac{1}{2}$ | | | $^{-}2$ |
| $y$ | 5 | $4\frac{1}{2}$ | | | 3 | | |

**c** Hence, using a suitable scale, plot the graph of $x + 2y = 8$.

**7** Using the method from Question **6**, plot the graphs of

**a** $x + 2y = 6$ **b** $2x + y = 10$
**c** $x - 3y = 9$ **d** $4x + 3y = 12$

for $x = {}^{-}2$ to $x = 4$.

**8** Represent the information as a graph and hence state whether or not the graph represents a linear function.

**a**

| Number of litres | Cost of fuel in cents |
|---|---|
| 1 | 22 |
| 2 | 44 |
| 3 | 66 |
| 4 | 88 |
| 5 | 110 |
| 6 | 132 |

**b**

| Number of km | Cost of hiring taxi in cents |
|---|---|
| 0 | 50 |
| 1 | 62 |
| 2 | 74 |
| 3 | 86 |
| 4 | 98 |
| 5 | 110 |
| 6 | 122 |

**c**

| Length of side of square in cm | Area of square in $cm^2$ |
|---|---|
| 1.2 | 1.44 |
| 1.7 | 2.89 |
| 2.1 | 4.41 |
| 2.5 | 6.25 |
| 3.0 | 9.00 |
| 3.6 | 12.96 |

**d**

| Time for journey in hours | Distance travelled in km |
|---|---|
| 1 | 32 |
| 2 | 64 |
| 3 | 100 |
| 4 | 128 |
| 5 | 160 |
| 6 | 200 |
| 7 | 224 |

**9** The information given represents a linear relation. Show this on a graph and hence complete the table.

**a**

| Time for journey in hours | Distance travelled in km |
|---|---|
| 1 | 83 |
| 2 | 166 |
| 3 | |
| 4 | 332 |
| 5 | |
| 6 | 498 |
| 7 | |

**b**

| Number of hectares | Cost of spraying in $ |
|---|---|
| 0 | 5 |
| 1 | 22.50 |
| 2 | |
| 3 | 57.50 |
| 4 | |
| 5 | |
| 6 | 110 |
| 7 | |

**c**

| Time taken in hours | Distance from home in km |
|---|---|
| 0 | |
| 1 | 172 |
| 2 | |
| 3 | 116 |
| 4 | |
| 5 | 60 |
| 6 | 32 |
| 7 | |

In Questions **5**, **6** and **7** of Exercise 14A, you were asked to rearrange equations to make $y$ the subject before working out coordinates and plotting the graphs.
There is an alternative method you can use.

To draw a straight line we only need two points – we usually work out more than two points because if we make an error in calculating the coordinates it can easily be seen from the points.

For an equation in the form $2x + 5y = 30$, it may be simpler and quicker to find where the graph crosses the axes.

- To find where the graph of an equation crosses the $y$-axis substitute for $x = 0$.
- To find where the graph of an equation crosses the $x$-axis substitute for $y = 0$.

Be careful using this method as you will have only worked out two points. You may want to check after drawing the line, using a third point.

**EXAMPLE 2**

**a** Describe the relationship between the $x$ and $y$ coordinates for the graph of $2x + 5y = 30$

**b** Draw the graph of $2x + 5y = 30$

---

**a** If you double the $x$-coordinate and add it to five times the $y$-coordinate you get 30

**b** The graph crosses the $y$-axis at $x = 0$

$2(0) + 5y = 30$

$5y = 30$

$y = 6$

$(0, 6)$ is a point on the graph.

The graph crosses the $x$-axis at $y = 0$

$2x + 5(0) = 30$

$2x = 30$

$x = 15$

$(15, 0)$ is a point on the graph.

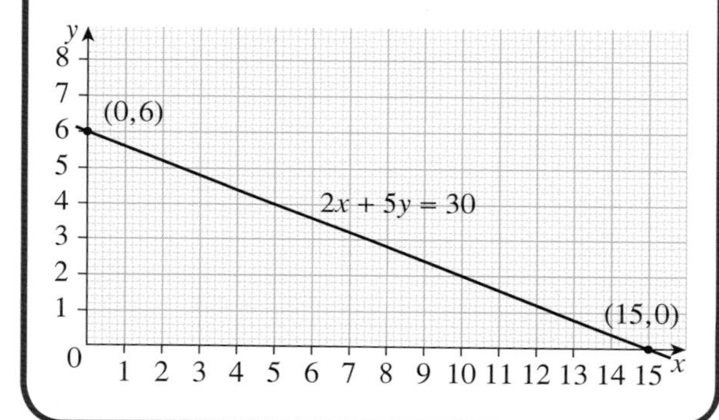


You can choose a point on the line to check. The point $(5, 4)$ is on the line drawn in Example 2.

Substituting for $x = 5$ and $y = 4$ into $2x + 5y = 30$ gives

$$\begin{aligned} 2(5) + 5(4) &= 10 + 20 \\ &= 30 \end{aligned}$$

as required, so we can be confident our line is correctly drawn.

Be careful with this check: the point needs to be exactly on the line. A point very close to but not exactly on the line will not work.

## Exercise 14B

**1** Draw the graphs of these equations using the method used in Example 2.

- **a** $3x + 2y = 18$
- **b** $x + 6y = 12$
- **c** $4x + 7y = 28$
- **d** $5x + 3y = 30$

**2** Repeat Question **1** but this time rearrange the equations to make $y$ the subject.

Which method is easier?

**3** Using the method from Example 2, repeat Exercise 14A, Question **7**.

Is this easier than the method you used the first time you answered the question?

**4** Draw the graphs of these equations using a method of your choice.

- **a** $5x + 2y - 20 = 0$
- **b** $3x = 24 - 4y$
- **c** $3y = 3(2x - 5)$

## 14.2 Solving simultaneous equations graphically

You will need graph paper.

In Chapter 8 you learned how to solve a pair of simultaneous equations using algebra. You can also solve them by drawing their graphs.

The values $x = 0, y = 0$ satisfy the equation $y = 2x$. These values do not satisfy the equation $x + y = 3$.

The values $x = 1, y = 2$ satisfy both equations.

In fact $x = 1, y = 2$ are the only values that satisfy both equations, and hence $(1, 2)$ is the only point that lies on both lines.

The pair of simultaneous equations

$y = 2x$ and $x + y = 3$

Simultaneous equations are equations which have a common solution.

have $x = 1$ and $y = 2$ as their common solution.

**EXAMPLE 3**

Plot the graphs of

$x + y = 6$
$x - y = 2$

Hence find the solution to this pair of equations.

Tables of values for $x = 0$ to $x = 6$:
$x + y = 6$

| x | 0 | 1 | 2 | 3 | 4 | 5 | 6 |
|---|---|---|---|---|---|---|---|
| y | 6 | 5 | 4 | 3 | 2 | 1 | 0 |

$x - y = 2$

| x | 0 | 1 | 2 | 3 | 4 | 5 | 6 |
|---|---|---|---|---|---|---|---|
| y | ⁻2 | ⁻1 | 0 | 1 | 2 | 3 | 4 |

Plotting both equations on the same axes gives:

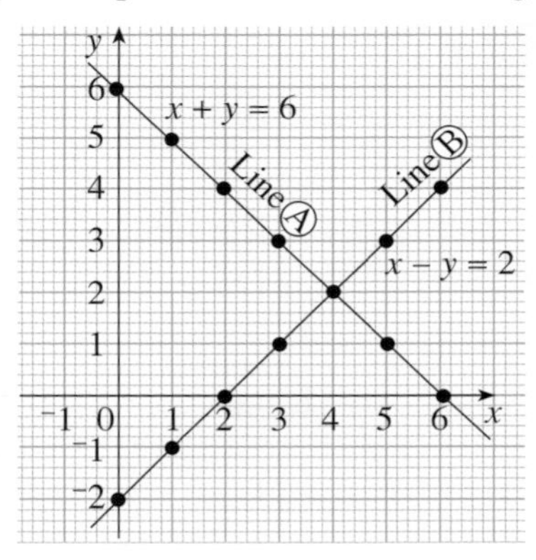


Line (A) gives the solutions to $x + y = 6$ and

Line (B) gives the solutions to $x - y = 2$.

So the points that lie on both lines are solutions to both equations.

Here there is only one point, (4, 2), on both lines.

So the solution to both equations is $x = 4, y = 2$

You need to be aware that this method isn't as exact as using an algebraic method, particularly if the coordinates are not integers. So we say that we can find the **approximate solutions** of a pair of simultaneous linear equations by finding the **point of intersection** of their graphs.

## Exercise 14C

**1** Plot $y = 2x - 1$ and $y = x + 1$ on the same axes. Hence find the solution to this pair of equations.

**2** Draw a graph for each pair of equations and find their solution.

**a** $y = x + 4$, $y = 6 - x$ **b** $y = 3 - x$, $y = 3 - 3x$

**c** $y = 2x$, $y = x + 1$ **d** $y - 2x = 1$, $y - x = {}^-5$

**e** $y = 2x$, $y = x + 2$ **f** $y - x = 3$, $y - 2x = 1$

**3** Draw each pair of lines on the same axes. Find the point of intersection. Show that this point satisfies both equations.

**a** $y = 2x$, $y = 9 - x$ **b** $y = 2x - 4$, $y = x + 1$

**c** $y = 3x - 2$, $y = x + 4$ **d** $y = 4x - 3$, $y = 2x + 2$

**4** Plot graphs for this pair of equations:
$y = 3x - 1$
$y = 3x + 2$
Why is there no solution to this pair of equations?

**5** Find an approximate solution for each pair of simultaneous equations by finding the point of intersection of their graphs.

**a** $2x + 5y = 20$
$3x - 2y = 6$

**b** $3x + 4y = 24$
$5x - 3y = 15$

**c** $5x + 8y = 40$
$3x - 4y = 12$

**d** ${}^-2x + 7y = 14$
$5x - 3y = 30$

**6** The graph shows $y = x - 1$ and $y = 10 - 3x$.

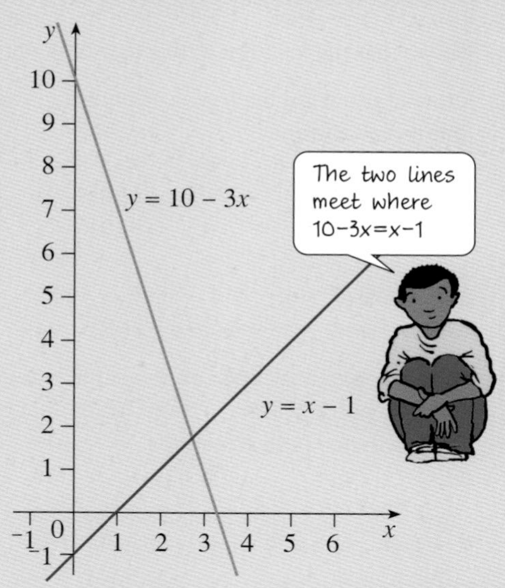


**a** Which of the following is the value of $x$ where the lines intersect?

$2\frac{1}{2}, 2\frac{3}{4}, 1\frac{7}{10}, 3\frac{1}{4}, 2\frac{1}{3}$

**b** Find the corresponding value of $y$.

**7** Find the $(x, y)$ equation that describes each set of ordered pairs. Check that the common ordered pair satisfies both equations.

**a** $\{(0,0),(1,2),(2,4),(3,6),(4,8),(5,10)\}$, $\{(0,6),(1,5),(2,4),(3,3),(4,2),(5,1),(6,0)\}$

**b** $\{(0,3),(1,5),(2,7),(3,9),(4,11),(5,13)\}$, $\{(0,{}^-2),(1,1),(2,4),(3,7),(4,10),(5,13)\}$

**8** Show each set of ordered pairs on the same axes.
Find the point of intersection of the two lines.
Write down the equation for each line.
Show that the common point satisfies both equations.

**a** $\{(1,6),(2,5),(3,4),(4,3),(5,2),(6,1)\}$, $\{(1,4),(2,7),(3,10),(4,13),(5,16),(6,19)\}$

**b** $\{(0,1),(1,3),(2,5),(3,7),(4,9),(5,11)\}$, $\{(0,{}^-4),(1,0),(2,4),(3,8),(4,12),(5,16)\}$

# 14.3 Equations of lines in the form $y = mx + c$

You will need graph paper.

Small gradient

Large gradient

## Gradients

You can measure the steepness of a hill, or a straight line, by its gradient.

- gradient of a line $= \dfrac{\text{vertical rise}}{\text{horizontal shift}}$

Some people use the equation: gradient $= \dfrac{\text{rise}}{\text{run}}$

**EXAMPLE 4**

Find the gradient of the line $y = 2x + 1$

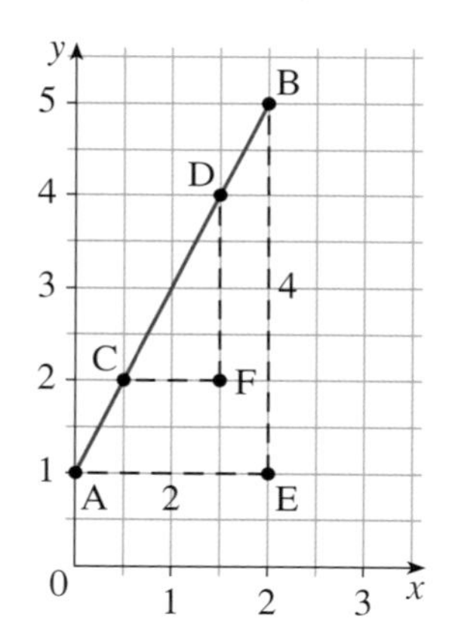


Choose two points on the line A (0, 1) and B (2, 5):

$$\text{gradient} = \frac{\text{vertical rise}}{\text{horizontal shift}} = \frac{BE}{AE}$$

$$= \frac{5-1}{2-0} = \frac{4}{2} = 2$$

**Note**: It does not matter which two points on the line are chosen.

If we choose instead the points C (0.5, 2) and D (1.5, 4), the right-angled triangle is CDF, and

$$\text{gradient} = \frac{DF}{CF} = \frac{4-2}{1.5-0.5} = \frac{2}{1} = 2$$

## Exercise 14D

**1** **a** On graph paper draw the graphs of these linear equations:

**i** $y = 3x + 1$ for $^{-}5 \leqslant x \leqslant 6$

**ii** $y = 5x - 2$ for $^{-}4 \leqslant x \leqslant 6$

**iii** $y = 4x - 3$ for $^{-}2 \leqslant x \leqslant 6$

**iv** $y = 7x + 1$ for $^{-}3 \leqslant x \leqslant 6$

**b** Use the method in Example 4 to find the gradient of each line drawn.

**c** Complete this table:

| Equation of line | Gradient |
|---|---|
| $y = 3x + 1$ | |
| $y = 5x - 2$ | |
| $y = 4x - 3$ | |
| $y = 7x + 1$ | |

**d** What do you notice about

**i** the coefficient of $x$ and the gradient

**ii** the constant in the equation and the point of intersection with the $y$-axis?

(**Hint**: The coefficient of $x$ is the number in front of $x$.)

**2** **a** On the same set of axes, plot the graphs of:

**i** $y = x$ **ii** $y = 2x$

**iii** $y = 3x$ **iv** $y = 4x$

**b** Which line has the largest gradient?

**3** **a** On the same set of axes, draw the graphs of:

**i** $y = 2x - 1$ **ii** $y = 2x$

**iii** $y = 2x + 1$

**b** What do you notice about the gradient of each?

**4** Draw the graph of each equation in the table, then copy and complete the table.

| Equation of line | Gradient |
|---|---|
| $2y = 4x + 8$ | |
| $4y - 12x = 12$ | |
| $3y = 15x - 9$ | |
| $2y - 8x = 16$ | |

### INVESTIGATION

By investigating equations in the form $y = mx + c$, where $m$ and $c$ can be either positive or negative, determine a link between the value of $m$ and whether the line slopes up or down.

In Question **1** of Exercise 14D, you should have found that the coefficient of $x$ (the number in front of $x$) gives you the gradient (or steepness) of the line. This only works for equations in the form $y = mx + c$. When equations are in this form we say $y$ is given **explicitly** in terms of $x$, and we can identify the gradient as the value of $m$. In Question **4** the equations were in forms in which the relationship between $x$ and $y$ is given **implicitly**. To find the gradient from these equations, we have to rearrange them to make $y$ the subject.

All of the gradients in Exercise 14D were positive. In the Investigation you learned that if the line slopes down instead of up, the value of $m$ in $y = mx + c$ will be negative; that is, the line will have a negative gradient.

Examples 5 and 6 involve finding the gradient of a line segment joining two points from the coordinates of those points, and show why the gradient of a line is positive or negative.

**(Note, however, that finding the gradient of a line from the coordinates of two points on it is not mentioned specifically in the Cambridge Secondary 1 curriculum framework, and is included here as extension material.)**

### EXAMPLE 5

Find the gradient of the line joining the points $(1, 3)$ and $(5, 11)$.

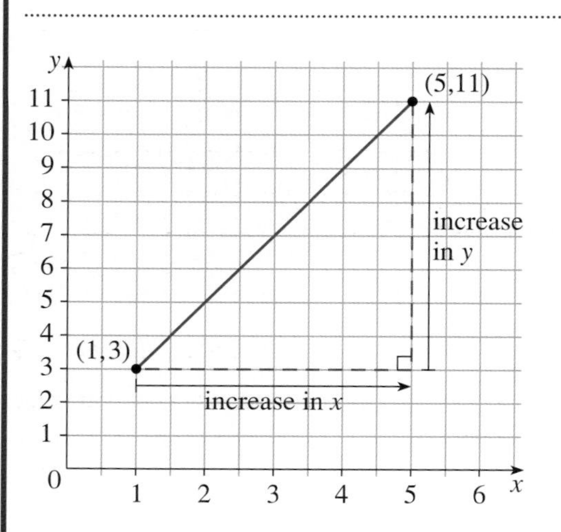


$$\text{gradient} = \frac{\text{vertical rise}}{\text{horizontal shift}}$$

$$= \frac{\text{increase in value of } y}{\text{increase in value of } x}$$

$$= \frac{11 - 3}{5 - 1} = \frac{8}{4} = 2$$

### EXAMPLE 6

Find the gradient of the line joining the points $(1, 5)$ and $(5, 3)$.

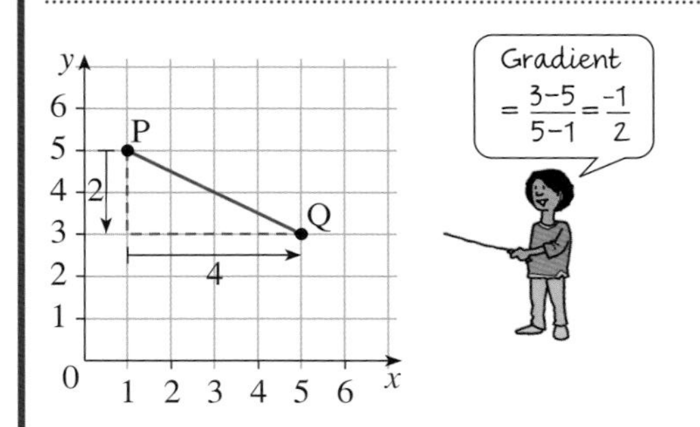


The line PQ slopes down to the right.

As the value of $x$ increases from 1 to 5 the value of $y$ *decreases* from 5 to 3. We say the *increase* in $y$ is $^{-}2$.

The gradient of the line PQ is:

$$\frac{\text{the increase in } y}{\text{the increase in } x} = \frac{^{-}2}{4} = \frac{^{-}1}{2}$$

The gradient of the line is negative because the line *slopes down to the right.*

- A line that slopes up to the right ⁄ has a positive gradient. A line that slopes down to the right ╲ has a negative gradient.

## Exercise 14E

**(Note: Questions 1, 2 and 5 are extension work.)**

**1** Find the gradients of the lines joining these pairs of points.

**a** $(4, 2)$ and $(8, 10)$ **b** $(3, 1)$ and $(4, 3)$
**c** $(^{-}3, ^{-}2)$ and $(4, 5)$ **d** $(4, 2)$ and $(3, ^{-}3)$
**e** $(^{-}2, 3)$ and $(3, 3)$ **f** $(4, 7)$ and $(4, 3)$

**2 a** Find the gradient of the line joining:
**i** $(1, 6)$ and $(3, 2)$ **ii** $(2, 5)$ and $(4, 1)$
**iii** $(2, 7)$ and $(3, 9)$ **iv** $(1, 8)$ and $(3, 7)$

**b** Which of these lines has a positive gradient?

**3** The Downtown Youth Club found that its committee meetings took longer if more members were present. This table shows how long some of the meetings took:

| **Number present ($x$)** | 5 | 10 | 15 | 20 |
|---|---|---|---|---|
| **Time taken in minutes ($y$)** | 25 | 40 | 55 | 70 |

**a** Draw a graph of the time taken against the number present. Does the graph show a linear relation?
**b** Where does the graph cut the $y$-axis?
**c** Find the gradient of the graph.
**d** What information does the gradient give us?

**4** In an experiment, a spring is stretched by hanging weights on its end. Here are the results:

| **Weight ($x$ kg)** | 5 | 10 | 20 | 25 |
|---|---|---|---|---|
| **Length of spring ($y$ cm)** | 22 | 24 | 28 | 30 |

**a** Draw a graph of the results of the experiment.
**b** What is the unstretched length of the spring? How can you tell?
**c** Find the gradient of the graph.
**d** By how much is the spring extended for each 1 kg?

**5** Use a diagram to show that the gradient of the line joining the points $(x_1, y_1)$ and $(x_2, y_2)$ is $\frac{y_2 - y_1}{x_2 - x_1}$.

Questions **3** and **4** of Exercise 14E gave you examples of the meaning of gradient in practical contexts.

In general:

- Gradient means the increase in the $y$-value for an increase of 1 in the $x$-value.

For example, in Question **3** of Exercise 14E, the gradient of 3 shows that each extra person (that is, each increase of 1 in the $x$-value) means a meeting will take another 3 minutes (that is, there will be an increase of 3 in the $y$-value).

## Parallel and perpendicular lines

In Question **3** of Exercise 14D you should have found that the graphs of the equations

$y = 2x - 1, y = 2x, y = 2x + 1$

all have gradient 2.

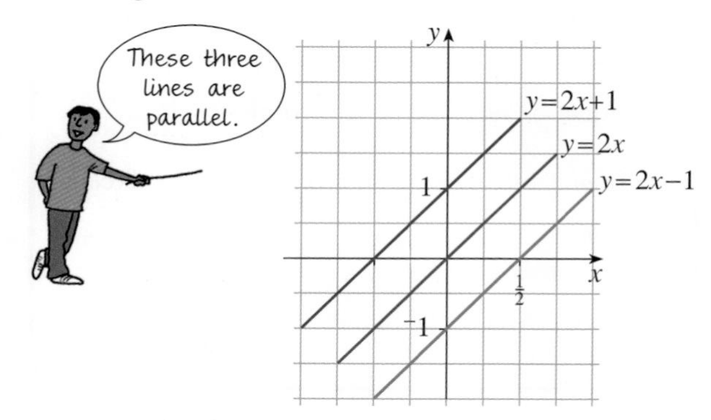


Notice that the coefficient of $x$ in each equation gives the gradient.

The three lines in the graph are all parallel.

- Parallel lines have the same gradient.

$y = 2x + 1$ and $y = 2x - 1$ are parallel because they both have a gradient of 2.

$y = 3x + 2$ and $y = {}^{-}3x + 1$ are *not* parallel. Their gradients are 3 and $^{-}3$, which are *not* the same.

The gradients of perpendicular lines are also connected, however.

Look at the diagram below. The line AB is at right angles to the line BC.

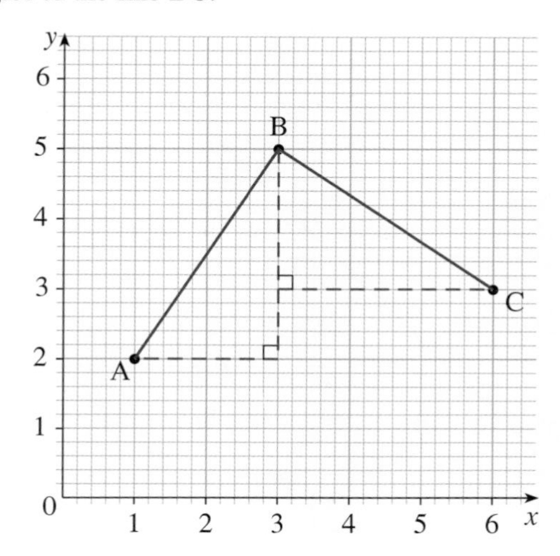


The gradient of AB is $\frac{5-2}{3-1} = \frac{3}{2}$

The gradient of BC is $\frac{3-5}{6-3} = \frac{^{-}2}{3}$

Notice that the product of the gradients is

$\frac{3}{2} \times \frac{^{-}2}{3} = {}^{-}1$

So if two lines are at right angles, the product of their gradients is $^{-}1$.

- If $y = mx + c$ and $y = nx + d$ are perpendicular lines, then $\boldsymbol{m \times n = {}^{-}1}$

## Exercise 14F

**(Note: Questions 1, 5 and 6 are extension work.)**

**1** Find the gradient of the line joining:

**a** (1, 2) and (3, 6) **b** (1, 3) and (3, 4)
**c** (1, 4) and (2, 2) **d** (1, 4) and (4, 10)
**e** (2, 5) and (6, 7) **f** (2, 7) and (5, 1)

**2** Which pairs of lines in Question **1**:

**a** are parallel
**b** have the same gradient as $y = {}^{-}2x + 3$?

**3** Which of these lines have the same gradient?

**a** $y = 2x + 3$ **b** $y = 4x + 3$
**c** $y = 4 + 2x$ **d** $y = 3 - 4x$
**e** $y + 2x = 3$ **f** $y - 4x = 4$

**4** Pick out the pairs of perpendicular lines:

**a** $y = 2x + 3$ **b** $y = 3x + 2$
**c** $y = \frac{^{-}1}{2}x + 1$ **d** $y = \frac{1}{4}x - 1$
**e** $y = {}^{-}4x + 5$ **f** $y = \frac{^{-}1}{3}x + 4$

**5** **a** Draw a diagram to show the points P (4, 1), Q (5, 4) and R (1, 2).
**b** Find the gradients of PQ and PR.
**c** Find the lengths of PQ, PR and RQ and show that RP is at right angles to PQ using Pythagoras' theorem.
**d** Do you agree that: (gradient of PR) × (gradient of PQ) = $^{-}1$?

**6** If P is (3, 1), Q is (7, 2) and R is (6, 6) find the gradients of PQ, QR, and PR. Find which two lines are perpendicular and check your answer using Pythagoras' theorem.

## The *y*-intercept

**Note: The *y*-intercept is not mentioned in the Cambridge Secondary 1 curriculum framework, but is included here to explain the '*c*' part of $y = mx + c$.**

- Any line with an equation in the form $y = mx + c$ crosses the *y*-axis at the point $(0, c)$.

  The point $(0, c)$ is called the ***y*-intercept**.

  The value of $m$ tells you the **gradient** of the line.

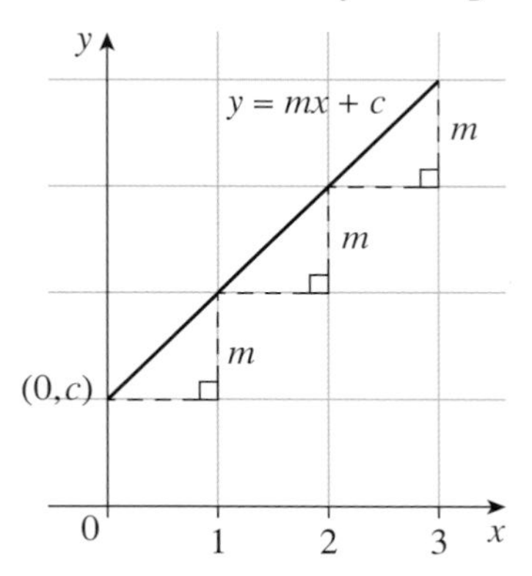


When you know the equation of a straight-line graph it is easy to write down its gradient and *y*-intercept.

It is possible to draw the graph of an equation using just the *y*-intercept and the gradient. You can plot the *y*-intercept $(0, c)$, then for every increase of 1 along the *x*-axis you can plot a step up of the gradient, $m$.

**EXAMPLE 7**

Find the *y*-intercept and gradient of the line

**a** $2y = x - 6$

**b** $3x + 2y = 6$

To do this each equation must be rewritten in the form $y = \square x + \square$.

**a** For $2y = x - 6$, this can be done by dividing each side of the equation by 2.

This gives $y = \frac{1}{2}x - 3$.
Therefore the *y*-intercept is $(0, {}^-3)$ and the gradient is $\frac{1}{2}$.

**b** $3x + 2y = 6$ must first be rewritten as $2y = {}^-3x + 6$.

Each side is then divided by 2, which gives $y = \frac{-3}{2}x + 3$.

The *y*-intercept is $(0, 3)$. The gradient is $\frac{-3}{2}$.

In the same way, if you know the gradient and the *y*-intercept of a line you can write down the equation of the line.

**EXAMPLE 8**

What is the equation of a line that passes through the point $(0, 3)$ and has gradient of 2?

The equation of a line has the form
$y = mx + c$
where $m$ is the gradient and $(0, c)$ is the *y*-intercept.

Here $m = 2$
and $c = 3$

So the equation of the line is
$y = 2x + 3$

### Exercise 14G (extension)

**1 a** By rewriting the equation in the form $y = mx + c$, find the *y*-intercept and gradient.

**i** $2y = 4x + 6$ **ii** $2y = 6x - 4$
**iii** $3y = 4x + 1$ **iv** $4x + 2y = 8$
**v** $3x + 6y = 2$ **vi** $2y - 4x = {}^-5$

**b** Use your answers to part **a** to draw the graph of each equation, first plotting $(0, c)$ then for every step along plotting a step up of $m$.

**2 a** Write down the coordinates of the points marked by triangles on each line in the diagram.

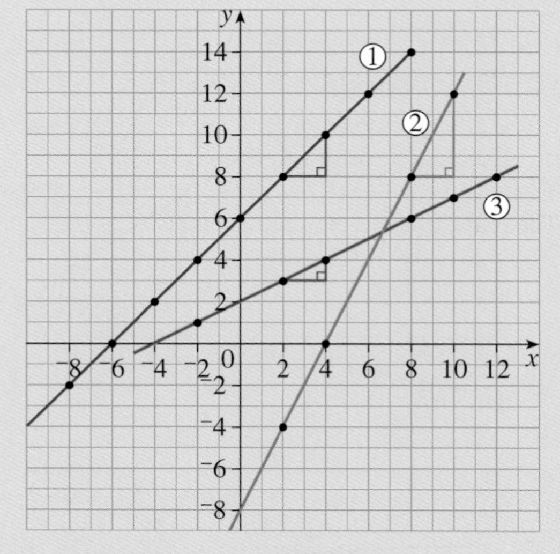


**b** For each line, write down

**i** the value of $c$, the *y*-intercept

**ii** the value of $m$, the gradient.

**3** **a** Write down the values of $m$ and $c$ in each equation

**i** $y = 3x + 4$ **ii** $y = 5x - 2$

**iii** $y = \frac{1}{2}x + 4$ **iv** $y = {}^{-}x + 3$

**b** Use the gradient and $y$-intercept to sketch each graph in part **a**.

**4** Write down the equation of a line parallel to $y = 2x + 3$ which has $y$-intercept

**a** $(0,0)$ **b** $(0,1)$ **c** $(0,5)$

**d** $(0,{}^{-}3)$ **e** $(0,\frac{3}{2})$

**5** Write down the equation of a line which has $y$-intercept $(0,2)$ and a gradient of:

**a** 3 **b** 5 **c** $\frac{1}{2}$ **d** $^{-}2$ **e** 0

**6** Write down the equation of a line which

**a** passes through $(0,5)$ and has a gradient of 3

**b** passes through $(0,{}^{-}2)$ and has a gradient of 4

**c** passes through $(0,1)$ and has a gradient of $^{-}1$.

**7** Find the gradient of a line which passes through

**a** $(0,2)$ and $(1,5)$ **b** $(0,4)$ and $(2,8)$

**c** $(0,{}^{-}1)$ and $(1,3)$ **d** $(0,{}^{-}3)$ and $(2,5)$

Hence write down the equation of each line.

## 14.4 Real-life functions

**EXAMPLE 9**

The following estimate is received for printing copies of a wedding programme:

| No. of copies | 50 | 100 | 200 | 300 |
|---|---|---|---|---|
| Cost in $ | 11.50 | 12.50 | 14.50 | 16.50 |

**a** Find an equation giving the cost $\$y$ of $x$ copies.

**b** Estimate the cost of 325 copies.

The graph of cost ($y$) against number of copies ($x$) is:

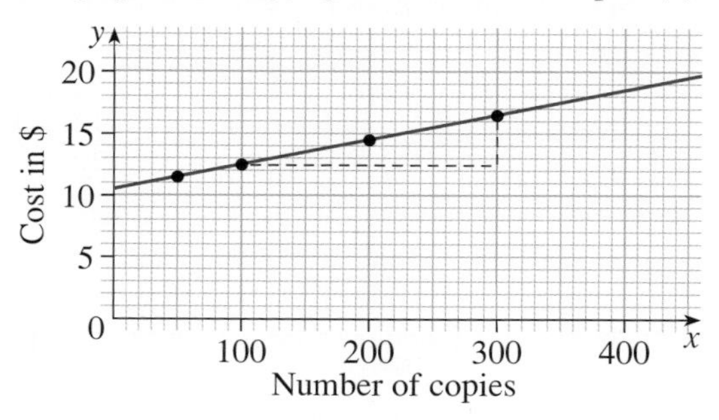


The graph is a straight line, so $x$ and $y$ are related linearly.

**a** The line cuts the $y$-axis at the point (0,10.5), so

$c = 10.5$

and

$$m = \frac{16.5 - 12.5}{300 - 100}$$

$$= \frac{4}{200} = 0.02$$

The linear relation is $y = 0.02x + 10.5$

This means there is a fee of $10.50 before any programmes are printed.

This means that each programme printed costs 2 cents.

**b** When $x = 325$,

$$y = 0.02\,(325) + 10.5$$
$$= 6.5 + 10.5$$
$$= 17$$

so the cost of 325 copies is $17.

### Exercise 14H

**1** The graph shows a set of measurements collected during an experiment. A line is drawn so that all the points are close to it (some on one side and some on the other).

This line of 'best fit' can be used to find the approximate linear relationship between $I$ and $V$. Use the values of the gradient and $V$-intercept to write down the relation.

**2** Three people exchanged US dollars ($) for euros (€) in a bank.
The table shows the amounts exchanged.

| $x | 200 | 80 | 140 |
|---|---|---|---|
| €y | 150 | 60 | 105 |

**a** Draw a graph showing this information with dollars on the $x$-axis and euros on the $y$-axis.
**b** Find the gradient of the graph.
**c** What does the gradient tell you?

**3** Show the information given in the tables on graphs.
Draw a line on each graph to show how the information approximates to a linear relation.
Use this line to state the relationship between $s$ and $t$ in each case.

**a**

| t | s |
|---|---|
| 1.2 | 16.4 |
| 2.5 | 34.1 |
| 2.9 | 39.5 |
| 3.4 | 46.4 |

**b**

| t | s |
|---|---|
| 1.8 | 6.2 |
| 2.3 | 8.1 |
| 3.1 | 10.8 |
| 3.8 | 13.4 |
| 4.7 | 16.5 |

For part **a** use a scale of:

2 cm to represent one unit on the $t$-axis
1 cm to represent five units on the $s$-axis.

For part **b** use a scale of:
2 cm to represent one unit on the $t$-axis
2 cm to represent five units on the $s$-axis.

**4** Mr Khan is a plumber. The table shows his fees for three jobs.

| Number of hours, x | 1 | 2 | 5 |
|---|---|---|---|
| Fee in dollars, y | 60 | 80 | 140 |

**a** Draw a graph to show this information with number of hours on the $x$-axis and fee in dollars on the $y$-axis.
**b** Find the gradient of the graph. What does this tell you?
**c** Find the $y$-intercept of the graph. What does this tell you?

Mr Brown is also a plumber. He makes a $20 call-out charge and has an hourly rate of $25.

**d** Draw a graph showing Mr Brown's fees on the same axes.
**e** What is the point of intersection of the two graphs?
**f** If I have a plumbing job that is estimated to take 3 hours, which plumber would be cheapest?

**5** The values of $C$ and $n$ are connected by the linear function $C = Pn + I$.

The data in the table was collected in an experiment. Show the information on a graph using a scale of 1 cm to represent 1 unit on the $n$-axis, and 1 cm to represent 10 units on the $C$-axis.

| n | C |
|---|---|
| 2.4 | 22.6 |
| 3.4 | 27.9 |
| 4.3 | 32.7 |
| 5.7 | 40.1 |
| 7.4 | 49.0 |
| 9.5 | 60.1 |
| 11.9 | 72.7 |
| 15.0 | 89.0 |

Hence find approximate values for $P$ and $I$.

**6** **a** Find, by drawing a graph, which of the values given in the table seem to indicate that an error was made when collecting data satisfying the linear relation
$E = 0.51W + 3$

| W | E |
|---|---|
| 12 | 9.2 |
| 15 | 10.6 |
| 19 | 11.1 |
| 23 | 14.7 |
| 28 | 17.3 |
| 31 | 18.8 |
| 36 | 21.4 |

Use a scale of 1 cm to represent 5 units on the $W$-axis and 4 cm to represent 10 units on the $E$-axis.
**b** Use your graph to find a more likely value for $E$.

**7** **a** Find, by drawing a graph, which two results given in the table seem to indicate that an error was made when collecting the data.
Use a scale of 4 cm to represent 1 unit on the $W$-axis and 1 cm to represent 1 unit on the $E$-axis.

| W | E |
|---|---|
| 0.2 | 8.8 |
| 0.5 | 7.7 |
| 0.9 | 6.4 |
| 1.1 | 6.0 |
| 1.6 | 3.9 |
| 1.9 | 2.8 |
| 2.4 | 1.1 |
| 2.7 | 0.6 |

**b** Assuming the data fits the linear relation $E = aW + b$, find from your graph approximate values for $a$ and $b$, and hence find a better value for the two inaccurate results.

**8** Factory A and Factory B both make electrical cables. The cost of a cable depends on its length. The delivery cost is the same no matter how long the cable is.

The formula for the cost of a cable from Factory A is $C = 10 + 5d$, where $C$ is the cost in dollars and $d$ is the length of the cable in metres.

**a** How does the formula show the cost of delivering the cable? What is the delivery charge?

**b** What is the cost of a cable per metre?

**c** Draw a graph to show this information.

Factory B has a lower delivery cost but the cable is more expensive per metre. The table shows the cost of cables from Factory B.

| Length in metres, $d$ | 2 | 3 | 6 |
|---|---|---|---|
| Cost in dollars, $C$ | 17 | 23 | 41 |

**d** Draw the graph for the cost of a cable from Factory B on the same axes as your graph for Factory A.

**e** Find the gradient of the line for Factory A. What does this tell you?

**f** Find the intercept of the line for Factory A. What does this tell you?

**g** Write down the formula for the cost in dollars, $C$, of a cable $d$ metres long from Factory B.

## 14.5 Direct proportion

In Chapter 13 you learned that the graph of two quantities which are in direct proportion is a straight line through the origin. Since the line passes through the origin, the $y$-intercept is 0 and so the equation is simply $y = mx$. You can use the graph or a table of values to find the value of $m$.

An example of two items in direct proportion used in Chapter 13 was the number of apples and the cost of apples:

| Number of apples, $x$ | 2 | 4 | 12 | 10 |
|---|---|---|---|---|
| Cost of apples, $y$ (in cents) | 60 | 120 | 360 | 300 |

$\times 30$

Notice that the multiplier going down the table is 30

It is clear that, when we plot these values as a graph, it will pass through the origin. If you have no apples it will cost nothing.

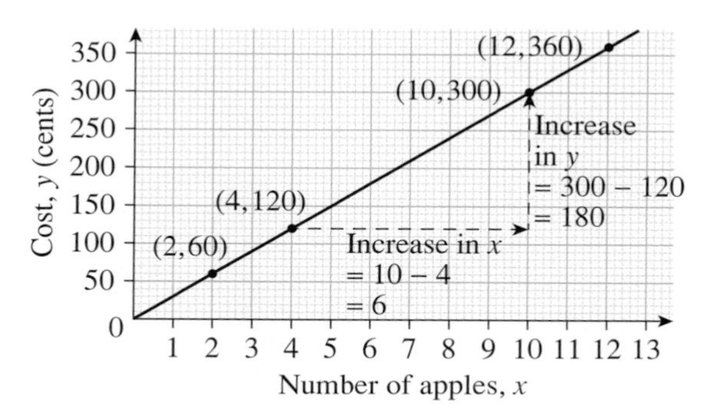


The gradient of the line is $\frac{\text{increase in } y}{\text{increase in } x} = \frac{180}{6} = 30$

Notice that the gradient of the line is equal to the multiplier going down the table.

So the equation of the line is $y = 30x$

This equation can be used to find the cost of, for example, 7 apples. Simply substitute $x = 7$ into the equation: $y = 30 \times 7 = 210$. So the cost of 7 apples is 210 cents.

### Exercise 14I

**1** $x$ and $y$ are directly proportional.

| $x$ | 2 | 5 | 8 |
|---|---|---|---|
| $y$ | 6 | 15 | 24 |

**a** Find the multiplier from $x$ to $y$.

**b** Plot the values in the table on a graph. Draw the line passing through the points.

**c** Extend the line drawn in part **b**. Does it pass through the origin?

**d** Find the gradient of the graph.

**e** Were your answers to parts **a** and **d** the same?

**f** Write down the equation of the graph in the form $y = mx$.

**g** Using the equation, find the value of $y$ when $x = 7$.

**h** Using the equation, find the value of $x$ when $y = 33$.

**2** The volume of paint and the area the paint covers are directly proportional. A 5-litre tin of paint covers 60 square metres.

**a** Draw a graph to show this information, with volume of paint, in $\ell$, on the $x$-axis and area covered, in $m^2$, on the $y$-axis.

**b** Use the graph to find the equation relating volume of paint to area covered, in the form $y = mx$.

**c** Use this equation to find the area you would expect 8 litres of paint to cover.

**d** Use the equation to find how much paint you would need to cover an area of 48 square metres.

**e** What area does 1 litre of paint cover?

**f** Compare your answer to part **e** to the gradient of the line.

**3** I change 80 US dollars ($) into 64 euros (€). Assuming the exchange rate doesn't change:

**a** Draw a graph to show this information, with euros on the $x$-axis and US dollars on the $y$-axis.

**b** Find the equation connecting number of euros and number of US dollars, in the form $y = mx$.

**c** Using the graph, find how many

**i** euros I would get for $73

**ii** US dollars I would get for €31.

**d** Repeat question **c** using the equation. Which method is more accurate?

**4** $x$ and $y$ are directly proportional

| x | 3 | 8 | 14 |
|---|---|---|---|
| y | 7.5 | 20 | 35 |

**a** Find the equation connecting $x$ and $y$ in the form $y = mx$.

(**Hint**: you may find it helpful to draw the graph.)

**b** Using the equation, find the value of

**i** $y$ when $x = 5$

**ii** $x$ when $y = 23$

**5** Barrel A and Barrel B are both cylindrical, but different sizes.

Barrel A

Barrel B

Water was poured into each barrel at a constant rate. The rate was the same for both barrels.

When the water was stopped, one of the barrels was completely full. The other was not.

The graph shows the height of water in each barrel, $h$, in centimetres, after $t$ seconds.

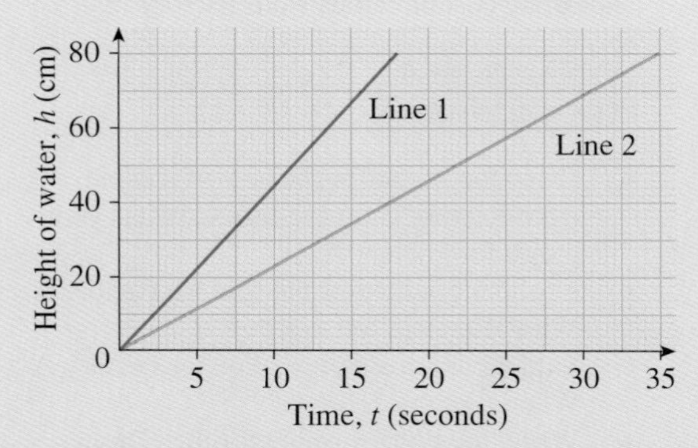


Using this information, write down as much as you can about what the graph tells you.

## 14.6 Inverse of a function

In the previous section, you used the equation $y = 30x$ to find the cost, $y$, of $x$ apples. If you want to find the number of apples, $x$, when you know the cost, $y$, you need to rearrange the equation, which gives $x = \frac{y}{30}$.

Using function notation, $f(x) = 30x$. This can also be shown as a flow chart:

$x \longrightarrow \boxed{\times 30} \longrightarrow 30x$

The inverse of $\times$ 30 is $\div$ 30. Reversing the flow chart gives:

$\frac{y}{30} \longleftarrow \boxed{\div 30} \longleftarrow y$

- The **inverse** of a function is its reverse. It 'undoes' the effect of the original function.

Using function notation, the inverse of the function $f(x)$ is written as $f^{-1}(x)$.

If $f(x) = 30x$ then $f^{-1}(x) = \frac{x}{30}$.

You can show simple linear functions as flow charts or function machines.

**EXAMPLE 10**

Draw a flow chart to show $f(x) = 3x - 2$

$x \rightarrow$ [$\times 3$] $\xrightarrow{3x}$ [$-2$] $\rightarrow 3x - 2$

You can find the inverse of a function by reversing a flow chart.

**EXAMPLE 11**

Find the inverse of $f(x) = 3(x + 2) - 1$
First show $f(x) = 3(x + 2) - 1$ as a function machine:

$x \rightarrow$ [$+2$] $\xrightarrow{x+2}$ [$\times 3$] $\xrightarrow{3(x+2)}$ [$-1$] $\rightarrow 3(x + 2) - 1$

To find the inverse, reverse the machine:

$\dfrac{(x+1)}{3} - 2 \leftarrow$ [$-2$] $\xleftarrow{\frac{x+1}{3}}$ [$\div 3$] $\xleftarrow{x+1}$ [$+1$] $\leftarrow x$

or $x \rightarrow$ [$+1$] $\xrightarrow{x+1}$ [$\div 3$] $\xrightarrow{\frac{x+1}{3}}$ [$-2$] $\rightarrow \dfrac{(x+1)}{3} - 2$

So $f^{-1}(x) = \dfrac{(x + 1)}{3} - 2$

There are occasions when the flow chart method will not work (when a letter appears more than once). You will learn more about this when you are older, but there is an alternative method that doesn't involve a flow chart. If you are confident about rearranging formulae, it is a better method to use because it will always work.

**EXAMPLE 12**

**a** Find the inverse of the function $f(x) = 2x + 5$
**b** Find $f(3)$
**c** Find $f^{-1}(11)$

**a** Write the function as an equation: $y = 2x + 5$
Swap $x$ and $y$: $x = 2y + 5$

Rearrange to make $y$ the subject: $x - 5 = 2y$

$\dfrac{x - 5}{2} = y$

Therefore the inverse function is $f^{-1}(x) = \dfrac{x - 5}{2}$

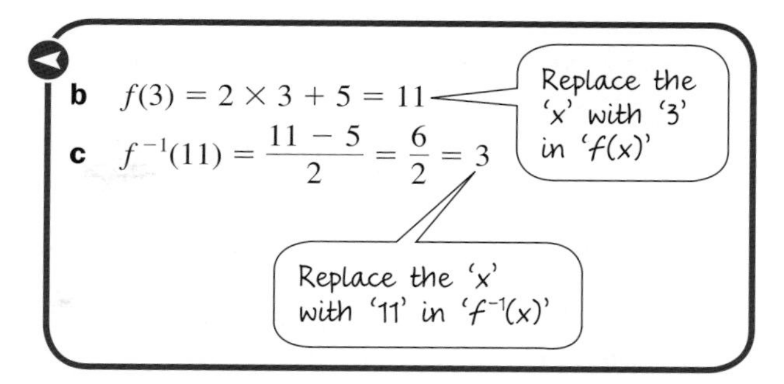


In Example 12, notice that $f(3) = 11$ and $f^{-1}(11) = 3$, demonstrating the 'undoing' nature of the inverse function.

## Exercise 14J

**1** Describe each flow chart as '$f(x) =$'.

**a** $\rightarrow$ [$\times 3$] $\rightarrow$ [$+2$] $\rightarrow$ **b** $\rightarrow$ [$\times 4$] $\rightarrow$ [$-3$] $\rightarrow$

Draw the inverse of each flow chart and write down the rule for the inverse.

**2** Describe each flow chart as '$f(x) =$'.

**a** $\rightarrow$ [$-5$] $\rightarrow$ [$\times 4$] $\rightarrow$ **b** $\rightarrow$ [$+2$] $\rightarrow$ [$\div 3$] $\rightarrow$

Draw the inverse of each flow chart and write down the rule for the inverse.

**3** Draw a flow chart to show:
**a** $f(x) = 4x + 7$ **b** $f(x) = 3x - 5$
**c** $f(x) = 5(x + 2)$ **d** $f(x) = 7(x - 5)$
Use your flow charts to help you write down the inverse of each rule.

**4** Using the method shown in Example 12, find the inverse function, $f^{-1}(x)$, for each of these functions.
**a** $f(x) = 6x - 1$ **b** $f(x) = 4x + 2$
**c** $f(x) = 2x$ **d** $f(x) = 3x - 7$

**5** Using a method of your choice, find the inverse function, $f^{-1}(x)$, for each of these functions.
**a** $f(x) = \dfrac{x}{3}$ **b** $f(x) = \frac{1}{2}x - 1$
**c** $f(x) = 2 - 3x$ **d** $f(x) = 4(3x + 6)$
**e** $f(x) = 3(2x - 7) + 1$
**f** $f(x) = \dfrac{x + 2}{5}$

**6** Find the inverse of each of these functions.
(**Note**: any letter can be used to denote a function, not just $f$.)
**a** $g(x) = 2x + 3$
**b** $h(x) = 4(x - 5)$
**c** $k(x) = 4(x + 3) - 2$
**d** $l(x) = 3(2x + 1)$
**e** $m(x) = \dfrac{(x + 1)}{3} - 2$

**7** Use the functions given in Question **6** and their inverses to write down:

**a** g(5) **b** h(7) **c** k(3)
**d** l(4) **e** $g^{-1}(13)$ **f** $h^{-1}(8)$
**g** $k^{-1}(22)$ **h** $l^{-1}(27)$

**8** $f(x) = 4x + 1$. Find:

**a** $f^{-1}(x)$ **b** $f(2)$
**c** $f^{-1}(9)$ **d** $f^{-1}(2)$

**9** $g(x) = 2(x + 5)$. Find:

**a** $g^{-1}(x)$ **b** $g(1)$
**c** $g^{-1}(12)$ **d** $g^{-1}(1)$

**10** $h(x) = (x - 3) \div 2$. Find:

**a** $h^{-1}(x)$ **b** h(3)
**c** $h^{-1}(0)$ **d** $h^{-1}(3)$

**11** $f(x) = 3(2x + 1) + 4$. Find:

**a** $f^{-1}(x)$ **b** $f(3)$
**c** $f^{-1}(25)$ **d** $f^{-1}(4)$

**12 a** Copy and complete the table.

| Function | $f(x) = 2x$ | $f(x) = x - 2$ | $f(x) = 3x + 2$ | $f(x) = 1 - x$ |
|---|---|---|---|---|
| Inverse function | $f^{-1}(x) =$ | $f^{-1}(x) =$ | $f^{-1}(x) =$ | $f^{-1}(x) =$ |

**b** Plot each function and its inverse on the same set of axes

**c** What do you notice?

**13** Write down as many functions as you can, in two minutes, with an inverse function which is the same as the function itself.

## 14.7 Sequences

### Term-to-term rules

A **sequence** is an ordered set of numbers following a rule or pattern. Each number in a sequence is called a **term**.

There are many different types of sequences. Here are some examples.

6, 10, 14, 18, 22, 26, …

A **term-to-term** rule tells you how to get from the current term to the next term in a sequence. The term-to-term rule here is *add 4*.

This is an example of an **arithmetic sequence**. A sequence is arithmetic when the term-to-term rule is adding the same number each time. For this sequence, the **common difference** – the difference between each pair of consecutive terms – is 4. All arithmetic sequences have a common difference, which can be either positive or negative.

For example, 25, 20, 15, 10, 5, 0, $^{-}5$, … is also an arithmetic sequence, this time with a common difference of $^{-}5$.

A **geometric sequence** is a different type of sequence, in which the term-to-term rule is to multiply by the same number each time. The number that you multiply by each time is called the **common ratio**.

For example, 2, 4, 8, 16, 32, … is a geometric sequence with a common ratio of 2. The term-to-term rule is *multiply by 2*.

Note that geometric sequences can get smaller. The geometric sequence 1000, 100, 10, 1, 0.1, … has the common ratio $\frac{1}{10}$ and the term-to-term rule *divide by 10*.

There are lots of other sequences. Here are three more examples:

- 0, 1, 1, 2, 3, 5, 8, 13, … is the Fibonacci sequence, with the term-to-term rule *add the previous two terms*. The sequence is named after a famous thirteenth-century mathematician.
- 1, 3, 6, 10, 15, … are the triangle numbers, with the term-to-term rule *add one more than you added the previous time*.
- 1, 4, 9, 16, 25, 36, … are the square numbers, with the term-to-term rule *add two more than you added the previous time*.

### Position-to-term rules

A **position-to-term** rule describes how to calculate a term in a sequence from its position in the sequence. This is often more useful than the term-to-term rule, particularly when you want to find a term which is a long way into the sequence (e.g. the $100^{th}$ term) without having to work out all the terms that come before it.

**EXAMPLE 13**

Find the first 5 terms of the sequence for which the position-to-term rule is *multiply by 2 then add 3*.

| Position | 1 | 2 | 3 | 4 | 5 |
|---|---|---|---|---|---|
| Term | $1 \times 2 + 3 = 5$ | $2 \times 2 + 3 = 7$ | $3 \times 2 + 3 = 9$ | $4 \times 2 + 3 = 11$ | $5 \times 2 + 3 = 13$ |

$\times 2 + 3$

## Exercise 14K

**1** Write down whether each of these sequences is **i** arithmetic **ii** geometric **iii** another type of sequence.

- **a** 7, 10, 13, 16, 19, …
- **b** 16, 9, 4, 1, 0, …
- **c** 1, 3, 9, 27, 81, …
- **d** 8, 4, 0, $^{-}4$, $^{-}8$, …
- **e** 10, 20, 40, 80, 160, …
- **f** 7, 12, 17, 22, 27, …
- **g** 64, 32, 16, 8, 4, …
- **h** 1, 1.5, 2, 2.5, 3, …
- **i** 4, 2, 1, $\frac{1}{2}$, $\frac{1}{4}$, …

**2** For each of the sequences in Question **1**, write down the next two terms.

**3** For each of the sequences in Question **1**, write down the term-to-term rule.

**4** For each of the sequences in Question **1**, write down the tenth term.

**5** Write down the first five terms of the following sequences:

- **a** The first term is 40, the term-to-term rule is *subtract 10.*
- **b** The first term is 2, the term-to-term rule is *multiply by 3.*
- **c** The **fourth** term is 7, the term-to-term rule is *add 6.*
- **d** The **third** term is 20, the term-to-term rule is *add 12.*
- **e** The **second** term is 4, the term-to-term rule is *multiply by 2.*
- **f** The **third** term is 4, the term-to-term rule is *divide by 2.*
- **g** The first term is 3, the term-to-term rule is *multiply by 3 then subtract 1.*

**6** What are the missing numbers, □, in these sequences?

- **a** 21, 27, 33, □, 45, 51, □, 63, …
- **b** □, □, 9, 14, 19, □, 29, …
- **c** $^{-}11$, □, $^{-}5$, □, □, 4, 7, 10, …
- **d** 7, □, □, $^{-}5$, $^{-}9$, $^{-}13$, □, …

**7** Write down the first five terms of these sequences.

- **a** The position-to-term rule is *subtract 3.*
- **b** The position-to-term rule is *add 4.*
- **c** The position-to-term rule is *multiply by 3 then add 1.*
- **d** The position-to-term rule is *multiply by 10 then subtract 3.*

## The *n*th term

When we use algebra to describe terms using the position-to-term rule, this is known as finding the ***n*th term.**

In Book 2 you learned various methods for finding the *n*th term. Here are two methods you used.

To find the *n*th term in the sequence
12, 16, 20, 24, 28, …
look at the differences:

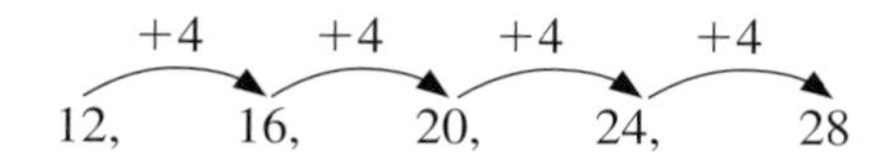


The common difference is + 4, so compare the sequence to the sequence 4*n*:

| Sequence 4*n* | 4, | 8, | 12, | 16, | 20 |
|---|---|---|---|---|---|
| Sequence | 12, | 16, | 20, | 24, | 28 |

+ 8

Each number in the sequence is 8 more than the sequence $4n$, so the sequence is $4n + 8$.

You also learned the 'position zero' method.

To find the *n*th term in the sequence 1, 4, 7, 10, 13, …
look at the differences:

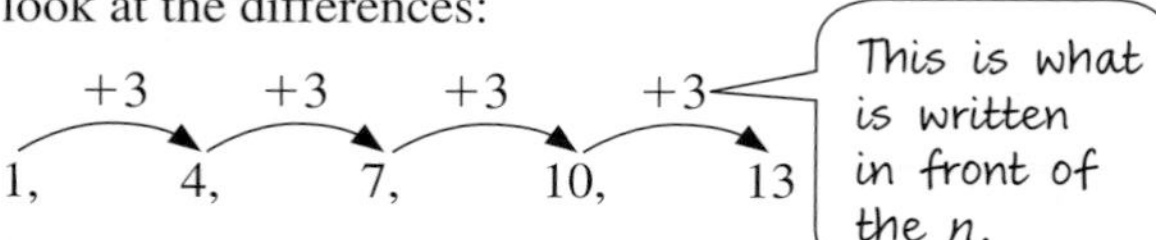


Notice you add 3 each time.

| Position | Term |
|---|---|
| 0 | $^{-}2$ |
| 1 | 1 |
| 2 | 4 |
| 3 | 7 |
| 4 | 10 |
| 5 | 13 |

− 3

+ 3

*n*th terms is $3n - 2$

Work backwards by subtracting 3 from the first term to get to the term in position zero. This is written after the 3*n*.

The 'position zero' method can be used to derive an expression to describe the $n$th term of a general arithmetic sequence. Consider an arithmetic sequence with first term $a$ and common difference $d$.

| Position | Term |
|---|---|
| 0 | $a - d$ |
| 1 | $a$ |
| 2 | $a + d$ |
| 3 | $a + 2d$ |
| 4 | $a + 3d$ |
| 5 | $a + 4d$ |

$-d$, $+d$

$n$th terms is $dn + a - d$

You can see from the table that the number in front of the $d$ is always one less than the position number. The 20th term (with position number 20) will be $a + 19d$. The 100th term will be $a + 99d$. So the $n$th term will be $a + (n - 1)d$.

Notice that $dn + a - d$, from the table, can be written as $a + (n - 1)d$ by putting the terms in $d$ together and taking $d$ out as a common factor.

- The $n$th term of an arithmetic sequence is $a + (n - 1)d$ where $a$ is the first term and $d$ is the common difference.

### EXAMPLE 14

**a** Find the $n$th term of the arithmetic sequence 4, 11, 18, 25, 32, … .

**b** Find the 40th term.

**c** Which term in the sequence is 158?

---

**a** Using $a + (n - 1)d$, the first term, $a = 4$ and the common difference, $d = 7$,

the $n$th term is: $4 + 7(n - 1)$

$= 4 + 7n - 7$ Multiply out brackets

$= 7n - 3$ Simplify

**b** The $n$th term is $7n - 3$, so the 40th term is $7(40) - 3 = 280 - 3 = 277$

**c** $7n - 3 = 158$

+ 3] $7n = 161$

÷ 7] $n = 23$

So 158 is the 23rd term in the sequence.

### EXAMPLE 15

Find the $n$th term of the arithmetic sequence 17, 15, 13, 11, 9, … .

---

Using $a + (n - 1)d$, the first term, $a = 17$ and the common difference, $d = {}^{-}2$,

the $n$th term is: $17 - 2(n - 1)$

$= 17 - 2n + 2$ Multiply out brackets

$= 19 - 2n$ Simplify



## Exercise 14L

**1** Using the $n$th term formula $a + (n - 1)d$, find the $n$th term for these arithmetic sequences.

**a** 4, 7, 10, 13, 16, …
**b** 3, 8, 13, 18, 23, …
**c** 2, 8, 14, 20, 26, …
**d** 20, 15, 10, 5, 0, …
**e** 7, 7.5, 8, 8.5, 9, …
**f** $4\frac{1}{4}, 4\frac{1}{2}, 4\frac{3}{4}, 5, 5\frac{1}{4}, \ldots$

**2** For each of the sequences in Question **1**, write down

**i** the 10th term
**ii** the 20th term.

**3 a** Find the $n$th term of the arithmetic sequence 8, 11, 14, 17, 20, … .
**b** Find the 50th term
**c** Which term in the sequence is 98?

**4 a** Find the $n$th term of the arithmetic sequence 1, 11, 21, 31, 41, … .
**b** Find the 100th term
**c** Is 205 in the sequence? How do you know?
**d** Which term in the sequence is 161?

**5** The 6th term of an arithmetic sequence is 43. The first term is 3.

**a** Find the first 5 terms of the sequence.
**b** Find the $n$th term.

**6** The 5th term of an arithmetic sequence is 23. The 8th term is 35.

**a** Find the first 5 terms of the sequence.
**b** Find the $n$th term.

# Consolidation

### Example 1

Plot the graph $y + 2x = 6$ for $x = {}^{-}1$ to $x = 4$.

Rearrange to make $y$ the subject:
$y = 6 - 2x$
The table of values is:

| $x$ | $^{-}1$ | 0 | 1 | 2 | 3 | 4 |
|---|---|---|---|---|---|---|
| 6 | 6 | 6 | 6 | 6 | 6 | 6 |
| $^{-}2x$ | 2 | 0 | $^{-}2$ | $^{-}4$ | $^{-}6$ | $^{-}8$ |
| $y$ | 8 | 6 | 4 | 2 | 0 | $^{-}2$ |

The graph is:

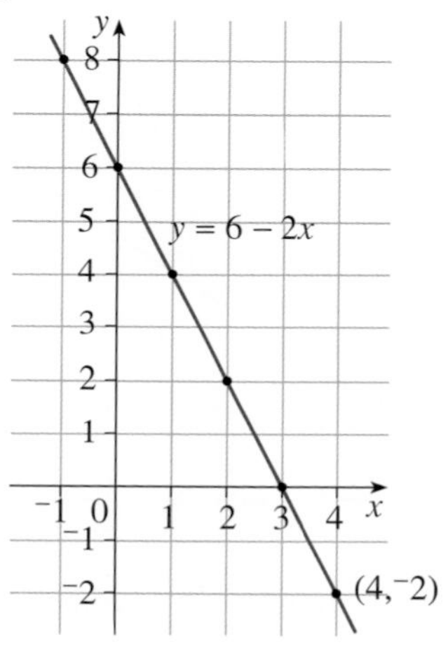


The gradient of this line is $^{-}2$, the coefficient of $x$.
The $y$-intercept is 6.

### Example 2

Find the gradient of the line joining the points A (2, 5) to B (6, 9).

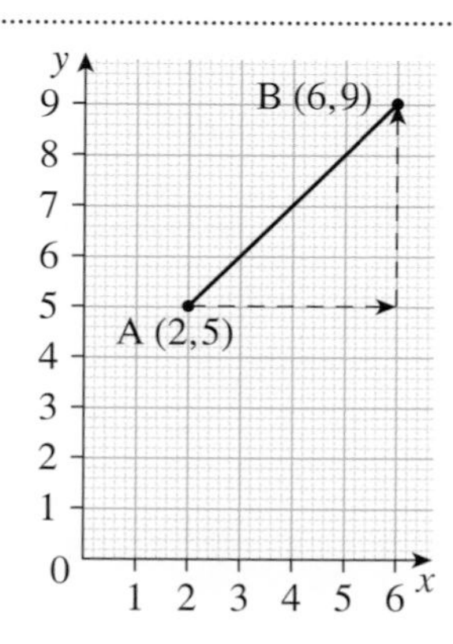


$$\text{Gradient} = \frac{\text{vertical rise}}{\text{horizontal shift}} = \frac{9-5}{6-2} = \frac{4}{4} = 1$$

### Example 3

Draw a graph to solve the simultaneous equations
$y - 3x = {}^{-}2 \quad y - 2x = 2$

Rearranging:
$y - 3x = {}^{-}2 \qquad y - 2x = 2$
$y = 3x - 2 \qquad y = 2x + 2$

The table of values for $y = 3x - 2$ (for $x = {}^{-}2$ to $x = 5$) is:

| $x$ | $^{-}2$ | $^{-}1$ | 0 | 1 | 2 | 3 | 4 | 5 |
|---|---|---|---|---|---|---|---|---|
| $y$ | $^{-}8$ | $^{-}5$ | $^{-}2$ | 1 | 4 | 7 | 10 | 13 |

and for $y = 2x + 2$ (for $x = {}^{-}2$ to $x = 5$) is:

| $x$ | $^{-}2$ | $^{-}1$ | 0 | 1 | 2 | 3 | 4 | 5 |
|---|---|---|---|---|---|---|---|---|
| $y$ | $^{-}2$ | 0 | 2 | 4 | 6 | 8 | 10 | 12 |

Plotting the graphs:

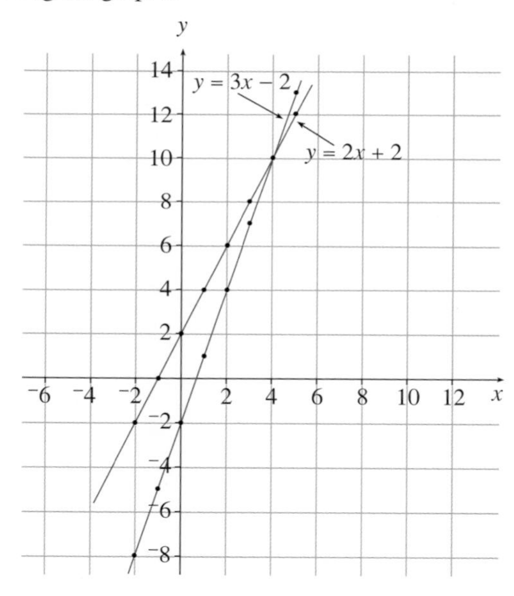


The two graphs intersect at (4, 10), so the solution to the two equations is $x = 4$, $y = 10$.

### Example 4

Find the inverse of the function $f(x) = 3x - 7$

Using a flow chart:

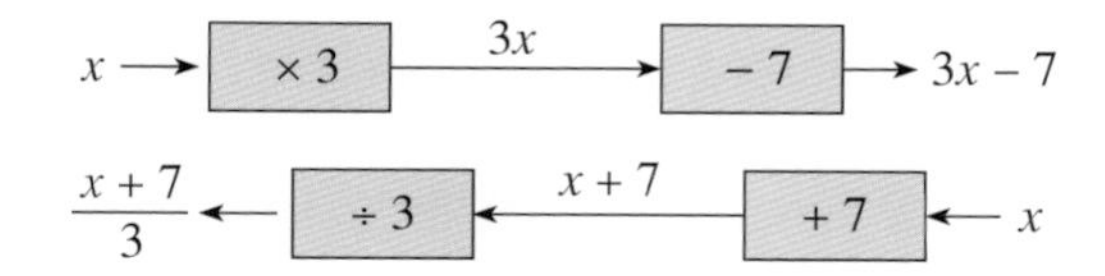

So $f^{-1}(x) = \dfrac{x + 7}{3}$

or

As an equation: $y = 3x - 7$

Swapping $x$ and $y$: $x = 3y - 7$

Rearranging: $x + 7 = 3y$

$$\frac{x + 7}{3} = y$$

So $f^{-1}(x) = \dfrac{x + 7}{3}$

**Example 5**

Mr Fox is an electrician. The table shows his fees for three jobs.

| Number of hours, $x$ | 1 | 2 | 4 |
|---|---|---|---|
| Fee in dollars, $y$ | 50 | 70 | 110 |

**a** Draw a graph to show this information with number of hours on the $x$-axis and fee in dollars on the $y$-axis.

**b** Find the gradient of the graph. What does this tell you?

**c** Find the $y$-intercept of the graph. What does this tell you?

**d** Are number of hours and fee in dollars directly proportional?

**a**

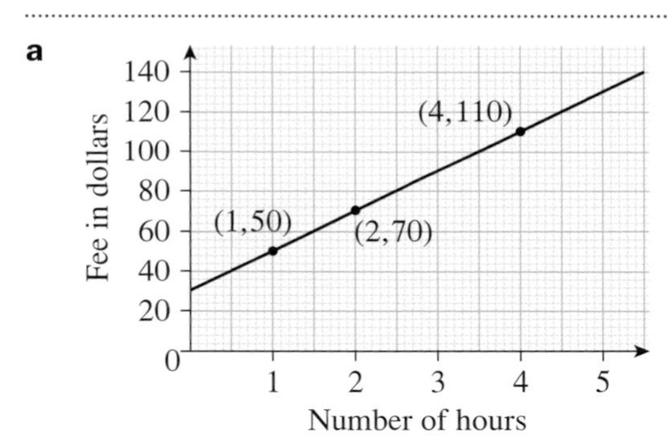


**b** Gradient $= \dfrac{110 - 70}{4 - 2} = \dfrac{40}{2} = 20$. This is the hourly rate.

**c** Intercept $= 30$. This is the call-out fee.

**d** Number of hours and fee in dollars are not directly proportional, as the graph does not pass through the origin and the two values do not increase at the same rate (that is, if you double the number of hours you do not double the fee).

**Example 6**

$x$ and $y$ are directly proportional.

| $x$ | 3 | 5 | 8 |
|---|---|---|---|
| $y$ | 4.5 | 7.5 | 12 |

**a** Find the equation relating $x$ and $y$ in the form $y = mx$

**b** Check your equation is correct by drawing the graph and finding the gradient and $y$-intercept.

**c** Find the value of

**i** $y$ when $x = 7$

**ii** $x$ when $y = 19.5$

**a** The multiplier going down the table is $4.5 \div 3 = 1.5$, so $y = 1.5x$

**b**

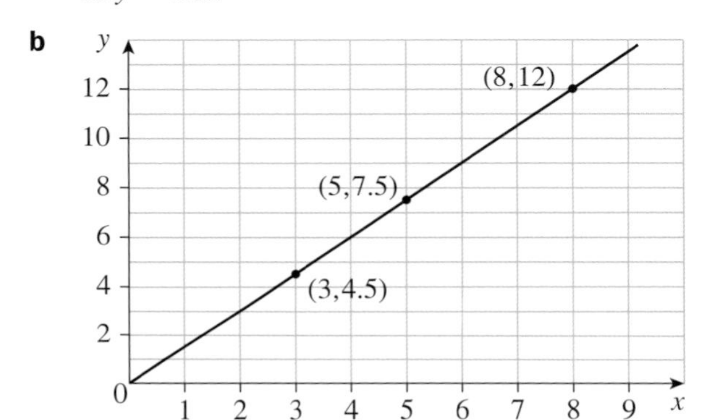


The graph is a straight line passing through the origin. This shows direct proportion.

The gradient is $\dfrac{7.5 - 4.5}{5 - 3} = \dfrac{3}{2} = 1.5$

The $y$-intercept is 0.

**c** **i** $y = 1.5x$ so $y = 1.5 \times 7 = 10.5$

**ii** $19.5 = 1.5x$ so $x = 19.5 \div 1.5 = 13$

**Example 7**

Find the first five terms of these sequences.

**a** The term-to-term rule is *divide by 4*, the first term is 512.

**b** The position-to-term rule is *multiply by 3 then subtract 1.*

**a**

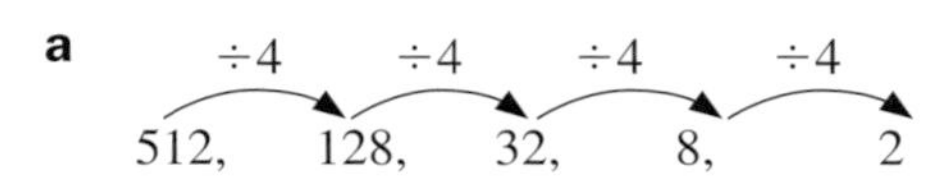


**b**

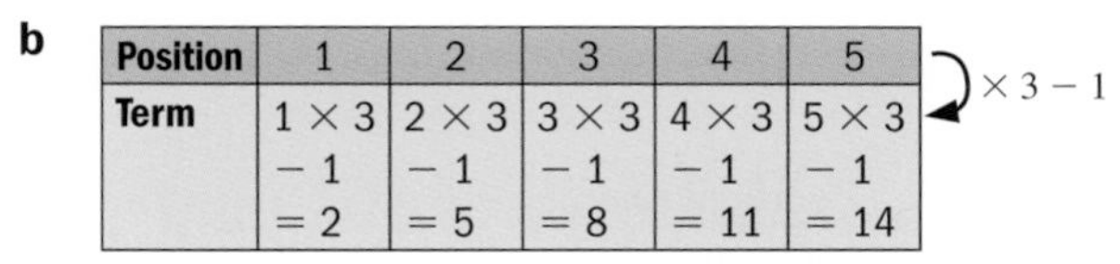

| Position | 1 | 2 | 3 | 4 | 5 |
|---|---|---|---|---|---|
| Term | $1 \times 3 - 1 = 2$ | $2 \times 3 - 1 = 5$ | $3 \times 3 - 1 = 8$ | $4 \times 3 - 1 = 11$ | $5 \times 3 - 1 = 14$ |

$\times 3 - 1$

## Exercise 14

**(Note: Question 9 is extension work.)**

**1** **a** Complete suitable tables of values for $x = {}^{-}2$ to $x = 3$ for the equations:

**i** $y + 2 = 3x$ **ii** $y = \frac{1}{2}x + 4$

**iii** $y + 3x = 4$

**b** Draw a coordinate graph for each equation.

**c** Write down the gradient and $y$-intercept for each graph.

**2** Plot graphs for each pair of equations, and find their solution.

**a** $y = 3x + 2$, $y = 6x - 1$ **b** $y = 3x - 2$, $y = 2x - 1$

**c** $y = x - 1$, $y = 3x - 5$ **d** $y = 3x - 4$, $y = 4x - 5$

**e** $y = 2x + 3$, $y = 3x + 5$ **f** $x + y = 4$, $2x - y = 2$

**g** $x - y = 3$, $3x - 2y = 10$ **h** $2x + 3y = 7$, $4x - 5y = 3$

**3** Write down the inverse functions $f^{-1}(x)$.

**a** $f(x) = x + 5$ **b** $f(x) = 3x$

**c** $f(x) = 6 - x$ **d** $f(x) = 3x - 5$

**e** $f(x) = \frac{x}{4} + 6$

**4** **a** Plot the graphs of

**i** $y = 2x + 1$

**ii** $y = 5 - 3x$

**iii** $2y = 3x - 6$

**iv** $4x + 2y = 5$

**b** Write down the gradient of each line.

**5** Given $f(x) = 2x$ and $g(x) = x + 2$, calculate:

**a** $g(2)$ **b** $g^{-1}(4)$

**c** $f^{-1}(4)$ **d** $f(2)$

**6** Write down the first five terms of these sequences.

**a** The first term is 10, the term-to-term rule is *subtract 2*.

**b** The position-to-term rule is *add 7*.

**c** The first term is 5, the term-to-term rule is *multiply by 2*.

**d** The position-to-term rule is *multiply by 4 then add 3*.

**e** The **third** term is 8, the term-to-term rule is *add 3*.

**f** The position-to-term rule is *multiply by 5 then subtract 3*.

**g** The **second** term is 8, the term-to-term rule is *multiply by 2*.

**7** $x$ and $y$ are directly proportional.

| x | 4 | 7 | 12 |
|---|---|---|---|
| y | 14 | 24.5 | 42 |

**a** Find the equation relating $x$ and $y$ in the form $y = mx$.

**b** Check your equation is correct by drawing the graph and finding the gradient and $y$-intercept.

**8** If $f(x) = 3x + 2$ and $g(x) = \frac{1}{2}x + 1$, calculate:

**a** $f(2)$ **b** $g(8)$

**c** $f^{-1}(8)$ **d** $g^{-1}(5)$

**9** Find the gradient of the line joining the points:

**a** $(3,5)$ and $(6,8)$

**b** $(1,6)$ and $(3,9)$

**c** $({}^{-}1,5)$ and $(4,9)$

**d** $({}^{-}2,{}^{-}5)$ and $(3,{}^{-}8)$

**e** $({}^{-}2,7)$ and $(2,5)$

**10** Find the inverses of these functions.

**a** $f(x) = x + 3$ **b** $f(x) = \frac{x}{3}$

**c** $f(x) = 2x + 1$ **d** $f(x) = 1 - x$

**e** $f(x) = 3x - 2$ **f** $f(x) = \frac{x + 3}{4}$

**11** Savings Cellular has a fixed charge on all mobile phone calls made. Otherwise charges depend on the duration, in seconds, of the call. A 2-minute call costs \$1.30 while a 4-minute call costs \$2.50.

**a** Show these two costs on a cost–time graph.

**b** Join these two points with a straight line. Where does the line intersect the $y$-axis?

**c** Find the gradient of the line.

**d** If $C$ is call cost in cents and $t$ is time in seconds, find the equation of the line.

**e** What does the intercept tell you? The gradient?

**12** Find the $n$th term of these arithmetic sequences.

**a** 1, 4, 7, 10, 13, …

**b** 15, 12, 9, 6, 3, …

**c** 14, 19, 24, 29, 34, …

**d** 3, 5.5, 8, 10.5, 13, …

**e** 1, 9, 17, 25, 33, …

**f** $15\frac{1}{4}$, 16, $16\frac{3}{4}$, $17\frac{1}{2}$, $18\frac{1}{4}$, …

**13 a** A gallon (about $4\frac{1}{2}$ litres) of petrol costs \$6.50. Copy and complete the table.

| Number of gallons ($g$) | 0 | 1 | 2 | 3 | 4 | 5 | 6 |
|---|---|---|---|---|---|---|---|
| Cost ($c$) \$ | 0 | 6.50 | 13.00 | | | | |

**b** Using a suitable scale, plot a graph to show this information.

**c** From your graph find

**i** the price of $4\frac{1}{2}$ gallons of petrol

**ii** how much petrol can be bought for \$20.

**d** Write down the equation of the graph in terms of $g$ and $c$.

**e** Are number of gallons and cost directly proportional? How do you know?

**14** Find the gradient and equation of the line joining the points

**a** $(0,0)$ to $(2,2)$ **b** $(0,1)$ to $(1,2)$

**c** $(^-2,0)$ to $(0,^-2)$

**15** Fahrenheit temperatures, $t$, can be converted to Celsius temperatures using the relationship

$$f(t) = \frac{5}{9}(t - 32)$$

**a** What is the temperature in degrees Celsius if the Fahrenheit temperature is 86°?

**b** Find $f^{-1}(t)$.

**c** Use your result to find the temperature in degrees Fahrenheit if the Celsius temperature is 25°.

# Summary

## You should know ...

**1** How to determine whether a set of points represent a linear relation by plotting a graph.

*For example:*

$(^-2,^-1), (^-1,2), (0,5), (1,8), (2,11)$

represent the linear relation $y = 3x + 5$

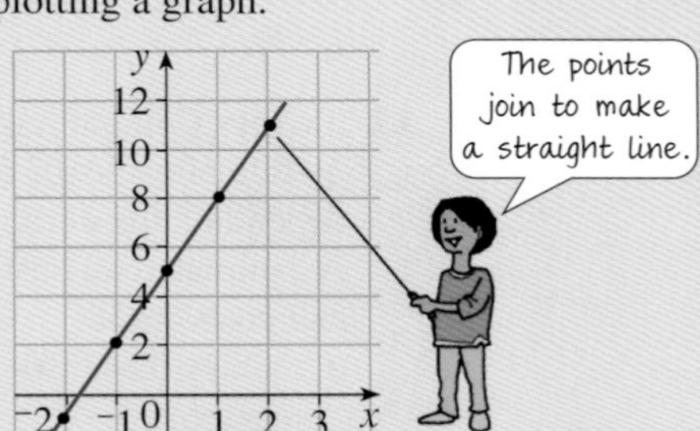


**2** How to construct tables of values and plot the graphs of linear functions where $y$ is given implicitly in terms of $x$.

*For example:*

$2y - 4x = 3$

$+ 4x]$ $2y = 4x + 3$

$\div 2]$ $y = 2x + 1.5$

| x | 0 | 1 | 2 | 3 |
|---|---|---|---|---|
| y | 1.5 | 3.5 | 5.5 | 7.5 |

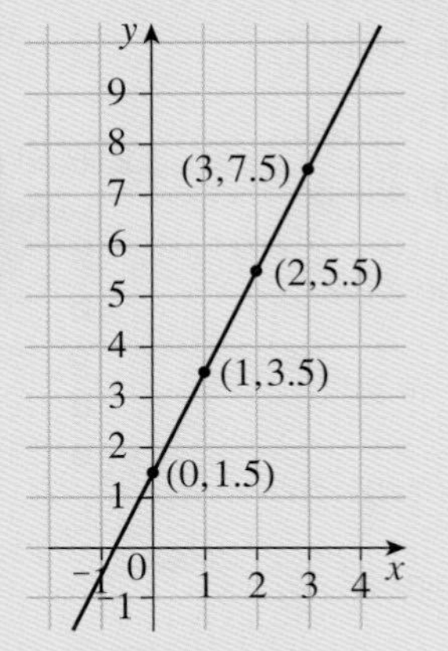


## Check out

**1** Determine which of the following sets of points represent a linear relation.

**a** $(^-3,^-5), (^-1,^-1), (1,3), (3,7)$

**b** $(^-2,^-2), (^-1,0), (0,1), (1,3)$

**2** Draw the graphs of these equations.

**a** $2y - 8x = 2$

**b** $2x - y = 3$

**c** $3y - 3x = 15$

**3** How to find the gradient of a line.
*For example:*
The gradient of the line joining
(4, 1) and (6, 5) is $\frac{5-1}{6-4} = \frac{4}{2} = 2$

**3** Find the gradient of the line joining:
**a** (4, 3) and (7, 7)
**b** (2, 4) and (6, ⁻2)
**c** (⁻4, 4) and (⁻6, ⁻2)

**4** How to find the equation of a line.
*For example:*

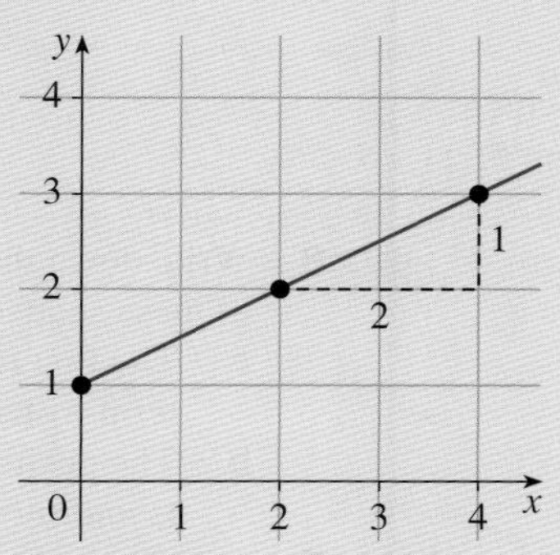


This line passes through (2, 2) and (4, 3).

Gradient of line $= \frac{3-2}{4-2} = \frac{1}{2}$

The equation of the line has the form $y = \frac{1}{2}x + c$
and it passes through (0, 1), so $c = 1$
so the line has equation $y = \frac{1}{2}x + 1$.

**4**

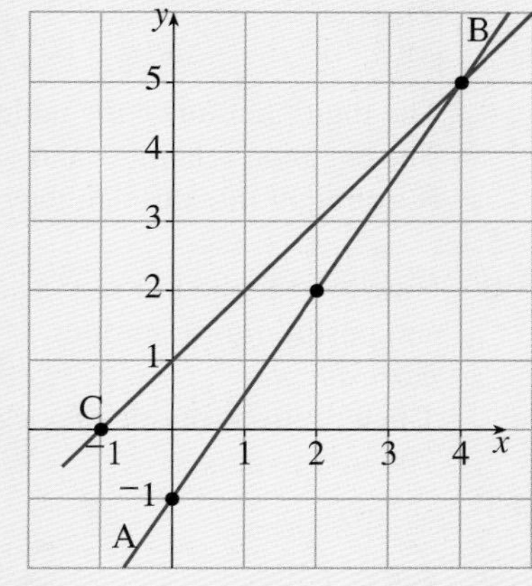


**a** Determine the equation of the line AB.
**b** CB

**5** How to solve simultaneous linear equations graphically
- Draw both equations on the same graph.
- The solution is the point of intersection of the two graphs.

**5** Draw graphs to solve
**a** $x + y = 6$
$x - 2y = 3$
**b** $2x - y = 3$
$3x + 2y = 22$

**6** The inverse of the function $f$ is written $f^{-1}$. You can find the inverse of a function using a flow chart.
*For example:*
If $f(x) = 3x + 4$

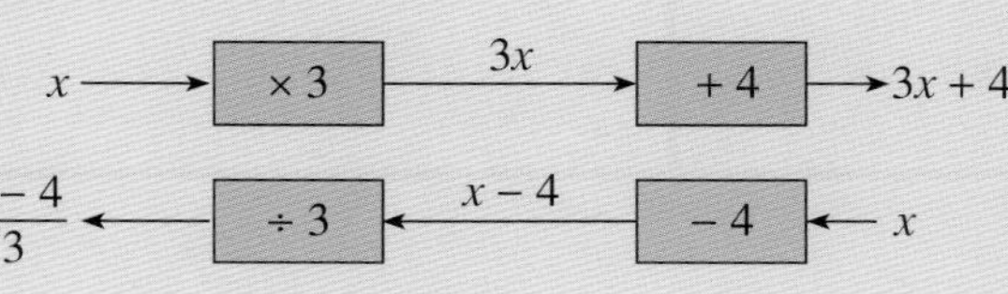


so $f^{-1}(x) = \frac{x-4}{3}$
Alternatively you can write the function as:
$y = 3x + 4$
Swap $x$ and $y$: $x = 3y + 4$
And rearrange it to make the subject:
− 4] $x - 4 = 3y$
÷ 3] $y = \frac{x-4}{3}$
so $f^{-1}(x) = \frac{x-4}{3}$

**6** Find the inverse of:
**a** $f(x) = x + 2$
**b** $f(x) = 3x$
**c** $f(x) = 4x - 5$
**d** $f(x) = 5(2x + 3)$

**7** How to use algebraic methods with quantities which are in direct proportion, and know the relationship with their graph.
*For example:*
$x$ and $y$ are directly proportional

| $x$ | 7 | 11 | 16 |
|---|---|---|---|
| $y$ | 35 | 55 | 80 |

× 5

The multiplier down the table is $35 \div 7 = 5$
The equation relating $x$ and $y$ is $y = 5x$, the graph of which is a straight line through the origin, (0, 0), with gradient 5.

**8** How to generate terms of a sequence using term-to-term and position-to-term rules.
*For example:*
The term-to-term rule of a sequence is *add 7*.
The first term is 3:

+7 +7 +7 +7
3, 10, 17, 24, 31

The position-to-term rule of a sequence is *multiply by 5 then subtract 2*:

| Position | 1 | 2 | 3 | 4 | 5 |
|---|---|---|---|---|---|
| Term | 3 | 8 | 13 | 18 | 23 |

**9** How to find the $n$th term of an arithmetic sequence using the formula
$n$th term $= a + (n - 1)d$
where $a$ is the first term and $d$ is the common difference.
*For example:*
For the arithmetic sequence 18, 16, 14, 12, 10, …
Using $a + (n - 1)d$, the first term, $a = 18$ and the common difference, $d = {}^{-}2$,
the $n$th term is: $18 - 2(n - 1)$
$18 - 2n + 2$ Multiply out brackets
$20 - 2n$ Simplify
The $n$th term is $20 - 2n$

**7** $x$ and $y$ are in direct proportion. Find the equation relating $x$ and $y$ and describe the graph of this equation.

| $x$ | 4 | 9 | 14 |
|---|---|---|---|
| $y$ | 18 | 40.5 | 63 |

**8** Write down the first five terms of these sequences.
- **a** The first term is 6, the term-to-term rule is *add 8*.
- **b** The position-to-term rule is *multiply by 2 then add 7*.
- **c** The first term is 0.5, the term-to-term rule is *multiply by 2*.
- **d** The position-to-term rule is *multiply by 6 then subtract 5*.
- **e** The **third** term is 22, the term-to-term rule is *add 5*.

**9** Find the $n$th term of these arithmetic sequences.
- **a** 12, 17, 22, 27, 32, …
- **b** 9, 13, 17, 21, 25, …
- **c** 52, 42, 32, 22, 12, …
- **d** 15, 13, 11, 9, 7, …

# 15 Transformations

## Objectives

- Use the coordinate grid to solve problems involving translations, rotations, reflections and enlargements.
- Transform 2D shapes by combinations of rotations, reflections and translations; describe the transformation that maps an object onto its image.
- Enlarge 2D shapes, given a centre and positive integer scale factor; identify the scale factor of an enlargement as the ratio of the lengths of any two corresponding line segments.
- Recognise that translations, rotations and reflections preserve length and angle, and map objects on to congruent images, and that enlargements preserve angle but not length.
- Know what is needed to give a precise description of a reflection, rotation, translation or enlargement.

## What's the point?

Transformations help describe the precise relationship between two similar shapes. Plans and maps are examples of the result of simple transformations.

## Before you start

### You should know ...

**1** How to plot points on graphs.

### Check in

**1** On suitable axes, plot these points.

**a** $(5,7)$ **b** $(^{-}3,5)$

**c** $(^{-}4,^{-}6)$ **d** $(8,^{-}5)$

**2** How to draw simple lines given their equations.
*For example:*

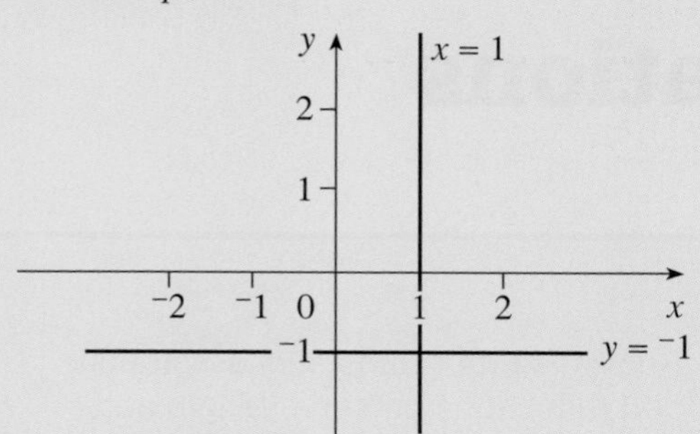


Points on the line $y = {}^{-}1$ always have a $y$-coordinate of $^{-}1$.

**2** Draw the graphs of these lines.

**a** $x = 2$ **b** $x = {}^{-}3$

**c** $y = 3$ **d** $y = {}^{-}2$

**3** How to use a pair of compasses to draw the perpendicular bisector of the line joining two points.
*For example:*

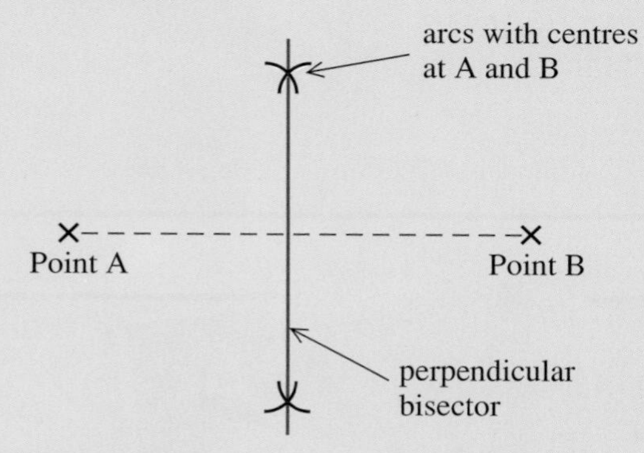


**3 a** Plot the points $(0, 1)$ and $(4, 6)$ on graph paper.

**b** Using a pair of compasses, draw the perpendicular bisector of the line joining the points in part **a**.

**4** How to solve ratio problems by writing them as fractions.
*For example:*
If $a:4 = 12:8$

then $\frac{a}{4} = \frac{12}{8}$

so $a = \frac{12}{8} \times 4 = 6$

**4** Find $x$ by converting these ratios to fractions:

**a** $x:5 = 9:15$

**b** $6:x = 3:4$

**c** $x:5 = (x + 1):10$

## 15.1 Translations

You will need graph paper.

A **translation** maps a shape to another position by a movement that is equivalent to sliding without turning.

In the diagram the quadrilateral ABCD has been translated to its **image**, A′B′C′D′.

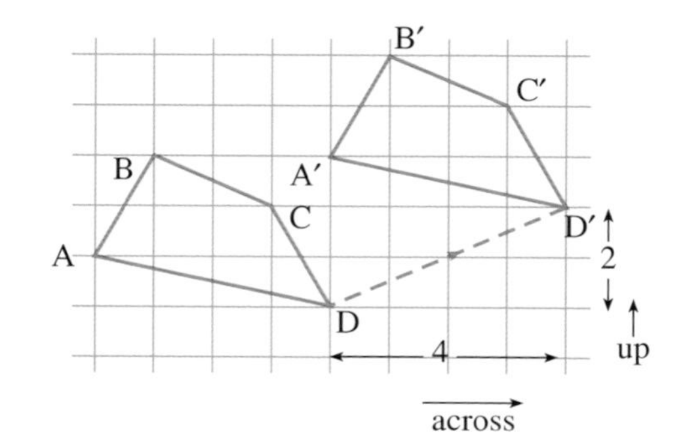

Notice that each point has been moved 4 squares across and 2 squares up.

This translation can be represented by the **vector T** $= \begin{pmatrix} 4 \\ 2 \end{pmatrix}$.

The **inverse translation** represented by

$\mathbf{T}^{-1} = \begin{pmatrix} ^-4 \\ ^-2 \end{pmatrix}$.

maps A′B′C′D′ back to ABCD.

Given the coordinates of a shape and a translation you can easily find its image.

Notice that ABCD and A′B′C′D′ are the same shape and size. All side lengths and angles in the image are the same as in the original shape. The two shapes are said to be **congruent**.

## EXAMPLE 1

Find the image A′B′C′ of triangle A(2, 1), B(5, 1), C(1, 4) under the translation $\begin{pmatrix} ^-5 \\ 3 \end{pmatrix}$.

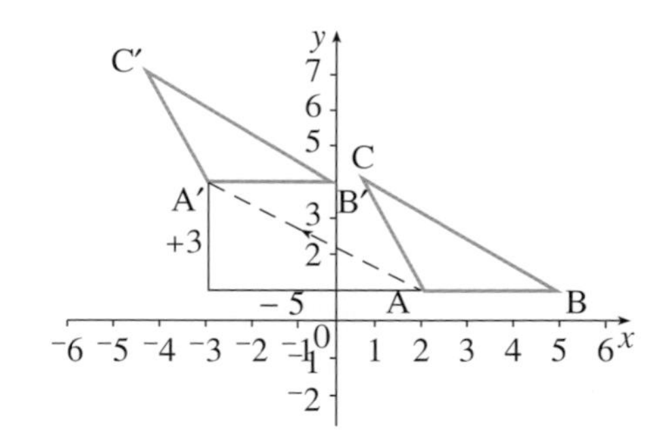


You can write:

$A\ (2,1) \xrightarrow{\begin{pmatrix} ^-5 \\ 3 \end{pmatrix}} A'(^-3,4)$

$B\ (5,1) \xrightarrow{\begin{pmatrix} ^-5 \\ 3 \end{pmatrix}} B'(0,4)$

$C\ (1,4) \xrightarrow{\begin{pmatrix} ^-5 \\ 3 \end{pmatrix}} C'(^-4,7)$

So the triangle ABC has been translated into triangle A′($^-$3,4), B′(0,4), C′($^-$4,7).

**Note:** triangle ABC and triangle A′B′C′ are the same shape and the same size.

Given the coordinates of the image and the translation vector, you can find the original shape (the **object**).

## EXAMPLE 2

The triangle ABC is mapped onto triangle A′B′C′ with vertices A′(3, 1), B′(1, $^-$2), C′(1, 2) under the translation $\mathbf{T} = \begin{pmatrix} 2 \\ 3 \end{pmatrix}$.

Find the vertices of triangle ABC.

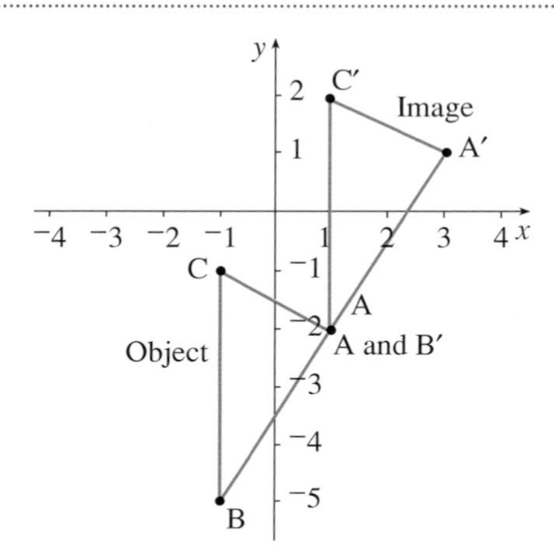


The translation, $\mathbf{T} = \begin{pmatrix} 2 \\ 3 \end{pmatrix}$ maps ABC → A′B′C′.

So the inverse translation, $\mathbf{T}^{-1} = \begin{pmatrix} ^-2 \\ ^-3 \end{pmatrix}$ will map A′B′C′ → ABC.

Thus $A'\ (3,1) \xrightarrow{\begin{pmatrix} ^-2 \\ ^-3 \end{pmatrix}} A\ (1,{}^-2)$

$B'\ (1,{}^-2) \xrightarrow{\begin{pmatrix} ^-2 \\ ^-3 \end{pmatrix}} B\ (^-1,{}^-5)$

$C'\ (1,2) \xrightarrow{\begin{pmatrix} ^-2 \\ ^-3 \end{pmatrix}} C\ (^-1,{}^-1)$

So triangle ABC has vertices A (1, $^-$2), B ($^-$1, $^-$5), C ($^-$1, $^-$1).

## Exercise 15A

**1** The vertices of triangle PQR are ($^-$4, 2), ($^-$2, 3) and ($^-$3, 5) respectively. Determine the vertices of the image triangle P′Q′R′ under the translation $\mathbf{T} = \begin{pmatrix} 4 \\ ^-3 \end{pmatrix}$.

**2** The vertices of a parallelogram ABCD are (1, 1), (4, 1), (5, 3) and (2, 3) respectively. Find the vertices of the image of the parallelogram under a translation $\mathbf{T} = \begin{pmatrix} ^-3 \\ ^-4 \end{pmatrix}$.

**3** The diagram shows a triangle ABC.

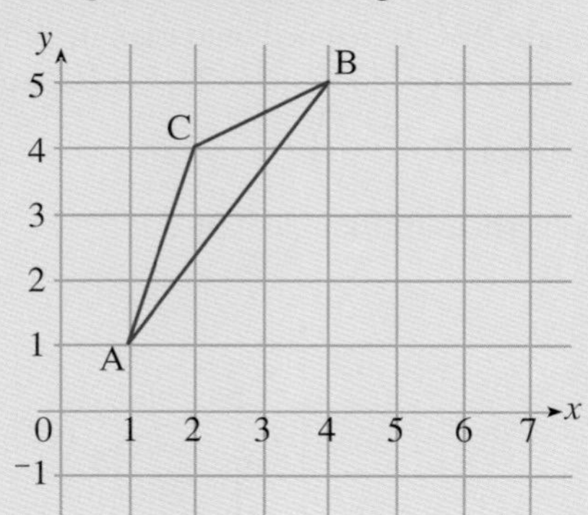


Draw a copy of the diagram.

On your copy draw also the triangles with vertices as listed in parts **a**–**e**.

In each case, if the triangle is a translation of ABC, write down the column vector of the translation.

**a** $(4, {}^{-}1), (7, 3), (5, 2)$
**b** $(2, 1), (4, 2), (5, 5)$
**c** $(0, 0), (0, 3), (2, 4)$
**d** $(0, 1), (1, 4), (3, 5)$
**e** $(2, 0), (3, 3), (5, 4)$

**4** **a** Are all the triangles in Question **3** the same shape and size as triangle ABC?
**b** If two triangles are the same shape and size as each other, must one be a translation of the other?

**5** The vector $\begin{pmatrix}4\\3\end{pmatrix}$ maps the triangle ABC to its image PQR. Point A maps to P, as shown in the diagram below.

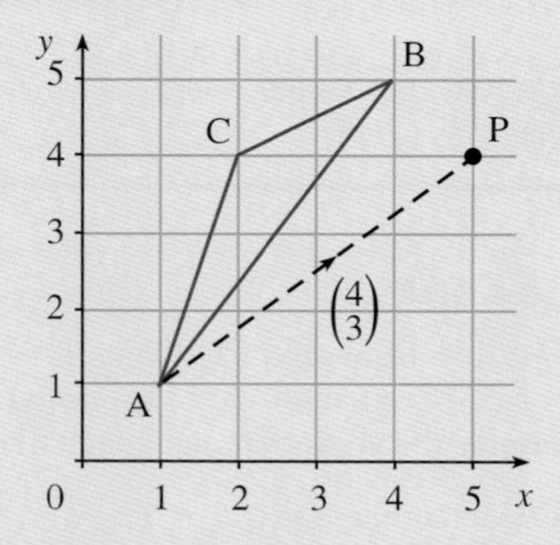


**a** What are the coordinates of A?
**b** What are the coordinates of P?
**c** Given the translation vector and the coordinates of A, explain how to find the coordinates of P without a diagram.

**6** **a** For Question **5**, copy and complete:

$$A \xrightarrow{\begin{pmatrix}4\\3\end{pmatrix}} P$$
$$(1, 1) \longrightarrow (\quad)$$

**b** Show the mappings $B \to Q$ and $C \to R$ in a similar way.
**c** On squared paper, draw a pair of axes marked from ${}^{-}1$ to 10. Mark on triangles ABC and PQR.
**d** Does a translation map all points the same distance and in the same direction?
**e** Is triangle ABC the same shape and size as its image, PQR?

**7** Another translation maps the triangle PQR from Question **5** to STU.

The vector for the translation is $\begin{pmatrix}1\\{}^{-}1\end{pmatrix}$

**a** Show STU on your diagram for Question **6**.
**b** Write down the coordinates of S, T and U.

**8** What is the translation vector that maps triangle ABC to triangle STU (Question **7**)?

Complete: $\begin{pmatrix}4\\3\end{pmatrix} + \begin{pmatrix}1\\{}^{-}1\end{pmatrix} = \begin{pmatrix}\ \end{pmatrix}$.

**9** X, Y and Z are three points. The vector $\begin{pmatrix}2\\7\end{pmatrix}$ maps X to Y and $\begin{pmatrix}1\\{}^{-}6\end{pmatrix}$ maps Y to Z.

**a** What single vector maps X to Z?
**b** What vector would map Z to X?

**10** A kite ABCD is mapped under a translation $\mathbf{T} = \begin{pmatrix}4\\5\end{pmatrix}$ to $A'(2,6), B'(4,8), C'(6,6)$ and $D'(4,2)$. Determine the vertices of the kite ABCD.

**11** On graph paper, draw triangle MNO with vertices M (1, 1), N (3, 3) and O (3, ${}^{-}1$).

Translate triangle MNO using the vector $\begin{pmatrix}2\\{}^{-}4\end{pmatrix}$. Label the vertices of the image $M'N'O'$.

Translate triangle $M'N'O'$ using the vector $\begin{pmatrix}{}^{-}1\\{}^{-}3\end{pmatrix}$. Denote this new image $M''N''O''$. Label the vertices of $M''N''O''$.

Give the vectors describing the translations which map:

**a** $\triangle MNO$ to $\triangle M''N''O''$
**b** $\triangle M'N'O'$ to $\triangle MNO$
**c** $\triangle M''N''O''$ to $\triangle MNO$.

**12** On graph paper, draw the rectangle ABCD with vertices A(2,1), B(2, $^{-}1$), C($^{-}2$, $^{-}1$) and D ($^{-}2$, 1). Translate ABCD with the vector $\mathbf{T} = \begin{pmatrix} ^{-}3 \\ 1 \end{pmatrix}$. Denote the image of ABCD as $A_1B_1C_1D_1$. Then translate $A_1B_1C_1D_1$ using the vector $\mathbf{T} = \begin{pmatrix} ^{-}2 \\ ^{-}2 \end{pmatrix}$. Denote this new image $A_2B_2C_2D_2$. State the vertices of $A_1B_1C_1D_1$ and $A_2B_2C_2D_2$. Give the vectors that translate:

**a** ABCD to $A_2B_2C_2D_2$

**b** $A_2B_2C_2D_2$ to ABCD

**c** $A_2B_2C_2D_2$ to $A_1B_1C_1D_1$.

**13** **T** is the translation $\begin{pmatrix} 4 \\ -3 \end{pmatrix}$. ABCD is a square with vertices at (2, 0), (2, 2) (4, 2) and (4, 0) respectively.

**a** Draw ABCD on graph paper.

**b** Draw the image of ABCD under **T**, that is, **T**(ABCD).

**c** Show that $\mathbf{T}^{-1} = \begin{pmatrix} ^{-}4 \\ 3 \end{pmatrix}$.

**d** Translate the image of ABCD under **T** once more, to get $\mathbf{T}^2$(ABCD).

## 15.2 Rotations

You will need squared and tracing paper.

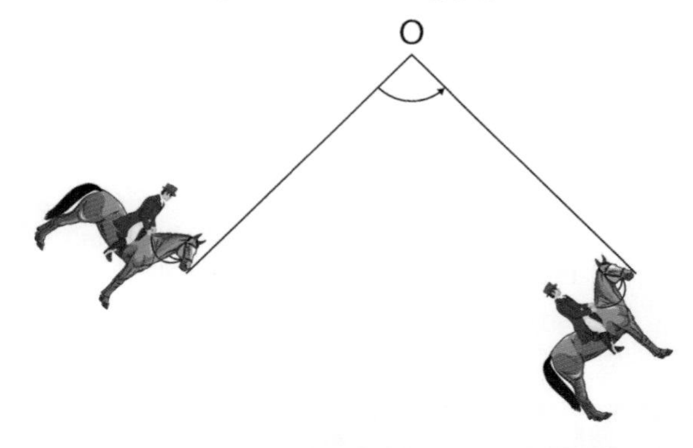


Here is a picture of a horseman and its image under a **rotation**.

A rotation has a **centre** and an **angle**.

Rotations can be clockwise or anticlockwise.

This rotation is 90° anticlockwise about centre O.

When describing a transformation we state whether it is clockwise or anticlockwise.

### EXAMPLE 3

The triangle with vertices at A(1, 1), B(4, 1) and C(1, 2) is rotated through 90° anticlockwise about the centre (0, 0). State the coordinates of the image vertices, A′, B′ and C′.

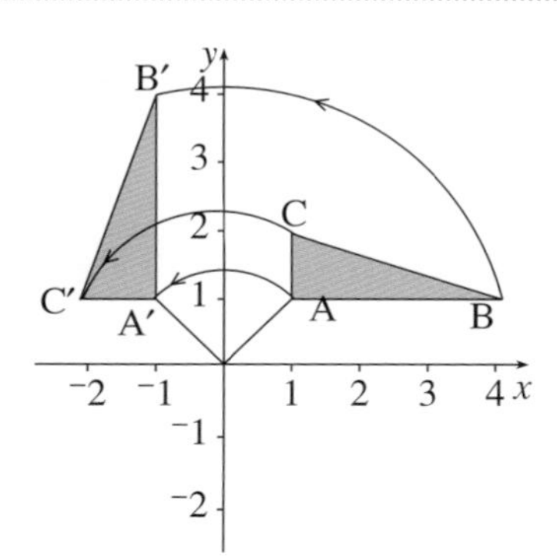


The vertices are A′($^{-}1$,1), B′($^{-}1$,4) and C′($^{-}2$,1).

Notice that triangles ABC and A′B′C′ in Example 3 are the same shape and size; that is, shape and size stay the same under rotation, and the triangles are congruent.

### Exercise 15B

You may use tracing paper throughout this exercise.

**1** The diagrams show a triangle ABC and its image after a rotation. In each case, write down the angle, direction and centre of rotation.

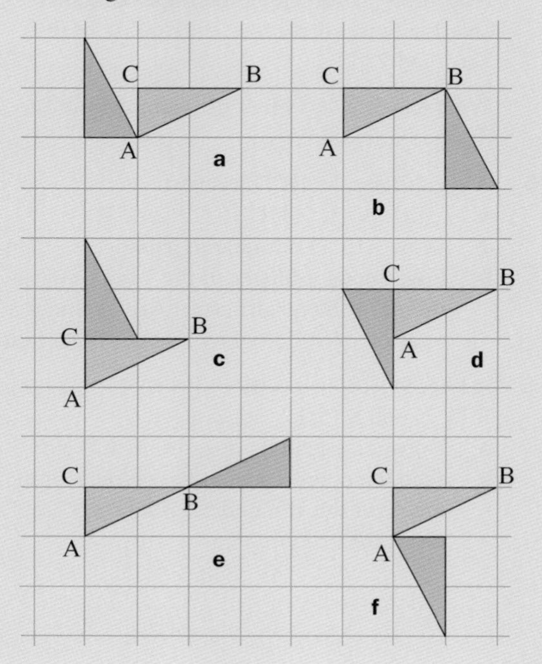

**2** In the diagram, the mapping of triangle ABC to triangle RBQ is a rotation of 90° anticlockwise, centre B.

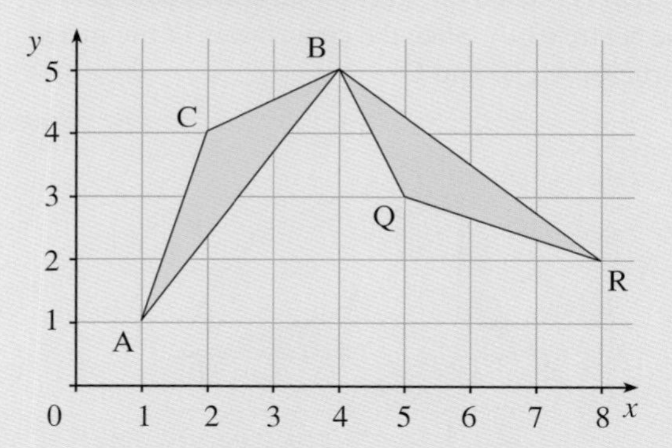


Draw a copy of triangle ABC using squared paper.

Now draw its image after a rotation of:

**a** 180°, centre B

**b** 90° anticlockwise, centre C

**c** 90° anticlockwise, centre A.

Each time, write down the coordinates of the vertices of the image of triangle ABC.

**3** Repeat Question **2** for a rotation of:

**a** 90° anticlockwise, centre (5, 4)

**b** 180°, centre (6, 5).

**4** Draw the trapezium with vertices (0, 0), (3, 0), (2, 1), (1, 1).

Now draw the image of the trapezium after anticlockwise rotations, centre (0, 0), of 90°, 180° and 270°.

What can you say about the symmetry of the finished drawing?

**5** A quadrilateral ABCD has vertices A $(1, {}^{-}3)$, B $(3, {}^{-}1)$, C $(6, {}^{-}2)$ and D $(6, {}^{-}5)$.

**a** Draw ABCD on squared paper.

**b** Find the coordinates of the vertices of the image of the quadrilateral ABCD under a rotation of 90° clockwise about the origin.

**6** A kite, PQRS, has vertices P $(3, {}^{-}4)$, Q (1, 1), R (3, 2) and S (5, 1).

**a** Draw, on squared paper, the kite PQRS.

**b** Find the coordinates of the vertices of the image of the kite under a rotation of 180° about the origin.

**7** **R** is a rotation centre $({}^{-}1, 1)$ through 90° clockwise and L is a triangle whose vertices are $({}^{-}1, 1)$, (3, 0) and (3, 3).

**a** Draw, on squared paper, the triangle L.

**b** Draw **R**(L), by rotating L once.

**c** Draw $\mathbf{R}^2$(L), by rotating L twice.

**8** **H** is a rotation centre (2, 0) through 180° and J is the triangle whose vertices are (1, 0), $(2, {}^{-}1)$ and (3, 4)

**a** Draw, on squared paper, the triangle J.

**b** Draw **H**(J).

**c** Draw $\mathbf{H}^2$(J).

**9** **R** is a rotation through 90° anticlockwise, centre (0, 0).
**Q** is a rotation through 90° clockwise, centre (3, 0).
L is a triangle whose vertices are (1, 0), (4, 0) and (4, 4) respectively.

On squared paper draw:

**a** L

**b** **R**(L)

**c** **Q**(L)

**d** What simple transformation maps **R**(L) onto **Q**(L)?

**10** On squared paper, draw the triangle A, which has vertices at (1, 1), (3, 4) and $(3, {}^{-}2)$.

**a** Triangle P is the image of triangle A under an anticlockwise rotation of 90° about the point (1, 1). Draw and label triangle P.

**b** Triangle Q is the image of triangle P under a clockwise rotation of 90° about the point $({}^{-}2, 0)$. Draw and label triangle Q.

## TECHNOLOGY

Have a look at the sign challenge at
www.figurethis.org/challenges/c05/challenge.htm
Can you create your own 'upside down' sign?

## 15.3 Reflections

You will need squared paper and compasses or a set square.

The image of the boat reflected in the water looks identical but is upside down.

The idea of reflection is used in mathematics.

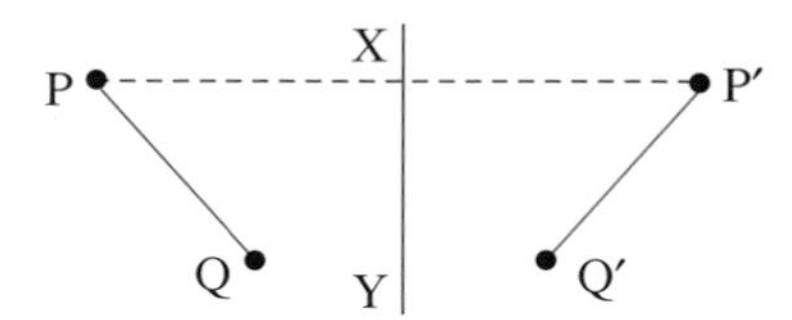


P′Q′ is the **reflection** of PQ in the line XY.

XY is called the **mirror line**.

P and P′ are the same distance from XY so PX = XP′ and PP′ is perpendicular to XY.

P′Q′ is the same length as PQ.

As with translation and rotation, the shape and size of the object are unchanged under a reflection: the image and the object are congruent.

It is often easy to see where a mirror line is when shapes are drawn on squared paper. But you can always find the mirror line by using the properties of reflections.

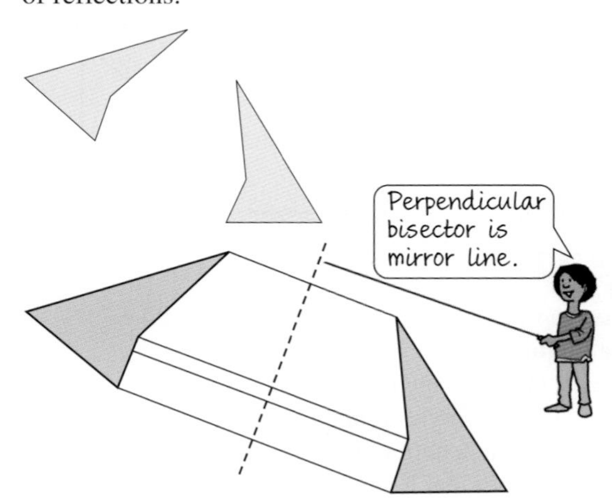


To find the mirror line, join corresponding points and find the midpoints of the lines.

### Exercise 15C

**1** What is the equation of the mirror line for this reflection of the image of a horseman?

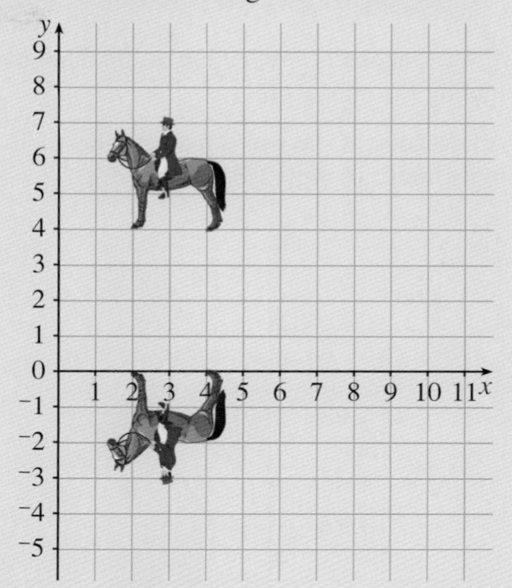


**2** Draw a line (to use as a mirror line) and mark two points P and Q anywhere on one side of the line. Use compasses or a set square to draw lines from P and Q perpendicular to the mirror line and so find the reflections P′ and Q′ by measurement.

Measure the lengths PQ and P′Q′. Are they the same?

**3** **M** is a reflection in the $x$-axis. L is the line joining $(4, 2)$ and $(^{-}3, ^{-}1)$.

**a** Draw the line L on squared paper.
**b** Draw **M**(L) by reflecting your line.

**4** **R** is a reflection in the line $x = 2$. J is the triangle with vertices $(^{-}1, ^{-}1)$, $(^{-}2, 2)$, $(^{-}3, 1)$.

**a** Draw the triangle J on squared paper.
**b** Draw **R**(J), the reflection of J.
**c** Draw $\mathbf{R}^2(\mathrm{J})$, the reflection of **R**(J).

**5** T is the triangle with vertices $(3, 1)$, $(2, 1)$ and $(2, 3)$. **R** is the reflection in the line $x = {}^{-}1$ and **N** is the reflection in the line $x = 4$.

**a** Draw the triangle T on squared paper.
**b** Draw **R**(T).
**c** Draw **N**(T).
**d** Draw **RN**(T), by reflecting **N**(T).
**e** Draw **NR**(T), by reflecting **R**(T).
**f** What single transformation maps **RN**(T) on to **NR**(T)?

**6** On squared paper plot two triangles with these vertices:
$(1.5, ^-2), (7, ^-1), (0.5, ^-9)$
$(^-1.5, 2), (1, 7), (^-8.5, 3)$

Show that one is a reflection of the other by finding the mirror line.

**7** A reflection in the $y$-axis maps triangle ABC to its image triangle PQR.

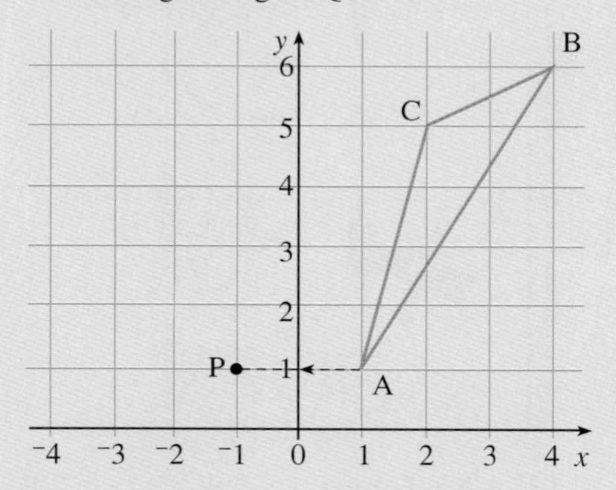


**a** The position of A is $(1, 1)$. What is the position of P?
**b** Copy and complete: $A(1, 1) \rightarrow P(\quad)$.
**c** Write down similar statements for $B \rightarrow Q$ and $C \rightarrow R$.
**d** If $(x, y)$ is any point, copy and complete this statement for a reflection in the $y$-axis: $(x, y) \rightarrow (\quad)$.

**8** Copy triangle ABC in Question **7** and show its image after a reflection in the line:
**a** $x = 3$ **b** $y = 3$ **c** $y = 5 - x$

Each time, write down the vertices of the image triangle.

**9** Under a reflection, are a shape and its image the same shape and size?

**10** Copy triangle ABC from Question **7**. Show its image XYZ with vertices at $(^-4, 6)$, $(1, 9)$ and $(0, 7)$. Find the equation of the mirror line.

**11** A reflection maps the following points.
$(5, 2) \rightarrow (5, 2)$
$(2, 5) \rightarrow (2, ^-1)$
$(0, 0) \rightarrow (0, 4)$

What is the equation of the mirror line of the reflection?

**12** Draw a copy of triangle ABC from Question **7**. Next, draw its image after a reflection in the line $x = 4$
Now reflect the image in the line $x = 6$
**a** What type of transformation maps the original triangle to its final image?
**b** What vector represents this transformation?

**13** Draw triangle ABC from Question **7** and its image after a reflection in $x = 5$, followed by a reflection in $x = 6$
**a** Could ABC map to its final image by a translation?
**b** What vector represents this translation?
**c** What is the connection between this vector and the distance between the lines $x = 5$ and $x = 6$?

**14** **a** What vector represents the translation that is equivalent to a reflection in $x = 1$ followed by a reflection in $x = 5$?
**b** Draw a triangle with vertices at $(1, 1)$, $(0, 3)$ and $(1, 5)$. By carrying out the reflections, check that your answer to part **a** is correct.

## 15.4 Enlargements

You will need squared paper.

In an **enlargement** the size of an object is either increased or reduced.

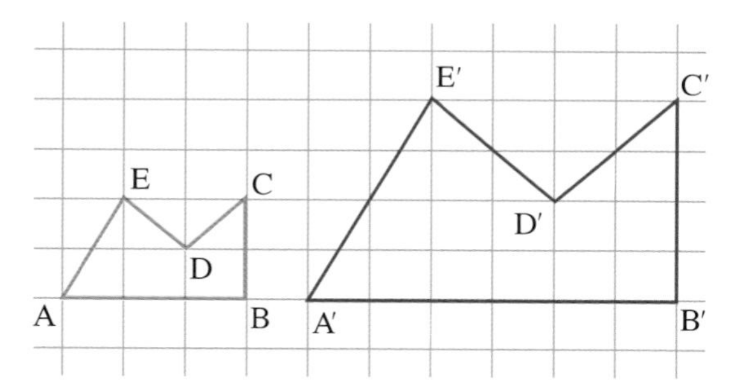


In the diagram the shape ABCDE is enlarged by **scale factor** 2 into A′B′C′D′E′.

**Note:** (1) the shape remains the same
(2) the angles remain the same
but (3) the length of every side of the image is original length × scale factor

The shapes ABCDE and A′B′C′D′E′ are **similar** but not congruent. The word 'similar' is a mathematical word meaning one shape is an enlargement of another, preserving shape and angles but increasing or decreasing side lengths.

## Exercise 15D

**1** Draw an enlargement of each of these shapes with scale factor 2.

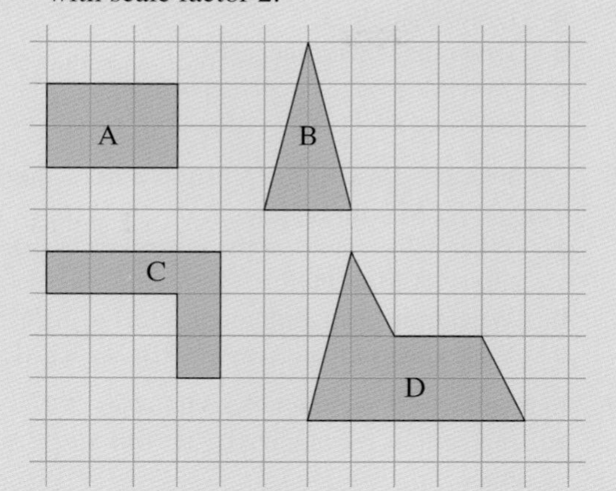


**2** Repeat Question **1**, but this time use scale factor 3.

**3** **a** In Question **1**, find the area, in square units, of shapes A, B, C and D.

**b** What is the new area of each shape after an enlargement with scale factor 2?

**c** How are the areas of the objects and images related?

**d** How would the areas be related if the scale factor was 3?

To completely define an enlargement you need to know the:

(1) **scale factor** of the enlargement
(2) **centre of enlargement**.

All parts of the image are enlarged from the centre of enlargement.

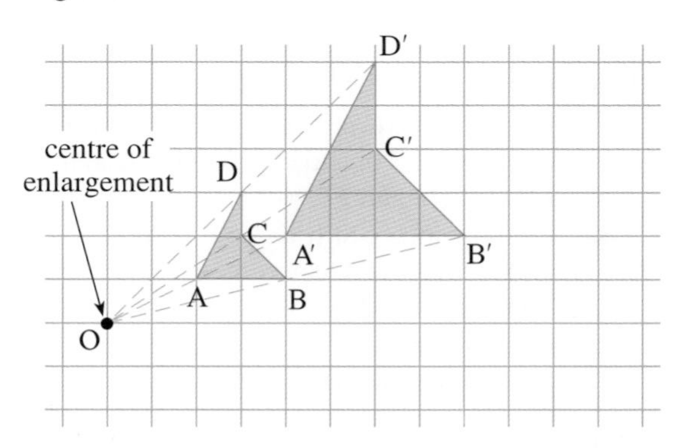


The diagram shows ABCD enlarged with scale factor 2 and centre of enlargement O.

### EXAMPLE 4

Find the image A′B′C′ of triangle A (2, 1), B (2, 3), C (1, 1) under an enlargement of scale factor 2 with centre of enlargement O (0, 0).

Join the vertices A, B, C of the triangle to the centre of enlargement, O. Extend OA to A′ so that

$OA' = 2 \times OA$

Repeat for B′ and C′.

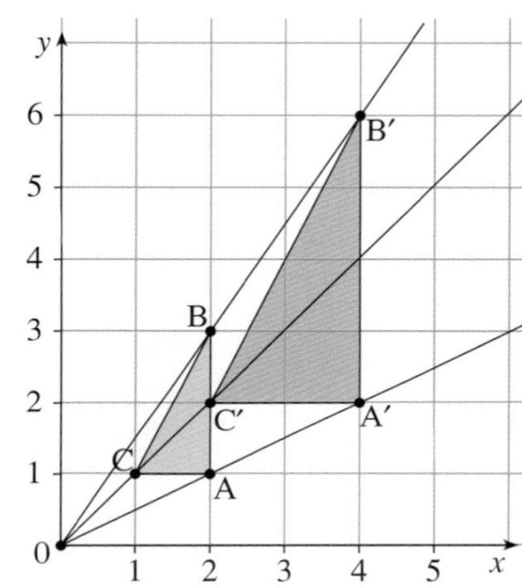


The coordinates of the enlarged triangle are $A'(4,2)$, $B'(4,6)$ and $C'(2,2)$.

Notice that each side of the original triangle is multiplied by 2, so triangle ABC is **similar** to triangle A′B′C′.

## Exercise 15E

**1** In the diagram, a rectangle has been enlarged by a scale factor 2, using the origin, O, as centre.

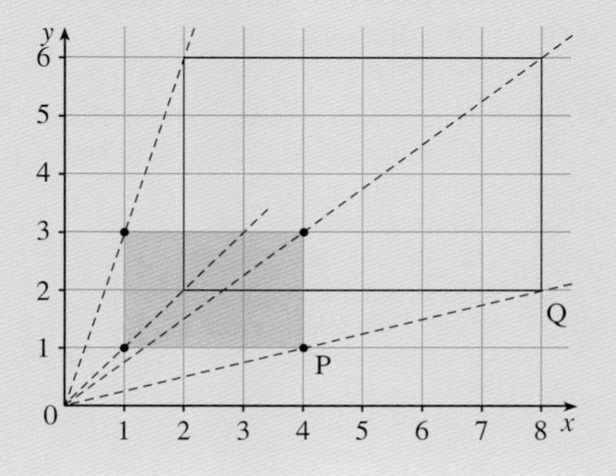

**a** In the enlargement, the image of P is Q. What can you say about the lengths of OP and OQ?
**b** What are the lengths of the sides of the small rectangle? What are the lengths of the sides of the large rectangle?
**c** How does an enlargement scale factor 2 change the lengths of the sides?

**2** In this diagram, triangle ABC is mapped to triangle DEF by an enlargement, centre S.

SD = 3SA and SE = 3SB

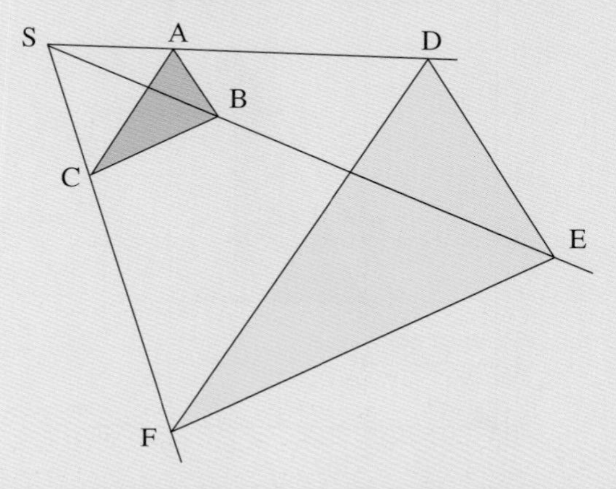


**a** What is the scale factor of the enlargement?
**b** Copy and complete: SF = . . .
**c** If AB = 3 units, what is the length of DE?
**d** If DF = 12 units, what is the length of AC?

**3** On squared paper, draw the triangles ABC and PQR with:
A (1, 1), B (4, 5), C (2, 4)
P (2, 0), Q (8, 8), R (4, 6)
**a** Is triangle PQR an enlargement of triangle ABC?
**b** Draw lines through AP, BQ and CR to find the centre of enlargement.
**c** What is the scale factor of the enlargement?

**4** T is a triangle with vertices (2, 1), (4, 1) and (4, 2). T is enlarged with centre (0, 0) and scale factor 2.
**a** Draw triangle T.
**b** Draw the enlarged triangle.
**c** Write down the coordinates of the enlarged triangle's vertices.

**5** A is the triangle with vertices (3, 1), (1, 2) and (1, 1). **E** is an enlargement with scale factor 2 and centre (3, 0).
**a** Draw triangle A.
**b** Draw the enlargement, **E**(A).
**c** Write down the coordinates of the enlarged triangle's vertices.

**6** On the same drawing as Question **3** draw the triangle KLM with K (5, 1), L (8, 5), M (6, 4).
**a** What is the centre of the enlargement that maps KLM to PQR?
**b** What transformation maps KLM to ABC?
**c** What transformation maps ABC to KLM?

**7** In the diagram, rectangle KLMN is an enlargement of rectangle ABCD with S as centre.

MN = 12 cm, CD = 4 cm, and BC = 3 cm.

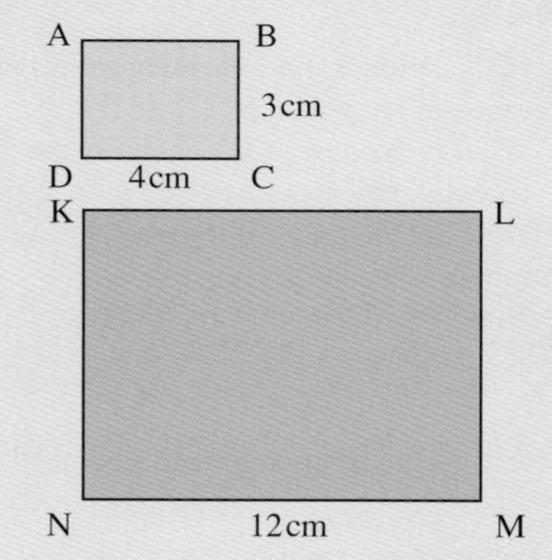


**a** What is the scale factor of the enlargement?
**b** What is the length of LM?
**c** IF SA = 3 cm, what is the length of SK?
**d** AC = 5 cm. Find the length of KM.

**8** **a** Construct a rectangle ABCD in which AB is 2 cm and BC is 4 cm. On BC construct a triangle EBC outside the rectangle so that EB = 2 cm and EC = 4 cm.
**b** Find the image of triangle EBC under an enlargement with centre A, given that the scale factor is 2.
**c** Find the image of triangle EBC under an enlargement with centre D, given that the scale factor is 2.
**d** What can you say about the two images in parts **b** and **c**?

## 15.5 Similar triangles

**Note that similar triangles are not mentioned specifically in the Cambridge Secondary 1 curriculum framework, and are included here as extension material.**

Two shapes are said to be similar if they are the *same shape*. They do not need to be the same size.

Two triangles are **similar** when all corresponding angles are equal.

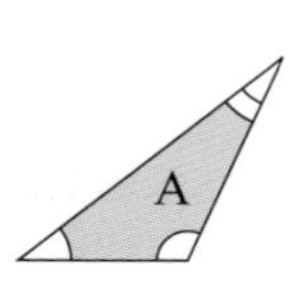


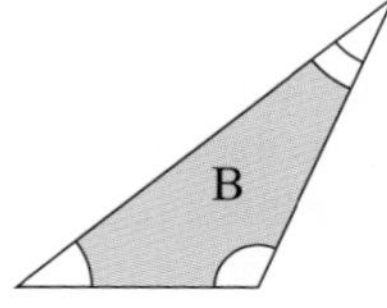


Triangle B is similar to triangle A.

Triangle B is an enlargement of triangle A.

The ratio of the sides in similar shapes gives the scale factor of the enlargement.

**EXAMPLE 5**

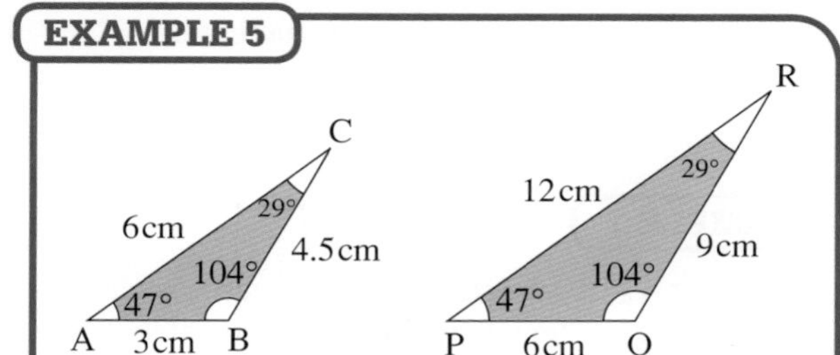


Show that triangles ABC and PQR are similar.
What is the scale factor of enlargement?

$\hat{A} = \hat{P} = 47°, \hat{B} = \hat{Q} = 104°, \hat{C} = \hat{R} = 29°$

Corresponding angles are equal so triangle ABC and triangle PQR are similar.

Ratio of sides:

$$\frac{PQ}{AB} = \frac{6}{3} = \frac{QR}{BC} = \frac{9}{4.5} = \frac{PR}{AC} = \frac{12}{6} = 2$$

Hence, the scale factor of enlargement is 2.

### Exercise 15F (extension)

**1** Is each pair of triangles similar?

**a**

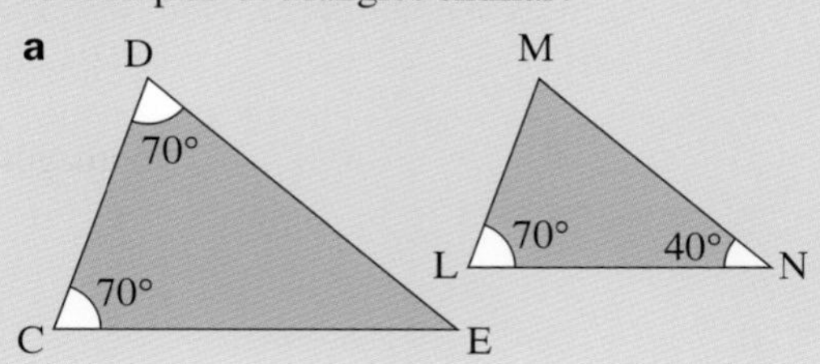


**b**

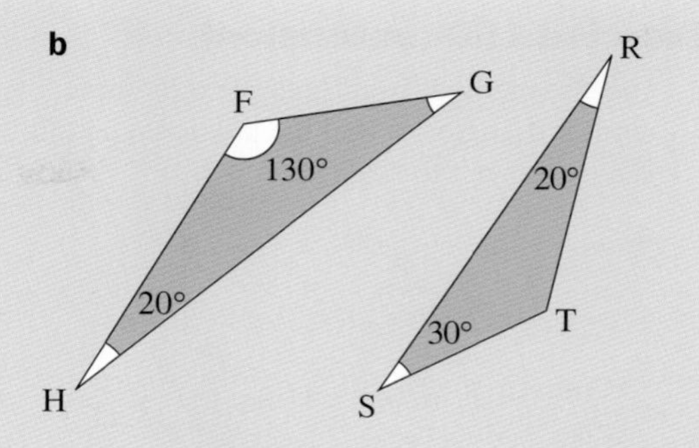


**2** Look at triangles OAB and OPQ.

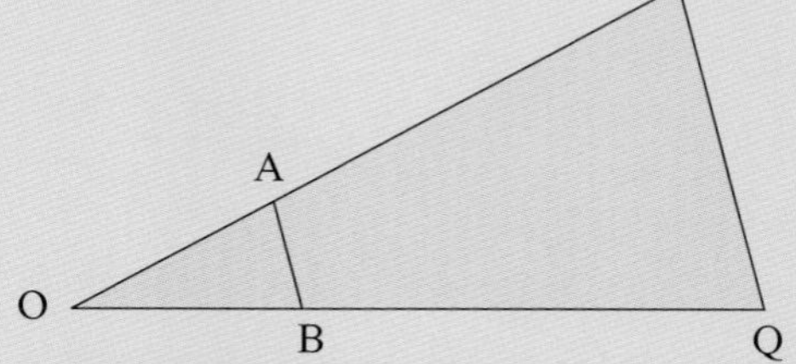


$O\hat{A}B$ corresponds to $O\hat{P}Q$. OA corresponds to OP.

**a** What corresponds to:
**i** $O\hat{B}A$ **ii** $P\hat{O}Q$ **iii** PQ **iv** OQ?

**b** Measure the angles of triangles OAB and OPQ. Are the two triangles similar? Why?

### Using ratios to find lengths of sides

**EXAMPLE 6**

Triangles LMN and STR are similar. (They are not drawn to scale.)

Find the value of $x$.

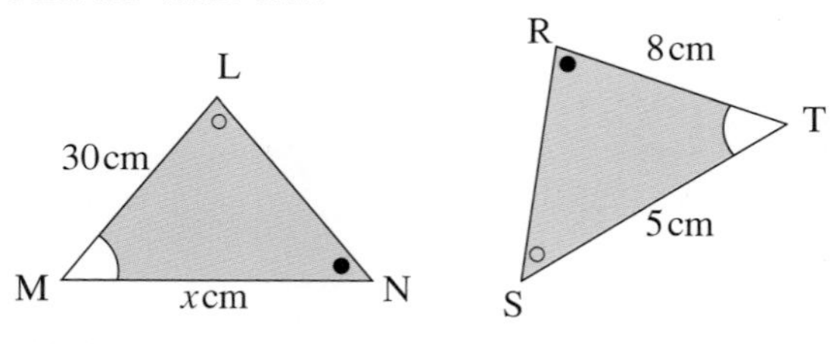


MN corresponds to TR.
LM corresponds to ST.

$$\frac{MN}{TR} = \frac{LM}{ST}$$

$$\frac{x}{8} = \frac{30}{5}$$

Multiplying both fractions by 8 gives:

$x = 48$

## Exercise 15G (extension)

**1** Use the method from Example 6 on these similar triangles to find the unknown length. (Diagrams not to scale.)

**a**

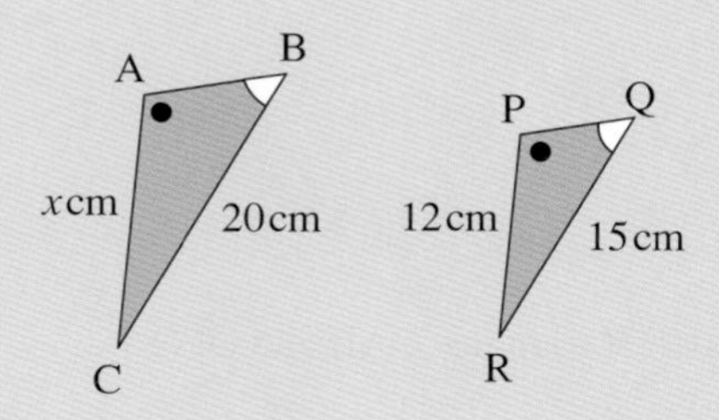


**b**

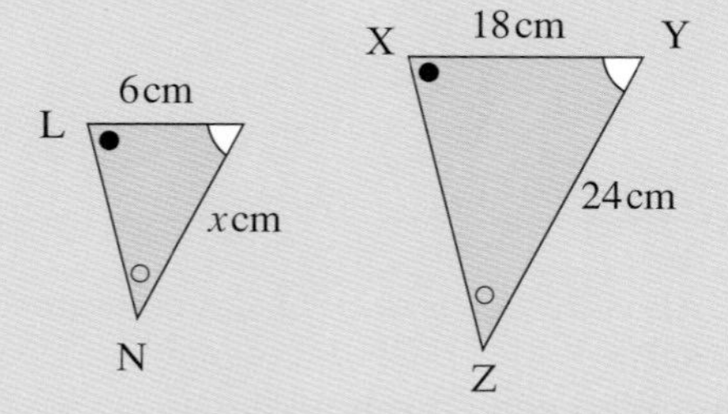


**2** Use ratios to find the lengths $a$, $b$, $c$, $d$, $e$ and $f$.

**a**

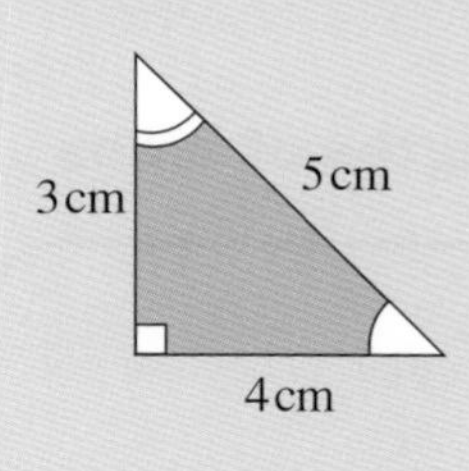


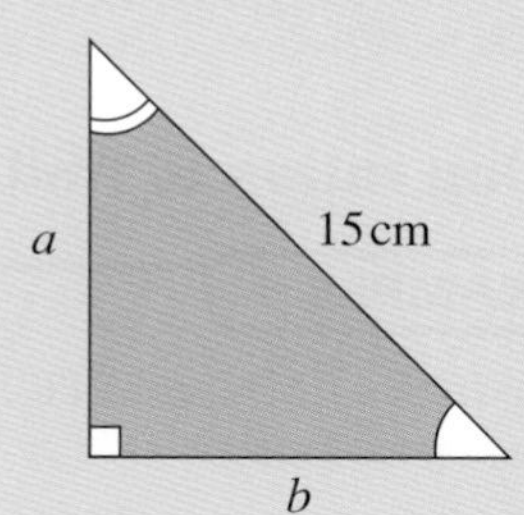


**b**

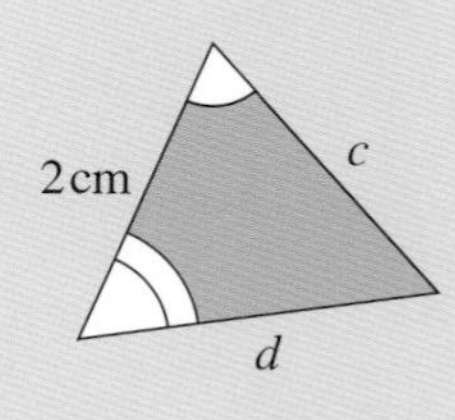


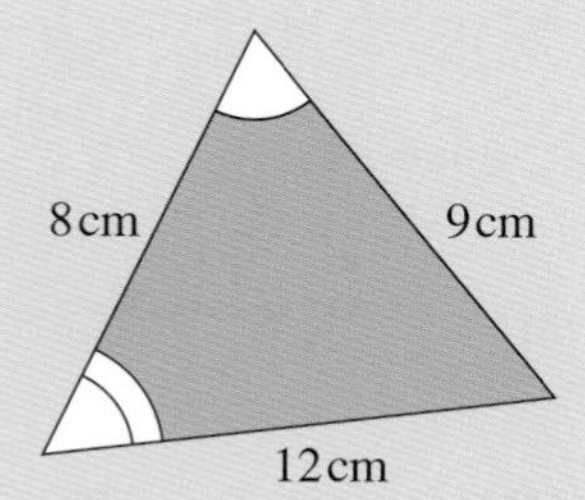


**c**

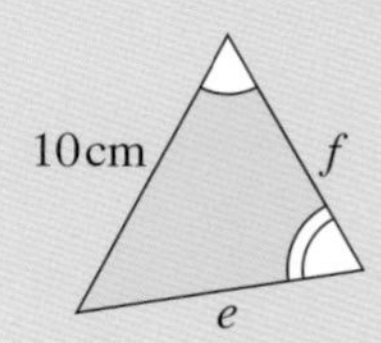


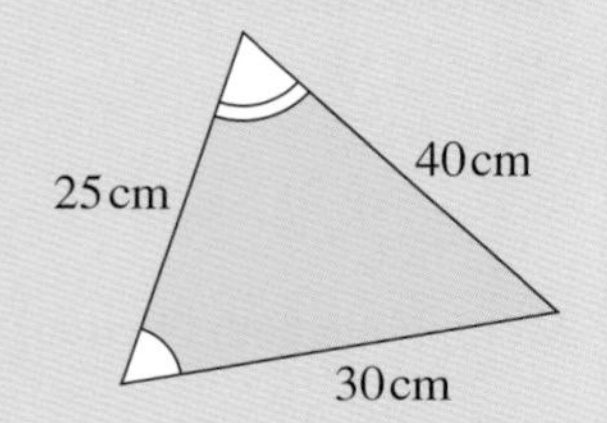


## 15.6 Describing transformations

You need to give a precise and full description of a transformation. First of all, you need to say which transformation it is: a reflection, rotation, translation or enlargement.

Then you need to give more detail:

- to describe a translation – give the vector
- to describe a rotation – give the angle and direction of the rotation and the coordinates of the centre of rotation
- to describe a reflection – give the equation of the mirror line
- to describe an enlargement – give the coordinates of the centre of enlargement and the scale factor of the enlargement.

### Finding the equation of the mirror line

If object and image points after a reflection are known, you can find the mirror line by constructing the perpendicular bisector of the two points.

Use the same scale on both the $x$- and $y$-axis.

**EXAMPLE 7**

The point A (1, 3) is reflected in a line to give an image B (5, 7). Draw the mirror line accurately and find its equation.

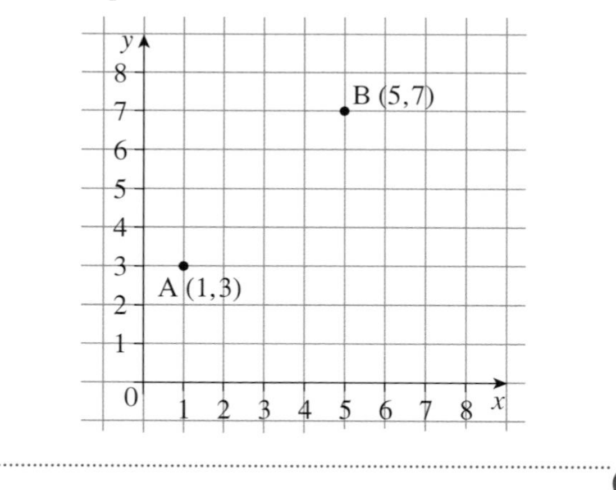

Join the two points. With your compasses construct the perpendicular bisector of the line AB.

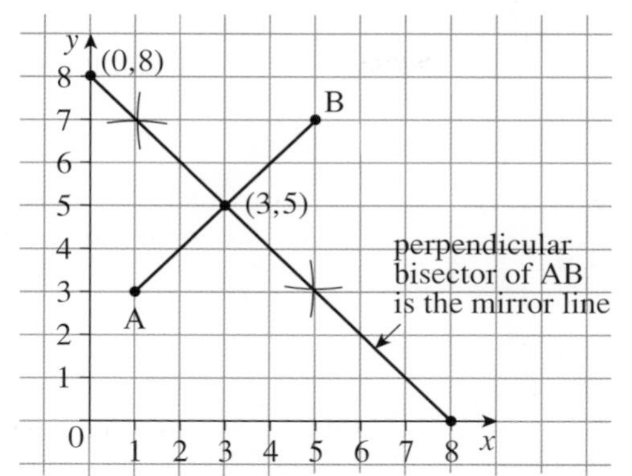


Two points on the mirror line are $(0, 8)$ and $(3, 5)$.

The $y$-intercept is 8 and the gradient $= \frac{5-8}{3-0} = \frac{^-3}{3} = {^-1}$

So the equation of the mirror line is $y = 8 - x$

Look back at Chapter 14 if you have forgotten about finding the equation of a line.

## Finding the centre of rotation

In the diagram, the line XY has been rotated anticlockwise to the line X′Y′.

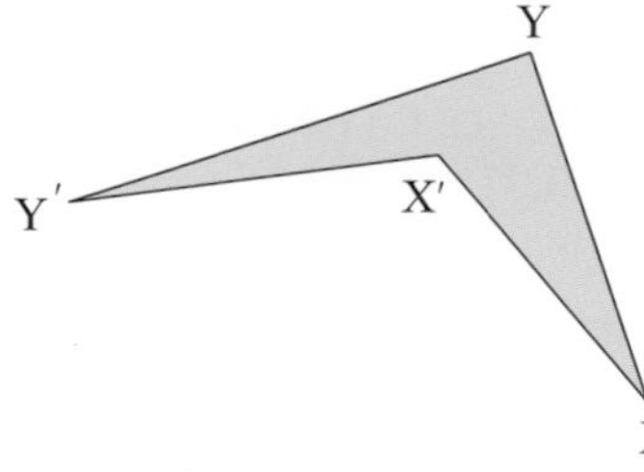


To find the centre of rotation, first join object and image points, X to X′ and Y to Y′.

Now find the perpendicular bisectors of XX′ and YY′.

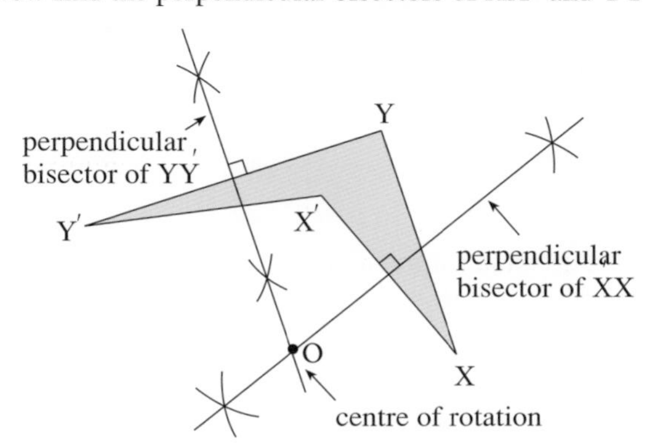


The perpendicular bisectors meet at the centre of rotation, O.

You can check by drawing arcs XX′ and YY′ using O as a centre.

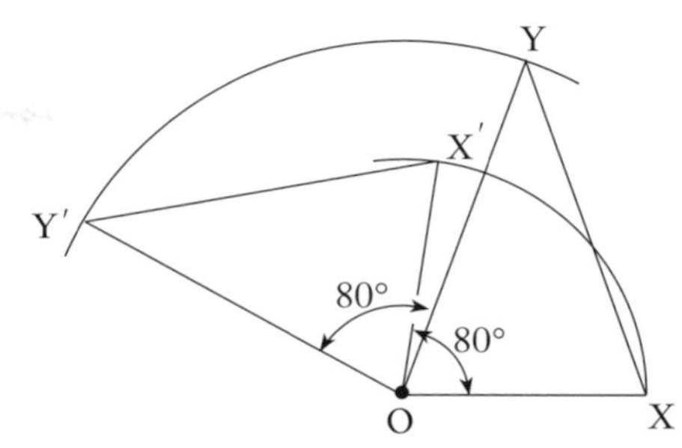


To find the angle of rotation measure the angle XOX′ or YOY′.

You can use the same method when the coordinates of the object and image are given.

**EXAMPLE 8**

Triangle P $(6, 4)$, Q $(5, 1)$, R $(3, 2)$ is rotated to triangle P′ $(0, 8)$, Q′ $(3, 7)$, R′ $(2, 5)$.

Construct and find the coordinates of the centre of rotation. Describe the transformation.

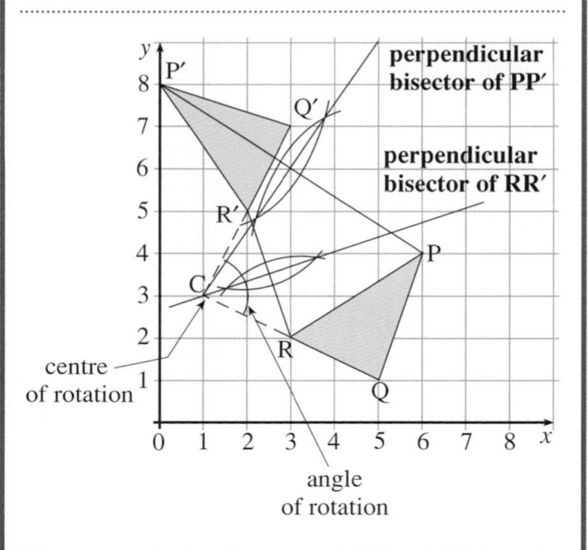


The perpendicular bisectors of PP′ and RR′ meet at C $(1, 3)$. So the centre of rotation is at $(1, 3)$.

Angle RCR′ is the angle of rotation, which is 90° anticlockwise.

The transformation is a rotation 90° anticlockwise about $(1, 3)$.

## Finding the centre of enlargement and scale factor

**EXAMPLE 9**

Describe the transformation from triangle BCD to triangle B′C′D′.

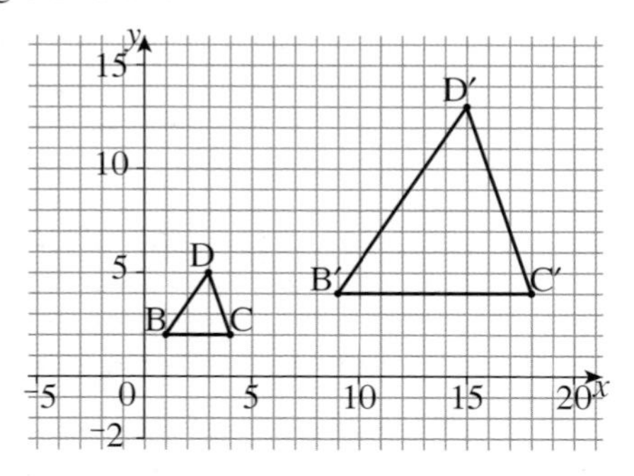


To find the scale factor, find the ratio of corresponding sides:

$C'B' \div CB = 9 \div 3 = 3$

So the scale factor is 3.

Join corresponding points on the object and the image. Extend the lines until they intersect. This is the centre of enlargement.

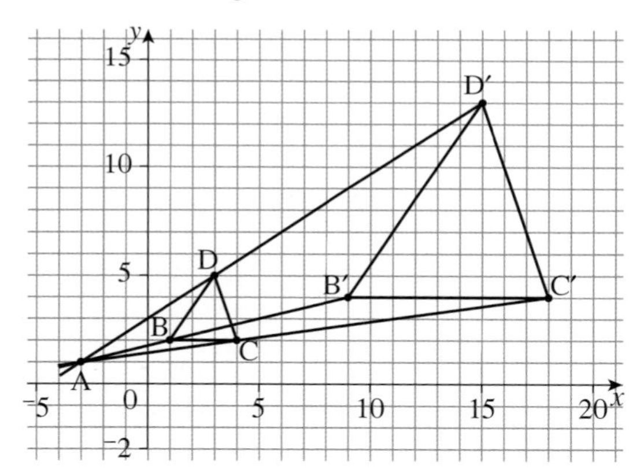


The centre of enlargement is the point A (⁻3, 1).

So the transformation is an enlargement with scale factor 3, centre (⁻3, 1).

### Exercise 15H

**1** In each case construct the mirror line.

| | *Object* | *Image* |
|---|---|---|
| **a** | (2, 3) | (5, 3) |
| **b** | (⁻1, 2) | (4, 4) |
| **c** | (3, 1) | (1, 3) |
| **d** | (⁻3, ⁻2) | (2, 3) |
| **e** | (1, 6) | (⁻3, ⁻1) |

**2** The triangle X (2, 3), Y (5, 4), Z (6, 1) becomes X′ (⁻3, ⁻2), Y′ (⁻4, ⁻5), Z′ (⁻1, ⁻6), after reflection in a certain line.

**a** Plot the two triangles on graph paper.
**b** Construct the mirror line.
**c** Write down the equation of the mirror line.

**3** Draw two equal lines XY and X′Y′ anywhere on your page. Find the centre and angle of rotation that maps XY → X′Y′.

**4** Copy Example 8 onto graph paper.

**a** Construct the perpendicular bisector of QQ′. Does it also pass through C?
**b** With your protractor measure the angles:
**i** $P\hat{C}P'$ **ii** $Q\hat{C}Q'$
Are they equal to $R\hat{C}R'$?

**5** For the lines AB with A (5, 6), B (3, 8) and A′B′ with A′ (6, 3), B′ (4, 1), find the centre and angle of rotation that maps AB → A′B′.

**6** After a transformation, triangle ABC with A (1, 2), B (4, 1), C (1, 5) becomes A′B′C′ with A′ (⁻9, ⁻2), B′ (0, ⁻5) and C′ (⁻9, 7). Describe the transformation.

**7** After a transformation quadrilateral ABCD with A (1, 2), B (2, 4), C (4, 3), D (5, 1) becomes A′B′C′D′ with A′ (4, ⁻2), B′ (5, 0), C′ (7, ⁻1), D′ (8, ⁻3). Describe the transformation.

**8** Find the equation of the mirror line for each of the following pairs of points.

| | *Object* | *Image* |
|---|---|---|
| **a** | (1, 5) | (1, 3) |
| **b** | (⁻3, 2) | (7, 2) |
| **c** | (⁻3, ⁻1) | (⁻3, 4) |
| **d** | (6, 2) | (2, 6) |
| **e** | (⁻4, 5) | (⁻5, 4) |

**9** After a transformation, triangle A (2, ⁻1), B (3, 2), C (4, 1) becomes triangle A′ (⁻1, 2), B′ (2, 3), C′ (1, 4). Describe the transformation.

**10** After a transformation, quadrilateral A (1, 1), B (⁻1, 4), C (5, 2), D (2, 2) becomes quadrilateral A′ (0, 5), B′ (⁻2, 8), C′ (4, 6), D′ (1, 6). Describe the transformation.

**11** After a transformation, triangle A (0, 0), B ($^-$4, 3), C ($^-$1, 5) becomes triangle A′ (2, $^-$3), B′ ($^-$6, 3), C′ (0, 7). Describe the transformation.

**12** Under a rotation, the line AB with A (7, 1), B (8, 4) is mapped onto A′B′ with A′ (4, 6), B′ (1, 7).

**a** Plot the points on graph paper.

**b** Construct the perpendicular bisectors of AA′ and BB′.

**c** What are the coordinates of the centre of rotation, C?

**d** With a protractor, measure the angle ACA′.

**e** What is the angle of rotation?

**13** After a transformation, triangle A (2, 1), B (4, 0), C (6, 2) becomes triangle A′ (7, 1), B′ (9, 0), C′ (11, 2). Describe the transformation.

## 15.7 Combinations of transformations

Transformations can be combined by performing one transformation followed by another transformation.

### EXAMPLE 10

The transformation, **P**, consists of a reflection in the $y$-axis, **Q**, followed by a translation $\mathbf{R} = \binom{0}{4}$.

Find the image of triangle A (1, 1), B (2, 1), C (2, 4) under **P**.

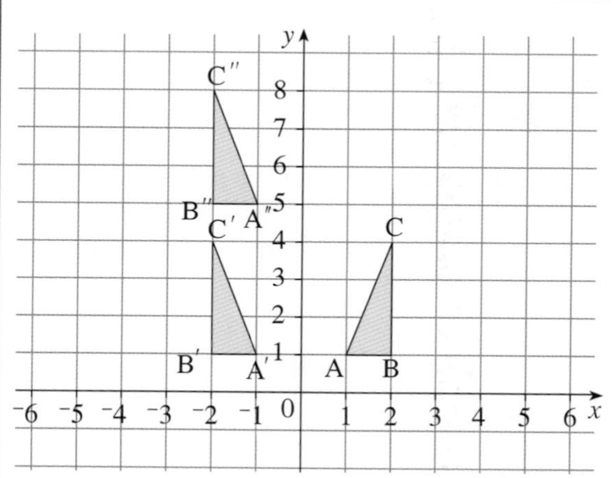


Under reflection **Q**:

A (1, 1) → A′ ($^-$1, 1)

B (2, 1) → B′ ($^-$2, 1)

C (2, 4) → C′ ($^-$2, 4)

Under translation **R**:

A′ ($^-$1, 1) → A″ ($^-$1, 5)

B′ ($^-$2, 1) → B″ ($^-$2, 5)

C′ ($^-$2, 4) → C″ ($^-$2, 8)

So, the image of ABC under the combined transformation **P** has vertices A″ ($^-$1, 5), B″ ($^-$2, 5), C″ ($^-$2, 8).

### EXAMPLE 11

**X** is a reflection in the line $y = 2$. **Y** is a reflection in the line $x = 3$. Find **XY**(P) where P is the point (4, 1).

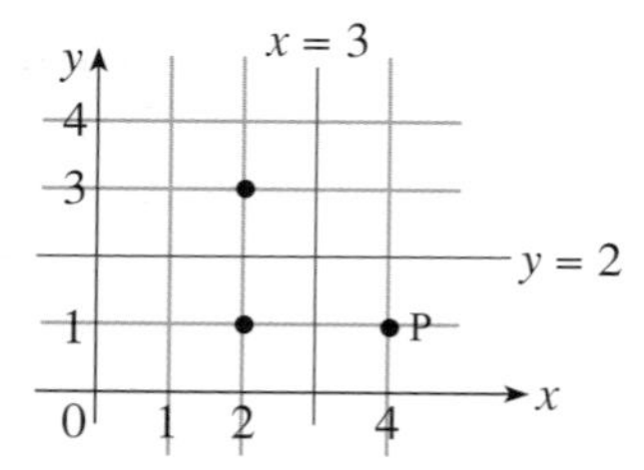


**XY**(P) = **XY** (4, 1) = **X** (2, 1) = (2, 3).

**Note:** In finding **XY**(P), the transformation **Y** is done first, followed by the transformation **X**.

### Exercise 15I

**1** Draw a diagram to show the reflection of a triangle with vertices A (1, 1), B (2, 4), C (1, 4) in the line $x = 2$ and the reflection of the image in the line $x = 4$. Describe the single transformation equivalent to these two.

**2** Draw a sketch to show the image of a triangle under transformations as follows:

*first* use a reflection in $x = 1$
*followed by* a reflection in $x = {}^-1$
*followed by* a reflection in $x = 2$.

Find the single transformation that is equivalent to these three.

**3 a** On a coordinate grid, mark point Z at ($^-$3, 1).

**b** Show its image, Z′, after reflection in the $y$-axis.

**c** Z″ is the image of Z′, after a rotation of 90° anticlockwise, centre (3, 6). Show Z″, and write down its coordinates.

# Consolidation

### Example 1

What is the image of the quadrilateral A $(^-2,3)$, B $(^-1,^-3)$, C $(2,^-4)$, D $(4,2)$ under the translation $\binom{2}{1}$?

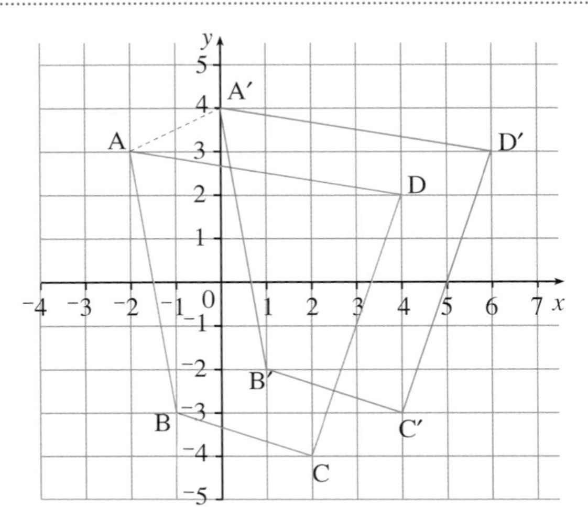


Under $\binom{2}{1}$:

A $(^-2,3) \rightarrow$ A′ $(0,4)$, B $(^-1,^-3) \rightarrow$ B′ $(1,^-2)$,
C $(2,^-4) \rightarrow$ C′ $(4,^-3)$, D $(4,2) \rightarrow$ D′ $(6,3)$

### Example 2

Find the image of the quadrilateral A $(0,3)$, B $(1,1)$, C $(3,0)$, D $(4,4)$ under an enlargement centred at the origin with a scale factor 2.

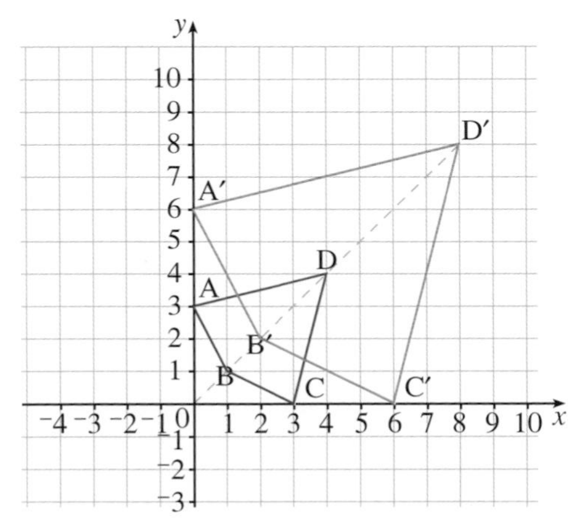


The coordinates of the enlarged quadrilateral are:

A′ $(0,6)$, B′ $(2,2)$, C′ $(6,0)$, D′ $(8,8)$

Notice that the distance from the origin to C′ is twice the distance from the origin to C.

### Example 3

The image of P $(1,^-2)$ after reflection in a mirror line is P′ $(3,4)$. Construct and hence find the equation of the mirror line.

First draw PP′. Then, using compasses, construct the perpendicular bisector of PP′.

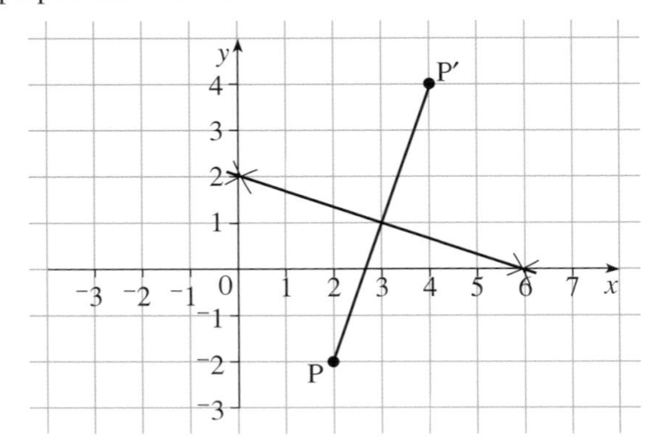


Two points on the mirror line are $(0,2)$ and $(3,1)$.
The $y$-intercept is 2 and the gradient $= \frac{1-2}{3-0} = \frac{^-1}{3}$
So the equation of the mirror line is $y = 2 - \frac{1}{3}x$

### Example 4

Under a rotation the line A $(3,1)$, B $(2,2)$ is sent to A′ $(0,4)$, B′ $(^-1,3)$. Find the angle and centre of rotation.

The centre of rotation is where the perpendicular bisectors of the lines AA′ and BB′ meet, which is at $(0,1)$. The angle of rotation is the angle BOB′ or AOA′ $= 90°$ anticlockwise.

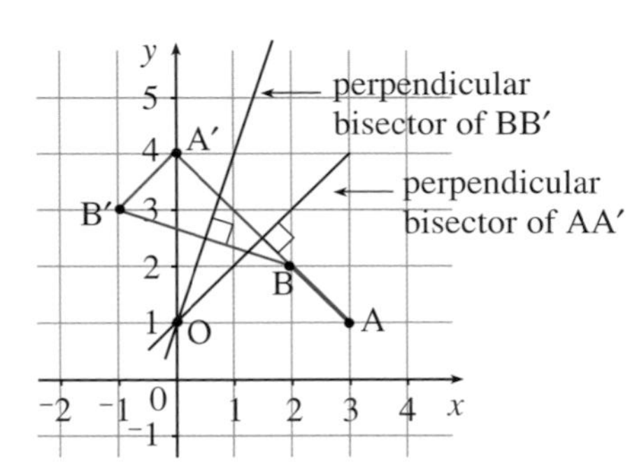

**Example 5**

What is the image of triangle A (1, 1), B (3, 2), C (2, 2) under the combined transformation **G** consisting of a translation $\begin{pmatrix} 0 \\ {}^-2 \end{pmatrix}$ followed by a reflection in the *y*-axis?

Under the translation:

$A\ (1,1) \rightarrow A'\ (1,{}^-1)$
$B\ (3,2) \rightarrow B'\ (3,0)$
$C\ (2,2) \rightarrow C'\ (2,0)$

Under the reflection:

$A'\ (1,{}^-1) \rightarrow A''\ ({}^-1,{}^-1)$
$B'\ (3,0) \rightarrow B''\ ({}^-3,0)$
$C'\ (2,0) \rightarrow C''\ ({}^-2,0)$

The image of triangle ABC under **G** is
$A''\ ({}^-1,{}^-1), B''\ ({}^-3,0), C''\ ({}^-2,0)$

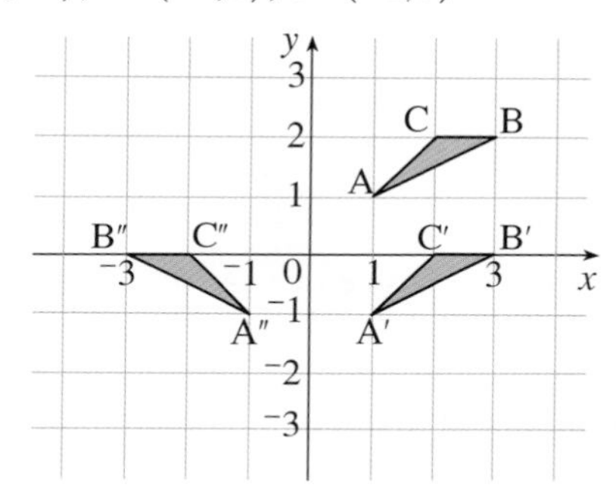


## Exercise 15

**(Note: Question 10 is extension work.)**

**1** The vertices of a triangle ABC are A (2, 1), B (4, 2) and C (4, 5).

**a** Find the vertices A′, B′ and C′ after a translation $\mathbf{T} = \begin{pmatrix} {}^-5 \\ {}^-6 \end{pmatrix}$.

**b** Find the vertices of image A″B″C″ of triangle A′B′C′ after a translation $\mathbf{U} = \begin{pmatrix} 0 \\ 6 \end{pmatrix}$.

**c** Find the vertices of the image of A″B″C″ after a translation $\mathbf{V} = \begin{pmatrix} 5 \\ 0 \end{pmatrix}$.

**2 a** Draw the trapezium with vertices A (1, 2), B (4, 2), C (3, 4) and D (2, 4).

**b** Draw the image of ABCD after a rotation of 90° anticlockwise about the origin.

**c** Draw the image of ABCD after a rotation of 180° about the origin.

**3** **S** is a rotation of 90° anticlockwise about $({}^-1, {}^-1)$. Triangle A has vertices at (3, 1), (6, 4), (4, 4).

**a** Draw, on squared paper, the triangle A.

**b** Draw **S**(A), by rotating triangle A.

**c** Draw $\mathbf{S}^2(A)$, by rotating triangle A twice.

**4 a** Draw the quadrilateral ABCD, which has vertices A (5, 1), B (7, 2), C (6, 5) and D (4, 4).

**b** Reflect ABCD in the line:

**i** $x = 0$

**ii** $y = 0$

**iii** $x = 2$

**5** On squared paper, draw rectangles ABCD and WXYZ with:

A (2, 1), B (2, 2), C (4, 2), D (4, 1)

W (3, 3), X (3, 5), Y (7, 5), Z (7, 3)

Describe the transformation of ABCD onto WXYZ.

**6 a** On squared paper, draw a triangle with vertices at (1, 3), (2, 4) and (2, 7).

**b** Draw the image of the triangle under an enlargement with centre (0, 0) and scale factor 2.

**7** These triangles are all congruent. Write down the sizes of all the unmarked angles and sides.

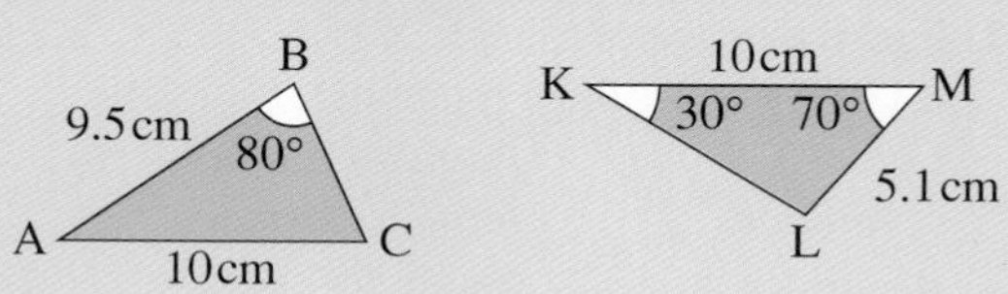


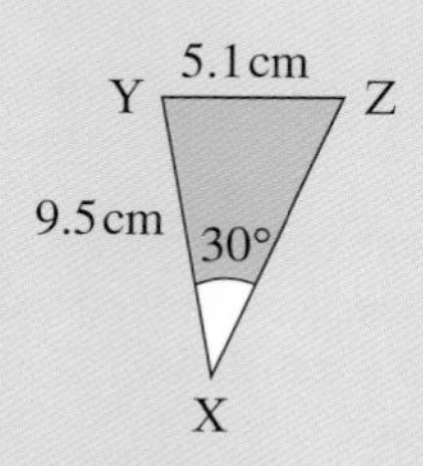


**8** A reflection maps these points onto the images given:

$(1, {}^-4) \rightarrow (4, {}^-1)$

$({}^-1, {}^-5) \rightarrow (5, 1)$

$({}^-1, {}^-7) \rightarrow (7, 1)$

What is the equation of the mirror line of the reflection?

**9** **a** On squared paper, draw triangle ABC with vertices A $(3, 2)$, B $(1, 1)$ and C $(1, 3)$.

**b** Translate ABC to A′B′C′, using the vector $\begin{pmatrix} ^-5 \\ ^-5 \end{pmatrix}$.

**c** Translate A′B′C′ to A″B″C″, using the vector $\begin{pmatrix} 1 \\ 7 \end{pmatrix}$.

**d** Give the vectors that translate:

**i** ABC to A″B″C″

**ii** A′B′C′ to ABC

**iii** A″B″C″ to ABC.

**10** ABC and A′B′C′ are similar triangles.

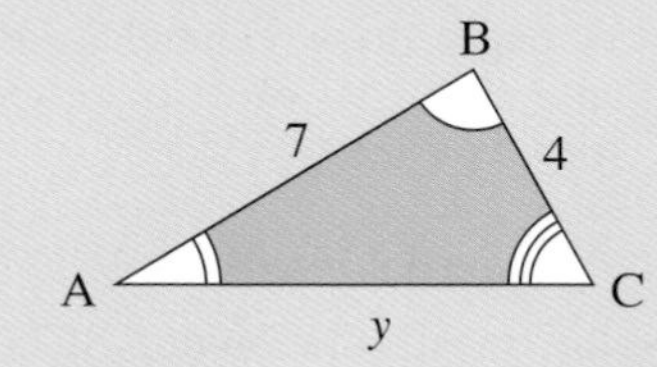


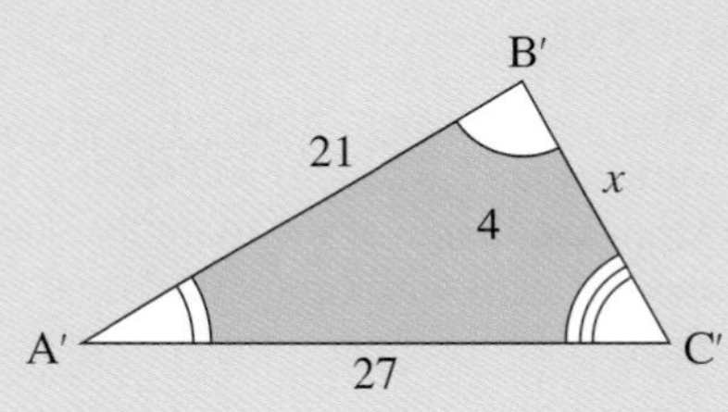


Work out $x$ and $y$.

**11** Find the image of triangle X $(^-1, 2)$, Y $(^-2, 2)$, Z $(^-3, 3)$ under these rotations.

| | *Centre of rotation* | *Angle of rotation (anticlockwise)* |
|---|---|---|
| **a** | $(0, 0)$ | 90° |
| **b** | $(0, 2)$ | 60° |
| **c** | $(^-1, 1)$ | 45° |
| **d** | $(3, 1)$ | 190° |

**12** Find the image of triangle P $(^-3, 4)$, Q $(^-2, 2)$, R $(^-1, 3)$ under a combined transformation consisting of a translation, **T,** and a reflection, **R,** as follows:

| | **T** | **R** |
|---|---|---|
| **a** | $\begin{pmatrix} ^-1 \\ 0 \end{pmatrix}$ | reflection in $y = 1$ |
| **b** | $\begin{pmatrix} 0 \\ 2 \end{pmatrix}$ | reflection in $y$-axis |
| **c** | $\begin{pmatrix} 0 \\ 4 \end{pmatrix}$ | reflection in $x = 2$ |
| **d** | $\begin{pmatrix} ^-1 \\ ^-1 \end{pmatrix}$ | reflection in $y = x$ |

**13** The triangle A $(2, 3)$, B $(4, 2)$, C $(5, 5)$ becomes triangle A′ $(^-4, 3)$, B′ $(^-6, 2)$, C′ $(^-7, 5)$ after reflection in a certain line.

**a** Draw ABC and A′B′C′ on squared paper.

**b** Construct the mirror line.

**c** Write down the equation of the mirror line.

**14** The transformation **R** is a reflection in the line $y = 1$. The transformation **S** is a translation $\begin{pmatrix} 3 \\ ^-1 \end{pmatrix}$.

**a** On squared paper, plot the triangle A $(1, 1)$, B $(2, 1)$, C $(2, 5)$.

**b** On your diagram show:

**i** the image A′B′C′ of ABC under **S**

**ii** the image A″B″C″ of A′B′C′ under **R.**

**c** Is **SR** = **RS**? Explain your answer.

**15** A trapezium with vertices W $(2, 3)$, X $(3, 5)$, Y $(5, 5)$ and Z $(6, 3)$ is transformed to a trapezium with vertices W′ $(^-3, 4)$, X′ $(^-5, 5)$, Y′ $(^-5, 7)$, Z′ $(^-3, 8)$.

Describe the transformation.

**16** Triangle ABC has vertices A $(0, 2)$, B $(2, 1)$, C $(1, 0)$.

**a** Draw a labelled diagram to show the images of ABC after the following transformations:

| | *Transformation 1* | *Transformation 2* |
|---|---|---|
| **i** | Enlargement scale factor 3, centre $(0, 0)$ | Translation $\begin{pmatrix} 1 \\ 2 \end{pmatrix}$ |
| **ii** | Rotation 90° anticlockwise about $(3, 2)$ | Reflection in line $x = 6$ |
| **iii** | Reflection in line $y = x$ | Translation $\begin{pmatrix} ^-3 \\ ^-3 \end{pmatrix}$ |

**b** In each case in part **a** describe a single (if possible) transformation that takes ABC to its final image.

**17** A quadrilateral has vertices P $(2, 1)$, Q $(3, 2)$, R $(5, 2)$, S $(5, 1)$.

**X** is a reflection in the $x$-axis.
**Y** is a rotation of 90° anticlockwise about $(0, 2)$.
**Z** is the translation $\begin{pmatrix} ^-2 \\ ^-1 \end{pmatrix}$.

Find the coordinates of the vertices of:

**a** **XY**(PQRS)

**b** **XZ**(PQRS)

**c** **YZ**(PQRS)

**18** A combined transformation **G** consists of a translation $\mathbf{T} = \begin{pmatrix} 5 \\ 0 \end{pmatrix}$ followed by a reflection in the $x$-axis.

**a** Find the image of a triangle with vertices A $(0, 2)$, B $(4, 3)$ and C $(4, 5)$ under **G**.

**b** If the triangle ABC was first reflected in the $x$-axis and then translated by $\mathbf{T} = \begin{pmatrix} 5 \\ 0 \end{pmatrix}$, would the image be the same?

# Summary

## You should know ...

**1** How to represent a translation by a column vector.
*For example:*

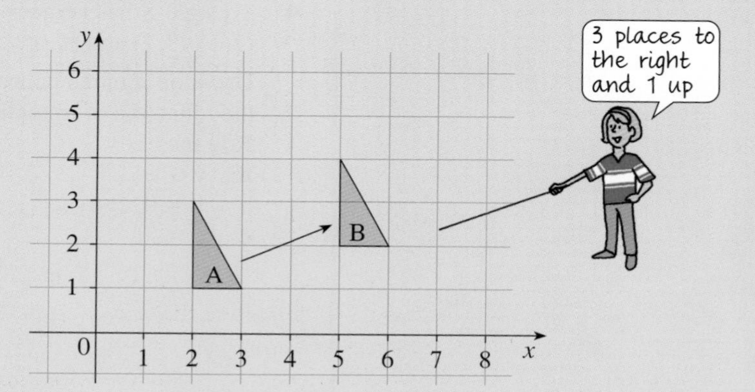


Triangle A is translated to triangle B by the vector $\begin{pmatrix} 3 \\ 1 \end{pmatrix}$.

**2** How to find the mirror line given the object and image point.
*For example*:
A is mapped to A′ by a reflection:

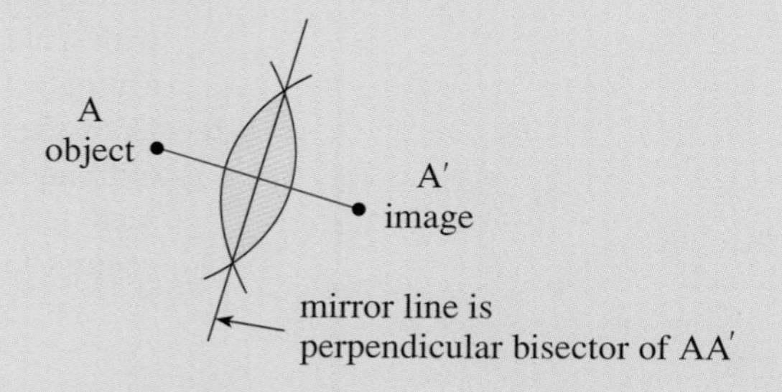


## Check out

**1** The vertices of a square are at $(1, 1)$, $(3, 1)$, $(3, 3)$ and $(1, 3)$.

Find the coordinates of the vertices after:

**a** the translation $\begin{pmatrix} 2 \\ 2 \end{pmatrix}$.

**b** the translation $\begin{pmatrix} ^-5 \\ ^-7 \end{pmatrix}$.

**2** **a** Find the image of the point $(3, 6)$ after reflection in the line $y = x$.

**b** The point $(^-2, 3)$, when reflected in a line, has image $(^-3, 2)$.
What is the equation of the mirror line?

**3** A rotation has a centre and an angle.
*For example:*

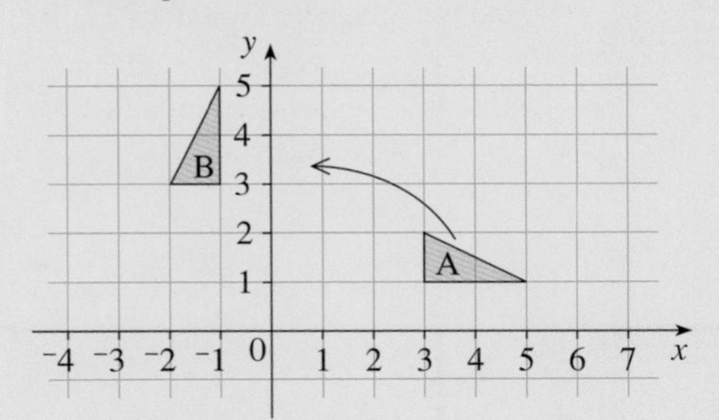


Triangle A is rotated 90° anticlockwise about the origin to triangle B.

**3** The triangle with vertices X ($^-2, 4$), Y ($^-2, 6$) and Z ($^-1, 5$) is rotated by 180° about the point (1, 0). Show, on squared paper, triangle XYZ and its image after the rotation.

---

**4** How to reflect a shape in a mirror line.
*For example:*

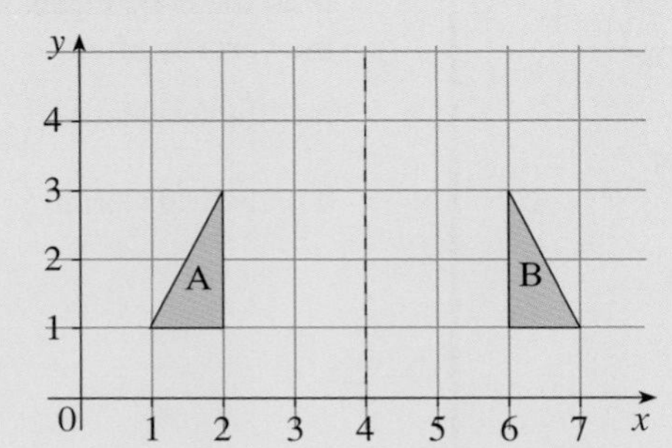


Shape B is the reflection of shape A in the line $x = 4$

**4** Triangle X has vertices at (2, 1), (3, 2) and (4, 0).

Draw, on squared paper, the reflection of triangle X in the line

**a** $y = 0$
**b** $x = 0$
**c** $x = {}^-1$

---

**5** An enlargement has a scale factor and a centre of enlargement.
*For example:*

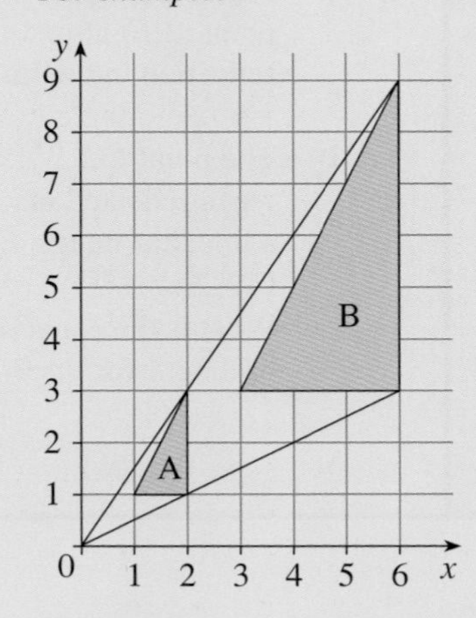


B is the enlargement of A by a scale factor of 3, with centre (0, 0).

**5** Triangle XYZ has vertices at (6, 1), (8, 1) and (8, 4) respectively.

**a** Draw the enlargement of triangle XYZ with scale factor 2, centre (8, 0).
**b** Draw the enlargement of triangle XYZ with scale factor 3 and centre (6, 0).

**6** Two triangles are similar if all corresponding angles are equal.
*For example:*

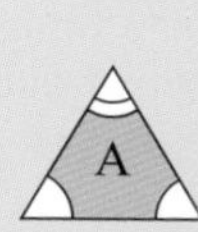


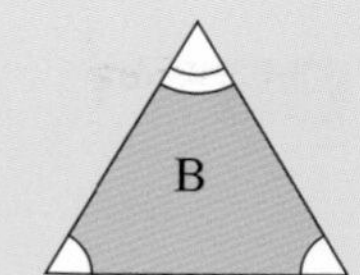


Triangle A is similar to triangle B. In this case, triangle B is an enlargement of triangle A.

The ratio of corresponding sides of similar triangles gives the scale factor of the enlargement.

Enlarged images will be similar to the object (the original shape). Rotated, reflected and translated images are congruent to their objects.
**(Note: This is extension work.)**

**6** Use the ratios of corresponding sides of these similar triangles to find

**a** the scale factor

**b** $x$ and $y$.

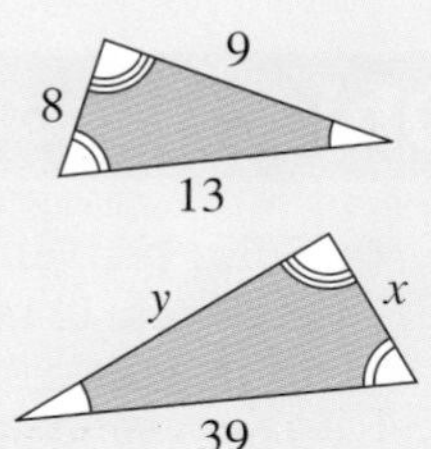


**(Note: This is extension work.)**

**7** How to find the centre of rotation given the object and image.
*For example*:
AB is mapped to A′B′ by a rotation:

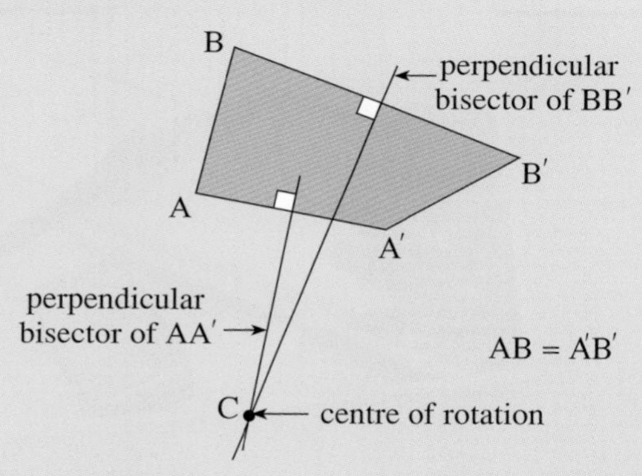


**7** X is the point $(6, 1)$ and Y is the point $(3, 2)$. Find the centre and angle of rotation that maps XY to X′Y′ with X′ $(^-3, ^-2)$ and Y′ $(^-4, ^-5)$.

**8** How to find the image of a point under two transformations.
*For example*:
For point P $(1, 1)$,
**R** is a reflection in $y = 2$
**S** is a reflection in $y = 5$

**SR**$(1, 1) =$ **S** (**R**$(1, 1)) =$ **S**$(1, 3) = (1, 7)$

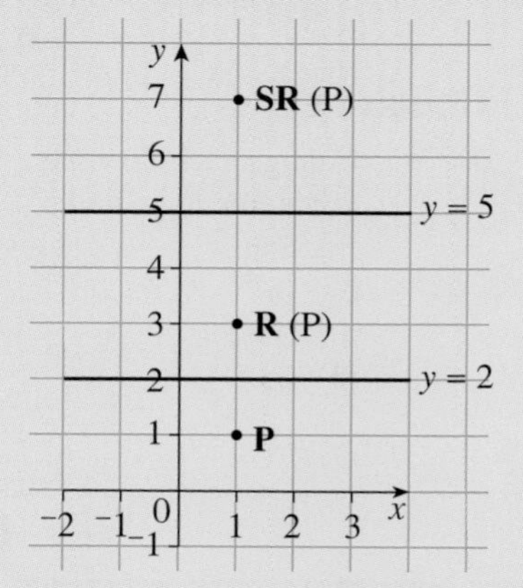


**8** **a** **M** is a reflection in the line $x = 3$ and J is the triangle with vertices $(2, 2)$, $(2, 6)$ and $(8, 1)$.

**i** Draw J.

**ii** Draw **M**(J), by reflecting J.

**iii** Draw **M²**(J), by reflecting J twice.

**b** If **N** is a reflection in the line $y = x$, draw

**i** **MN**(J)

**ii** **NM**(J)

# 16 Fractions, decimals and percentages

## Objectives

- Solve problems involving percentage changes, choosing the correct numbers to take as 100% or as a whole, including simple problems involving personal or household finance, e.g. simple interest, discount, profit, loss and tax.
- Recognise when fractions or percentages are needed to compare different quantities.

## What's the point?

Profit, loss, discount, tax and interest are important to any business. These are often expressed as fractions, decimals or percentages. Without an understanding of these it would be difficult to run a company. It is important to understand taxation. Any business trading in different countries must deal with the different tax laws in those. And everyone has to pay tax in some way – most things you buy have sales tax added, and most people earning a wage have to pay tax, the amount usually depending on how much they earn.

## Before you start

### You should know ...

**1** How to convert between decimals, percentages and fractions.
*For example*:

To convert 0.4 to

- a percentage: multiply by 100, to give 40%
- a fraction: the 4 is in the tenths column, so $0.4 = \frac{4}{10}$, which cancels down to $\frac{2}{5}$

### Check in

**1** **a** Change each decimal to a percentage and a fraction.
- **i** 0.7
- **ii** 0.62
- **iii** 0.08

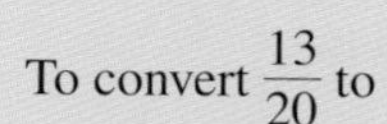

To convert $\frac{13}{20}$ to

- a decimal: rewrite the fraction as an equivalent fraction over 100, to give $\frac{65}{100}$, then $65 \div 100 = 0.65$; or simply do $13 \div 20$
- a percentage: multiply by 100, to give $\frac{13}{20} \times 100 = 65\%$

To convert 18% to

- a decimal: divide by 100, to give 0.18
- a fraction: write the percentage with a denominator of 100 and simplify the fraction, so $\frac{18}{100} = \frac{9}{50}$

**b** Change each fraction to a percentage and a decimal.

**i** $\frac{3}{10}$

**ii** $\frac{7}{20}$

**iii** $\frac{4}{5}$

**c** Change each percentage to a decimal and a fraction.

**i** 24%

**ii** 6%

**iii** 85%

**2** How to write a given number as a fraction or percentage of another number.

*For example:*

To write 24 g as **a** a fraction **b** a percentage of 160 g:

**a** Using equivalent fractions, $\frac{24}{160} = \frac{3}{20}$

**b** Using equivalent fractions, $\frac{24}{160} = \frac{3}{10} = \frac{15}{100}$, and $\frac{15}{100} = 15\%$

Or, if you are allowed your calculator, $(24 \div 160) \times 100 = 15\%$

**2** Write the first amount as a fraction and as a percentage of the second amount.

**a** 66 mm, 120 mm

**b** 20 g, 50 g

**c** 90 ml, 200 ml

**d** 18 minutes, an hour

**3** How to find a percentage of a quantity.

*For example*:

To find 15% of 240 km,

$0.15 \times 240 = 36$

So 15% of 240 km is 36 km.

Change the percentage to a decimal then multiply.

**3** Work out:

**a** 24% of 350 m

**b** 16% of \$520

**c** 72% of 3800 g

**d** 4% of 820 $cm^2$

## 16.1 Profit and loss

If you sell something for more money than it has cost you, you make a **profit**.

The picture shows Ken, who bought a bike for $220 and sold it for $300.

$$\text{Ken's profit} = \$300 - \$220 = \$80$$

- Profit = selling price − cost price. If the cost price is more than the selling price a **loss** is made.

### Exercise 16A

1 For each item he sold last week, Johnny wrote down how much it cost him (**cost price**) and how much he sold it for (**selling price**).

For each item:

a find the difference between the selling price and the cost price.

b say whether Johnny made a profit or loss.

| Item | Cost price | Selling price |
|---|---|---|
| Table | $320 | $365 |
| Four chairs | $288 | $240 |
| Bed | $495 | $570 |
| Fridge | $2163 | $2437 |
| Gas stove | $1400 | $1077 |

2 Copy and complete this table.

| Cost price | Selling price | Profit |
|---|---|---|
| $3.00 | $5.00 | |
| $1.50 | | $1.50 |
| | $6.40 | $1.40 |
| $3.00 | $3.00 | |
| | $6.20 | $1.20 |

3 Copy and complete this table.

| Cost price | Selling price | Loss |
|---|---|---|
| $4.00 | | $1.00 |
| $3.50 | $3.25 | |
| $1.00 | | $0.75 |
| | $3.50 | $1.20 |
| | $10.00 | $3.50 |

4 Look again at the list in Question **1**.

a Altogether, by the end of the week, had Johnny made a profit or loss on his sales?

b How did you work out the answer?

Some profit and loss problems are more complicated.

**EXAMPLE 1**

Khalid bought 150 pairs of shoes for $6000. He sold 100 of them for $50 each and the remaining 50 for $55 each. How much profit did he make altogether?

Khalid sold 100 pairs of shoes for $50:

$$100 \times \$50 = \$5000$$

Khalid sold the other 50 pairs for $55:

$$50 \times \$55 = \$2750$$

$$\text{Selling price of 150 shoes} = \$5000 + \$2750 = \$7750$$

Khalid bought the shoes for $6000

$$\text{So Khalid's profit} = \$7750 - \$6000 = \$1750$$

### Exercise 16B

1 Ian bought 200 shirts for $6000. He sold 120 of them for $32.50 each. He sold the remaining 80 for $35 each. How much profit did he make altogether?

2 Albert bought 25 calculators for $162.50. He sold them for $7.25 each.

a How much did each calculator cost Albert?

b How much profit did Albert make on each calculator?

c How much profit did Albert make altogether?

3 Belle bought 30 pairs of trousers for $1560. She sold them all at the same selling price and made a loss of $180.
   a How much did each pair of trousers cost Belle to buy?
   b How much loss did she make on each pair of trousers?
   c What was the selling price of each pair?

4 Belle also bought 12 pairs of men's sandals. She sold them for $34.95 each and made a total profit of $51.
   a How much profit did she make on each pair of sandals?
   b How much did each pair cost her?

5 Yasmina bought a bag of 300 mangoes for $225. When she sorted then she found she had

   90 top quality mangoes
   186 premium mangoes

   and the rest were bad.
   She sold the top quality mangoes for $1.10 each and her premium fruits for 90 cents each.
   She threw away the bad ones.
   What profit, if any, did she make?

## Buying and selling

The profit made when selling something can be written as a percentage. This percentage is called the **percentage profit**.

**EXAMPLE 2**

Jim bought a motorbike for $800 and sold it for $1120. Find his percentage profit.

Actual profit = $1120 − $800
= $320

Profit as a fraction of cost price

$= \frac{320}{800} = \frac{40}{100}$

Profit as a percentage = 40%

In general:

- **Percentage profit or loss** $= \dfrac{\textbf{profit or loss}}{\textbf{cost price}} \times 100$

### Exercise 16C

1 Copy and complete these tables.

a

| Item | Cost price | Selling price | Profit | Percentage profit |
|---|---|---|---|---|
| Shoes | $80 | $100 | | |
| Table | $600 | $750 | | |
| Stove | $750 | $1000 | | |
| Glass | $5 | $6 | | |
| Mat | $3 | $6 | | |
| Dish | $7.50 | $10 | | |

b

| Item | Cost price | Selling price | Loss | Percentage loss |
|---|---|---|---|---|
| Car | $16 000 | $4000 | | |
| Camera | $150 | $120 | | |
| Coat | $200 | $120 | | |

2

Tom bought an old van. It cost him $8000. What percentage profit would he make if he sold it again for:
   a $8800   b $9600   c $10 000?

3 Sugar can be bought wholesale for $3.60 per kilogram. What is the percentage profit if it is sold for
   a $4.05 per kg   b $4.50 per kg?

4 Maduka bought a new car for $80 000. He sold it three years later for $20 000.
   a What was Maduka's loss?
   b Write this loss as a fraction of the cost price.
   c What was his percentage loss?

5 Abraham bought a refrigerator for $2500. What was the percentage loss if he sold it for:
   a $2000   b $1500   c $2150?

## Calculating selling prices

If you know the cost of an item and the percentage profit made by selling it, you can find its selling price.

**EXAMPLE 3**

A craftsman makes a chair for \$80 and sells it to make a 20% profit on its cost price. What is his selling price?

$$\begin{aligned}\text{Profit made} &= 20\% \text{ of } \$80\\ &= \tfrac{20}{100} \times \$80\\ &= \tfrac{1}{5} \times \$80\\ &= \$16\\ \text{Selling price} &= \text{cost price} + \text{profit}\\ &= \$80 + \$16\\ &= \$96\end{aligned}$$

In general:

- **Selling price = cost price + profit**

In Book 2 you learned how to increase and decrease by a given percentage by adding to or subtracting from 100%, then changing the percentage to a decimal before multiplying. Here is a reminder of this method:

To increase 480 by 32%:
Find 132% of 480

100% + 32% = 132%

$1.32 \times 480 = 633.6$

To decrease 320 by 25%:
Find 75% of 320

100 - 25% = 75%

$0.75 \times 320 = 240$

You can use this method to find selling price from cost price and the percentage profit or loss. If there is a percentage profit, this is the same a percentage increase.

So Example 3 can be done in a shorter way:

Cost price = \$80
Profit = 20%, so selling price is 120% of the cost price
120% of \$80 = $1.2 \times \$80$ = \$96

If there is a percentage loss, you work out selling price using the percentage decrease method.

## Exercise 16D

1 Copy and complete the table.

| Item | Cost price | Profit as % of cost price | Selling price |
|---|---|---|---|
| Iron | \$70 | 25% | |
| Chair | \$240 | 15% | |
| Mug | \$3.60 | 10% | |
| Vase | \$48 | 8% | |
| Radio | \$220 | 9% | |

2 Selling price = 120% of \$290

Jason sells his stereo for a 20% profit. If he bought it for \$290, what did he sell it for?

3 What is the selling price of a television made for \$1000 and sold at a 25% profit?

4 Workers in a glove factory earn \$150 a week. The management offers them a wage increase of 10%.
  **a** How much would the workers earn if they accepted the management offer?
  **b** The workers want an increase of 15%. How much would that give them?

5 A restaurant adds a service charge of 8% onto every bill. What is the total bill if a meal costs
  **a** \$40  **b** \$65  **c** \$9.50?

6 Achim sells his car at a 30% loss. What is his selling price if he bought it for \$30 000?

7 Beverly buys six marbles for 60 cents. She sells them to her friend for a 20% loss.
  **a** For how much did she buy each marble?
  **b** How much did she lose on each marble she sold?
  **c** What did she sell each marble for?

8 

Davina buys a goat for \$75. What is her selling price for the goat if she sells it for
  **a** a loss of 20%  **b** a loss of 15%?

**9** A car bought for \$50 000 depreciates in value by 20% each year.
(**Note:** 'depreciates' means 'goes down in value'.)

**a** What is its value after one year?
**b** What is its value after two years?

**10** **a** Find the value of the car in Question **9** after five years.
**b** What would be its value after five years if the depreciation rate was only 10%?

## 16.2 Discounts and sales

20% DISCOUNT
ON ALL SALES !!

The amount that you subtract from a bill is called the **discount**.

A discount is usually a percentage of your total bill.

**EXAMPLE 4**

A television marked at \$1400 is sold at a 15% discount. What is the discount price?

**Method 1**

Discount = 15% of \$1400

$= \frac{15}{100_1} \times \$1\overset{14}{400}$

= \$210

Discount price = cost price − discount

= \$1400 − \$210

= \$1190

**Method 2**

Discount = 15% of cost price, so discount price is 100% − 15% = 85% of the cost price

Discount price = 0.85 × \$1400 = \$1190

### Exercise 16E

**1** Find the discounted prices of the items shown.

**a**

**b**

SALE!
20% Discount
on marked prices
of JEANS
\$85

**c**

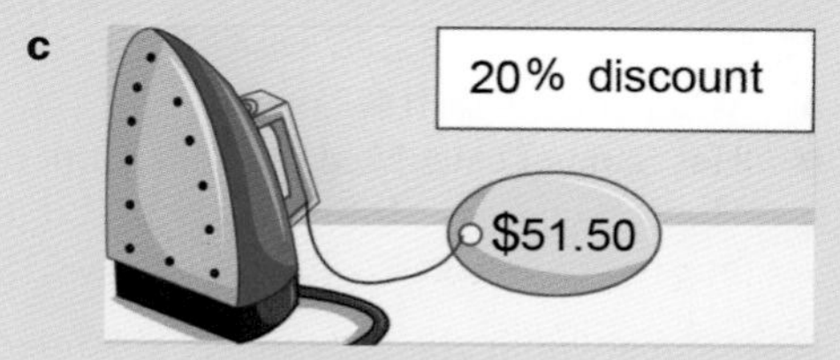


**d**

**e**

**2** What would be the selling prices of the items in Question **1** if the discount was 30%?

**3** The sign in Samuel's paint shop reads

10% OFF, if you spend $50 or more!

Mr Singh bought some paint and paint brushes in Samuel's shop.
Here is the bill he got.

| | $ |
|---|---|
| 2.5 litres white gloss paint | 19.50 |
| 2.5 litres white undercoat | 16.30 |
| 2 5-cm paint brushes | 7.20 |
| 1 10-cm paint brush | 4.50 |
| 1 tin white spirit | 6.50 |
| Total | 54.00 |
| less 10% | 5.40 |
| | $ 48.60 |

**a** What does 'less 10%' mean?
**b** Has Samuel calculated the 10% properly?
**c** How much does Mr Singh pay for his brushes and paint?

**4** How much would you have to pay at Samuel's paint shop, from Question **3**, if your bill total before discount was
**a** $205 **b** $84.15 **c** $196.84
**d** $923 **e** $303.50 **f** $416.03?

**5** Find how much you would have to pay on a bill of $65 if a discount was given of either
**a** 7% or **b** 9%

**6** Here is a bill sent out by the Cool & Casual Clothing Company, to Mrs Akin's shop.

THE COOL & CASUAL CLOTHING CO.

*To:* Akin Clothing
*Address:* 401 Shoreham Street, Kettering
*Date:* 20th January 2013

| No. | Article | Price each | Cost |
|---|---|---|---|
| 12 | Dress | $165.00 | $1980.00 |
| 10 | Evening dress | $315.00 | $3150.00 |
| 24 | Blouse | $115.25 | |
| 10 | Trousers | $123.75 | |
| | | Total | |

Note: Customers are reminded that a discount of 10% is given for payment within 15 days of the above date.

**a** Copy and complete the bill.
**b** By what date must Mrs Akin pay this bill, in order to get a discount?
**c** After discount, how much will she have to pay?
**d** Why do you think some companies offer a discount if you pay your bill quickly?

## Sales

Here is a shop window. There is a **sale** going on in the shop. This means the shop is offering a discount on the usual prices.

## Exercise 16F

1. Why do shops have sales?
   Try to think of three reasons.
2. **a** What is the meaning of 'big reductions'?
   **b** What is the meaning of 'bargains'?
3. Look again at the shop window. A 10% reduction is given on all marked prices. Find the sale prices of all the items in the window.
4. What would be the sale prices of the items in the shop window if the discount given was 30%?
5. Malik owns a men's clothing shop. In his sale, he gives a discount of 22% on all selling prices. Copy and complete the table below for items on sale in Malik's shop.

| Item | Usual price | Sale price |
|---|---|---|
| Shirt | $46.50 | |
| T-shirt | $19.00 | |
| Trousers | $54.00 | |
| Pair of socks | $12.50 | |
| Belt | $14.50 | |

If the cost price and the sale price are given, you can work out the percentage discount.

Remember, to convert a fraction to a percentage, you multiply the fraction by 100%.

Find the percentage discount on the dress shown.

Discount $= \$120 - \$80$

$= \$40$

Discount as fraction of cost price

$$= \frac{40}{120} = \frac{1}{3}$$

Percentage discount $= \frac{1}{3} \times 100\%$

$= 33\frac{1}{3}\%$

It is important that you know when to use fractions and when to use percentages to compare different quantities. If you are comparing discounts and all the original cost prices are equal, you can compare them easily by giving them as fractions. It is a lot harder to compare discounts if you give them as a fraction of the cost price and the cost prices are different. This is because the denominators of the fractions will be different. Changing all the discounts to percentages makes comparison much easier.

## Exercise 16G

1. Calculate the percentage discount on these items.

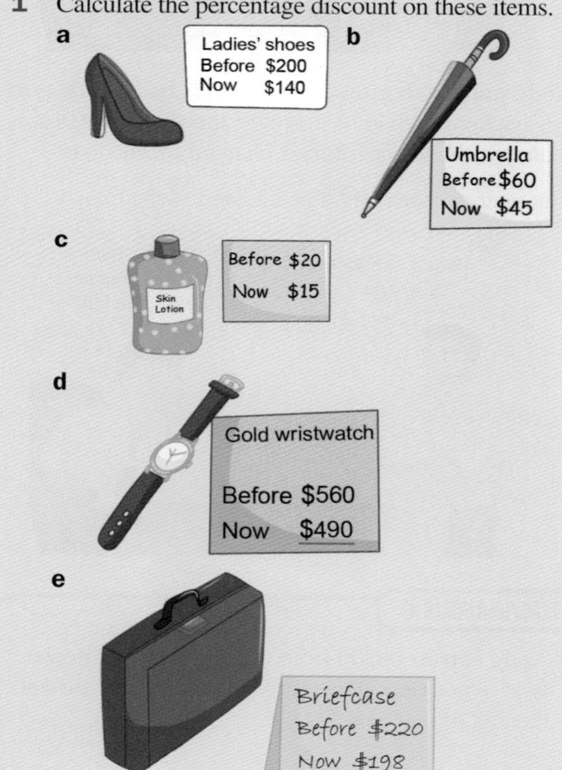


2. Which is the better buy,
   9 music albums for $53.10 or
   12 music albums for $76.80 with a 10% discount?
3. An insect spray is sold in 600-ml and 450-ml containers. The 600-ml container costs $19.50 and has a 10% cash discount. The 450-ml container sells for $13.22.

   Which is the better buy? Why?
4. A restaurant offers a 10% discount for senior citizens. If a senior citizen pays $59.40 for a meal, what would he have paid if there was no discount?

# 16.3 Loans and savings

## Loans

Andy wants to borrow $200 from his bank. He has to pay back more than he borrows, though. The extra money is the fee or **interest** charged for the loan.

The interest is usually a percentage of the loan.

**EXAMPLE 6**

Andy borrows $200 from the bank. If the interest is 15% per year, how much interest will he owe at the end of one year?

Loan = $200

Interest = 15% of $200

$= \frac{15}{100} \times \$200$

= $30

So Andy will owe $30 interest.

Altogether he will owe $230.

The total can be worked out as
115% of $200 = 1.15 × $200 = $230

## Savings

Andy's brother Tom has saved $200. He puts his money in a savings account at his bank. This time Tom loans the bank $200. Now the bank has to pay Tom interest on his loan.

To **deposit** money just means to put it in your account at the bank.

The fee that banks pay you when you lend money to them is also called **interest**.

The amount of money that you borrow or save is called the **principal**. You can borrow or save money for more than one year. If the interest is worked out as a percentage of the principal in the first year this is called **simple interest**. If you are saving, the interest earned is not reinvested. This means that the amount of interest earned each year does not change.

There is another type of interest, called 'compound interest', which you will learn about at Cambridge IGCSE® level. This is worked out a different way: the interest earned each year is reinvested, so the interest earned each year increases.

If the investment is only for one year the simple interest and the compound interest are the same amount, so there is no need to say which sort of interest is involved. If interest is calculated for more than one year then it is important to make clear which type of interest is being earned.

The formula for calculating simple interest is

$$I = \frac{PRT}{100}$$

where $I$ is the interest, $P$ is the principal, $R$ is the percentage rate per year and $T$ is the time in years.

**EXAMPLE 7**

Find the simple interest earned on a deposit of \$300 saved for 4 years at a rate of 5% per year.

$$I = \frac{PRT}{100}$$

$$= \frac{300 \times 5 \times 4}{100}$$

$$= 60$$

So the simple interest earned is \$60.

After 4 years, the savings will be worth \$360.

## Exercise 16H

**1** How much interest would Andy have to pay at the end of the year, if the interest rate on his loan of \$200 was:
- **a** 10% per year
- **b** 20% per year
- **c** 7% per year?

**2** In Question **1**, how much money must be paid back to the bank at the end of year, altogether?

**3** If the interest charged is 15% per year, how much interest will Andy pay on year-long loans of
- **a** \$100
- **b** \$500
- **c** \$5000?

**4** How much interest will the bank pay Tom if he deposits \$200 for one year and the interest rate is
- **a** 3% per year
- **b** 4% per year
- **c** 5% per year?

**5** The interest rate at the People's Bank is 4%. How much will Anika have in the bank after one year if she deposits
- **a** \$50
- **b** \$200
- **c** \$250
- **d** \$500
- **e** \$10 000
- **f** \$50 000?

**6** Layla wants to buy a washing machine for \$2500. Her bank charges 12% interest on loans. How much will her washing machine cost her if she uses a loan to buy it and pays the bank back after one year?

**7** Alan needs a loan of \$8000 to buy a piece of land. If the bank's interest rate is 10%, find the simple interest payable on his loan
- **a** after one year
- **b** after two years
- **c** after five years.

**8** A Credit Union offers a 5% interest rate on deposits. What amount of simple interest will have been paid on a deposit of \$400 after
- **a** one year
- **b** two years
- **c** five years?

**9** Mr Khan put \$6000 in his bank. He withdraws it all four years later. How much does he withdraw if the simple interest rate was 4% per year?

**10** Find the simple interest paid on the following loans
- **a** \$200 for 4 years at 15% per year
- **b** \$500 for 4 years at 15% per year
- **c** \$1000 for 2 years at 15% per year
- **d** \$3000 for 2 years at 12% per year
- **e** \$4500 for 3 years at 9% per year

**11**
- **a** What principal will earn \$360 in 3 years at a rate of 6% simple interest per year?
- **b** If \$192 is earned in simple interest on a principal of \$600 invested for 4 years, what is the rate of interest per year?
- **c** How many years would an investment take to earn \$36 in simple interest on a principal of \$300 at 4% interest per year?
- **d** What principal will earn \$105 in 5 years at 3% simple interest?
- **e** What rate of interest will earn \$45 in simple interest on a principal of \$750, invested for 2 years?
- **f** How many years would a principal of \$540 earning 5% interest per year take to earn \$216 in simple interest?

**12** The interest rate at a bank is 2% per year. George has \$1000. How long will he have to keep his money in the bank before he gets \$200 in simple interest?

**13** Debbie puts \$500 in the First Street Bank. Four years later her money is worth \$600.
- **a** What amount did the bank pay her each year in simple interest?
- **b** What was the bank's interest rate?

14 **a** Write down two reasons why people save money in banks.
**b** Try to find out how banks use the money that people deposit.

## INVESTIGATION

Visit banks near where you live on go online to find the answers to the following questions:

a What does a bank do with the extra money it receives as interest on loans?
b Who can get loans from a bank? What are the conditions?
c Find out the interest rates on deposits and loans with different banks.
Are they all the same?
Which, do you think, is the best bank?
Explain why.
Discuss your answer with your friends.
d Find out about compound interest. Do you think it is better or fairer than simple interest?

## TECHNOLOGY

Try out the puzzle about percentage increase and codes at this web page:
www.nrich.maths.org/4799

## 16.4 Taxes

Where does the money come from to pay for

- roads
- the police
- hospitals
- airports?

The answer is: from **taxes**.

Taxes are usually payments made to government to pay for the cost of services.

The main taxes used are:

- import duties
- sales tax, somtimes called 'value added tax'
- income tax.

The first two of these taxes are usually levied on goods and services. Income tax is levied on people.

Taxes on goods and services are usually charged as a percentage of the cost.

**EXAMPLE 8**

In the UK sales tax is 20%. What is the sale price of a mobile phone which costs \$95 before sales tax?

Sales tax = 20% of \$95
= 0.2 × \$95
= \$19

Sale price of cell phone = \$95 + \$19
= \$114

### Exercise 16I

**1** The prices of these items before sales tax are shown below.

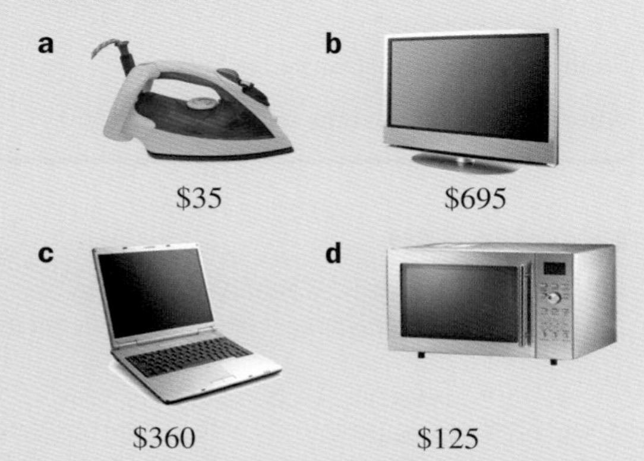


What are the sale prices if sales tax is charged at 15%?

**2** Import duty on luxury items is charged at 40%. What is the duty on:
**a** a sports car valued at \$145 000
**b** perfume valued at \$203
**c** a DVD player valued at \$460
**d** an electric razor valued at \$115?

**3** Sales tax in a country is 16.5%. What will the selling prices of these three items be, given the prices before tax shown?
**a** Chair \$132
**b** Coffee table \$29
**c** Chest of drawers \$215

**4**

A double room at the Grand Hotel costs \$220 per night. James and his wife stay at the hotel for 4 nights. What will the bill be if a 10% hotel tax and a 7.5% service charge are added to their bill?

**5** What is the rate of sales tax if a radio is priced at \$130 before tax and \$151.45 after tax?

**6** What is the cost price of a chair which is priced at \$330 including sales tax at 15%?

## Income tax

Income tax is the tax paid to government on the money you earn.

Only part of your annual income is taxed. The part that is not taxed is called **tax-free income**.

The amount of tax-free income is decided by government, and is usually a fixed amount.

**Taxable** or **chargeable income** is the money you earn in excess of the tax threshold.

In the UK the basic rate of income tax is 20%. Different countries have different rates.

### EXAMPLE 9

Anuradha earns \$72 000 each year. The tax allowance and rate in her country are:

| Tax allowance | Tax rate |
|---|---|
| \$60 000 | 25% |

**a** What is Anuradha's taxable income?
**b** How much tax does she pay?

---

**a** Taxable income = annual income − tax allowance
= \$72 000 − \$60 000
= \$12 000

**b** Tax payable = 25% of taxable income
= 25% of \$12 000
$= \frac{25}{100} \times \$12\,000$
= \$3000

In Example 9, Anuradha earns \$72 000. This is his **gross pay**. She pays \$3000 in income tax each year. Her take-home annual salary is \$69 000; this is her **net pay**.

### Exercise 16J

**1** **a** Find out about the tax allowances for earned income in your country.
**b** What are the income tax rates?

Use this tax allowance and rate for Questions **2–5**:

| Tax allowance | Tax rate |
|---|---|
| \$60 000 | 25% |

**2** Dakshi earns \$115 000 each year.
**a** What is his chargeable income?
**b** How much income tax does he pay?
**c** What is his net annual income?

**3** How much income tax will Tarek pay if his annual salary is:
**a** \$60 000 **b** \$600 000 **c** \$26 000?

**4** In Question **3**, what is Tarek's monthly net salary?

**5** Melinda paid a total of \$12 000 in income tax last year.
**a** What was her chargeable income?
**b** What was her net salary?
**c** What was her net monthly salary?
**d** What was her gross salary?

Use this table of tax rates and allowances for Questions **6** and **7**.

| Tax allowance | Tax rates | |
|---|---|---|
| \$10 000 | First \$20 000 | 10% |
| | Next \$20 000 | 20% |
| | Excess | 40% |

**6** Anil has an annual salary of \$48 500. Find
**a** his chargeable income
**b** the amount of tax he pays
**c** his net income.

**7** Find Anna's net income if her annual salary is
**a** \$8000 **b** \$80 000 **c** \$115 000

## 16.5 Percentage change

So far you have looked at lots of different percentage problems, involving, for example, percentage profit, percentage loss, percentage discount, percentage interest or percentage tax. You have used quite a few different formulae for these problems. Many of these can be summarised in one formula:

- $\text{Percentage change} = \dfrac{\text{actual change}}{\text{original amount}} \times 100$

where the 'change' is profit, loss, appreciation, depreciation, increase, decrease, discount, error, and so on.

(**Note: appreciation** means 'go up in value', while **depreciation** means 'go down in value'.)

It is important, when working with such percentages, to choose the correct number to use as 100% or the whole; that is, as the 'original amount'. You need to read the question carefully to make sure you choose correctly. In addition, you may be given the change, or you may have to use subtraction to find it.

### Exercise 16K

**1** A piece of paper 150 cm long is trimmed to 120 cm. What percentage has been removed?

**2** A spring 60 cm long is stretched to 69 cm. What percentage of the original length is the increase?

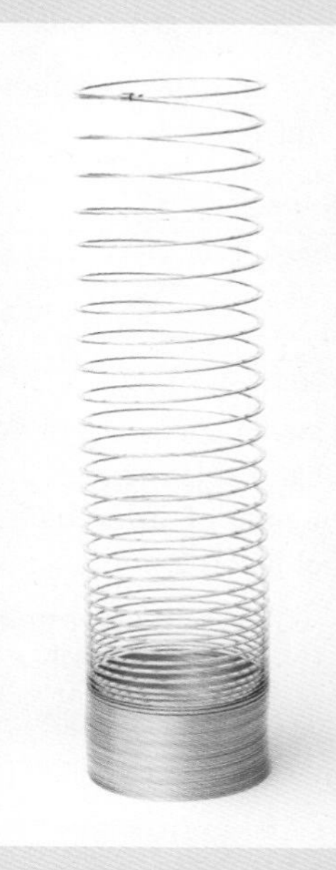

**3** After a storm, 135 trees out of 180 were left standing. What is the percentage loss in the number of trees?

**4** A coat with an original price of $300 was reduced to $180 in the sale. What percentage discount was given on it?

**5** A car which cost $4000 now has a value of $2400. What is the percentage depreciation in its value?

**6** A man had a mass of 70 kg. After six months on a diet his mass was 63 kg. What is the percentage reduction in his mass?

**7** A painting was bought for $12 000 and sold 3 years later for $22 200. What was the percentage profit in the sale?

**8** There are 195 litres of water left in a tank which originally held 300 litres. What is the percentage decrease?

**9** Gemma's wage was $300 per week. Following a pay rise, she now earns $366. What is the percentage increase in her wage?

**10** Sam measured a length of wood and found it to be 300 mm long. He made an error in his measuring, however, and the real length is 297 mm. What percentage error is there in Sam's measuring?

**11** A plant is 400 cm tall on Monday. By Thursday it is 424 cm tall. What is the percentage increase in height between Monday and Thursday?

**12** A house originally cost $120 000. Ten years later, it is now valued at $186 000. What is the percentage appreciation in the value of the house?

**13** In April there were 85 mm of rain. In May there was 68 mm. What is the percentage change in rainfall between April and May?

**14** Copy and complete with the correct percentage increase or decrease:

**a** 290 is ......... % greater than 200.
**b** 308 is ......... % greater than 175.
**c** 324 is ......... % less than 400.
**d** 143 is ......... % greater than 65.

**15** The temperature in the desert at 1 am was 4°C. By 1 pm it had risen to 48°C. What is the percentage rise in temperature?

**16** In one month the price of fuel increased by 20%. Then, over the next month, it decreased by 10%. What was the overall percentage change in the price of fuel over the two-month period? State whether it is a decrease or increase.

# Consolidation

**Example 1**

A car costing \$4800 is sold two years later for \$3000. What is

**a** the profit or loss

**b** the percentage profit or loss?

---

**a** The selling price is lower than the cost price so this is a loss.
Loss = \$4800 − \$3000 = \$1800

**b** $\text{Percentage change} = \frac{\text{actual change}}{\text{original amount}} \times 100$

$\text{Percentage loss} = \frac{1800}{4800} \times 100 = 37.5\%$

**Example 2**

What is the selling price of a bed priced at \$540 if 15% sales tax is then added?

---

After sales tax, the bed costs 100% + 15% = 115% of its original price.

115% of \$540 = 1.15 × \$540
= \$621

This is a percentage increase problem.

**Example 3**

What is the cost of a car priced at \$8400 if a 12% discount is given?

---

After the discount, the car costs 100% − 12% = 88% of its original price.

88% of \$8400 = 0.88 × \$8400
= \$7392

This is a percentage decrease problem.

**Example 4**

What is the simple interest to be paid on a loan of \$40 000 at a rate of 8% per year for 5 years?

---

$I = \frac{PRT}{100}$

where $I$ = interest, $P$ = Principal, $R$ = rate and $T$ = time.

$I = \frac{40\,000 \times 8 \times 5}{100} = 400 \times 8 \times 5$

$= 16\,000$

The interest is \$16 000.

## Exercise 16

**1** Mr Mulling is having a 10% discount sale in his shop. Copy and complete the table of sale items.

| Marked price | 10% of marked price | Sale price |
|---|---|---|
| \$50 | \$5 | \$45 |
| \$80 | \$8 | \$72 |
| \$20 | \$2 | |
| \$250 | | |
| \$180 | | |
| \$30 | | |
| \$10 | | |

**2** **a** What is the cost of a refrigerator priced at \$2000 if a sales tax of 3% is then charged?

**b** How much would the price rise if the government increased sales tax to 5%?

**3** A tin of beans normally holds 450 g of beans. A new tin holds 20% more.

**a** What is the mass of beans in the new tin?

**b** What should the price of the new tin be, if the original tin sold for \$0.48?

**4** Vishan has \$800 in his bank account. How much interest will he get after one year if the simple interest rate is

**a** 2% **b** 3% **c** 5%?

**5** In Question **4**, how much money will Vishan have in his account altogether after one year?

**6** The room rate at the Hastings Hotel is \$210 per night. Justin spends two nights at the hotel. What is his bill if

**a** a $7\frac{1}{2}$% service charge is added?

**b** a 10% government tax is added?

**c** both the $7\frac{1}{2}$% service charge and the 10% government tax are added?

**7** Find the total simple interest to be paid on these loans:

**a** \$5000 at 8% per year for 6 years

**b** \$40 000 at 7% per year for 3 years

**c** \$7050 at 6% per year for 4 years

**d** \$24 225 at $7\frac{1}{2}$% per year for 8 years

**8** Copy and complete the table based on simple interest rates.

| Principal | Rate | Time | Interest |
|---|---|---|---|
| \$6000 | 11% | 4 years | |
| \$3500 | 10% | | \$1050 |
| | $7\frac{1}{2}$% | 6 years | \$4308.75 |

**9** Natalia deposits \$8500 in the bank. How long must she wait for the total amount to become \$10 000 if the simple interest rate is:

**a** 5% **b** 3%?

**10** Copy and complete these tables.

**a**

| Cost price | Selling price | Percentage profit |
|---|---|---|
| \$130 | \$156 | |
| \$2.50 | \$2.70 | |
| \$482 | \$843.50 | |
| \$24 | \$48 | |

**b**

| Cost price | Selling price | Percentage loss |
|---|---|---|
| \$86 | \$73.10 | |
| \$4200 | \$4074 | |
| \$50 | \$25 | |
| \$140 | \$100.80 | |

**11** Sales tax of 15% is added to all goods by a government. What is the selling price of

**a** an MP3 player priced at \$80
**b** a television priced at \$350
**c** a book priced at \$6.40
**d** a pen priced at \$3.50
**e** a tin of biscuits priced at \$2.30
**f** a box of soap powder priced at \$2.80
**g** a bicycle priced at \$98
**h** a wardrobe priced at \$218?

**12** Mr Ramsaroop earns \$750 per week. He then gets a rise of 8%.

**a** By how much does his weekly wage increase?
**b** What is his new weekly wage?

**13** Una paid \$460 for a television. What is the percentage loss if she sells the television for \$299?

**14** Which of the following gives the better return on \$8000?

**a** $7\frac{1}{2}$% simple interest for 3 years
**b** 9% simple interest for 2 years

# Summary

## You should know ...

**1** Profit = selling price − cost price

Digital radio cost price \$96 Selling price \$120

$$\text{Profit} = \$120 - \$96 = \$24$$

If the selling price is lower than the cost price there is a loss.

**2** How to find the selling price if you know the percentage profit or percentage loss.

*For example*:

A salesman buys a watch for \$60 and sells it for a 25% profit. What is his selling price?

$$\begin{aligned}\text{Profit} &= 25\% \text{ of } \$60\\ &= \frac{25}{100} \times \$60\\ &= \frac{1}{4} \times \$60\\ &= \$15\\ \text{Selling price} &= \text{cost price} + \text{profit}\\ &= \$60 + \$15\\ &= \$75\end{aligned}$$

## Check out

**1** Find the profit or loss on.

**a** A car bought for \$75 000 and sold for \$50 000.
**b** 12 bags of sugar bought for \$95 each and sold for \$1200 altogether.

**2** Copy and complete:

| Item | Cost price | Percentage profit | Selling price |
|---|---|---|---|
| Stove | \$150 | 10% | |
| Computer | \$600 | 15% | |
| Clock | \$75 | 5% | |

**3** Discount is the amount you subtract from a bill.
It works the same as percentage decrease.

Sale price is $100\% - 20\% = 80\%$ of original price.

$$\begin{aligned}\text{Sale price} &= 80\% \text{ of } \$105 \\ &= 0.8 \times 105 \\ &= \$84\end{aligned}$$

**3 a** A store gives 5% discount on items. What is the discounted cost of a sofa marked at $1250?

**b** What would be the price of the sofa if a 3% sales tax was then added?

---

**4** The simple interest, $I$, on a principal, $P$, is given by the formula

$$I = \frac{PRT}{100}$$

where $R$ is the rate of interest per year and $T$ is the time in years.

**4 a** Find the simple interest on $600 kept in a bank for 5 years at interest of 4% per year.

**b** How long would you keep $750 in a bank that offers 5% per year simple interest before you received $112.50 in interest?

---

**5** The purpose of taxes and the types of taxes that are levied.

**5** Explain in your own words the meaning of:

**a** income tax **b** sales tax
**c** gross pay **d** net pay.

---

**6** How to work out percentage change.

$$\text{Percentage change} = \frac{\text{actual change}}{\text{original amount}} \times 100$$

The 'change' may be an increase or decrease and may reflect, for instance, profit, loss, appreciation, depreciation, discount or error.

*For example:*

A javelin thrower has a personal best throw of 60 m. He then throws 63 m. What is the percentage increase in his personal best?

$$\begin{aligned}\text{Percentage increase} &= \frac{\text{actual increase}}{\text{original amount}} \times 100 \\ &= \frac{63 - 60}{60} \times 100 \\ &= \frac{3}{60} \times 100 \\ &= 5\%\end{aligned}$$

**6 a** Last year there were 1200 children in a school. This year there are 1248. What is the percentage increase in the number of pupils?

**b** A piece of wood 700 mm long is cut down to 602 mm long. What is the percentage decrease in its length?

**c** A pair of shoes were priced at $80. They have been reduced to $44 in a sale. What is the percentage discount?

**d** A car depreciates in value from $8000 to $6800 in a year. What is the percentage depreciation?

# 17 Area, perimeter and volume

## Objectives

- Solve problems involving the circumference and area of circles, including by using the $\pi$ key of a calculator.
- Calculate lengths, surface areas and volumes in right-angled prisms and cylinders.

## What's the point?

What is the capacity of your freezer?
Which is the bigger container?
Which bottle holds more?
Volumes and capacities can be worked out to help market a product.

## Before you start

### You should know ...

1 How to find the areas of

**a** a square: $b^2$

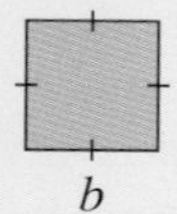

**b** a rectangle: $l \times w$

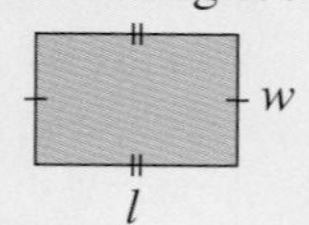

**c** a triangle: $\frac{1}{2} b \times h$

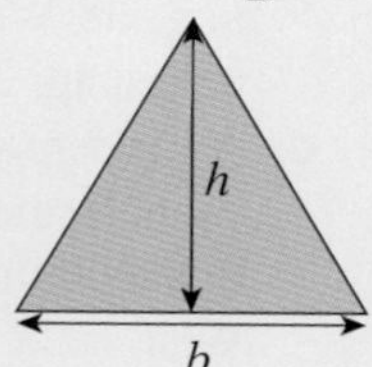

### Check in

1 Find the areas of these shapes.

**a**

15 cm

15 cm

**b**

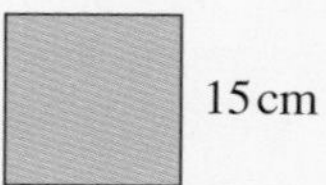

**c**

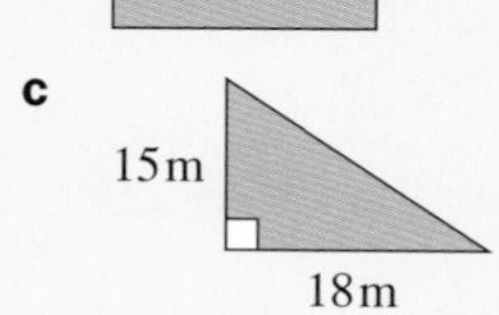

2 The parts of a circle.

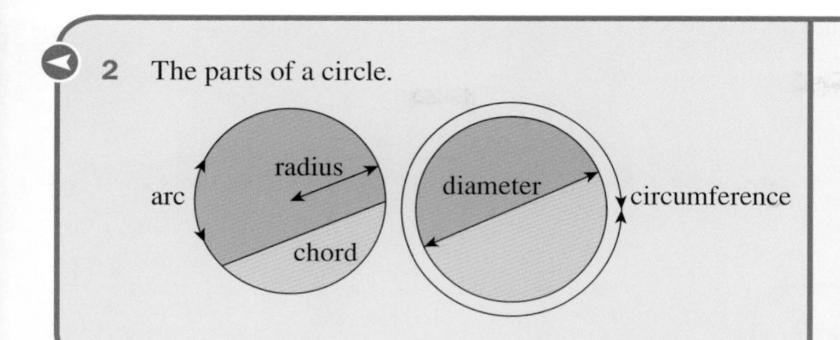


2 **a** What is the name for the perimeter of a circle?
**b** How do you find the diameter when you know the radius?
**c** How do you find the radius when you know the diameter?

## 17.1 Circles

The formula for the area of a circle is

- Area of circle $= \pi r^2$

where $r$ is the radius and $\pi = 3.142$

*For example:*

The area of a circle, $A$, with radius 8 cm is

$A = \pi r^2 = 3.142 \times 8\,\text{cm} \times 8\,\text{cm}$
$= 201.088\,\text{cm}^2$

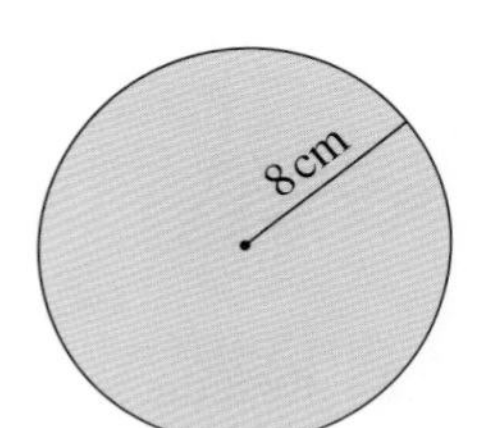


On most calculators you will find a pi ($\pi$) key. Look for it on your calculator. Ask your teacher if you can't find it. On some calculators you may need to press the shift or 2ndf key first. You will get the most accurate answer if you use this pi key in all your calculations.

Repeating the previous question using the pi key on your calculator gives:

$A = \pi \times 8\,\text{cm} \times 8\,\text{cm} = 201.06\,\text{cm}^2$ to 2 decimal places

This is close to the answer of 201.088 $\text{cm}^2$, which used $\pi = 3.142$. In fact both answers are the same to 4 significant figures.

Some people use other approximations. The table shows the outcome of the same calculation using different values for pi.

| Value used for pi | Area of circle with radius 8 cm ($\text{cm}^2$) | Percentage error |
|---|---|---|
| $\pi$ key | 201.06 (2 d.p.) | 0% |
| 3.142 | 201.088 | 0.01% |
| 3.14 | 200.96 | 0.05% |
| 3 | 192 | 4.51% |
| $\frac{22}{7}$ | 201.14 (2 d.p) | 0.04% |

The percentage error will be different for different radii, but the table clearly shows that the best values to use for pi are the $\pi$ key or 3.142. Using $\frac{22}{7}$ was common in the days before calculators and still can be a useful approximation for pi, as can 3, if you do not have a calculator.

The formula for the circumference of a circle is

- Circumference $= \pi d$

where $d$ is the diameter.

If you are given the radius instead of the diameter you can double the radius to get the diameter before using the formula, or you can use the alternative formula

- Circumference $= 2\pi r$

*For example:*

A circle with radius 8 cm has diameter 16 cm.

The circumference, $C$, is

$C = \pi d$
$= \pi \times 16\,\text{cm}$
$= 50.3\,\text{cm}$ (to 3 significant figures)

### Arcs and sectors

Look at this part of a circle:

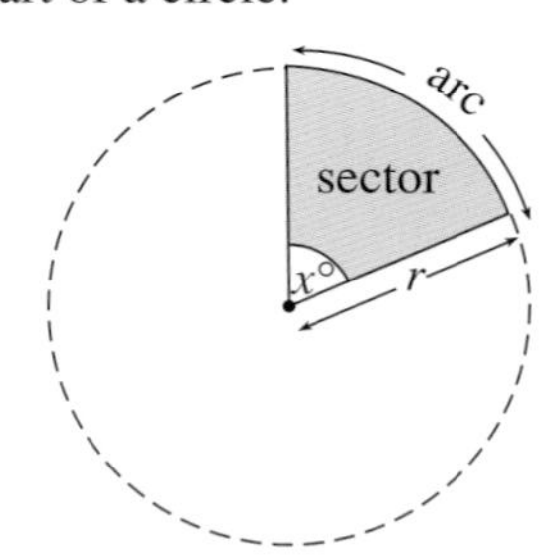


An **arc** is any piece of the circumference of a circle.

The arc length is a fraction of the circumference.

In the diagram

- Arc length $= \frac{x}{360} \times \pi d$

A **sector** of a circle is the shape enclosed by an arc and two radii of the circle.

The area of a sector is a fraction of the circle's area.

In the diagram

- Sector area $= \frac{x}{360} \times \pi r^2$

**EXAMPLE 1**

Find the arc length and sector area of the shape, to 3 significant figures.

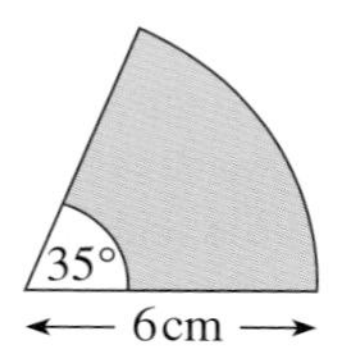


Length of arc $= \frac{x}{360} \times \pi d$

$= \frac{35}{360} \times \pi \times 12\,\text{cm}$

$= \frac{35}{360} \times 37.699\,\text{cm}$

$= 3.67\,\text{cm}$ (3 s.f.)

If you need the answer to 3 s.f., use values that are exact or to at least 4 s.f. in the working.

Sector area $= \frac{x}{360} \times \pi r^2$

$= \frac{35}{360} \times \pi \times 6\,\text{cm} \times 6\,\text{cm}$

$= \frac{35}{360} \times 113.1\,\text{cm}^2$

$= 11.0\,\text{cm}^2$ (3 s.f.)

Use values to at least 4 s.f.

## Exercise 17A

**1** Find

**i** the area

**ii** the circumference of these circles.

**a**

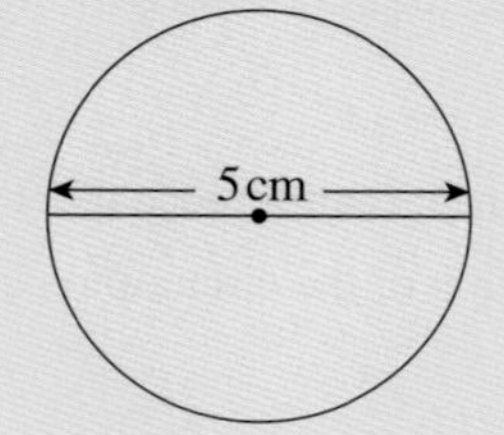


**b**

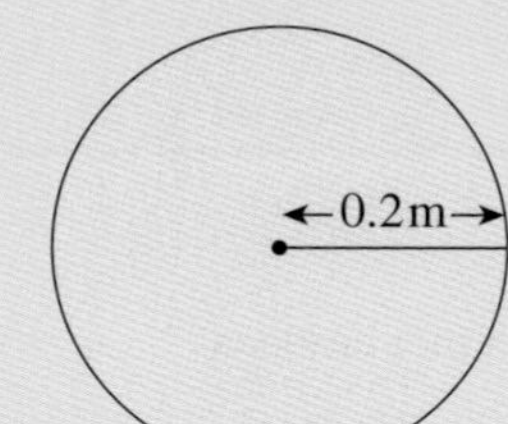


**c**

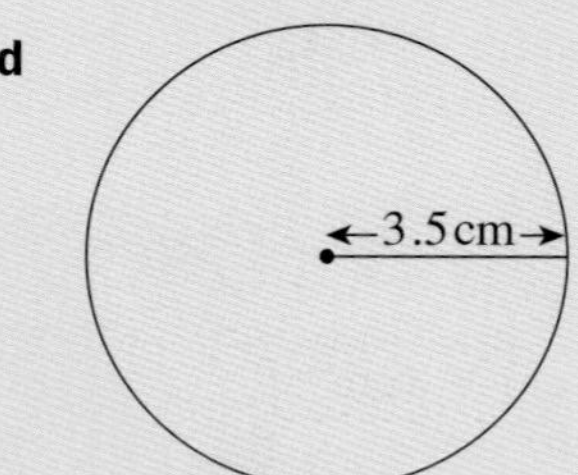


**d**

3.5cm

**2** Find the arc length and sector area of these shapes.

**a** **b**

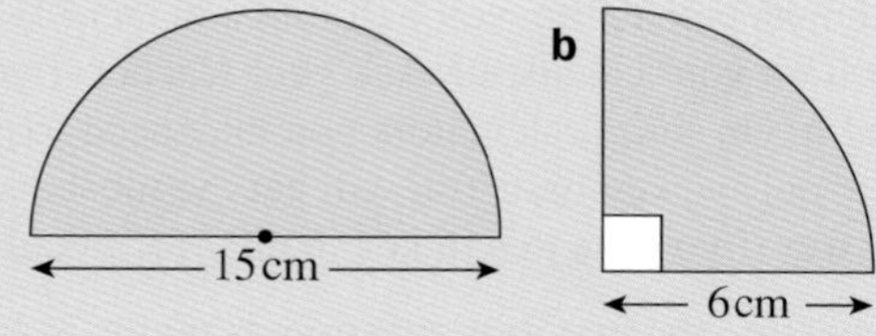


**c**

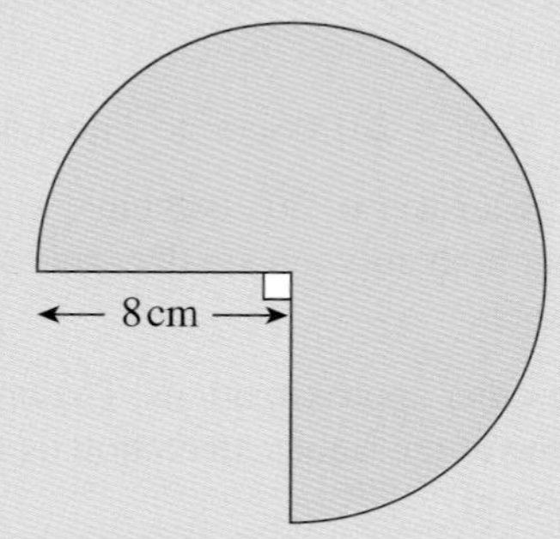


**3** Find the arc length and sector areas of these shapes.

**a**

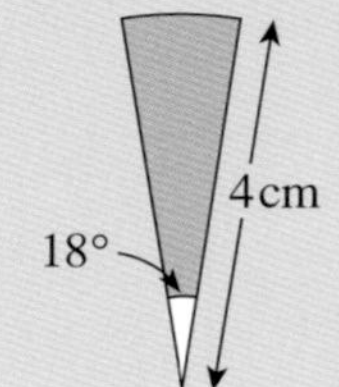

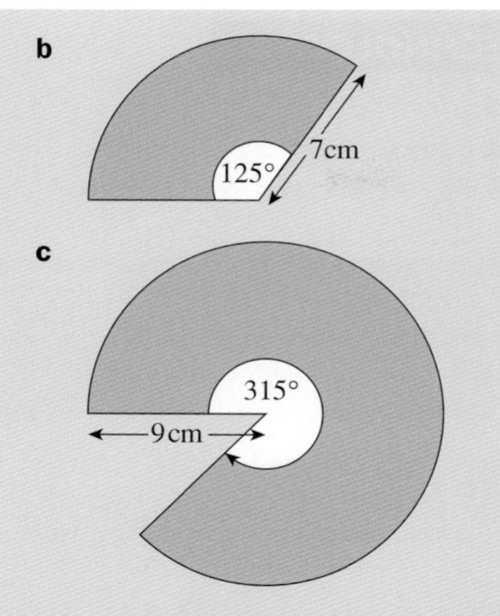


4 A circular watch has a minute hand that is 2.3 cm long.

a What distance does the tip of the hand move through in 20 minutes?

b What area of the watch face is covered by the minute hand in 25 minutes?

5 Calculate the areas and perimeters of these compound shapes.

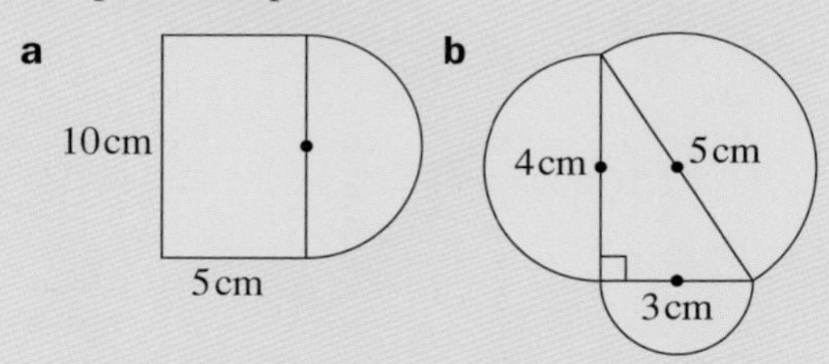


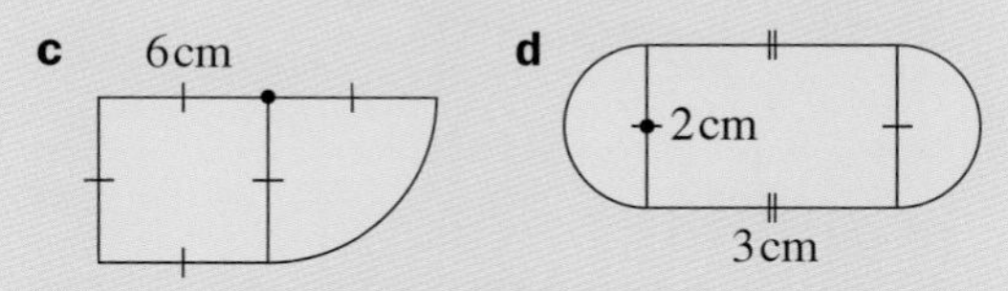


6 Find the area of the shaded parts of these shapes.

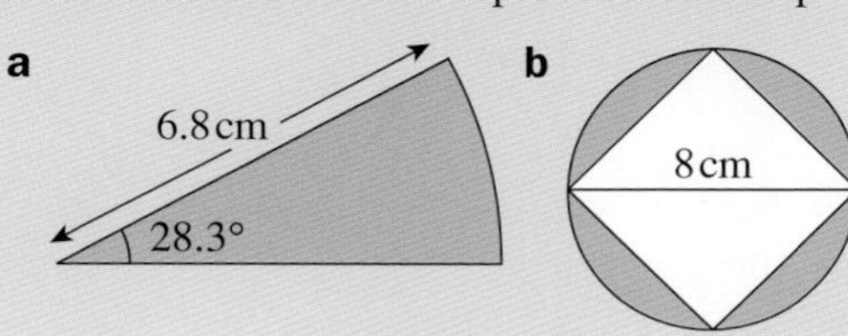


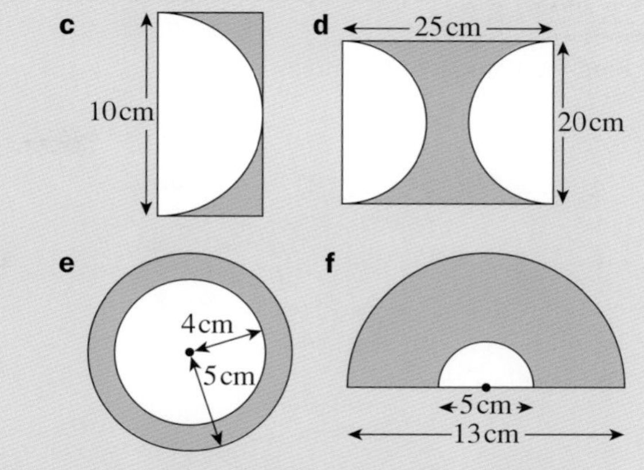


7 The radius of a wheel is 28 cm. How many revolutions will it make in travelling 3.52 km?

8 A sector of a circle has area 45 $cm^2$. What is the angle of the sector if the radius is 14.2 cm?

9 A circular tabletop has a diameter of 2.1 m.

a What is its area?

b The tabletop is to be painted. One tin of paint can cover an area of 1.75 $m^2$ and costs $5.50.

i How many tins of paint must be bought?

ii How much will it cost for materials and labour if the cost of labour is $\frac{2}{3}$ of the cost of materials?

## 17.2 Volume of prisms and cylinders

### Volume of a prism

A cuboid is one example of a **prism**.

Here are some more:

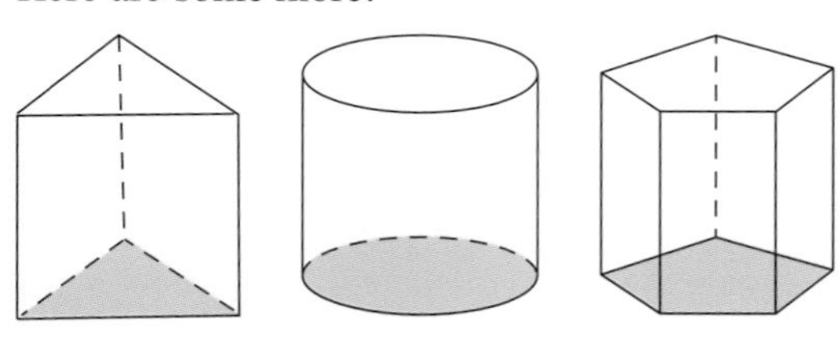

- A **prism** is a three-dimensional shape with a constant cross section.

Volume of a prism = area of cross section × height

- $V = A \times h$

**EXAMPLE 2**

Find the volume of this prism.

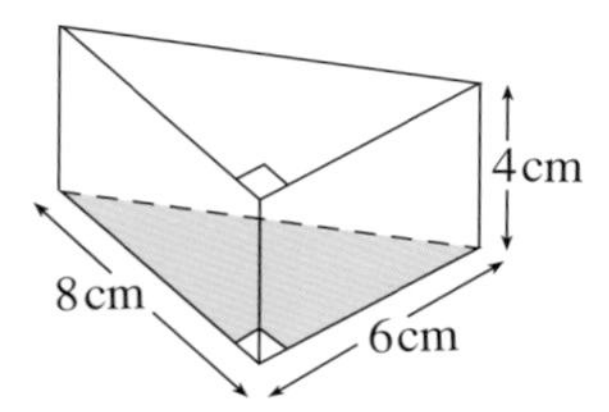


The base is a triangle.

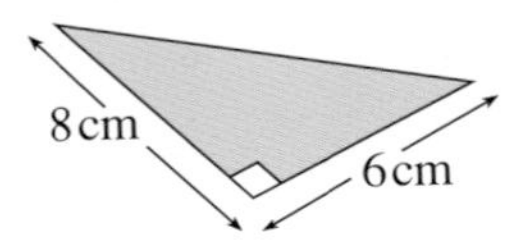


Area of base triangle, $A = \frac{1}{2}b \times h$

$= \frac{1}{2} \times 6\,\text{cm} \times 8\,\text{cm}$

$= 24\,\text{cm}^2$

Volume of prism $= A \times h$

$= 24\,\text{cm}^2 \times 4\,\text{cm}$

$= 96\,\text{cm}^3$

You can easily find missing side lengths for a prism when you know its volume and cross-sectional area.

**EXAMPLE 3**

Calculate the missing side length, $x$ cm, for this triangular prism with volume $336\,\text{cm}^3$.

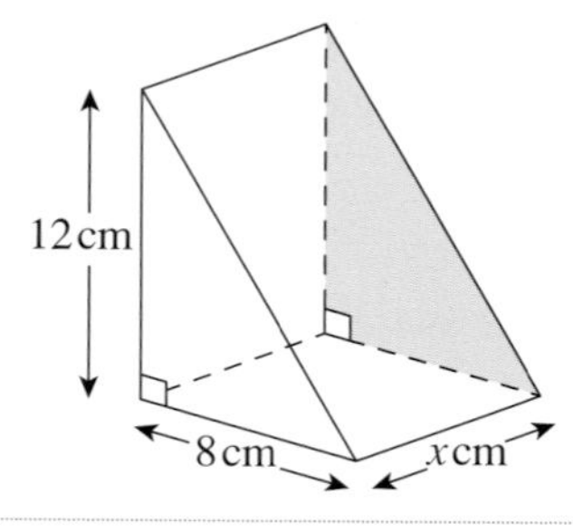


Cross-sectional area,

$$A = \frac{1}{2}b \times h = \frac{1}{2} \times 8\,\text{cm} \times 12\,\text{cm} = 48\,\text{cm}^2$$

Volume of prism, $V = A \times h$

Substituting $V = 336$, $A = 48$, $h = x$:

$336 = 48x$ [Solve for $x$

$x = 7$ [$\div 48$

So the missing side length is 7 cm.

**TECHNOLOGY**

Visit the website

www.onlinemathlearning.com

and follow the links to Geometry and Volume of Prisms.

Study what it says about finding volumes of prisms. Watch the videos!

When you are through, you are ready for Exercise 17B.

## Exercise 17B

1 Find the volumes of these triangular prisms.

a

b

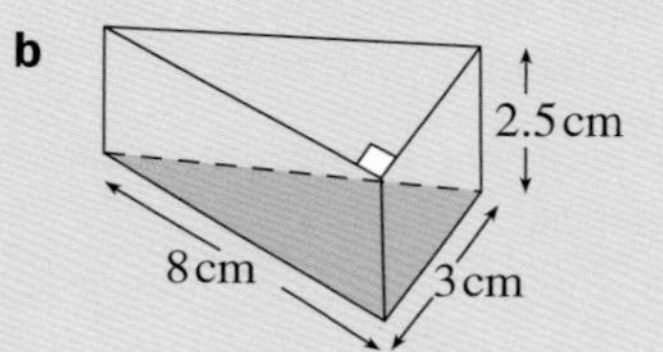


c

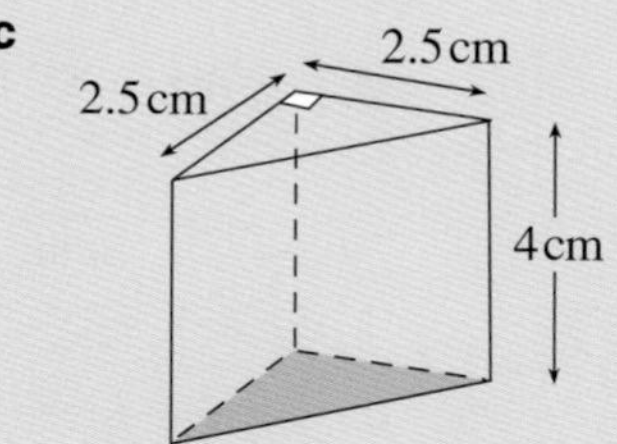


d

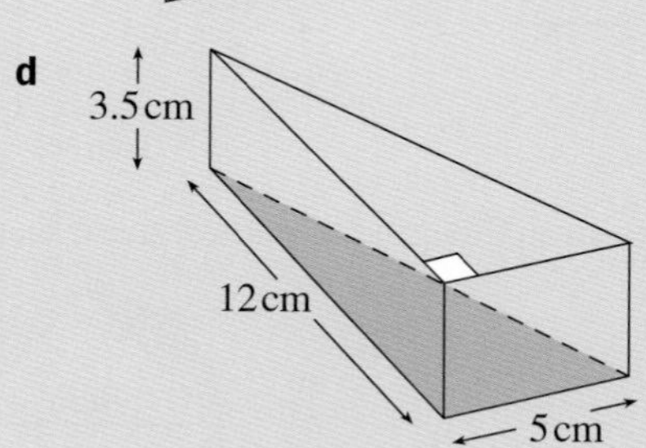

**2** **a** Find the area of this shape.

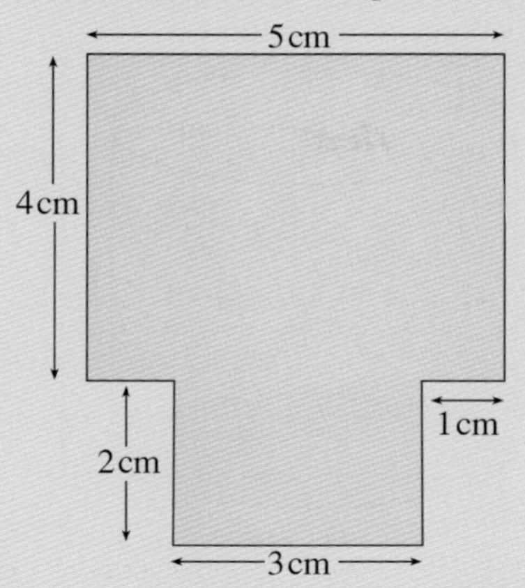


**b** Using your answer to part **a**, find the volume of this prism.

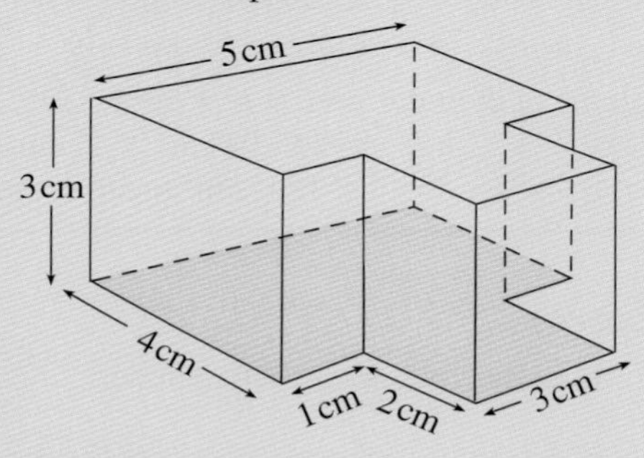


**3** **a** Calculate the missing side length for this cuboid with volume 180 cm$^3$.

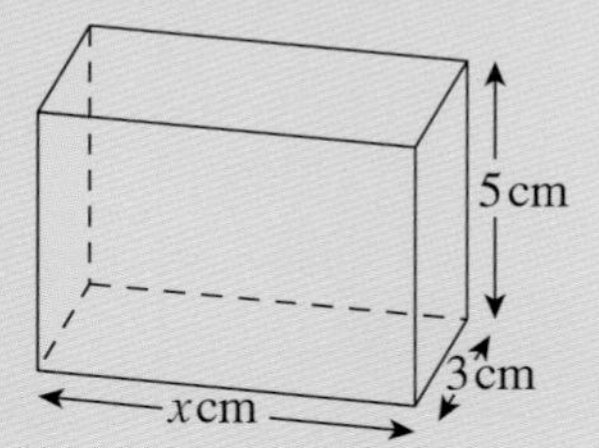


**b** Calculate the missing side length for this triangular prism with volume 48 m$^3$.

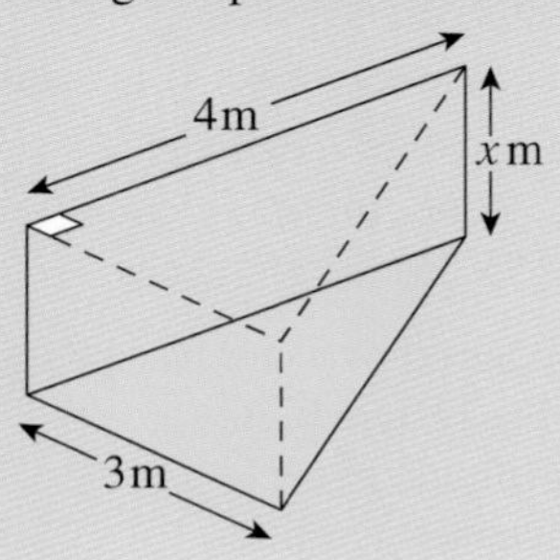


**4** Find the volume of each prism by first finding the area of its base.

**a**

**b**

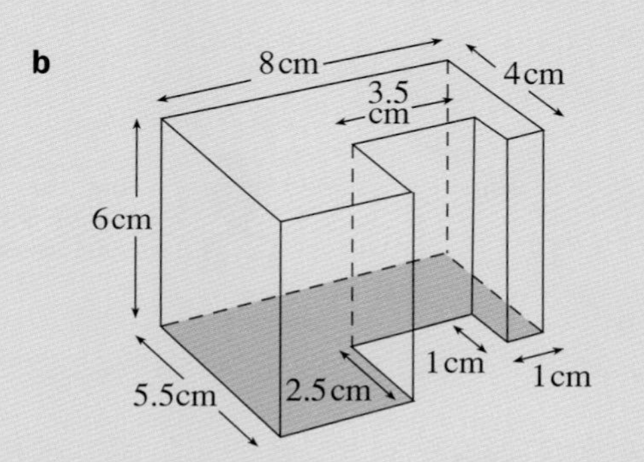


**5** The drawing shows the end of a steel girder 2 m long.

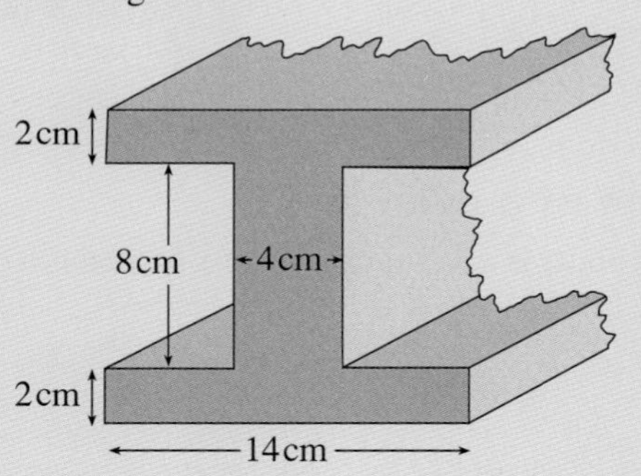


**a** Find the area of the end of the girder.
**b** Find the volume of the girder in cm$^3$.

**6** The total surface area of a cube is 150 cm$^2$. Find its volume.

**7**

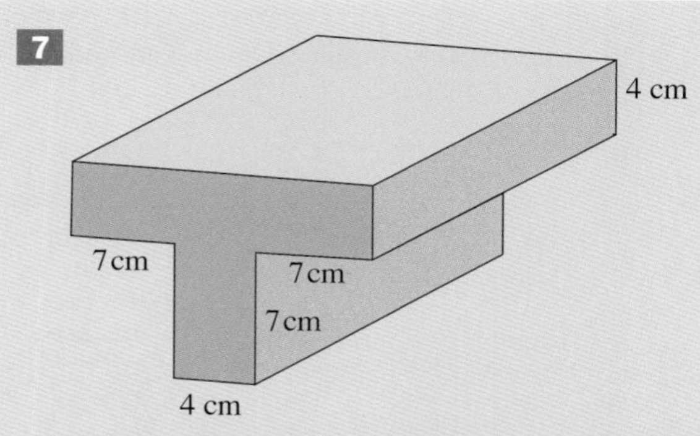


The volume of the steel girder in the diagram is $1200\,\text{cm}^3$.
Find its length.

**8** Find the volume of each prism.

**a**

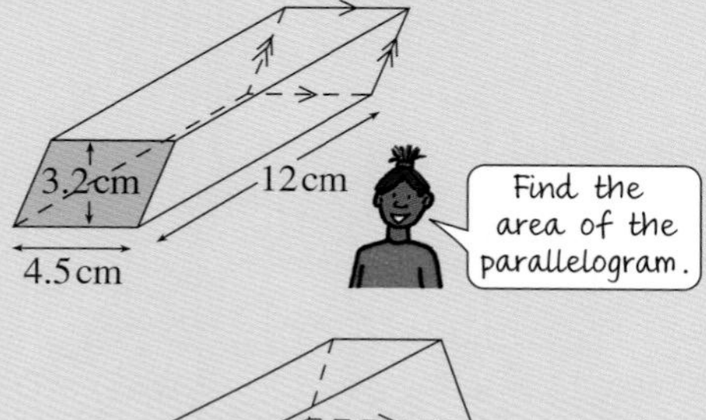


**b**

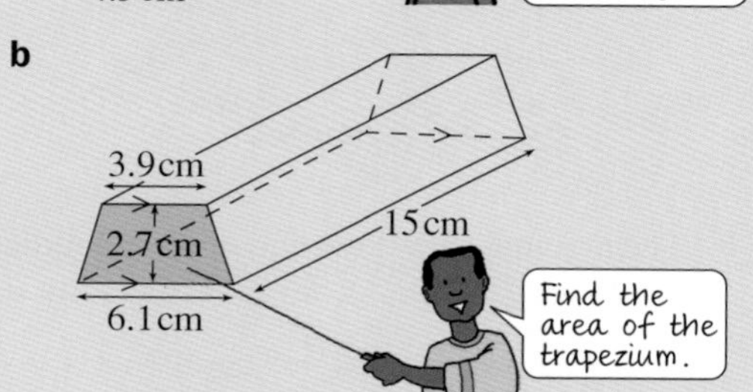


## Volume of a cylinder

- A cylinder is a prism whose cross section is a circle.

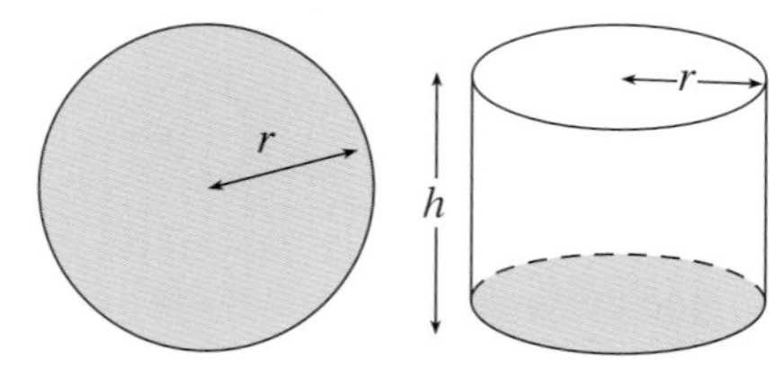


Area of a circle $= \pi r^2$
Volume of a cylinder = area of cross section × height
$= \pi r^2$

- The volume of a cylinder is $\boldsymbol{V = \pi r^2 h}$.

**EXAMPLE 4**

Find the volume of this cylinder.

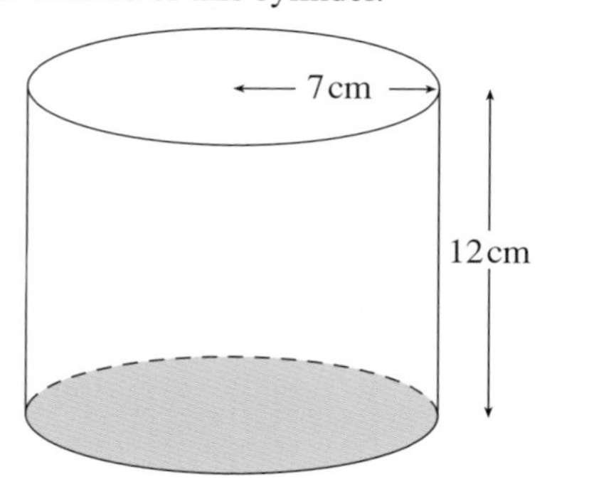


The base is a circle of radius 7 cm.

Area of circle $= \pi r^2$
$= \pi \times 7\,\text{cm} \times 7\,\text{cm}$
$= 153.94\,\text{cm}^2$ (2 d.p.)

Volume of cylinder $= A \times h = \pi r^2 h$
$= 153.94\,\text{cm}^3 \times 12\,\text{cm}$
$= 1847\,\text{cm}^3$ (nearest $\text{cm}^3$)

### Exercise 17C

**1** Find the volume of each cylinder.

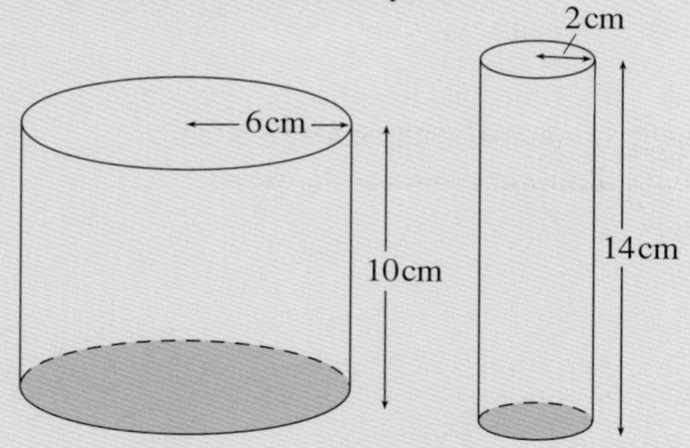


**2** **a** Find the volume of a cylinder if
  **i** $r = 3\,\text{cm}, h = 6.4\,\text{cm}$
  **ii** $r = 3.2\,\text{cm}, h = 4\,\text{cm}$
  Give your answers correct to 1 decimal place.

**b** Find the height of a cylinder if
  **i** $V = 37.68\,\text{cm}^3, r = 1\,\text{cm}$
  **ii** $V = 75.36\,\text{cm}^3, r = 2\,\text{cm}$

**3** A cylindrical barrel of water has a base radius of 30 cm and is 110 cm high.
  **a** Find its volume in $\text{cm}^3$.
  **b** Find its capacity in litres
  (**Hint:** $1\,\text{cm}^3 = 1\,\text{ml}$)
  **c** Find the number of bottles of water that can be filled from the barrel if each bottle holds 0.73 litres.

**4** A wooden cylinder of radius 2.5 cm and length 32 cm has a square hole of edge 1.5 cm removed from it, as shown in the diagram.

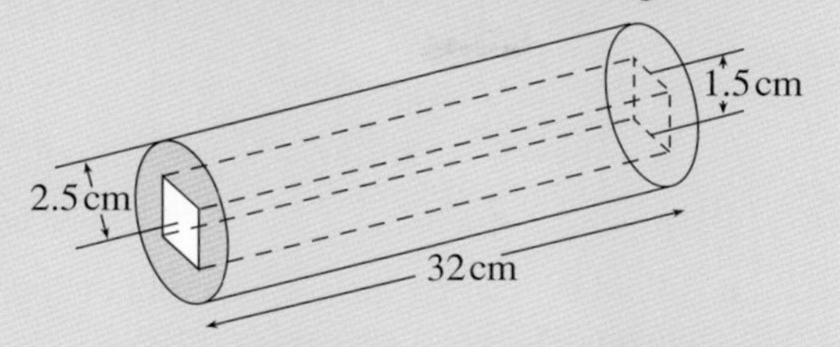


**a** Find the volume of the original cylinder.
**b** Find the volume of wood removed.
**c** Calculate the volume of the remaining wood.

**5** A pipe has an inner radius of 4 cm and an outer radius of 5 cm.

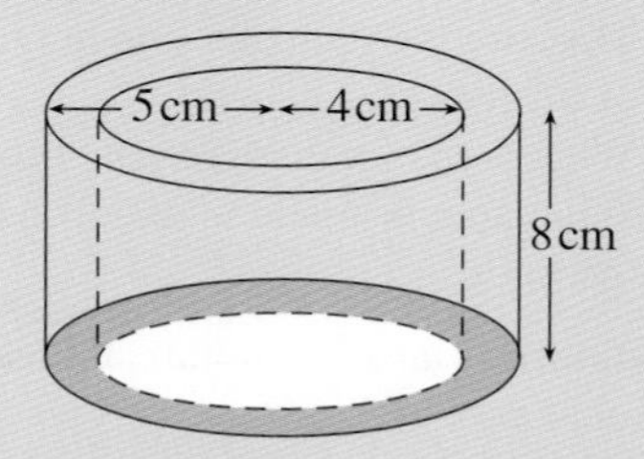


**a** Find the base area of the pipe.
**c** What is the volume of material used to make the pipe?

Review what you have learnt about volumes of cylinders by visiting the website
www.onlinemathlearning.com
and following the links to Geometry and Volume of Cylinders.
Watch the video examples.

## 17.3 Surface area of prisms and cylinders

The surface area of any prism is found by adding up the areas of the prism's faces.
For example, a triangular prism has five faces:

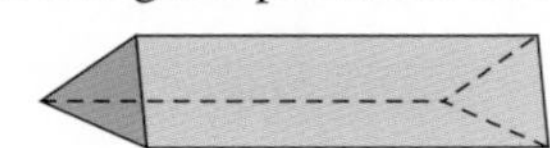

Two of its faces are triangles.
Three of its faces are rectangles.

To find the surface area you have to add the area of the two triangles to that of the three rectangles.

**EXAMPLE 5**

Find the surface area of this triangular prism.

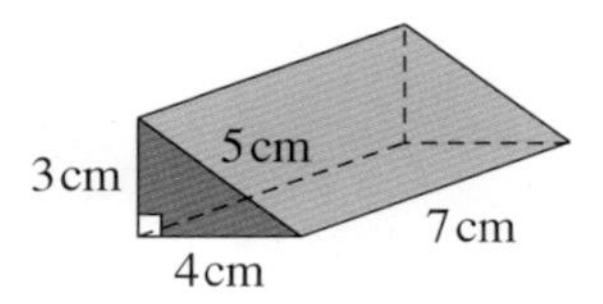


Area of triangular face $= \frac{1}{2} \times \text{base} \times \text{height}$
$= \frac{1}{2} \times 4\text{ cm} \times 3\text{ cm}$
$= 6\text{ cm}^2$

Area of base rectangle $= \text{length} \times \text{width}$
$= 7\text{ cm} \times 4\text{ cm}$
$= 28\text{ cm}^2$

Area of upright rectangle $= 7\text{ cm} \times 3\text{ cm}$
$= 21\text{ cm}^2$

Area of sloping rectangle $= 7\text{ cm} \times 5\text{ cm}$
$= 35\text{ cm}^2$

Surface area
= area of triangular face × 2 + area of rectangles
$= 6\text{ cm}^2 \times 2 + 28\text{ cm}^2 + 21\text{ cm}^2 + 35\text{ cm}^2$
$= 12\text{ cm}^2 + 28\text{ cm}^2 + 21\text{ cm}^2 + 35\text{ cm}^2$
$= 96\text{ cm}^2$

A cylinder has three faces – two flat circular faces and one curved face:

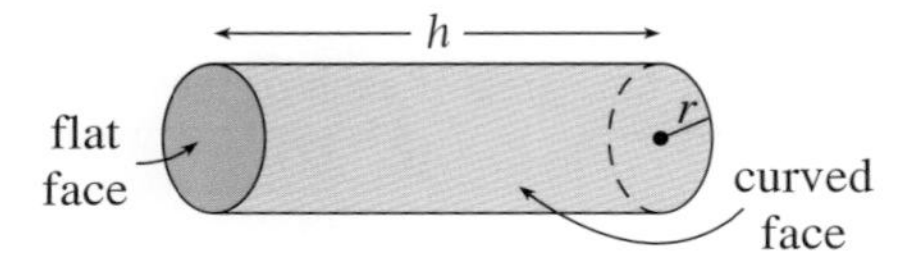


The area of the curved surface of the cylinder = distance around end × height
= circumference of flat face × $h$
$= \pi dh$

Area of flat face = area of circle
$= \pi r^2$

Total surface area of cylinder $= \pi dh + 2\pi r^2$

**EXAMPLE 6**

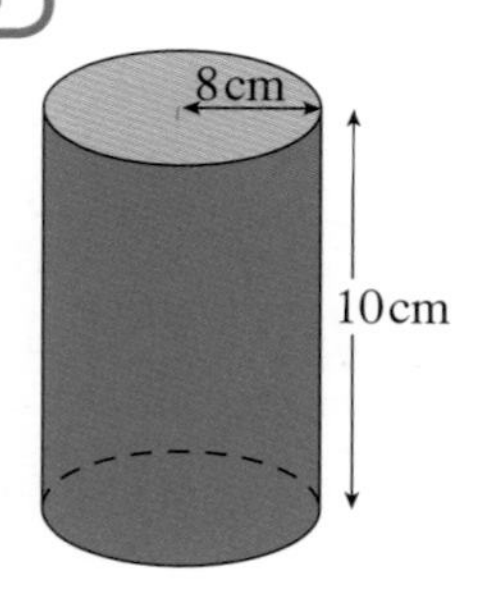


Find

**a** the curved surface area

**b** the total surface area of the cylinder.

Give your answers to 1 decimal place.

**a** Curved surface area $= \pi dh$

$= \pi \times 16\,\text{cm} \times 10\,\text{cm}$

$= 502.65\,\text{cm}^2$ (2 d.p.)

$= 502.7\,\text{cm}^2$

**b** Area of circular face $= \pi r^2$

$= \pi \times 8\,\text{cm} \times 8\,\text{cm}$

$= 201.06\,\text{cm}^2$ (2 d.p.)

Total surface area $= 2 \times 201.06\,\text{cm}^2 + 502.65\,\text{cm}^2$

$= 904.8\,\text{cm}^2$ (1 d.p.)

To get an answer to 1 d.p., remember to calculate using numbers to at least 2 d.p.

## Exercise 17D

**1** Find the surface area of each of these prisms.

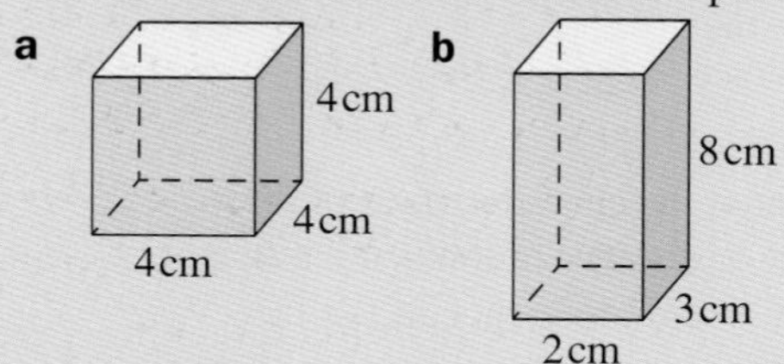


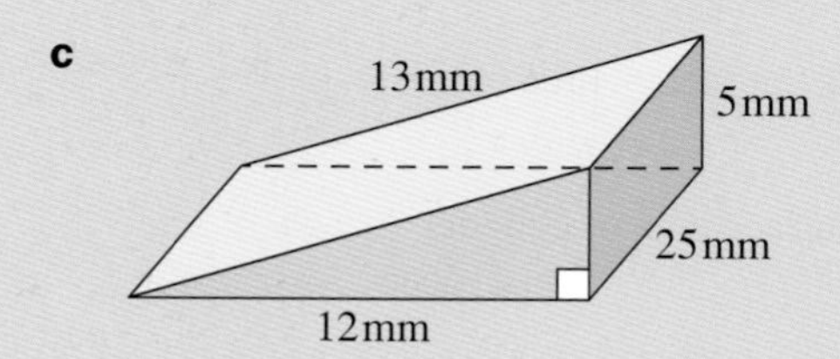


**2** Find the surface area of these shapes.

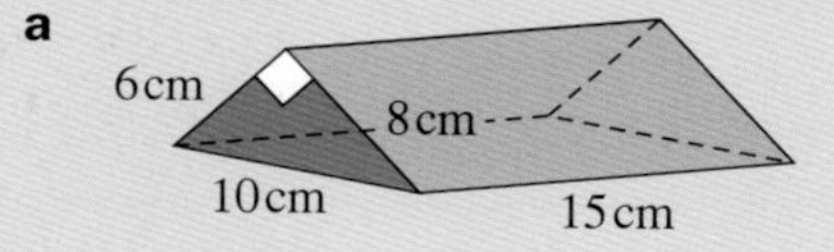


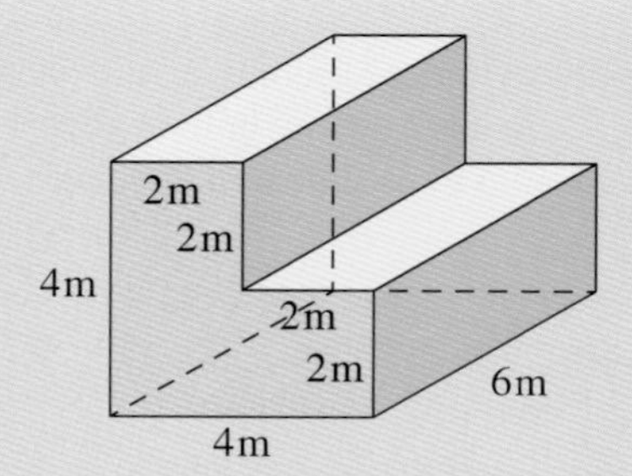


**3** **a** Find the curved surface area of a cylinder with

**i** $r = 5\,\text{cm}, h = 10\,\text{cm}$

**ii** $r = 3\,\text{cm}, h = 6\,\text{cm}$

**iii** $r = 7.5\,\text{cm}, h = 12\,\text{cm}$

**b** What is the total surface area of each of these cylinders?

**4** **a** Measure the diameter and height of a tin of beans.

**b** What area of paper is needed for the label?

**c** What area of metal is required to make the tin?

**5** A cylindrical tank is 2 m tall and has a cross-sectional area of $3\,\text{m}^2$.

**a** What is the curved surface area of the tank?

**b** How many litres of paint will be needed to paint the curved surface of the tank if one litre of paint covers $11.25\,\text{m}^2$?

## INVESTIGATION

A soft drink manufacturer wants to make a cylindrical can to hold $350\,\text{cm}^3$ of soft drink.
To reduce costs they wish to use the least amount of metal possible to make the can. That is, they wish to construct a cylindrical can with the smallest surface area that holds $350\,\text{cm}^3$.
What would the height of the can be if its radius was 2 cm?
What would its surface area be?
Copy and complete the table.

| Radius (cm) | Height (cm) | Surface area (cm²) ($\pi dh + 2\pi r^2$) |
|---|---|---|
| 2 | | |
| 3 | | |
| 4 | | |
| 5 | | |

Plot a graph of the radius of the can against its surface area.
From your graph determine the dimensions that use the least amount of metal.

# Consolidation

### Example 1

Find the area of the sector and the length of the arc of this shape.

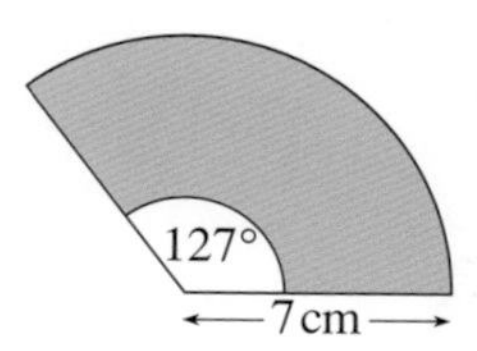


$$\text{Area of sector} = \frac{x°}{360°} \times \pi r^2$$
$$= \frac{127°}{360°} \times \pi \times 7\,\text{cm} \times 7\,\text{cm}$$
$$= \frac{127°}{360°} \times 153.94\,\text{cm}^2$$
$$= 54.3\,\text{cm}^2\ (1\text{ d.p.})$$

$$\text{Length of arc} = \frac{x°}{360°} \times \pi d$$
$$= \frac{127°}{360°} \times \pi \times 14\,\text{cm}$$
$$= \frac{127°}{360°} \times 43.98\,\text{cm}$$
$$= 15.5\,\text{cm}\ (1\text{ d.p.})$$

### Example 2

Find the volume of the cylinder.

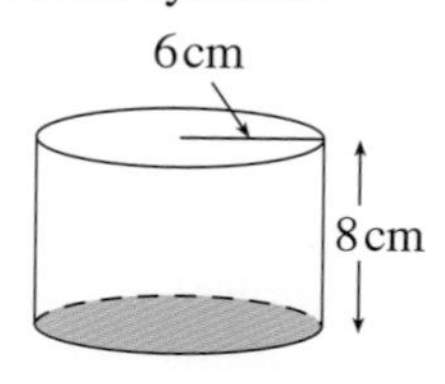


$$\text{Volume of cylinder} = \text{base area} \times \text{height}$$
$$= \pi r^2 \times h$$
$$= \pi \times 6^2 \times 8\,\text{cm}^3$$
$$= 904.8\,\text{cm}^3\ (1\text{ d.p.})$$

### Example 3

Find the surface area of these prisms.

**a**

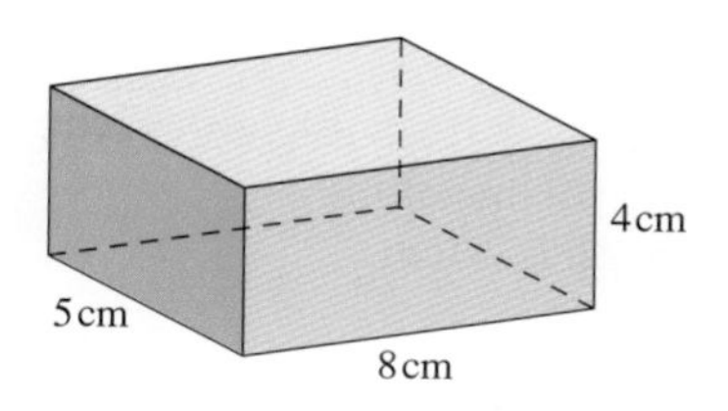


Surface area is the total area of the six faces.

$$\text{Area of front face} = \text{length} \times \text{width}$$
$$= 8\,\text{cm} \times 4\,\text{cm}$$
$$= 32\,\text{cm}^2$$

$$\text{Area of side face} = 5\,\text{cm} \times 4\,\text{cm}$$
$$= 20\,\text{cm}^2$$

$$\text{Area of top face} = 8\,\text{cm} \times 5\,\text{cm}$$
$$= 40\,\text{cm}^2$$

$$\text{Total surface area} = 2 \times 32\,\text{cm}^2 + 2 \times 20\,\text{cm}^2 + 2 \times 40\,\text{cm}^2$$
$$= 64\,\text{cm}^2 + 40\,\text{cm}^2 + 80\,\text{cm}^2$$
$$= 184\,\text{cm}^2$$

**b**

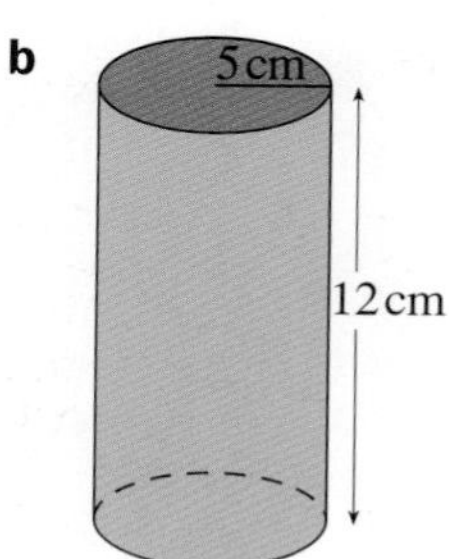


$$\text{Curved surface area} = \pi dh$$
$$= \pi \times 10 \times 12\,\text{cm}^2$$
$$= 376.99\,\text{cm}^2$$

$$\text{Area of circle} = \pi r^2$$
$$= \pi \times 5 \times 5\,\text{cm}^2$$
$$= 78.54\,\text{cm}^2$$

$$\text{Total surface area} = \text{area of circle} \times 2 + \text{curved surface area}$$
$$= 2 \times 78.54 + 376.99\,\text{cm}^2$$
$$= 534.1\,\text{cm}^2\ (1\text{ d.p.})$$

### Example 4

Find the length, $x$ cm, of this prism, if cross-sectional area = $82\,\text{cm}^2$ and volume = $615\,\text{cm}^3$.

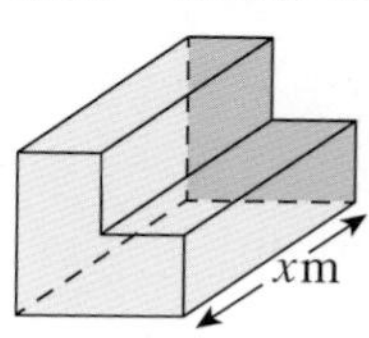


Volume = cross-sectional area × $x$

Substituting:

$615 = 82x$ [Solve for $x$

$x = 7.5$ [÷ 82

So the length is 7.5 cm.

## Exercise 17

**1** Find **i** the perimeter **ii** the area of these shapes.

**a**

32m

28m

**b**

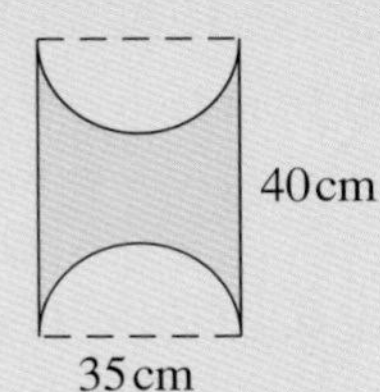


**c**

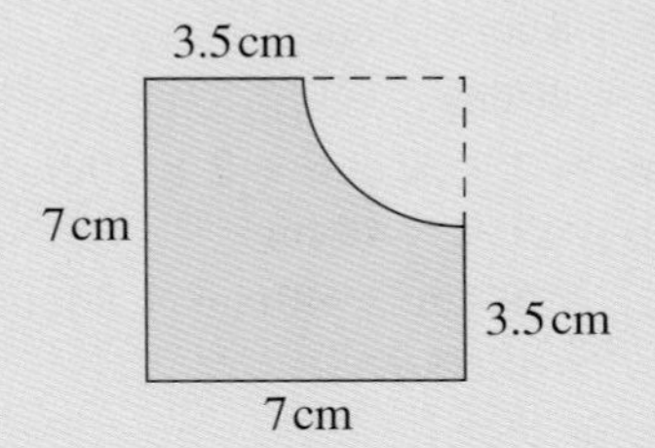


**d**

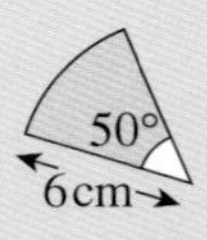


**e**

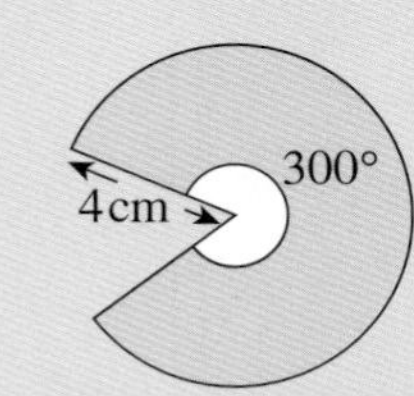


**2** Find the volume of the L-shaped solid. The dimensions given are in centimeters.

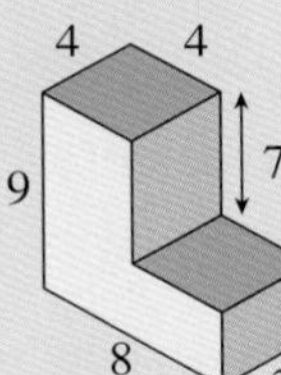


**3** The end face of a prism of length 10 cm is a right-angled triangle with sides 3 cm, 4 cm and 5 cm.

**a** Draw a sketch of the prism.

**b** Find the volume of the prism.

**4** A circular pond of diameter 9 m is surrounded by a path of width 1.5 m.
Calculate the area of

**a** the pond

**b** the pond and path together

**c** the path alone.

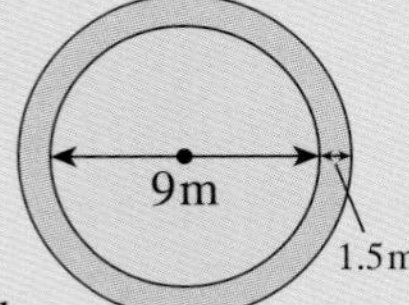


**5** The diagram shows a triangular prism.

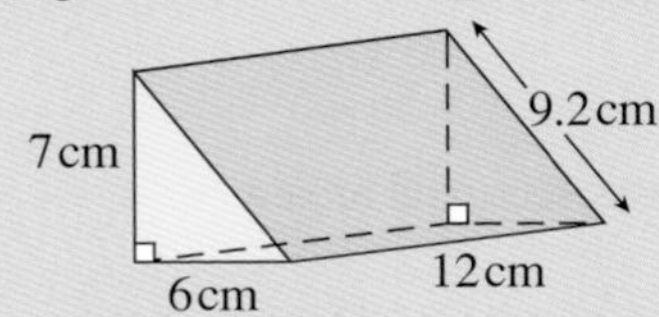


Calculate **a** the volume **b** the surface area of the prism.

**6** Find the volume of a cylinder with

**a** $r = 1\text{ cm}, h = 2\text{ cm}$

**b** $r = 7\text{ cm}, h = 10\text{ cm}$

**c** $r = 5\text{ mm}, h = 12\text{ mm}$

**d** $r = 0.15\text{ m}, h = 1\text{ m}$

**7** Find the volume of each of these prisms.

**a**

4.6cm

3cm

14cm

8cm

2cm

**b**

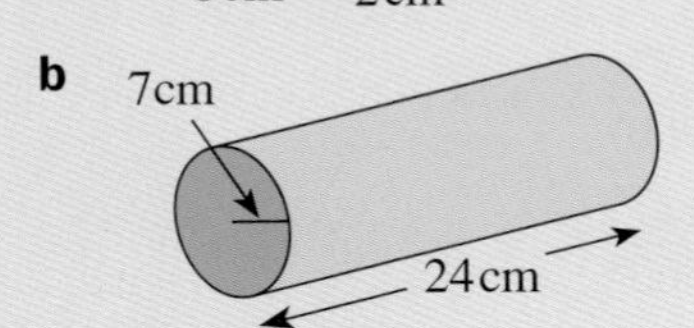


**8** Find **a** the volume

**b** the total surface area of this prism.

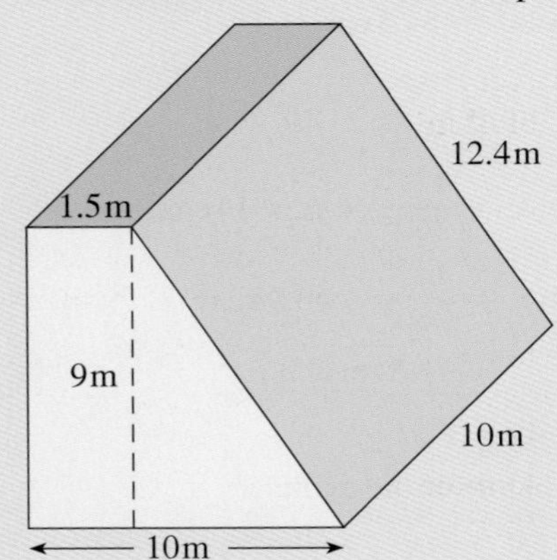


**9** A coin has diameter 2 cm and thickness 2 mm.

**a** Write down the radius and height of the cylinder in centimetres.

**b** What volume of metal is needed to produce 100 coins?

**10** How many cylindrical jars of diameter 16 cm and height 28 cm can be filled from a cylindrical vessel of diameter 60 cm and height 40 cm?

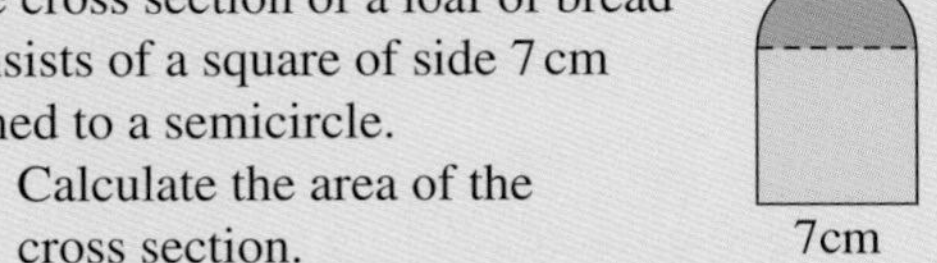

**11** The cross section of a loaf of bread consists of a square of side 7 cm joined to a semicircle.

**a** Calculate the area of the cross section.

**b** Calculate the volume of the loaf if its length is 30 cm.

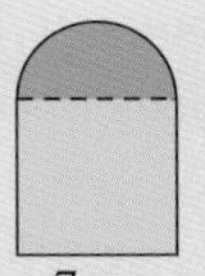

**12** A gardener's shed has the dimensions shown in the diagram. Find the volume of the shed.

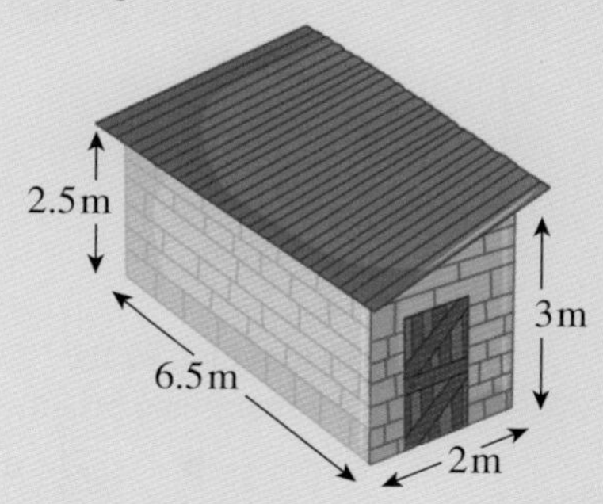


**13** The diameter of a bicycle wheel is 42 cm.
- **a** Find the circumference of the wheel.
- **b** Find the number of revolutions that the wheel makes in travelling a distance of 1.98 km.

**14** The circumference of a tyre is 54.95 cm. Find its diameter.

**15**
- **a** A room is 3 m high and has a volume of 45 $m^3$.
  Calculate the area of the floor of the room.
- **b** A dormitory is 10 m × 6 m × 3 m. How many people can sleep in the dormitory if each person needs 18 $m^3$ of air space?

**16** The minute hand of a clock is 49 mm long. How far does the tip of the hand travel in
- **a** 15 minutes
- **b** 45 minutes
- **c** 50 minutes?

**17** A battery-operated toy aeroplane is attached to the end of a cord 2 m long.
- **a** What is the radius of the largest circle that the aeroplane can fly?
- **b** The aeroplane flies at a speed of 3.2 m/s. How long will it take to fly the largest possible circle?

**18** The diagram shows a cylinder of height 11 m and radius 5 cm.
Calculate the volume of the cylinder
- **a** in $cm^3$
- **b** in $m^3$.

**19** A concrete block is made by pouring 720 $cm^3$ of concrete into a 12 cm × 10 cm rectangular tray.

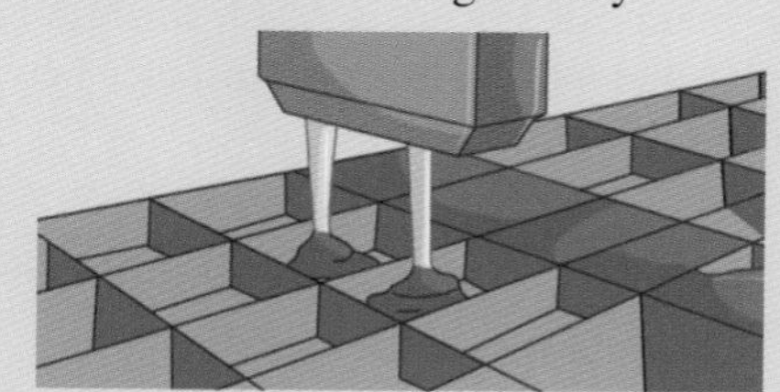

How thick is the block?

**20** The hour hand of a clock is 10 cm long. What area does it pass over in 3 hours?

**21** 33 litres of water are poured into a rectangular tank 55 cm long and 48 cm wide.
Calculate the depth of the water in the tank.
(**Hint:** 1 ml = 1 $cm^3$)

## Summary

### You should know ...

**1**
- **a** Part of the circumference of a circle is called an arc.
- **b** The area bounded by two radii and an arc is called a sector.

*For example:*

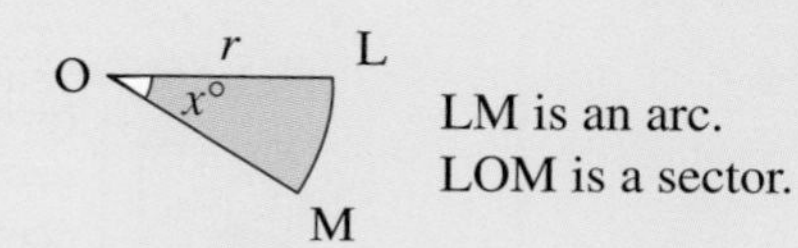


LM is an arc.
LOM is a sector.

Length of arc LM $= \frac{x}{360} \times \pi d$

Area of sector LOM $= \frac{x}{360} \times \pi r^2$

### Check out

**1**
- **a** Find
  - **i** the length of arc AB
  - **ii** the area of sector AOB
  - **iii** the perimeter of sector AOB.

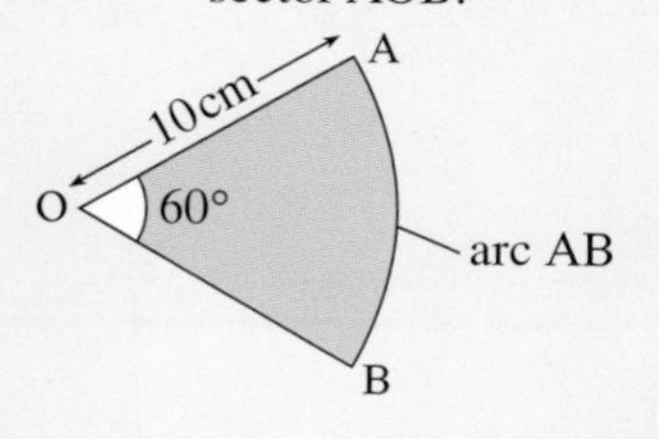

**b** What are the area and perimeter of this shape?

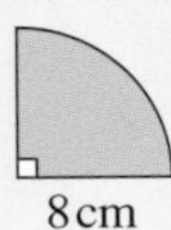


---

**2** A prism is a solid shape with a constant cross section.
*For example:*

 is a prism.

**2** State which of the following is a prism.

**a** 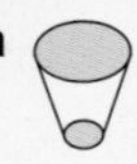 **b**  **c**

---

**3** The volume of a prism is $A \times h$ where $A$ is the area of the cross section and $h$ is the height or distance between the flat faces.
*For example:*

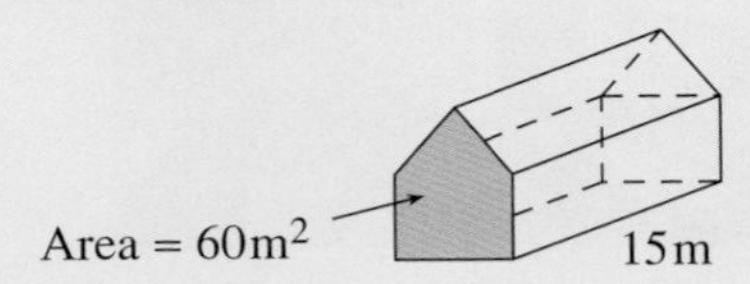


Volume $= 60 \times 15\,\text{m}^3$
$= 900\,\text{m}^3$

**3** Find the volume of this prism.

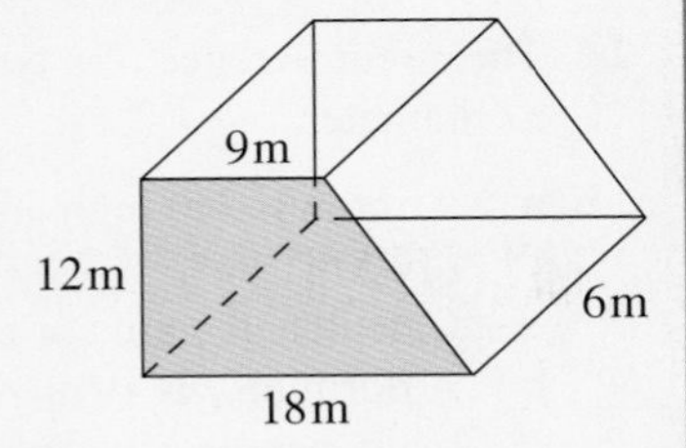


---

**4** The total surface area of a prism is the sum of the areas of all its faces.
*For example:*

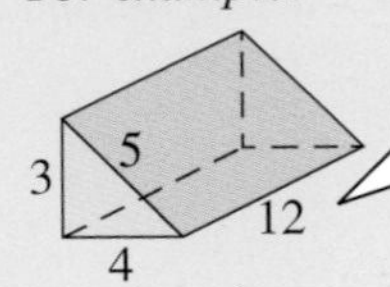


Surface area $= 2(\frac{1}{2} \times 4 \times 3) + (12 \times 4) + (12 \times 5)$
$+ (12 \times 3)\,\text{cm}^2$
$= 156\,\text{cm}^2$

**4** Find the total surface area of the prism. All measurements are in centimetres.

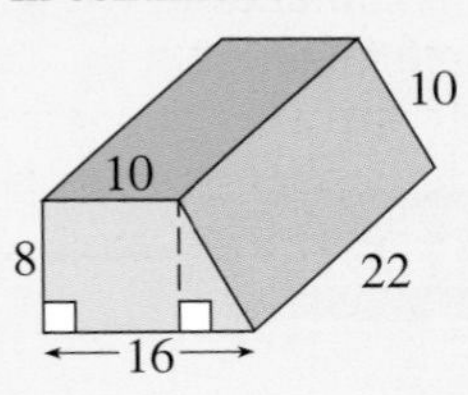


---

**5** Total surface area of a cylinder $= \pi dh + 2\pi r^2$

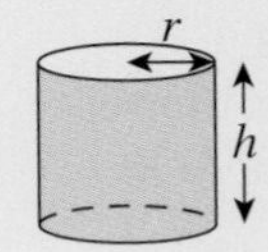


**5** Find the total surface area of this cylinder.

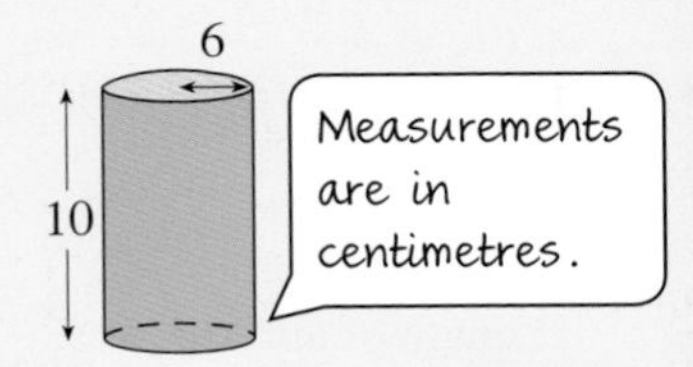

# 18 Probability

## Objectives

- Know that the sum of probabilities of all mutually exclusive outcomes is 1 and use this when solving probability problems.
- Find and record all outcomes for two successive events in a sample space diagram.
- Understand relative frequency as an estimate of probability and use this to compare outcomes of experiments in a range of contexts.

## What's the point?

Predicting the weather is based upon probabilities. Weather forecasters look at lots of evidence and collect lots of data, for example about atmospheric pressure, and combine this with their experience of the weather when conditions were similar in the past to make their forecast. They might be able to say, for example, that there is a 70 per cent chance of rain tomorrow, and even predict when during the day that rain is likely to fall.

## Before you start

### You should know ...

**1**
- The probability of an event happening must lie between 0 (impossible) and 1 (certain).
- Probability can be written as a fraction, percentage or decimal. It should not be written as a ratio or in words.
- If the probability of an event occurring is $p$, then the probability of it not occurring is $1 - p$.

### Check in

**1** **a** If the probability of picking a red sweet is $\frac{3}{5}$, what is the probability of not picking a red sweet?

**b** If the probability of it raining tomorrow is 0.6, what is the probability of it not raining tomorrow?

*For example:*
If the probability of picking a yellow sweet is $\frac{5}{8}$, then the probability of not picking a yellow sweet is $1 - \frac{5}{8} = \frac{3}{8}$

Do not write '3 : 8' or '3 out of 8' or '3 in 8'

**2** How to add and subtract decimals.
*For example:*
$0.3 + 0.45 = 0.75$
$1 - 0.58 = 0.42$

**3** How to add and subtract fractions by using a common denominator.
*For example:*

$\frac{1}{2} + \frac{2}{5} + \frac{1}{10} = \frac{5}{10} + \frac{4}{10} + \frac{1}{10} = \frac{10}{10} = 1$

$1 - \frac{3}{7} = \frac{7}{7} - \frac{3}{7} = \frac{4}{7}$

**c** If the probability of winning a cricket match is 53%, what is the probability of not winning the cricket match?

**d** Which of these are possible correct answers to a probability question?

**i** 0.6 **ii** 3 : 5
**iii** 62% **iv** 1.4
**v** $1\frac{3}{7}$ **vi** 2 out of 3

**2** Work out:
**a** $0.41 + 0.2$
**b** $0.3 + 0.25 + 0.1$
**c** $1 - 0.75$
**d** $1 - 0.08$

**3** Work out:
**a** $\frac{1}{4} + \frac{2}{5}$ **b** $\frac{1}{8} + \frac{2}{5} + \frac{1}{10}$
**c** $1 - \frac{4}{5}$ **d** $1 - \frac{3}{10}$

## 18.1 Successive events

When rolling ordinary six-sided dice, there are six **possible outcomes**: 1, 2, 3, 4, 5 or 6. Each of these are **equally likely outcomes** if the dice are fair. They are also **mutually exclusive** outcomes. This means that they cannot happen at the same time.

In Book 2 you learned that if the probability of an event occurring is $p$, then the probability of it not occurring is $1 - p$. This can be extended to all mutually exclusive outcomes.

- The sum of the probabilities of all mutually exclusive outcomes of an event is 1.

### EXAMPLE 1

There are four colours of sweets in a bag: red, green, yellow and purple. The probabilities of choosing a red, green or yellow sweet are shown in the table. What is the missing probability, for choosing a purple sweet?

| Colour of sweet | Red | Green | Yellow | Purple |
|---|---|---|---|---|
| Probability of choosing colour | $\frac{3}{8}$ | $\frac{1}{5}$ | $\frac{3}{10}$ | |

There are only four outcomes, because there are no other colours of sweets. The outcomes are mutually exclusive, as each sweet can only be one colour, so the probabilities of all the outcomes must add up to 1.

$$\begin{aligned} \text{P(red or green or yellow)} &= \frac{3}{8} + \frac{1}{5} + \frac{3}{10} \\ &= \frac{15}{40} + \frac{8}{40} + \frac{12}{40} \\ &= \frac{35}{40} = \frac{7}{8} \end{aligned}$$

$$\text{P(purple)} = 1 - \frac{7}{8} = \frac{1}{8}$$

An **event** is a particular outcome to an experiment. For the experiment of rolling a dice and the event of 'getting an even number', there are three possible successful outcomes: 2, 4 or 6. The theoretical probability of an event is found using the formula

$$\text{P(event)} = \frac{\text{number of successful outcomes}}{\text{number of possible outcomes}}$$

The theoretical probability of getting an even number when rolling a dice is

$$\text{P(even number)} = \frac{3}{6} = \frac{1}{2}$$

**Successive events** are events that happen one after the other. It is important to know how many possible outcomes of successive events there are, in order to use the probability formula. You can write down a list of all the possible outcomes. This list of all possible outcomes is called the **sample space**. Often a diagram helps us to organise the list in a logical way, to make sure we don't miss out any outcomes. This is called a **sample space diagram**. Two-way tables and tree diagrams are two examples of sample space diagrams.

**EXAMPLE 2**

A coin is tossed and then a tetrahedral (four-sided) dice is rolled. The coin has heads (H) on one side and tails (T) on the other. The dice is numbered from 1 to 4.

Draw two different sample space diagrams to show all the possible outcomes. Use the diagrams to work out the probability of

**a** getting tails and a 3

**b** getting heads and an even number

**c** not getting heads and an even number.

Two-way table:

| | | Dice | | | |
|---|---|---|---|---|---|
| | | 1 | 2 | 3 | 4 |
| Coin | H | H1 | H2 | H3 | H4 |
| | T | T1 | T2 | T3 | T4 |

Tree diagram:

| Coin | Dice | Outcome |
|---|---|---|
| H | 1 | H1 |
| | 2 | H2 |
| | 3 | H3 |
| | 4 | H4 |
| T | 1 | T1 |
| | 2 | T2 |
| | 3 | T3 |
| | 4 | T4 |

**a** There are 8 possible outcomes, one of which is a successful outcome: T3.
So P(tails and a 3) $= \frac{1}{8}$

**b** There are 8 possible outcomes, 2 of which are successful outcomes: H2, H4.
So P(heads and even number) $= \frac{2}{8} = \frac{1}{4}$

**c** P(not heads and even number) $= 1 - \frac{1}{4} = \frac{3}{4}$

## Exercise 18A

**1** Two tetrahedral dice are rolled. Each dice is numbered 1 to 4.

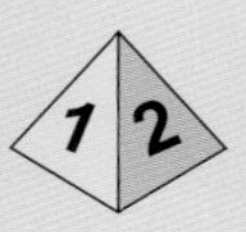


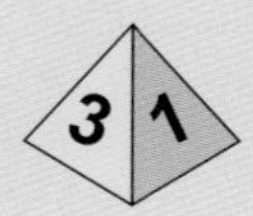


**a** Copy and complete this two-way table to show all the possible outcomes.

| | | Dice 1 | | | |
|---|---|---|---|---|---|
| | | 1 | 2 | 3 | 4 |
| Dice 2 | 1 | | | | 4, 1 |
| | 2 | | | 3, 2 | |
| | 3 | | | | |
| | 4 | | 2, 4 | | |

**b** What is the probability of the score on the two dice being the same?

**c** What is the probability of the score on the two dice not being the same?

**2** **a** Two coins, each with heads (H) and tails (T), are tossed. Copy and complete this tree diagram to show all the possible outcomes.

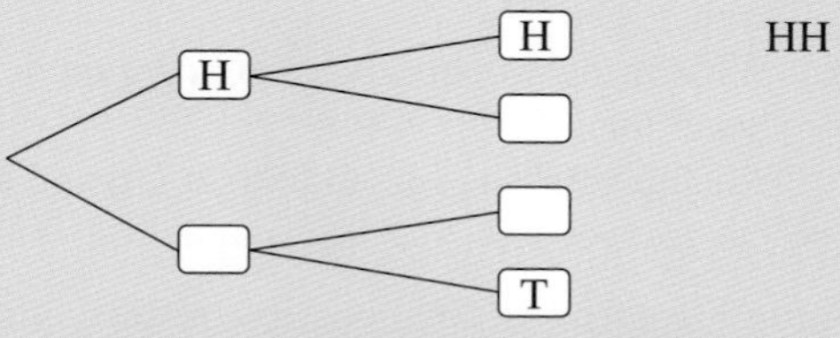


**b** What is the probability of getting tails on both coins?

**c** What is the probability of not getting tails on either coin?

**3** There are four colours of pens in a tin: red, green, blue and black. The probabilities of choosing a red, green or blue pen are shown in the table. What is the missing probability, for choosing a black pen?

| Colour of pen | Red | Green | Blue | Black |
|---|---|---|---|---|
| Probability of choosing colour | $\frac{3}{10}$ | $\frac{1}{5}$ | $\frac{1}{4}$ | |

**4** Two ordinary dice are rolled. Each dice is numbered 1 to 6.

The scores on the two dice are added together.

**a** Copy and complete this two-way table to show all the possible outcomes.

| | | Dice 1 | | | | | |
|---|---|---|---|---|---|---|---|
| | | 1 | 2 | 3 | 4 | 5 | 6 |
| Dice 2 | 1 | | 3 | | | | |
| | 2 | | | | | | |
| | 3 | | | | | | |
| | 4 | | | | | 9 | |
| | 5 | | | | | | |
| | 6 | | | | | | |

**b** What is the probability of rolling
- **i** a total of 8
- **ii** a total of 2
- **iii** a total of 11?

**c** Which total score has the highest probability of being rolled?

**d** What is the probability of rolling a total higher than 8?

**e** What is the probability of rolling a total lower than 8?

**f** Add up your answers to parts **b i**, **d** and **e**. Do you get 1?

**5** A rugby match has three possible results: win, lose or draw. Based on data from previous matches, the probability that Jim's team will win their next match is 0.4, while the probability they will lose their next match is 0.22. What is the probability they will draw their next match?

**6** Mr Wood and Mrs Boyd teach the same year in a primary school. Tomorrow is the year photograph. The teachers don't want to appear in the photograph wearing the same colour clothes. Mrs Boyd has blue, white and red blouses. Mr Wood has blue, purple, white and green shirts.

**a** Copy and complete the sample space diagram to show all the possible colour combinations.

| | | Mrs Boyd's blouses | | |
|---|---|---|---|---|
| | | Blue | White | Red |
| Mr Wood's shirts | Blue | | | |
| | Purple | | PW | |
| | White | | | |
| | Green | GB | | |

**b** Assuming that the teachers do not plan what they are wearing together, what is the probability that
- **i** Mrs Boyd wears red and Mr Wood wears white
- **ii** they both wear blue
- **iii** they do not wear the same colour as each other?

**7** These two spinners are spun and the scores added together.

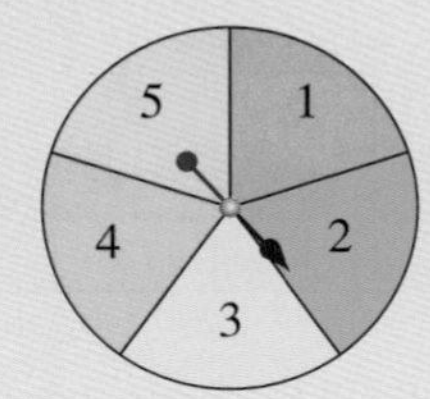

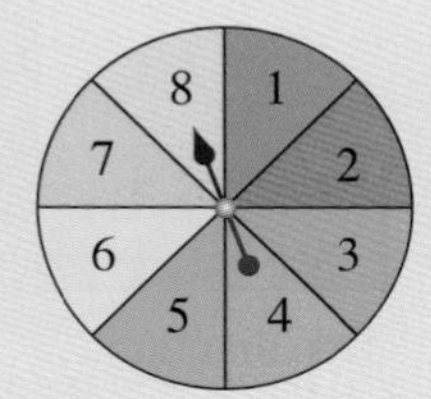


Draw a two-way table to show the sample space.

**a** What is the highest possible score?

**b** How many possible outcomes are there?

**c** What is the probability that the score is 5?

**d** What is the probability that the score is greater than 10?

**e** What is the probability that the score is less than 4?

**8** A letter is picked at random from the word HAT, then another letter is picked at random from the word CAP.

**a** Draw a tree diagram to show the sample space.

**b** What is the probability of
- **i** both letters being A
- **ii** only one of the letters being A
- **iii** at least one letter picked being H or P?

**9** A biased spinner has four colours: red, green, blue and white. The table shows the probability of spinning red, blue or white.

| Colour | Red | Green | Blue | White |
|---|---|---|---|---|
| Probability of spinning colour | $\frac{1}{4}$ | | $\frac{5}{12}$ | $\frac{1}{6}$ |

**a** Work out the probability of spinning green.

**b** Using compasses, a protractor and coloured pencils, draw the spinner.

**c** Jack was trying to draw a spinner that his friend Usuf said had these probabilities:

| Colour | Red | Yellow | Blue |
|---|---|---|---|
| Probability of spinning colour | $\frac{3}{8}$ | $\frac{1}{2}$ | $\frac{1}{4}$ |

It is not possible to draw this spinner. Why not?

**10** A blue and a red card are chosen at random from the set of cards shown. What is the probability that at least one of the cards is even?

| 2 | 3 | 1 | 1 | 5 | 3 | 4 | 2 |
|---|---|---|---|---|---|---|---|

## 18.2 Relative frequency and probability

The probabilities you have looked at so far have been **theoretical probabilities**, which means you can calculate them based on what you would expect to happen in theory, using the ratio

$$\frac{\text{(number of successful outcomes)}}{\text{(number of possible outcomes)}}.$$

Sometimes, though, the theoretical probability is not known – for example, you may be interested in the probability that a new design of light bulb will last longer than 500 hours or that a particular dice is biased.

When theoretical probabilities are not known we can work out **experimental probabilities**. This means an experiment is carried out and the results are used to work out the **relative frequency**, which is the ratio

$$\frac{\text{number of times an outcome occurs}}{\text{total number of trials}}.$$

For example, if 10 light bulbs are tested and 9 of them last longer than 500 hours, then the experimental probability of a light bulb lasting longer than 500 hours is $\frac{9}{10}$. You can see that you would not want to repeat an experiment like this – in which you are doing what is called 'testing to destruction' – with too many light bulbs. However, you learned in Book 2 that the more trials of you do of an experiment, the more accurate your experimental probability will be, and the closer it will get to the theoretical probability.

### EXAMPLE 3

Liam carries out a survey on the colours of cars passing his school. His results are shown in the table.

| Colour of car | Red | Green | Blue | White | Black | Other |
|---|---|---|---|---|---|---|
| Number | 92 | 45 | 84 | 69 | 47 | 63 |

Estimate the probability that the next car to pass will be blue.

Here, the total number of trials is the total number of cars seen.

Total number of trials = $92 + 45 + 84 + 69 + 47 + 63 = 400$

Relative frequency = $\frac{\text{number of times blue car occur}}{\text{total number of cars}} = \frac{84}{400} = 0.21$

So the estimated probability of the next car being blue is 0.21.

### Exercise 18B

**1** Use the survey results from Example 3 to estimate the probability that the next car will be

**a** red

**b** green

**c** not green

**d** white or black.

**2** Rashid rolled an ordinary six-sided dice 12 times. He rolled a 6 three times, so the experimental probability of rolling a 6 was $\frac{1}{4}$. Rashid said the dice must be biased because the probability of rolling a 6 should be $\frac{1}{6}$. Explain why he is wrong.

**3** Jess conducted a survey to find out the favourite ice-cream flavours of people buying her ice cream. The results are shown in the table.

| Favourite flavour | Vanilla | Chocolate | Strawberry | Raspberry | Other |
|---|---|---|---|---|---|
| Number of people | 64 | 45 | 52 | 38 | 51 |

**a** Use the results of Jess's survey to estimate the probability that the next person's favourite flavour will be

**i** vanilla

**ii** chocolate or strawberry

**iii** not raspberry.

**b** Which flavour of ice cream should Jess order the most of?

**c** Can you tell from this table which flavour she should order least of? Explain your answer.

**4** Mrs Brett suspects that a six-sided dice is biased. She rolls the dice 300 times and gets these results:

| Score | 1 | 2 | 3 | 4 | 5 | 6 |
|---|---|---|---|---|---|---|
| Frequency | 77 | 45 | 52 | 48 | 23 | 55 |

**a** How many times would you expect each score to occur if the dice is fair?

**b** None of the results equals the number of occurrences expected if the dice is fair. Does this mean the dice is biased?

**c** Do you think the dice is biased or fair? Explain your answer.

**5** An experiment is conducted which involves throwing a drawing pin up in the air 50 times and recording whether it lands point up or point down. The results are shown in the table.

| Outcome | Point up | Point down |
|---|---|---|
| Frequency | 27 | 23 |

**a** What is the experimental probability of the drawing pin landing point up? Give your answer as a decimal.

The experiment is then repeated. This time the drawing pin is thrown 1000 times, with these results:

| Outcome | Point up | Point down |
|---|---|---|
| Frequency | 420 | 580 |

**b** For this experiment, what is the experimental probability of the drawing pin landing point up? Give your answer as a decimal.

**c** Out of the two probabilities for parts **a** and **b**, which do you think is closest to the theoretical probability? Why?

**6** A bag contains an unknown number of balls. The balls are coloured blue, red and green. One ball is taken from the bag at random and its colour is recorded, then the ball is replaced in the bag. This is repeated 200 times. The results are shown in the table.

| Colour | Blue | Red | Green |
|---|---|---|---|
| Frequency | 80 | 52 | 68 |

If there are 50 balls in the bag altogether, how many of each colour of ball would you expect there to be?

## TECHNOLOGY

There are a lot of new words to learn when studying the topic of probability. Look at this web page for explanations of some of the terms used:

www.stats.gla.ac.uk/steps/glossary/probability.html

Note that the page uses set notation – you may want to have a look at Book 1, Chapter 19 to help you with this.

**Also, if you want to look at some extension work, the page goes beyond what is required at Cambridge Secondary 1 level.**

## ACTIVITY

**Probability crossword puzzle**

Ask your teacher for a copy of the probability crossword puzzle from their Teacher Book. You can use the web page just mentioned to help with some of the clues.

# Consolidation

### Example 1

The table shows the probabilities of spinning different scores on a biased spinner.

| Score | 1 | 2 | 3 | 4 | 5 |
|---|---|---|---|---|---|
| Probability | 0.2 | 0.13 | 0.4 | | 0.08 |

What is the missing probability, for spinning the number 4?

There are 5 mutually exclusive outcomes.
The sum of the probabilities of all mutually exclusive outcomes is 1.

P(1 or 2 or 3 or 5) = 0.2 + 0.13 + 0.4 + 0.08 = 0.81

P(4) = 1 − 0.81 = 0.19

### Example 2

A coin is tossed and then a dice is rolled. The coin has heads (H) on one side and tails (T) on the other. The dice is numbered 1 to 6. Draw two different sample space diagrams to show all the possible outcomes.

Two-way table:

| | | Dice | | | | | |
|---|---|---|---|---|---|---|---|
| | | 1 | 2 | 3 | 4 | 5 | 6 |
| Coin | H | H1 | H2 | H3 | H4 | H5 | H6 |
| | T | T1 | T2 | T3 | T4 | T5 | T6 |

Tree diagram:

Coin | Dice | Outcome

H → 1, 2, 3, 4, 5, 6 → H1, H2, H3, H4, H5, H6

T → 1, 2, 3, 4, 5, 6 → T1, T2, T3, T4, T5, T6

### Example 3

Jen carries out a survey on the favourite fruits of students in her year at school. Her results are shown in the table.

| Favourite fruit | Pears | Apples | Bananas | Other |
|---|---|---|---|---|
| Number of students | 33 | 48 | 52 | 17 |

Estimate the probability that the next student will choose pears.

Total number of trials = 33 + 48 + 52 + 17 = 150

$$\text{Relative frequency} = \frac{\text{number of times pears occurs}}{\text{total number of students}}$$

$$= \frac{33}{150} = 0.22$$

So the estimated probability that the next student will choose pears is 0.22.

## Exercise 18

**1** A blue and a red card are chosen at random from the set of cards shown.

**a** Draw a sample space table showing all the possible outcomes.

**b** What is the probability of

**i** both cards being 2

**ii** a red 8 and a blue 3

**iii** both cards being odd?

2 A bag contains an unknown number of coloured sweets. The colours are yellow, red, black and orange. One sweet is taken from the bag at random and the colour recorded, then the sweet is replaced in the bag. This is repeated 10 times. The results are shown in the table.

| Colour | Yellow | Red | Black | Orange |
|---|---|---|---|---|
| Frequency | 3 | 2 | 1 | 4 |

a Estimate the probability the next sweet chosen will be orange.
The experiment is repeated 190 more times. The final data is shown in the table.

| Colour | Yellow | Red | Black | Orange |
|---|---|---|---|---|
| Frequency | 43 | 59 | 60 | 38 |

b Using the second table, estimate the probability the next sweet chosen will be orange.

c Which of your answers to parts **a** and **b** do you think is more reliable? Give a reason for your answer.

3 A biased dice with sides numbered 1 to 4 is rolled. The table shows the probability of rolling different scores.

| Score | 1 | 2 | 3 | 4 |
|---|---|---|---|---|
| Probability | 0.3 | 0.24 | | 0.1 |

a What is the missing probability, for scoring 3?

b What is the probability of rolling a 1 or a 2?

c What is the probability of not rolling a 4?

4 These two spinners are spun, one after the other.

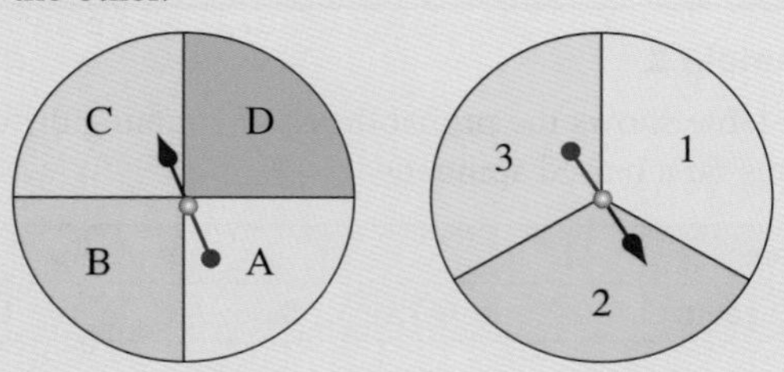


a Draw a tree diagram to show all the possible outcomes.

b What is the probability of spinning

i a D and a 2

ii a vowel and an odd number?

5 A magazine compares two different models of laptop computer, A and B, to see what the battery life is like. The results are shown in the table.

| | Time battery lasts (hours) | | | |
|---|---|---|---|---|
| | 0–2 | 3–5 | 6–8 | More than 8 |
| Laptop A | 3 | 18 | 13 | 6 |
| Laptop B | 12 | 48 | 25 | 15 |

a Find the experimental probability of **i** Laptop A **ii** Laptop B having a battery life of 3–5 hours.

b Repeat part **a** for 6–8 hours.

c If you want a laptop with a battery that lasts more than 8 hours, which of these laptops would you choose? Give a reason for your answer.

# Summary

## You should know ...

1 The sum of the probabilities of all mutually exclusive outcomes of an event is 1.

*For example:*

The possible results of a cricket match are: win, lose or draw.

| Result | Win | Lose | Draw |
|---|---|---|---|
| Probability | | 0.4 | 0.05 |

Using the probabilities in the table, what is the probability of a win?

P(lose or draw) = 0.4 + 0.05 = 0.45

P(win) = 1 − 0.45 = 0.55

## Check out

1 There are three colours of sweets in a bag: red, green and black. The probabilities of choosing a black or green sweet are shown in the table. What is the missing probability, for choosing a red sweet?

| Colour of sweet | Red | Green | Black |
|---|---|---|---|
| Probability of choosing colour | | $\frac{9}{20}$ | $\frac{1}{5}$ |

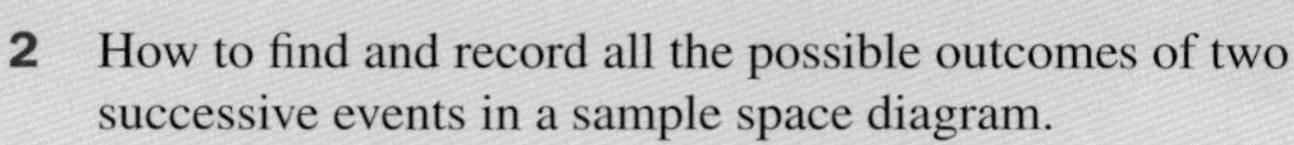



**2** How to find and record all the possible outcomes of two successive events in a sample space diagram.

*For example:*

A coin is tossed and then a spinner is spun. The coin has heads (H) on one side and tails (T) on the other. The spinner is numbered from 1 to 3. Draw two different sample space diagrams to show all the possible outcomes.

Two-way table:

| | | Spinner | | |
|---|---|---|---|---|
| | | **1** | **2** | **3** |
| **Coin** | **H** | H1 | H2 | H3 |
| | **T** | T1 | T2 | T3 |

Tree diagram:



**2** A letter is picked at random from the word HEN and a spinner is spun. The spinner is numbered from 1 to 3

Draw two different sample space diagrams to show all the possible outcomes.
(**Hint:** a possible outcome is, for example, N2.)

**3** That relative frequency is an estimate of probability. The more times the experiment is repeated the more reliable the estimate.

$$\text{relative frequency} = \frac{\text{number of times an outcome occurs}}{\text{total number of trials}}$$

*For example:*

A bag contains an unknown number of coloured pens. The colours are blue, red and black. One pen is taken from the bag at random and the colour recorded, then the pen is replaced in the bag. This is repeated 50 times. The results are shown in the table.

| **Colour** | Blue | Red | Black |
|---|---|---|---|
| **Frequency** | 17 | 10 | 23 |

Estimate the probability the next pen chosen will be red.

$$\text{Relative frequency} = \frac{\text{number of times red occurs}}{\text{total number of trials}} = \frac{10}{50} = \frac{1}{5}$$

So the estimated probability of the next pen being red is $\frac{1}{5}$.

**3** Using the data in the table on the left, estimate the probability that the next pen chosen will be

**a** blue

**b** black.

# Review C

**1** The Home Store has a sale.

**a** Find the sale price of the items shown.

**b** What would the actual price of the items be if a 15% sales tax then had to be included?

**2** Copy this diagram. The quadrilateral on the left is a mirror image of the quadrilateral on the right.

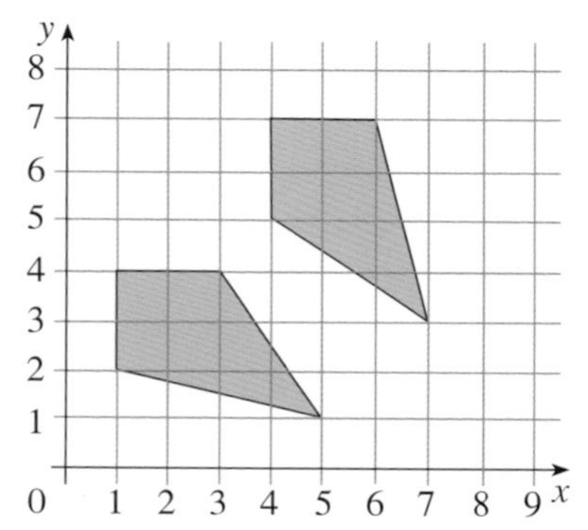


**a** On your diagram, draw in the mirror line.

**b** Write down the coordinates of three points on that line.

**c** Write down the equation of the mirror line.

**3** Simplify these ratios

**a** 240 : 420

**b** 105 : 75

**c** 84 : 70 : 112

**4** Mr Masood is a carpenter. The table shows his fees for three different jobs.

| Number of hours, $x$ | 1 | 2 | 5 |
|---|---|---|---|
| Fee in dollars, $y$ | 25 | 40 | 85 |

**a** Draw a graph to show this information, with number of hours on the $x$-axis and fee in dollars on the $y$-axis.

**b** Find the gradient of the graph. What does this tell you?

**c** Find the $y$-intercept of the graph. What does this tell you?

**d** Are number of hours and fee in dollars directly proportional? Explain your answer.

**5** A cylindrical water tank has a diameter of 6 m and a height of 2 m.

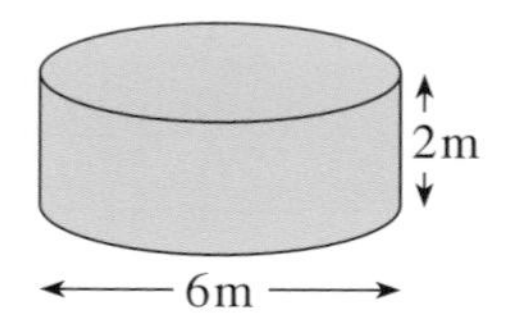


**a** What is the radius of the base of the tank?

**b** What is the area of the base of the tank?

**c** What is the surface area of the tank?

**d** What is the volume of the tank?

**6** Mohammad buys a car for $25 000. He sells the car three years later for $16 000. What is his percentage loss?

**7** A purple paint is a mix of blue paint and red paint in the ratio 7 : 4. How much blue and red paint is used to make 1650 ml of the purple paint?

**8** **i** $2y = 3x - 7$

**ii** $2x = 5 - 3y$

**iii** $2y - 3x = {}^{-}8$

**a** Write the three equations **i**–**iii** in the form $y = mx + c$.

**b** State the gradient and $y$-intercept of each line.

**c** Which two lines are parallel?

**d** Which two lines are perpendicular?

**e** Draw all three lines on the same set of axes to check your answers to parts **b**–**d**.

**9** Copy and complete the table.

| Usual price | Discount | Sale price |
|---|---|---|
| $400 | 15% | |
| $275 | 8% | |
| $1995 | $7\frac{1}{2}$% | |
| $3085 | $5\frac{1}{2}$% | |

**10** Describe fully these transformations.

**a**

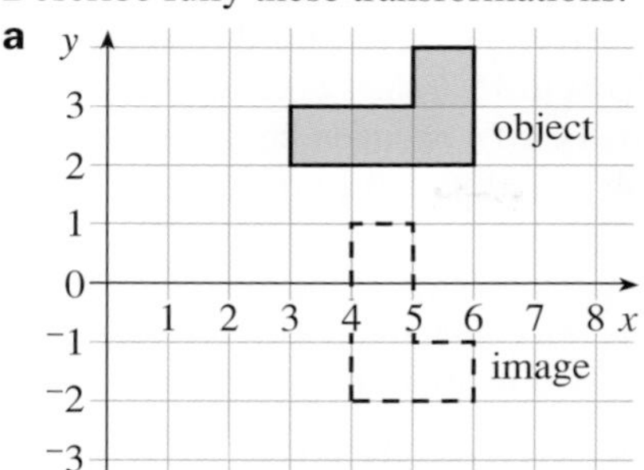


**b**

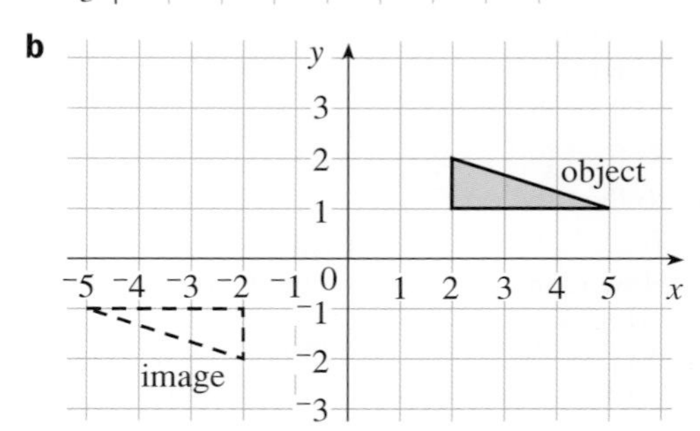


**11** A coin with heads (H) and tails (T) is tossed and a spinner is spun. The spinner has three equal-sized sections numbered 1 to 3.

**a** Copy and complete the tree diagram to show the sample space of all possible outcome.

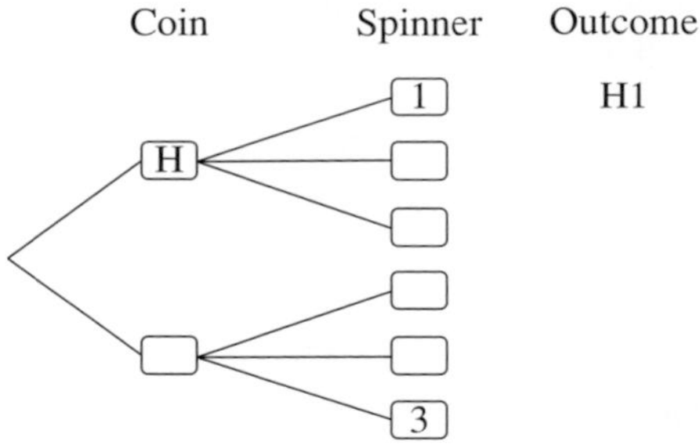


**b** What is the probability of getting

**i** heads and an even number

**ii** tails and an odd number?

**12** Share

**a** \$126 in the ratio 2 : 7

**b** 160 g in the ratio 3 : 5

**c** 0.8 m in the ratio 2 : 1 : 5

**13** A triangle T has vertices (0, 0), (2, 0), (0, 1). T is mapped to T′ by a combined transformation consisting of a translation $\begin{pmatrix} 4 \\ 0 \end{pmatrix}$ and a reflection in the line $y = 0$.

T′ is mapped to T″ by a combined transformation consisting of a translation $\begin{pmatrix} 0 \\ 3 \end{pmatrix}$ and a reflection in the line $x = 2$

**a** Draw a diagram to show T, T′ and T″

**b** What single transformation maps T to T″

**14**

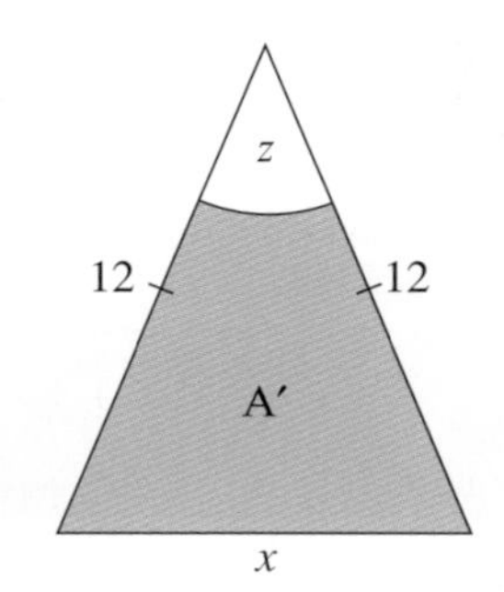


Triangle A is enlarged to make Triangle A′.

**a** What is the scale factor of the enlargement?

**b** Find the side length $x$.

**c** What can you say about angles $y$ and $z$?

**15** Find the **a** volume **b** surface area of this prism.

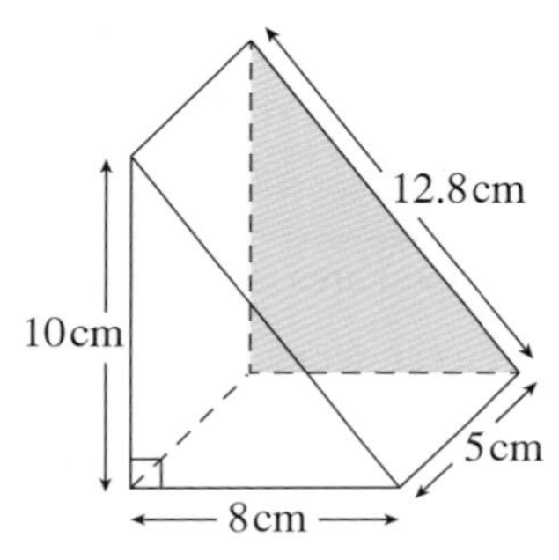


**16** If $f(x) = 7x - 5$, find

**a** $f^{-1}(x)$

**b** $f^{-1}(18)$

**17** Copy and complete:

To describe fully a

- reflection, give the equation of the ..............
- rotation, give the ........... and ............... of the rotation and the coordinates of the ...........
- translation, give the .....................
- enlargement, give the coordinates of the ..................... and the ..................... of the enlargement.

**18** Find the total simple interest on a loan of \$17 500 if the interest rate is 10.5% and the loan is taken for:

**a** 6 years **b** $2\frac{1}{2}$ years **c** 8 years.

**19** Describe the transformations that map these triangles to their images.

**a** A (2, 0), B (2, 2), C (0, 0) → A′ (3, 1), B′ (3, 3), C′ (1, 1)

**b** A (⁻1, 2), B (1, ⁻1), C (⁻2, ⁻2) → A′ (1, 3), B′ (3, 0), C′ (0, ⁻1)

**c** A (3, ⁻1), B (⁻2, 2), C (⁻1, 4) → A′ (1, ⁻2), B′ (⁻4, 1), C′ (⁻3, 3)

**20** There are four colours of balls in a bag: blue, green, yellow and white. The probabilities of choosing a blue, yellow or white ball are shown in the table. What is the missing probability, for choosing a green ball?

| Colour | Blue | Green | Yellow | White |
|---|---|---|---|---|
| Probability of choosing colour | $\frac{1}{8}$ | | $\frac{2}{5}$ | $\frac{3}{10}$ |

**21** Write each of these as a ratio in its simplest whole-number form:

**a** 2.7 : 9 **b** $\frac{1}{3} : \frac{2}{5}$ **c** 60% : 0.5 : $1\frac{2}{5}$

**22** This shape is made from a rectangle and two semicircles.

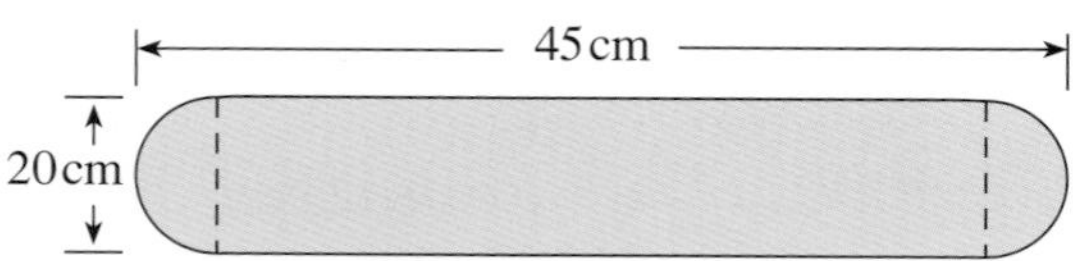


Calculate the total area of the shape.

**23** The value of a car depreciates by 10% each year. James buys a new car for $50 000.
What will the value of the car be after:
**a** one year
**b** two years
**c** three years?

**24** Triangle P with vertices (7, 0), (7, 2), (9, 3) is mapped to triangle Q with vertices (11, 0), (11, 4), (15, 6), by a transformation.
**a** Draw the two triangles.
**b** Describe the transformation

**25** Shamma buys some phones at a cost of $48 each. She sells them for $69.60 each. What is her percentage profit?

**26** A letter is picked at random from the word BOOKS, then another letter is picked at random from the word WORK.
**a** Draw a sample space diagram to show the possible outcomes.
**b** What is the probability of
**i** picking a B and a W
**ii** both letters being O?

**27** I change 250 US dollars ($) into 195 euros (€). At the same exchange rate, how many
**a** euros would I get for $200
**b** dollars would I get for €390?

**28** The price of a reconditioned car is $45 000.
**a** How much will Carl pay for the car if all sales are subject to 15% sales tax?
**b** What will Carl's payment be if there is a 7% cash discount on the price that includes sales tax?
**c** If the car depreciates at the rate of 20% per year from the price including sales tax, how much will it be worth after one year?

**29** On squared paper, plot the points ($^-2$, 4) and ($^-2$, 0). Now, using a coloured pencil:
**a** Join ($^-2$, 0) to (0, 0) and join ($^-2$, 4) to ($^-2$, 0)
**b** Reflect the shape in the $x$-axis.
**c** Reflect this image in the $y$-axis.
**d** Reflect the original shape in the $y$-axis.
**e** What letter have you made?
**f** How many lines of symmetry does it have?

**30** Calculate the volume of copper used to make this rod with radius 2 cm and length 2.5 m.

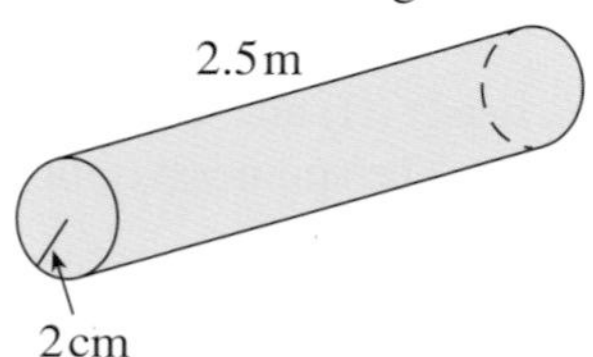


**31** Plot the graph of each equation. Use the graphs to find the solution to each pair of simultaneous equations.

**a** $x - 2y = 0$
$3x + 2y = 8$

**b** $x = 4$
$x - 2y = {}^-2$

**c** $y = x$
$3x - 2y = 10$

**d** $3x - y - 3 = 0$
$2x - y - 7 = 0$

**32** Copy and complete this table showing items sold in a bookstore

| Item | Cost price | Selling price | Profit or loss | Percentage profit or loss |
|---|---|---|---|---|
| Book | $4 | $6 | Profit | 50% |
| Bookmark | $0.30 | $0.54 | | |
| Game | $16 | $12 | | |
| Audio book | $12 | $15 | | |

**33** Students in a primary school were asked to vote for a school captain. The votes were in a ratio of 14 : 9 in favour of Rashid. 168 children voted for Rashid. How many children voted altogether?

**34 a** The simple interest rate at a bank is 4%. How much will Ashton have in the bank after 1 year if he deposits:

**i** \$600 **ii** \$750 **iii** \$5000
**iv** \$14 250 **v** \$82?

**b** For each of the deposits in part **a**, what will the total amount be, including the original deposit, after:

**i** 3 years
**ii** 8 years?

**35** A kite ABCD has vertices A $(1, ^-4)$, B $(3, ^-3)$, C $(1, 5)$ and D $(^-1, ^-3)$
Find the image of the kite under an anticlockwise rotation of:
**a** 180° **b** 270° about the origin.

**36** If 7 boxes of sweets contain 196 sweets in total, calculate how many sweets 5 boxes will contain.

**37** Write down the equation of the straight line with gradient $m = \frac{1}{2}$ which passes through the point $(0, 7)$.

**38** The triangle P $(2, 1)$, Q $(2, 3)$, R $(1, 3)$ is mapped to P′ $(4, 1)$, Q′ $(4, 5)$, R′ $(2, 5)$ by an enlargement.
**a** Draw PQR and its image on squared paper.
**b** By joining PP′, QQ′ and RR′ find the centre of enlargement. Write down its coordinates.
**c** What do you notice about the sides and angles of triangle PQR and its image?

**39 a** What is the name of the cuboid with all edges the same?
**b** Copy and complete this table, which shows the dimensions of two cuboids

| Length | Width | Height | Volume |
|---|---|---|---|
| 1 cm | 1 cm | 1 cm | |
| 10 mm | 10 mm | 10 mm | |

**Remember:** 1 cm = 10 mm

**c** Study your table. Now copy and complete:
$1\,\text{cm}^3 = \square\,\text{mm}^3$

**40** There are 184 litres of water left in a tank which originally held 400 litres. What is the percentage loss of water?

**41** If $x$ and $y$ are directly proportional, copy and complete these tables.

**a**

| $x$ | 5 | | 90 |
|---|---|---|---|
| $y$ | 35 | 14 | |

**b**

| $x$ | 2 | 15 | |
|---|---|---|---|
| $y$ | 32 | | 120 |

**42** On squared paper, draw the triangle with vertices $(3, 0)$, $(3, 4)$ and $(4, 4)$. Taking the origin as the centre of enlargement, draw the enlargements of the triangle using scale factors of:
**a** 2 **b** 3

**43** For each of these functions find the inverse function $f^{-1}(x)$.
**a** $f(x) = 2x$ **b** $f(x) = x - 5$
**c** $f(x) = 3x + 1$ **d** $f(x) = 7(x - 3)$

**44** At the Vacation Inn, the rates are advertised as:
Double room \$150 per night
Single room \$120 per night
**a** What would the bill be for Mr Alan if he stays for one night and there is a 10% government room charge and a 5% service charge?
**b** What would the bill be for Mr and Mrs Trotman if they stay for 3 nights and the same charges apply?

**45** Copy and complete:

| | Transformation | | | |
|---|---|---|---|---|
| | Rotation | Enlargement | Translation | Reflection |
| Object and image are | Congruent | Similar | | |
| Side lengths preserved? | | | Yes | |
| Angles preserved? | Yes | | | |

(**Note:** 'preserved' means 'stay the same'.)

**46** Compare these quantities using ratio.
**a** 300 mm and 35 cm
**b** 7.6 ℓ and 800 ml
**c** 6.24 kg and 460 g
**d** 23 cm² and 6900 mm²

**47** ABC is the triangle with vertices $(1, 0)$, $(2, 0)$, $(1, 2)$.
**a** Draw a diagram to show triangle ABC and its image following a rotation of 180° about $(0, 0)$. Use tracing paper to help you if necessary.
**b** The image is rotated 90° clockwise about the point $(0, 1)$ Draw this second image.
**c** What single rotation would take triangle ABC to this second image? State its centre and angle.

**48** Find the total length of wire used to make a wire circle set inside a wire square of side 9 cm, as shown in the diagram.

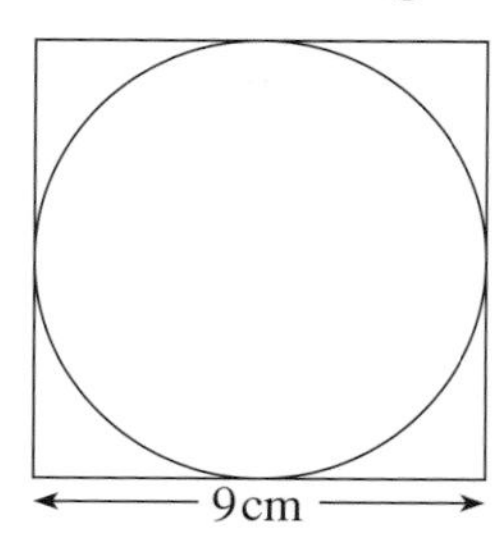

**49 a** The information in the table represents a linear relation. Show this information on a graph, with distance on the vertical axis and time on the horizontal axis. Hence copy and complete the table.

| Distance from home, $d$ (km) | Time of journey, $t$ (hours) |
|---|---|
| 17 | 1 |
| 19 | 2 |
| | 3 |
| 23 | 4 |
| | 5 |
| 37 | |

**b** Find the gradient of the line. Interpret its meaning.

**c** State the intercept on the vertical axis. Interpret its meaning.

**d** Hence, determine the equation of the straight line $d = mt + c$.

**50** Harry carries out a traffic survey on the types of vehicle passing his school. His results are shown in the table.

| Type of vehicle | Car | Bike | Lorry | Van | Bus/ Coach | Other |
|---|---|---|---|---|---|---|
| Number | 92 | 23 | 16 | 15 | 23 | 31 |

Estimate the probability that the next vehicle to pass will be a

**a** car

**b** lorry.

**51** A triangle has vertices at X $(3, {}^-1)$ Y $(1, 3)$ and Z $(5, 1)$.
It is transformed to give an image with vertices at X′ $(9, 5)$ Y′ $(5, 7)$ and Z′ $(7, 3)$.

**a** Draw a sketch to show this.

**b** Describe the transformation mapping XYZ to X′Y′Z′.

**52** $x$ and $y$ are directly proportional.

| $x$ | 3 | 5 | 8 |
|---|---|---|---|
| $y$ | 7.5 | 12.5 | 20 |

**a** Find the equation connecting $x$ and $y$ in the form $y = mx$.

**b** Check your answer to part **a** is correct by drawing the graph and finding the gradient and $y$-intercept.

**c** Find the value of

**i** $y$ when $x = 7$

**ii** $x$ when $y = 42.5$

**53** A coin is tossed and then a spinner is spun. The coin has heads (H) on one side and tails (T) on the other. The spinner has five equal sections numbered from 1 to 5. Draw a tree diagram to show the sample space of all the possible outcomes. Use the diagram to work out the probability of

**a** getting heads and a 5.

**b** getting tails and an even number

**c** not getting tails and an even number.

**54 a** An arithmetic sequence has first term $a$, and common difference $d$. Derive an expression to describe the $n$th term of this arithmetic sequence.

**b** Find the $n$th term of the arithmetic sequence 17, 24, 31, 38, 45, …

**c** Find the 20th term

**d** Which term in the sequence is 234?

**55** Find the image of triangle A $(1, 1)$, B $(2, 2)$, C $(2, 4)$ under an enlargement with

**a** scale factor 2 and centre $(0, 0)$

**b** scale factor 2 and centre $(1, 0)$

**c** scale factor 3 and centre $(1, 1)$.

**56** Amanda and Donna share some money in the ratio 4 : 7. Donna gets $54 more than Amanda. How much do they get each?

**57** Use the value of $m$ and $c$ to sketch the graph of each equation.

**a** $y = 2x + 1$ **b** $y = {}^-2x - 1$

**58** Find the image of triangle A $({}^-3, 1)$, B $({}^-1, {}^-1)$, C $(2, 2)$ under the translation

**a** $\begin{pmatrix} 1 \\ 0 \end{pmatrix}$ **b** $\begin{pmatrix} 0 \\ 1 \end{pmatrix}$ **c** $\begin{pmatrix} 2 \\ 3 \end{pmatrix}$

**d** $\begin{pmatrix} {}^-2 \\ 3 \end{pmatrix}$ **e** $\begin{pmatrix} 2 \\ {}^-3 \end{pmatrix}$ **f** $\begin{pmatrix} {}^-2 \\ {}^-3 \end{pmatrix}$

**59** Find the area and perimeter of this sector.

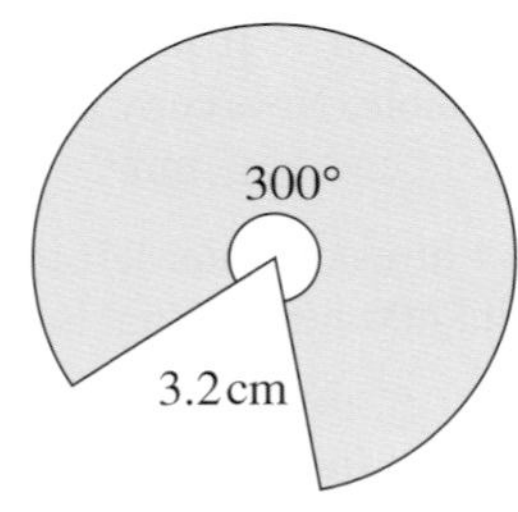


**60** Find the first five terms of these sequences.

**a** The term-to-term rule is *divide by 3*, the first term is 1215.

**b** The position-to-term rule is *multiply by 4 then add 3*.

# Quadratics

## Objectives

- Draw quadratic graphs.
- Expand the product of two linear expressions of the form $ax \pm n$ and simplify the corresponding quadratic expression.
- Use quadratics to help with arithmetic.
- Factorise quadratics, including the difference of two squares.
- Solve quadratic equations by factorising.
- Form quadratics based on word problems and solve them.

The work in this chapter is not in the Cambridge Secondary 1 Mathematics curriculum framework. It is not in the Checkpoint tests. This work is in the Cambridge IGCSE® maths curriculum and is here for you to try if you have completed the work from the other chapters.

## What's the point?

Living around 3000 years ago, the Babylonians were one of the world's first civilisations. They came up with some great ideas, like agriculture, irrigation and writing. They are also credited for the invention of the taxman. The Babylonian taxman asked farmers for a certain amount of their harvest in tax. Babylonians ended up solving quadratic equations – the quantity of a crop that can be grown in a field is proportional to side length of the field squared, which gives a quadratic equation – to work out how to maximise their harvest and what to give to the taxman. We still use quadratics in maximum/minimum problems today.

## Before you start

### You should know ...

**1** How to simplify algebraic expressions by collecting like terms.
*For example:*
$3y + x + 2x + y = 3x + 4y$

### Check in

**1** Simplify:
**a** $6x - 2x$
**b** $x^2 - 3y^2 + 2x^2$
**c** $ab - b + 6ab - 3b$

2 How to solve linear equations.
*For example:*

$$5x - 2 = 6 - 2x$$
+2x] $7x - 2 = 6$
+2] $7x = 8$
÷7] $x = 1\frac{1}{7}$

2 Solve:
**a** $3x - 5 = 10$
**b** $2 - x = 4 + x$
**c** $3x - 2 = 4 - 3x$

## 19.1 Graphs of quadratics

An equation with $x^2$ as the highest term is known as a **quadratic** equation.
For example

$y = x^2 + 3x - 2$
$x^2 - 4 = 0$

are quadratic equations.

To plot the graph of a quadratic equation you need first to complete a table of values, in the same way as you did for a linear equation (see Chapter 14).

### EXAMPLE 1

Plot the graph of $y = x^2 - 1$ for values of $x$ between $^-3$ and 3.

The table of values is:

| x | $^-3$ | $^-2$ | $^-1$ | 0 | 1 | 2 | 3 |
|---|---|---|---|---|---|---|---|
| $x^2$ | 9 | 4 | 1 | 0 | 1 | 4 | 9 |
| −1 | $^-1$ | $^-1$ | $^-1$ | $^-1$ | $^-1$ | $^-1$ | $^-1$ |
| y | 8 | 3 | 0 | $^-1$ | 0 | 3 | 8 |

Plot the points:

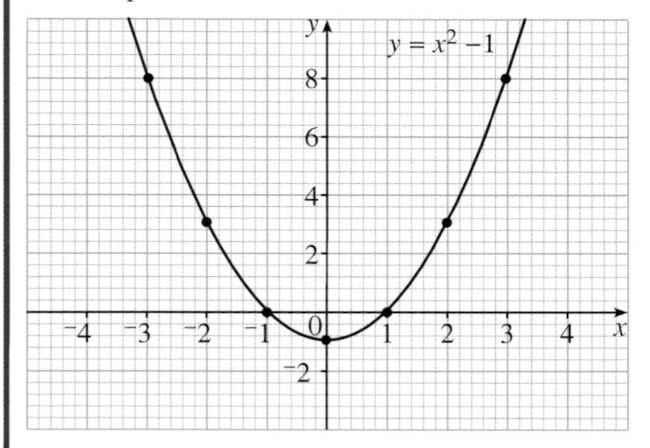


Notice that the graph is symmetrical and the curve is smooth.

For more complex quadratic functions the table of values has more rows.

### EXAMPLE 2

Draw the graph of $y = x^2 - 3x + 4$ for $x = {}^-3$ to $x = 3$.

The table of values is:

| x | $^-3$ | $^-2$ | $^-1$ | 0 | 1 | 2 | 3 |
|---|---|---|---|---|---|---|---|
| $x^2$ | 9 | 4 | 1 | 0 | 1 | 4 | 9 |
| −3x | 9 | 6 | 3 | 0 | $^-3$ | $^-6$ | $^-9$ |
| +4 | 4 | 4 | 4 | 4 | 4 | 4 | 4 |
| y | 22 | 14 | 8 | 4 | 2 | 2 | 4 |

Choose a suitable scale to plot the points.
Plotting the points gives:

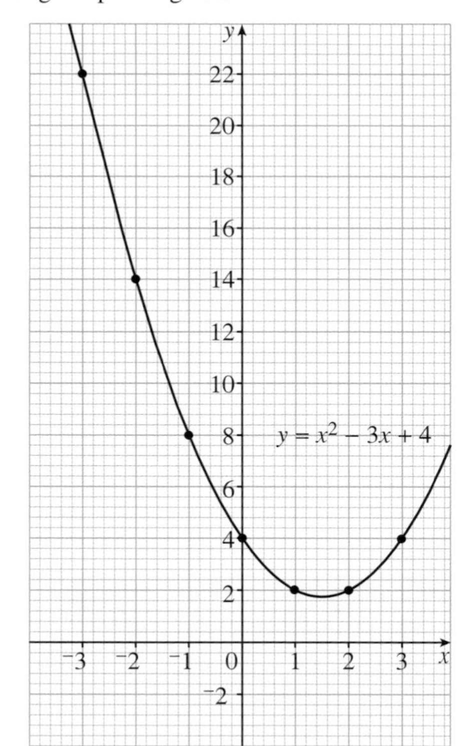


Make sure you draw a smooth curve to connect the points.

Your choice of scale is important. In Example 2, if the scale on the $x$-axis is too narrow, the graph will look like a hairpin!

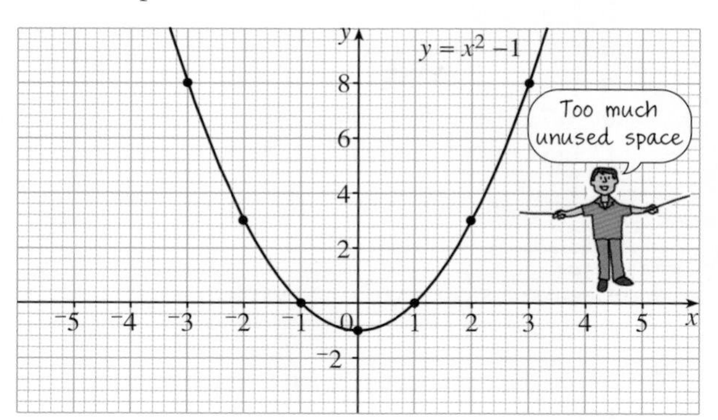


It is also important to join the points with a smooth curve – do not use a ruler!

WRONG

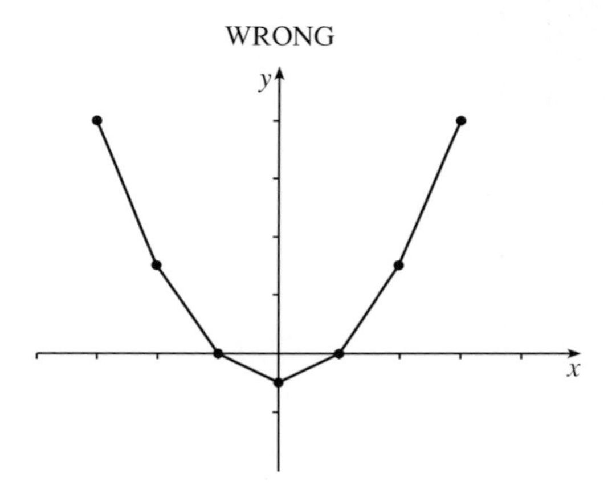

RIGHT

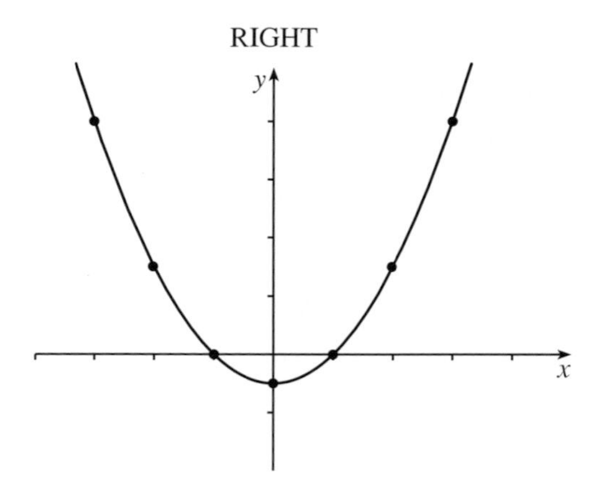

## Exercise 19A

**1** **a** Copy and complete the table of values for the equation $y = x^2 + 3$.

| $x$ | −3 | −2 | −1 | 0 | 1 | 2 | 3 |
|---|---|---|---|---|---|---|---|
| $x^2$ | | | 1 | | | | 9 |
| $+3$ | | 3 | | | | 3 | |
| $y$ | | | | 3 | | | |

**b** Plot the values to form a graph.

**c** Join the points with a smooth curve.

**d** Where does the graph cut the $y$-axis?

**2** Copy and complete the table for the equation $y = x^2 + 1$.

| $x$ | −3 | −2 | −1 | 0 | 1 | 2 | 3 |
|---|---|---|---|---|---|---|---|
| $y$ | | 5 | | 1 | | | |

Draw the graph of $y = x^2 + 1$.

**3** Draw a graph of $y = x^2 + 2$ by first making a table of values for $x$ from −3 to 3.

**4** Draw a graph of $y = x^2 - 4$ by first making a table of values for $x$ from −3 to 3.

**5** Draw the graph of $y = 6 - x^2$ for values of $x$ from −2 to 4.
The table of values has been started for you:

| $x$ | −2 | −1 | 0 | 1 | 2 | 3 | 4 |
|---|---|---|---|---|---|---|---|
| $x^2$ | 4 | 1 | 0 | 1 | 4 | | |
| $y = 6 - x^2$ | 2 | 5 | 6 | | | | |

In what way is the graph of $y = 6 - x^2$ different from the graphs in Questions **1**–**4**?

**6** **a** Copy and complete the table of values for the equation $y = x^2 - 3x$.

| $x$ | −3 | −2 | −1 | 0 | 1 | 2 | 3 |
|---|---|---|---|---|---|---|---|
| $x^2$ | | 4 | | | 1 | | |
| $-3x$ | 9 | | | | | −6 | |
| $y$ | | | 4 | | | | |

**b** Using suitable axes, plot the points on a graph. Join the points with a smooth curve.

**c** Where does the curve $y = x^2 - 3x$ cut the $x$-axis?

## INVESTIGATION

Find out about the uses of quadratics.
Search some websites to get information on their use.
Present your findings to your class.

## 19.2 Expanding two brackets

In Chapter 2 you learned how to expand the product of two linear expressions of the form $x \pm n$ and simplify the corresponding quadratic expression.
In this chapter this is extended to harder expressions, in the form $ax \pm n$.

In Chapter 2 you learned

$$(x + 3) \times (x + 2)$$
$$= x^2 + 2x + 3x + 6$$
$$= x^2 + 5x + 6$$

More complex expressions can be multiplied in the same way.

**EXAMPLE 3**

Simplify $(3x - 4) \times (2x + 6)$

$$= 3x \times 2x + 3x \times 6 - 4 \times 2x - 4 \times 6$$
$$= 6x^2 + 18x - 8x - 24$$
$$= 6x^2 + 10x - 24$$

### Exercise 19B

**1** By considering the areas of the rectangles, simplify the products:

**a**

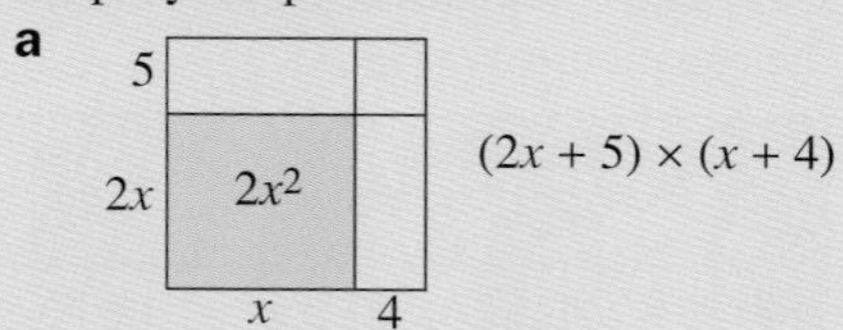


$(2x + 5) \times (x + 4)$

**b**

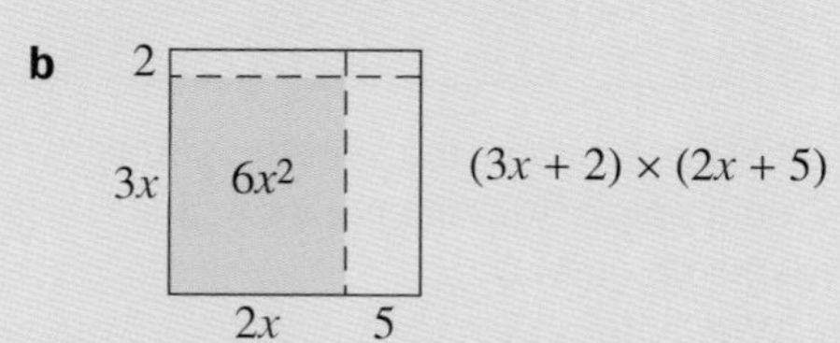


$(3x + 2) \times (2x + 5)$

**c**

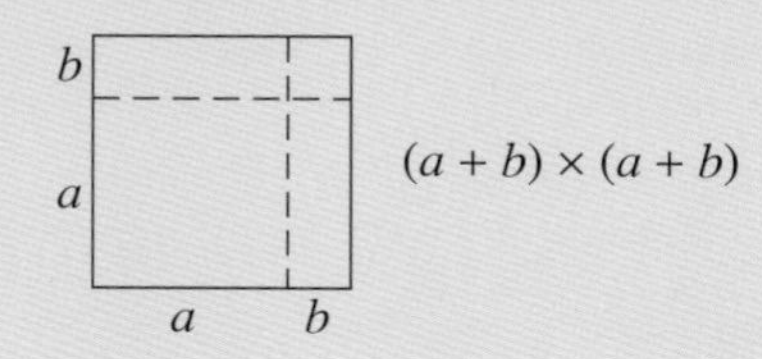


$(a + b) \times (a + b)$

**2** Expand and simplify:

**a** $(2x + 3)(x + 9)$
**b** $(4x + 2)(3x + 5)$
**c** $(5x + 1)(3x - 8)$
**d** $(6x - 2)(4x - 3)$
**e** $(7x + 8)(10x - 1)$
**f** $(8x - 7)(6x - 11)$

**3** Check that your answers for Question **2** are correct, by letting $x = 2$.

**4** Use the distributive law to explain why
$(x + a) \times (x + b) = x^2 + (a + b)x + ab$

**5** Write the product $(a + b)(c + d)$ without brackets.

**6**

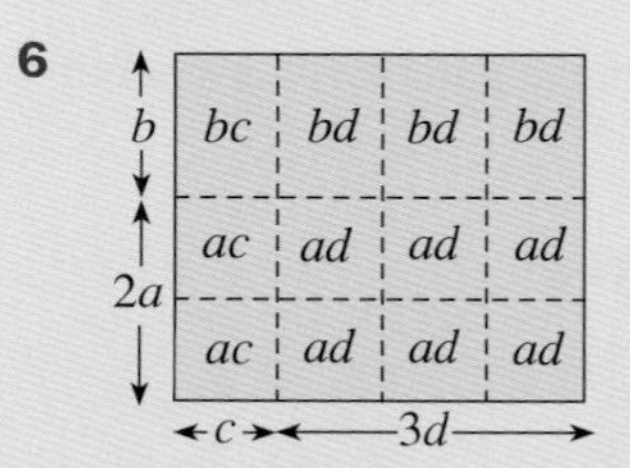


$(2a + b) \times (c + 3d)$
$= 2a \times c + 2a \times 3d$
$\quad + b \times c + b \times 3d$
$= 2ac + 6ad + bc + 3bd$

**a** The rectangle shown has sides of length $(2a + b)$ and $(c + 3d)$. Do you agree that its area is $(2a + b)(c + 3d)$?
**b** Do you agree that both methods above show that $(2a + b)(c + 3d)$ can be written as $2ac + 6ad + bc + 3bd$?

**7** Simplify these expressions:

**a** $(2p + q)(r + 3s)$
**b** $(a + 3b)(2c + d)$
**c** $(3x + y)(6 + 5m)$
**d** $(2p + 5q)(l + m)$
**e** $(5a + b)(4c + d)$
**f** $(x + 3y)(p + 7q)$

## 19.3 Difference between two squares and using quadratics

### Some important results

Look at these two diagrams:

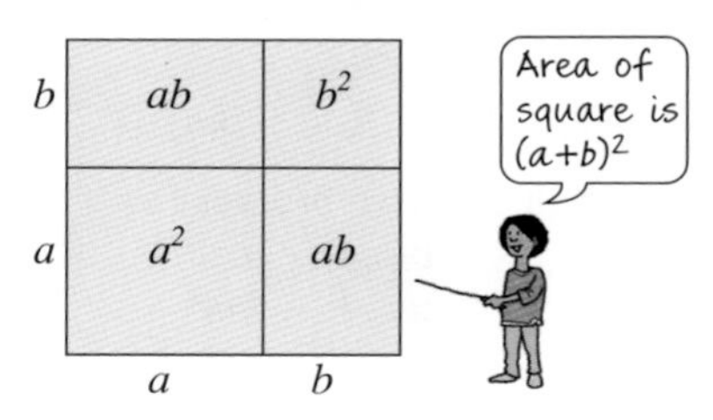


$(a + b)^2$
$= a^2 + ab + ab + b^2$
$= a^2 + 2ab + b^2$

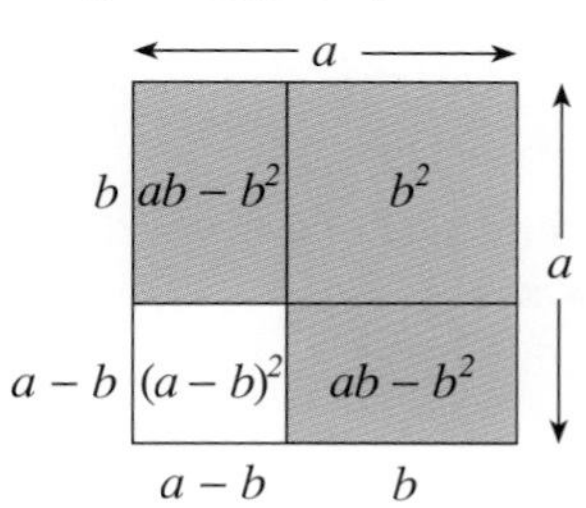


$(a - b)^2$
$= a^2 - 2(ab - b^2) - b^2$
$= a^2 - 2ab + b^2$

- The diagrams show that:
  $\boldsymbol{(a + b)^2 = a^2 + 2ab + b^2}$
  and $\boldsymbol{(a - b)^2 = a^2 - 2ab + b^2}$

Another interesting result is the product
$(a - b)(a + b) = a^2 + ab - ab - b^2$
$= a^2 - b^2$

That is, the **difference between two squares** can be expressed as a product:

$$a^2 - b^2 = (a - b)(a + b)$$

difference of square roots (points to $a - b$); sum of square roots (points to $a + b$)

**EXAMPLE 4**

Work out $73^2 - 27^2$

$73^2 - 27^2 = (73 + 27)(73 - 27)$
$= 100 \times 46$
$= 4600$

### Exercise 19C

**1** Show that $(a + b)^2 = a^2 + 2ab + b^2$ when:
**a** $a = 3, b = 4$ **b** $a = 9, b = 11$
**c** $a = 20, b = 1$ **d** $a = \frac{1}{3}, b = \frac{1}{2}$

**2** Do you agree that
$31^2 = (30 + 1)^2 = 30^2 + 2 \times 30 \times 1 + 1^2$
$= 900 + 60 + 1$?

Use this method to find:
**a** $41^2$ **b** $51^2$ **c** $23^2$ **d** $72^2$

**3** Show that $(a - b)^2 = a^2 - 2ab + b^2$ when:
**a** $a = 5, b = 3$ **b** $a = 15, b = 5$
**c** $a = 20, b = 1$ **d** $a = 1.4, b = 1.7$

**4** Do you agree that
$29^2 = (30 - 1)^2 = 30^2 - 2 \times 30 \times 1 + 1^2$
$= 900 - 60 + 1$?

Use this method to find:
**a** $39^2$ **b** $49^2$ **c** $27^2$ **d** $68^2$

**5** Show that $a^2 - b^2 = (a + b)(a - b)$ when:
**a** $a = 5, b = 4$ **b** $a = 7, b = 3$
**c** $a = 21, b = 20$

**6** Copy and complete the statements.
**a** $8^2 - 3^2 = (\square + \square) \times (\square - \square) = \square$
**b** $37^2 - 27^2 = (\square + \square) \times (\square - \square) = \square$

**7** Find:
**a** $61^2 - 60^2$ **b** $55^2 - 45^2$

**8** Copy and complete:
**a** $21^2 = (20 + 1)^2 =$
**b** $52^2 = (50 + 2)^2 =$
**c** $29^2 = (30 - 1)^2 =$
**d** $77^2 = (80 - 3)^2 =$

**9** Copy and complete:
**a** $7^2 - 3^2 = (7 - 3)(\quad) =$
**b** $14^2 - 6^2 = (14 - 6)(\quad) =$
**c** $89^2 - 11^2 = (89 - 11)(\quad) =$
**d** $28^2 - 18^2 = (28 - 18)(\quad) =$
**e** $53^2 - 33^2 = (\quad)(53 + 33) =$
**f** $72^2 - 42^2 = (\quad)(72 + 42) =$

**10** First write your answers as a product and then find:
**a** $(5.7)^2 - (4.3)^2$
**b** $(8.1)^2 - (1.9)^2$
**c** $(2.6)^2 - (1.4)^2$
**d** $(5.7)^2 - (3.7)^2$
**e** $(6.8)^2 - (4.2)^2$
**f** $(7.9)^2 - (2.9)^2$
**g** $(5.16)^2 - (4.84)^2$
**h** $(8.29)^2 - (1.71)^2$

## INVESTIGATION

Can you work out $45^2$ in your head? How about $75^2$? You can't?

Stun your friends by finding squares of multiples of 5 in your head. Look at the examples below:

| $45^2$ | | $65^2$ | |
|---|---|---|---|
| $4 \times 5$ | $5 \times 5$ | $6 \times 7$ | $5 \times 5$ |
| $= 20$ | $= 25$ | $= 42$ | $= 25$ |
| $45^2 = 2025$ | | $65^2 = 4225$ | |

Can you see how they work?

Show that this method always works.

(**Hint:** consider $45^2 - 5^2$.)

Since $a^2 - b^2 = (a - b)(a + b)$, the difference between two squares can be factorised.

### EXAMPLE 5

Factorise $25p^2 - 4q^2r^4$

$$25p^2 - 4q^2r^4$$
$$= (5p)^2 - (2qr^2)^2$$
$$= (5p - 2qr^2)(5p + 2qr^2)$$

You can use this idea to solve problems without having to square numbers.

### EXAMPLE 6

Find the volume of concrete in a pipe 6.3 m long with an external diameter of 1.7 m and an internal diameter of 1.5 m.

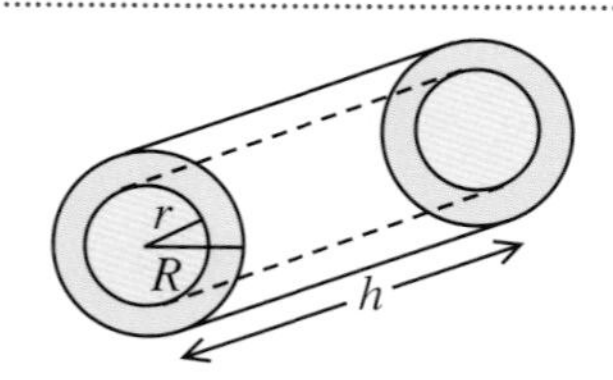


$$V = \pi R^2h - \pi r^2h = \pi h(R^2 - r^2)$$
that is, $V = \pi h(R - r)(R + r)$
so $V = \pi \times 6.3 \times (1.7 - 1.5) \times (1.7 + 1.5)$
$V = \pi \times 6.3 \times 0.2 \times 3.2$
$V = 4.032\pi$

Taking $\pi$ as 3.14, the volume of concrete is $12.7\,\text{m}^3$ to 3 s.f.

## Exercise 19D

**1** Copy and complete:

**a** $p^2 - q^2 = (p - q)(\quad)$
**b** $l^2 - m^2 = (\quad)(l + m)$
**c** $p^2 - 4q^2 = (p - 2q)(\quad)$
**d** $9l^2 - m^2 = (\quad)(3l + m)$
**e** $9p^2 - 4q^2 = (3p - 2q)(\quad)$
**f** $16l^2 - 9m^2 = (\quad)(4l + 3m)$

**2** Write as the product of two brackets:

**a** $x^2 - y^2$ **b** $s^2 - t^2$
**c** $4a^2 - b^2$ **d** $l^2 - 9m^2$
**e** $16a^2 - b^2$ **f** $p^2 - 25q^2$
**g** $16a^2 - 25b^2$ **h** $49p^2 - 16q^2$

**3** Write as the product of two brackets:

**a** $b^2 - c^2$ **b** $81x^2 - y^2$
**c** $p^2 - 100q^2$ **d** $100a^2 - 81b^2$
**e** $(xy)^2 - z^2$ **f** $a^2 - b^2c^2$
**g** $4p^2 - q^2r^2$ **h** $9l^2m^2 - n^4$

**4** Find the length of the third side of a right-angled triangle ABC where:

**a** AC = 13 cm, AB = 12 cm
**b** AC = 25 cm, AB = 7 cm
**c** AC = 41 cm, BC = 9 cm

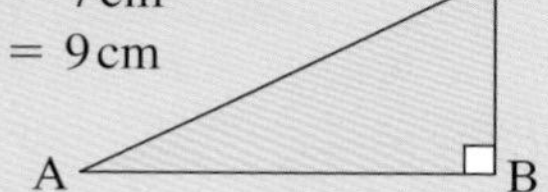


**5** PR is the hypotenuse of a right-angled triangle PQR. Find QR when:

**a** PR = 61 cm, PQ = 11 cm
**b** PR = 85 cm, PQ = 13 cm
**c** PR = 3.9 cm, PQ = 1.5 cm
**d** PR = 3.2 cm, PQ = 1.2 cm

**6** Look at the diagram.

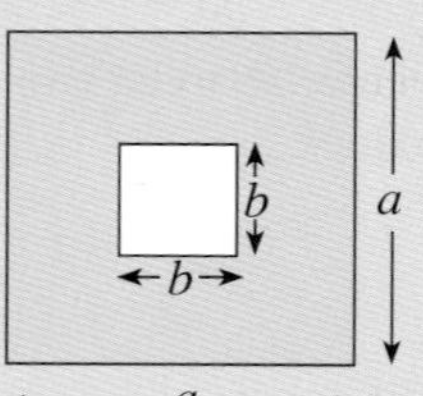


Find the area of the shaded part of the diagram where:

**a** $a = 5.7$ cm, and $b = 4.3$ cm
**b** $a = 13.2$ cm, and $b = 6.8$ cm
**c** $a = 7.9$ cm, and $b = 4.7$ cm

## 19.4 Factorising quadratic expressions

A quadratic expression is one which contains a term in $x^2$, but no higher power of $x$.

Examples are: $x^2 + 2x - 3$
$x^2 - 1$

Some quadratic expressions can be written as a product of two algebraic expressions. This is called **factorising** the quadratic expression.

*For example,*

$x^2 - y^2 = (x - y)(x + y)$

$x^2 + 5x + 6 = (x + 3)(x + 2)$

To factorise an expression like

$x^2 + 7x + 12$

you can write

$x^2 + 7x + 12 = (x + \square)(x + \triangle)$

which is the same as

$x^2 + 7x + 12 = x^2 + (\square + \triangle)x + (\square \times \triangle)$

where the product

$\square \times \triangle = 12$

and the sum

$\square + \triangle = 7$

Look at the factors of 12

The only factors of 12 whose sum is 7 are 3 and 4.

So $\square = 3$ and $\triangle = 4$

That is, $x^2 + 7x + 12 = (x + 3)(x + 4)$

**EXAMPLE 7**

Factorise $x^2 + 10x + 24$

The expression has to be written in the form
$(x + \square)(x + \triangle)$
where $\square \times \triangle = 24$
and $\square + \triangle = 10$

That is, you must find the factor pairs of 24 whose sum is 10.

$24 = 24 \times 1$ $\quad 24 + 1 = 25$
$\phantom{24} = 12 \times 2$ $\quad 12 + 2 = 14$
$\phantom{24} = 8 \times 3$ $\quad 8 + 3 = 11$
$\phantom{24} = 6 \times 4$ $\quad 6 + 4 = 10$

So $x^2 + 10x + 24 = (x + 6)(x + 4)$
You can check the answer by multiplying the brackets.

With negative terms, factorisation is a little more tricky.

**EXAMPLE 8**

Factorise $x^2 - 3x - 18$

$x^2 - 3x - 18 = (x + \square)(x + \triangle)$
where $\square \times \triangle = {}^{-}18$
and $\square + \triangle = {}^{-}3$

Look at the factors of $^{-}18$. Since $^{-}18$ is negative, one factor must be negative and one positive.

Now $9 \times {}^{-}2 = {}^{-}18$, but $9 + {}^{-}2 = 7$
$6 \times {}^{-}3 = {}^{-}18$, but $6 + {}^{-}3 = 3$

So try:

$3 \times {}^{-}6 = {}^{-}18$ and $3 + {}^{-}6 = {}^{-}3$

This works!
Hence, $x^2 - 3x - 18 = (x + 3)(x - 6)$

### Exercise 19E

**1** Copy and complete:

**a** $x^2 + 7x + 6 = (x + 6)(x + \triangle)$
**b** $x^2 + 16x + 15 = (x + 1)(x + \triangle)$
**c** $x^2 + 8x + 15 = (x + 5)(x + \triangle)$
**d** $x^2 + 9x + 20 = (x + 5)(x + \triangle)$
**e** $x^2 + 5x + 6 = (x + 3)(x + \triangle)$

**2** Copy and complete:

**a** $x^2 - 6x + 8 = (x - 2)(x - \triangle)$
**b** $x^2 - 8x + 15 = (x - 3)(x - \triangle)$
**c** $x^2 - 5x + 4 = (x - 1)(x - \triangle)$
**d** $x^2 - 9x + 8 = (x - 8)(x - \triangle)$
**e** $x^2 - 6x + 9 = (x - 3)(x - \triangle)$
**f** $x^2 - 5x + 6 = (x - 2)(x - \triangle)$

**3** Factorise, and check that your answer is correct.

**a** $x^2 + 10x - 11 = (x + \square)(x - \triangle)$
**b** $x^2 + 4x - 21 = (x + \square)(x - \triangle)$
**c** $x^2 - 2x - 35 = (x - \square)(x + \triangle)$
**d** $x^2 - 4x - 21 = (x + \square)(x - \triangle)$
**e** $x^2 + 32x + 60 = (x + \square)(x + \triangle)$
**f** $x^2 - x - 2 = (x + \square)(x - \triangle)$

**4** Factorise:

**a** $x^2 + 9x - 22$ **b** $x^2 + 4x - 12$
**c** $x^2 + 16x - 17$ **d** $x^2 - 8x - 33$
**e** $x^2 - 5x - 24$ **f** $x^2 + 0x - 49$

Factorising expressions like

$3x^2 - 14x - 5$

is more difficult, but the same method is used.

**EXAMPLE 9**

Factorise $12x^2 + 8x + 1$

This time try:
$12x^2 + 8x + 1 = (ax + 1)(bx + 1)$

which is the same as:
$12x^2 + 8x + 1 = abx^2 + (a + b)x + 1$
where $ab = 12$ and $a + b = 8$

So, you must find factors of 12 whose sum is 8.

That is, $a$ and $b$ are 6 and 2:
$12x^2 + 8x + 1 = (6x + 1)(2x + 1)$

**EXAMPLE 10**

Factorise $15x^2 - 38x + 11$

This time try:
$15x^2 - 38x + 11 = (ax - b)(cx - d)$

which is the same as:
$15x^2 - 38x + 11 = acx^2 - (ad + bc)x + bd$
where $ac = 15, bd = 11$ and $ad + bc = 38$
Now, 11 only has two factors, so
$b = 11$ and $d = 1$
That leaves
$ac = 15$ and $a + 11c = 38$
Looking at factors of 15 that satisfy
$a + 11c = 38$ gives
$a = 5, c = 3$
So, $15x^2 - 38x + 11 = (5x - 11)(3x - 1)$

## Exercise 19F

**1** Copy and complete:

**a** $3x^2 + 13x + 4 = (3x + 1)(x + \triangle)$
**b** $5x^2 + 7x + 2 = (5x + 2)(x + \triangle)$
**c** $6x^2 + 11x + 3 = (3x + 1)(2x + \triangle)$
**d** $6x^2 + 10x - 4 = (3x - 1)(2x + \triangle)$
**e** $15x^2 + 29x + 12 = (5x + \square)(3x + \triangle)$
**f** $6x^2 - 10x - 4 = (2x - \square)(3x + \triangle)$

**2** Factorise:

**a** $15x^2 + 8x + 1$ **b** $11x^2 + 12x + 1$
**c** $24x^2 + 10x - 1$ **d** $16x^2 + 6x - 1$
**e** $30x^2 - x - 1$ **f** $49x^2 - 14x + 1$

## 19.5 Solving quadratic equations

### Using factors

Any equation of the form $ax^2 + bx + c = 0$ is called a **quadratic equation**.

One way of solving a quadratic equation is by factorisation.

**EXAMPLE 11**

Solve $x^2 - 5x + 6 = 0$

This can be written as $(x - 2)(x - 3) = 0$
Since the right-hand side is zero, one bracket must be zero:

*either* $(x - 2) = 0$ so $x = 2$
*or* $(x - 3) = 0$ so $x = 3$

The solutions of this equation are $x = 2$ and $x = 3$.

Harder equations may be solved in the same way.

**EXAMPLE 12**

Solve $6x^2 - x - 15 = 0$

This can be written as $(3x - 5)(2x + 3) = 0$

*either* $3x - 5 = 0$ so $x = \frac{5}{3}$
*or* $2x + 3 = 0$ so $x = -\frac{3}{2}$

Notice that a quadratic equation usually has **two** solutions.

## Exercise 19G

**1** For what values of $x$ is:

**a** $(x - 3)(x - 4) = 0$
**b** $(x - 5)(x - 6) = 0$
**c** $(x - 5)(x + 2) = 0$
**d** $(x - 2)(x + 5) = 0$
**e** $(x + 4)(x - 7) = 0$
**f** $(x + 6)(x - 3) = 0$?

**2** Write down the two solutions for each quadratic equation.

**a** $(x - 3)(x - 4) = 0$
**b** $(x + 5)(x - 2) = 0$
**c** $(x + 2)(x + 9) = 0$
**d** $(2x - 1)(x - 3) = 0$
**e** $(4x - 1)(5x - 1) = 0$
**f** $x(x - 3) = 0$
**g** $(7x + 1)(4x + 1) = 0$
**h** $(2x - 3)(3x - 2) = 0$

**3** Find the solutions for each equation by first factorising the left-hand side.

**a** $x^2 - 6x + 5 = 0$
**b** $x^2 - 10x + 24 = 0$
**c** $x^2 - 15x + 56 = 0$
**d** $x^2 - 6x + 8 = 0$
**e** $x^2 - 12x + 32 = 0$
**f** $x^2 - 9x + 20 = 0$

**4** Solve these equations.

a $x^2 - 7x + 10 = 0$
b $x^2 - 7x + 12 = 0$
c $x^2 - 10x + 21 = 0$
d $x^2 - 13x + 42 = 0$
e $x^2 - 13x + 36 = 0$
f $x^2 - 4x + 4 = 0$

**5** Write down the two solutions:

a $(x + 3)(x - 7) = 0$
b $(3x - 1)(x + 4) = 0$
c $x(3x + 1) = 0$
d $(7x - 2)(3x + 4) = 0$
e $(5 - x)(2 - x) = 0$
f $(4 + x)(3 - x) = 0$

**6** Factorise the left-hand side of the equation and then find the solution.

a $x^2 - 7x - 8 = 0$
b $x^2 + 2x - 15 = 0$
c $6x^2 - 5x + 1 = 0$
d $5x^2 - 6x + 1 = 0$
e $3x^2 - 8x + 5 = 0$
f $5x^2 - 12x + 7 = 0$
g $2x^2 + 3x - 9 = 0$
h $7x^2 - 10x - 8 = 0$
i $4x^2 - 51x - 13 = 0$
j $5x^2 + 13x - 18 = 0$
k $9x^2 - 6x + 1 = 0$
l $25x^2 + 20x + 4 = 0$
m $9x^2 - 8x - 1 = 0$
n $4x^2 - 20x + 25 = 0$

**7** Solve the quadratic equation by first factorising.

a $x^2 - 4x + 4 = 0$
b $x^2 - 2x + 1 = 0$
c $x^2 - 6x + 9 = 0$
d $x^2 - 10x + 25 = 0$
e $x^2 + 8x + 16 = 0$
f $x^2 + 12x + 36 = 0$
g $x^2 + 4x + 4 = 0$
h $x^2 + 6x + 9 = 0$
i $4x^2 - 4x + 1 = 0$
j $9x^2 + 6x + 1 = 0$

**8** The equation $x^2 - 2x = 35$ can be written as $x^2 - 2x - 35 = 0$
so that $(x + 5)(x - 7) = 0$ and $x = {}^-5$ or 7.
Rewrite these equations and solve:

a $x^2 + 7x = {}^-12$ b $x^2 + 5x = 14$
c $x^2 + 5x = {}^-6$ d $x^2 - 8x = 9$
e $x^2 + 7x = {}^-6$ f $x^2 - 2x = 24$

**9** Collect the like terms together and solve the quadratic equation.

a $10 + 4x - x + x^2 = 28$
b $3x^2 + 40 + 8x = 2x^2 + 89 + 8x$
c $2x^2 - 60 + 4x + 2x^2 = 3x^2 + 3x - 40$
d ${}^-x^2 = 8x - 11x - 88$
e $92 - x^2 = 11x + 32$
f $500 + 13x = 34x + x^2 + 610$

**10** $2x^2 + 6x + 4 = 0$ $x^2 + 3x + 2 = 0$

a Show that both equations are the same.
b Solve each. Which one is the easier to solve? Are your answers the same?

## 19.6 Word problems

You can solve word problems by forming and solving equations.

**EXAMPLE 13**

The length of a rectangle is 5 cm more than its width and the area is $24\,\text{cm}^2$.
Find the length and the width of the rectangle.

A diagram usually helps.

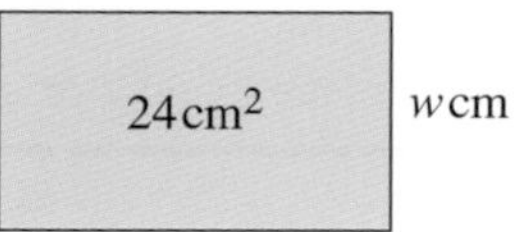


First, define the unknowns:
Let $w$ cm be the width of the rectangle, then $(w + 5)$ cm is the length.

Next, form the equation:

Area = length × width

so $24 = w(w + 5)$

Now, solve the equation:

$w^2 + 5w - 24 = 0$
$(w - 3)(w + 8) = 0$
$w = 3$ or $w = {}^-8$

Finally, interpret the solution:
$w = {}^-8$ is impossible as a rectangle cannot have a negative width.
Thus, rectangle width is 3 cm and length is 8 cm.

A diagram is not always possible, but the major steps in solving many word problems are:

- Read the problem carefully.
- Define unknown(s).
- Form or write down an equation involving your unknowns.
- Solve your equation.
- Interpret the solution – does it make sense?

**EXAMPLE 14**

Find two consecutive numbers whose squares add up to 85.

- Define the unknowns:
  Let $x$ be the first number, so $(x + 1)$ is the second number.
- Write an equation:
  $x^2 + (x + 1)^2 = 85$
- Solve the equation:
  $x^2 + x^2 + 2x + 1 = 85$
  $2x^2 + 2x - 84 = 0$
  $x^2 + x - 42 = 0$
  $(x + 7)(x - 6) = 0$
  so $x = 6$ or $^-7$
- Interpret the solution:
  The two numbers are 6 and 7 or $^-7$ and $^-6$.

## Exercise 19H

**1** One side of a rectangle is 10 cm longer than the other. If the area of the rectangle is 56 cm², show this information as a quadratic equation and hence find the length of each side.

**2** The three sides of a right-angled triangle are as shown in the diagram.

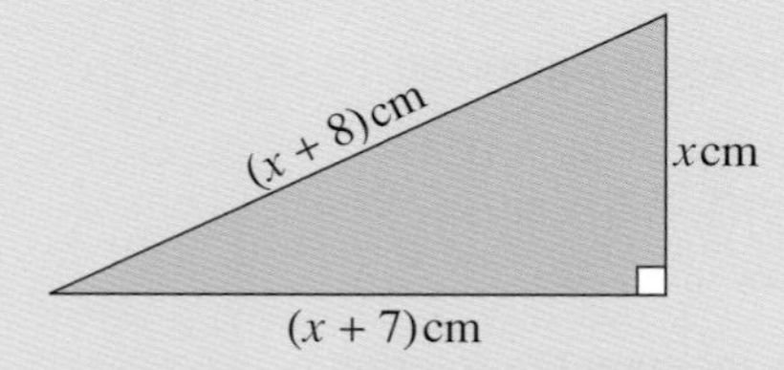


Use Pythagoras' theorem to form a quadratic equation, then by factorising find the lengths of the three sides.

**3** The hypotenuse of a right-angled triangle is 1 cm longer than the second side of the triangle and 2 cm longer than the third side. Using $x$ cm as the length of the shortest side, write down a quadratic equation to describe the triangle. (Use Pythagoras' theorem.) Solve the equation to find the length of the sides.

**4** The hypotenuse of a right-angled triangle is 1 cm longer than one side and 8 cm longer than the other. Find the length of each side.

**5** The squares of two consecutive odd numbers add up to 74. Show this information as a quadratic equation and hence find the numbers.

**6** The sum of the squares of three consecutive numbers is 77. If $x$ is the smallest number, show this information as a quadratic equation. Solve the equation to find the numbers.

**7** The sum of the squares of three consecutive numbers is 110. Show this information as a quadratic equation and hence find the numbers.

**8** A man makes a daily journey of 50 km. When he increases his normal speed by 10 km/h he finds he takes 10 minutes less than usual. Use speed = distance ÷ time to find his normal speed.

# Consolidation

**Example 1**

Simplify:

**a** $(x + 3)(x - 2)$
$= x(x - 2) + 3(x - 2)$
$= x^2 - 2x + 3x - 6$
$= x^2 + x - 6$

**b** $(3x - 2)(4x - 3)$
$= 3x(4x - 3) - 2(4x - 3)$
$= 12x^2 - 9x - 8x + 6$
$= 12x^2 - 17x + 6$

**Example 2**

Factorise these quadratics:

**a** $81 - 4x^2$
**b** $x^2 + 2x - 15$

---

**a** $81 - 4x^2 = 9^2 - (2x)^2 = (9 - 2x)(9 + 2x)$

Use the difference of two squares.

**b** $x^2 + 2x - 15 = (x + 5)(x - 3)$

Look for two numbers whose product is −15 and sum is 2: 5 and $^-3$

**Example 3**

Solve $x^2 - 4x - 12 = 0$

---

$x^2 - 4x - 12 = 0$
$(x - 6)(x + 2) = 0$
so $x = 6$ or $^-2$

## Exercise 19

**1** Plot these graphs for $^-3 \leq x \leq 3$:
**a** $y = x^2 - 2x + 5$
**b** $y = 8 - x^2$
**c** $y = 2x^2 - 3x + 4$

**2** Use the distributive law to work out:
**a** $(x + 2)(x + 3)$
**b** $(x - 1)(x - 7)$
**c** $(2x + 6)(x - 9)$
**d** $(4x - 3)(2x + 8)$

**3** Write as the product of two brackets:
**a** $x^2 - y^2$ **b** $36x^2 - y^2$
**c** $p^2 - 9q^2$ **d** $4p^2 - q^4$

**4** Factorise:
**a** $x^2 + 4x + 3$ **b** $x^2 - 4x + 3$
**c** $x^2 - 2x - 3$ **d** $x^2 + 2x - 3$

**5** Solve by factorisation:
**a** $x^2 - 5x + 6 = 0$
**b** $x^2 - 5x + 4 = 0$
**c** $x^2 - 3x - 10 = 0$
**d** $2x^2 - 5x - 3 = 0$

**6** Simplify:
**a** $(x + 2)(x + 3)$
**b** $(x + 3)(x + 5)$
**c** $(x - 3)(x + 4)$
**d** $(x + 3)(x - 5)$
**e** $(x - 3)(x - 4)$
**f** $(x + 3)(x - 3)$
**g** $(2x - 1)(x + 2)$
**h** $(2x + 1)(2x - 1)$
**i** $(3x + 2)(2x - 3)$
**j** $(3x + 4)(4x - 1)$
**k** $(3x - 4)(3x - 1)$
**l** $(3x - 7)(7 - 2x)$

**7** Solve these equations.
**a** $x^2 + 3x + 2 = 0$
**b** $x^2 - 7x + 6 = 0$
**c** $x^2 - 7x + 10 = 0$
**d** $x^2 + 7x + 10 = 0$

**8** **a** Factorise $x^2 - a^2$.
**b** Use the result in part **a** to work these out quickly:
**i** $99^2 - 1^2$ **ii** $998^2$ **iii** $9997^2$

**9** **a** Factorise:
**i** $81 - 25r^2$
**ii** $12x^2 - 23x + 10$
**iii** $a + b - bc - ac$
**b** Solve $2x^2 + 3x - 5 = 0$

**10** The sum of the squares of two consecutive even numbers is 164.
What are the numbers?

# Summary

## You should know ...

**1** How to draw quadratic graphs by working out coordinates.
*For example:*
Plot the graph of $y = x^2 + 1$ for $^{-}3 \leqslant x \leqslant 3$.

| **x** | $^{-}3$ | $^{-}2$ | $^{-}1$ | 0 | 1 | 2 | 3 |
|---|---|---|---|---|---|---|---|
| $x^2$ | 9 | 4 | 1 | 0 | 1 | 4 | 9 |
| **+1** | 1 | 1 | 1 | 1 | 1 | 1 | 1 |
| **y** | 10 | 5 | 2 | 1 | 2 | 5 | 10 |

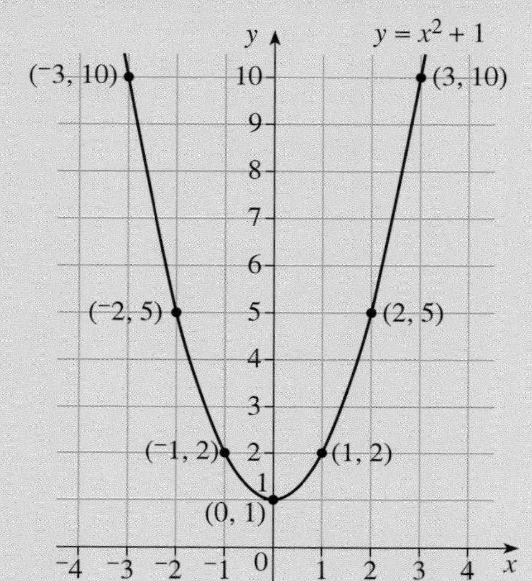


## Check out

**1** Plot the graph of
**a** $y = x^2 - 3$ for $^{-}3 \leqslant x \leqslant 3$
**b** $y = x^2 + 2x$ for $^{-}4 \leqslant x \leqslant 2$

---

**2** How the product of two brackets can be found using the distributive law.
*For example:*

$(3x - 2)(5x + 6)$

$= 3x \times 5x + 3x \times 6 - 2 \times 5x - 2 \times 6$
$= 15x^2 + 18x - 10x - 12$
$= 15x^2 + 8x - 12$

**2** Use the distributive law to work out:
**a** $(x + 2)(x + 3)$
**b** $(x - 2)(x - 2)$
**c** $(2x - 1)(x + 5)$
**d** $(x - 4)^2$
**e** $(3x - 5)^2$

---

**3** How to factorise a quadratic expression.
*For example:*
$x^2 + 7x - 8 = (x + a)(x + b)$
where $ab = {}^{-}8$ and $a + b = 7$
so, $a$ and $b$ are 8 and $^{-}1$:
$x^2 + 7x - 8 = (x + 8)(x - 1)$

$25x^2 - 36 = (5x)^2 - 6^2$
$= (5x - 6)(5x + 6)$

**3** Factorise:
**a** $x^2 + 4x + 3$
**b** $x^2 + 12x - 13$
**c** $x^2 - 10x + 25$
**d** $2x^2 + 5x - 12$
**e** $100m^2 - 49$

---

**4** How to solve a quadratic equation by factorisation.
*For example:*
$x^2 + x - 20 = 0$
$(x + 5)(x - 4) = 0$
so $x = {}^{-}5$ or 4

**4** Solve:
**a** $x^2 + 3x + 2 = 0$
**b** $x^2 - 8x + 16 = 0$
**c** $6x^2 - x - 1 = 0$

# Index